i	imaginary unit [6.3]
$\sqrt{-b}$	$i\sqrt{b}$ $(b>0)$ [6.3]
$a + bi$	complex number [6.3]
$\pm$	plus or minus [6.4]
(x,y)	ordered pair of numbers; first component is x and second component is y [7.1]
$R \times R$	Cartesian product of R and R [7.1]
f, g, etc.	names of functions [7.3]
$f(x)$, etc.	f of x, or the value of f at x [7.3]
f^{-1}	f inverse or the inverse of f [8.6]
$\log_b x$	logarithm to the base b of x [9.2]
$\text{antilog}_b x$	antilogarithm to the base b of x [9.4]
$\ln x$	logarithm to the base e of x [9.4]
s_n	nth term of a sequence [11.1]
S_n	sum of n terms of a sequence [11.1]
Σ	the sum [11.1]
S_∞	infinite sum [11.1]
$n!$	n factorial or factorial n [11.5]
$P(n,n)$	permutations of n things taken n at a time [11.6]
$P(n,r)$	permutations of n things taken r at a time [11.6]
$C(n,r)$ or $\binom{n}{r}$	combinations of n things taken r at a time [11.7]
$\begin{bmatrix} a_1 & b_1 \\ a_2 & b_2 \end{bmatrix}$	second-order matrix [B.1]
$\begin{bmatrix} a_1 & b_1 & c_1 \\ a_2 & b_2 & c_2 \\ a_3 & b_3 & c_3 \end{bmatrix}$	third-order matrix [B.1]
$\begin{vmatrix} a_1 & b_1 \\ a_2 & b_2 \end{vmatrix}$	second-order determinant [B.2]
δA or $\delta(A)$	determinant of A [B.2]
$\begin{vmatrix} a_1 & b_1 & c_1 \\ a_2 & b_2 & c_2 \\ a_3 & b_3 & c_3 \end{vmatrix}$	third-order determinant [B.3]

Project initiated by Harold Parnes
Mathematics Editor: Peter W. Fairchild
Production: Greg Hubit Bookworks
Cover photo: John Drooyan

Printed in the United States of America

4 5 6 7 8 9 10—88 87 86

ISBN 0-534-01433-X

Library of Congress Cataloging in Publication Data

Drooyan, Irving.
  Intermediate algebra.

  Rev. ed. of: Intermediate algebra / William Wooton,
Irving Drooyan. 5th ed. c1980.
  Includes index.
  1. Algebra.   I. Drooyan, Irving.   II. Wooton, William.
Intermediate algebra. 5th ed.   III. Title.
QA152.2.W66 1984   512.9   83–6800
ISBN 0-534-01433-X

INTERMEDIATE ALGEBRA

SIXTH EDITION

IRVING DROOYAN
Los Angeles Pierce College

WILLIAM WOOTON

Wadsworth Publishing Company
Belmont, California
A Division of Wadsworth, Inc.

CONTENTS

PREFACE

This edition of *Intermediate Algebra* retains the basic point of view of the fifth edition with respect to subject matter and pedagogy. However, we have rewritten parts of the text to improve clarity, made some changes in organization and content, and strengthened the exercise sets.

Retentions from earlier editions

The first six chapters review material normally encountered in some form in a beginning algebra course. The last five chapters are organized around the function concept, with heavy emphasis on graphing. In particular, linear and quadratic relations and functions and their graphs are explored in some detail; variation is treated from the function standpoint; logarithms are developed from a consideration of the inverse of the exponential function; and sequences and series are discussed through the use of function concepts. The last three chapters and the appendices are independent of each other and any one may be omitted for a short course.

To aid in making assignments, the exercise sets have been graded. Exercises labeled A provide routine practice and are sufficient for a basic course. Exercises labeled B contain more challenging problems.

Answers to the odd-numbered problems, including graphs, are provided at the end of the book. In addition, a solutions manual and a study guide covering topics in the text are available for student use. Additional ancillary materials are available to the instructor.

Changes in this edition

Interval notation (Section 1.3 in the fifth edition) is introduced where it is first used in Section 4.5. Chapter 1 now includes only topics that most students studied in elementary algebra. Hence, Chapter 1 may be used as a review or omitted.

Two additional laws of exponents involving powers (Section 5.1 in the fifth edition) are now included in Section 2.3, where the law of exponents pertaining to products of powers is first introduced.

Sign arrays are introduced in Section 6.8 to motivate the procedure used to solve quadratic inequalities.

Section 8.6 on inverse functions has been rewritten so that consideration is first given to one-to-one functions.

Chapter 9 on exponential and logarithmic functions now includes separate sections on using tables (Section 9.4) and using calculators (Section 9.5). Either section may be used in order to complete the following sections in the chapter. Whether instructors prefer to use tables or prefer to use calculators, the emphasis in this chapter continues to be on the properties of logarithms and on solving exponential equations.

Because the use of a calculator enables students to obtain values of e^x and $\ln x$ easily and because important applications of exponential equations involve the base e, as much attention is given to powers with base e as is given to powers with the base 10. Tables for e^x and $\ln x$, as well as the traditional $\log_{10} x$, are included for students using tables. Both Section 9.6, a new section on solving exponential equations, and Section 9.7, which treats a variety of applications of exponential equations, give equal emphasis to powers with the base 10 and the base e.

A new section (A.3) on graphing rational functions has been added to the appendix immediately following the section on graphing polynomial functions.

Many exercise sets have been revised. In particular, new word problems have been added and the number of "B" type exercises has been increased. The Chapter Reviews now also include "B" exercises.

New pedagogical features

Marginal annotations are used to highlight the major topics in each section.

Procedures and algorithms that require several steps and may be difficult for some students to follow are highlighted by displaying them in boxes. Common errors are also highlighted in the text and exercise sets.

Acknowledgments

We wish to thank the following people for reviewing the manuscript and for offering helpful suggestions: Harriett Beggs, State University of New York Agricultural and Technical College; C. Leary Bell, Columbus College; Dick J. Clark, Portland Community College; Steven Davis, El Camino College; Gerald H. James, Glendale Community College; Glenn Kindle, Community College of Denver; Mark Phillips, Cypress College; William H. Price, Middle Tennessee State University; Sheldon Rothman, C. W. Post College, Long Island University; Jo Steig, Mesa Community College; Eline Svendsen, Lake Land College; Thomas Wood, Central Connecticut State College.

In addition we give special thanks to our friends in the mathematics department of Pierce College who generously contributed valuable suggestions for this revision.

Irving Drooyan
William Wooton

1. REAL NUMBERS

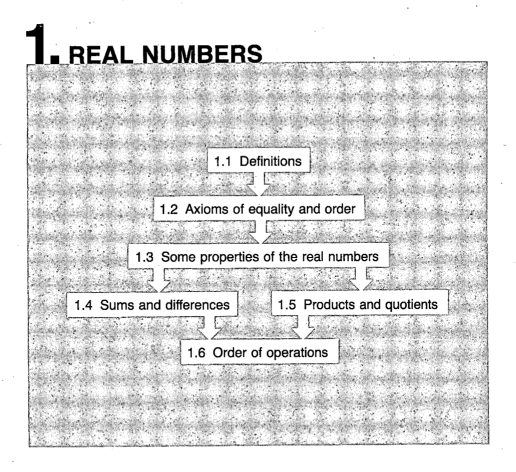

1.1 Definitions

1.2 Axioms of equality and order

1.3 Some properties of the real numbers

1.4 Sums and differences

1.5 Products and quotients

1.6 Order of operations

1.1

DEFINITIONS

We will start our study of intermediate algebra in this chapter by reviewing properties of numbers that are usually introduced in elementary algebra courses.

Sets and symbols

A **set** is a collection of objects. In algebra, we work primarily with collections, or sets, of numbers. Each item in a set is called an **element** or a **member** of the set. For example, the numbers 1, 2, 3, ... (the three dots mean "and so on") are the elements of a set we call the set of **natural numbers**. Because the natural numbers never end—we can never reach a *last* number—we refer to this set as an **infinite set**. A set whose elements can be counted is called a **finite set**.

We indicate sets by means of capital letters, such as *A*, *I*, or *R*, or by means of braces, {}, used together with words or symbols.

1

Examples   **a.** {natural numbers less than 4} is read "the set whose elements are natural numbers less than 4."

b. If $A = \{$natural numbers less than 7$\}$, then $A = \{1, 2, 3, 4, 5, 6\}$.
The set A is a finite set.

c. If $B = \{$natural numbers greater than 7$\}$, then $B = \{7, 8, 9, \ldots\}$.
The set B is an infinite set.

We say that two sets are **equal** if they have the same elements.

Examples   **a.** $\{2, 3, 4\} = \{$natural numbers between 1 and 5$\}$

b. $\{5, 6, 7, 8\} = \{7, 6, 5, 8\}$

We call a set containing no members the **empty set** or **null set**, and we regard it as a finite set. The empty set is denoted by $\varnothing$ or $\{\}$. Note that the symbol $\varnothing$ (read "the empty set") is not enclosed by braces.

Examples   **a.** {odd numbers exactly divisible by 2} $= \varnothing$

b. {natural numbers less than 1} $= \varnothing$

When we want to indicate an unspecified element in a set, we use a lowercase letter, such as a, b, c, x, y, or z. Such symbols are called **variables**. If the given set, called the **replacement set** of the variable, is a set of numbers (as it will be throughout this book), then the variable represents a number. A symbol used to denote a known, or specified, element of a set is called a **constant**.
We use the symbol $\in$ to denote membership in a set.

Examples   **a.** $x \in N$ is read "x is an element of N." The variable is x. The replacement set is N.

b. $x \in \{5\}$. Since x can be only one value, 5, x is a constant.

The slash, $/$, is a handy device for indicating the negation of a symbol.

Examples   **a.** $\{2, 3, 4\} \neq \{3, 4, 5\}$ is read "the set whose elements are 2, 3, and 4 *is not equal to* the set whose elements are 3, 4, and 5."

b. $2 \notin \{3, 4, 5\}$ is read "2 *is not an element of* the set whose members are 3, 4, and 5."

One way we can describe sets is by a convenient notation called **set-builder notation**. Notice how, in the following examples, the vertical line is read "such that."

Examples   **a.** $\{x | x \in A$ and $x \notin B\}$ is read "the set of all x *such that* x is an element of A and x is not an element of B."

b. $\{x | x \in A$ and $x \neq 3\}$ is read "the set of all x *such that* x is an element of A and x is not equal to 3."

Number relationships

In this book we are going to be concerned primarily with the following sets of numbers:

- The set N of natural numbers, also called **counting numbers**, whose elements consist of 1, 2, . . . , 7, . . . , 235, . . . , and so on, without ever coming to an end. Note that natural numbers are always positive numbers.

- The set W of **whole numbers**, whose elements consist of the natural numbers and zero. Examples of whole numbers are 5, 110, and 0.

- The set J of **integers**, whose elements consist of the natural numbers, their negatives, and zero. Examples are -7, -3, 0, 5, and 11.

- The set Q of **rational numbers**, whose elements are all the numbers that can be represented in the form a/b, where a and b are integers and b does not equal zero. Examples are $-\frac{3}{4}$, $\frac{18}{27}$, 3, and -6. (The numbers 3 and -6 are rational numbers because they can be written in the form $\frac{3}{1}$ and $-\frac{6}{1}$.) All rational numbers can be written as terminating or repeating decimals. For example, $-\frac{3}{4}$ is equivalent to the terminating decimal -0.75, and $\frac{2}{3}$ is equivalent to the repeating decimal 0.666. . . .

- The set H of **irrational numbers**, whose elements are the numbers with decimal representations that are nonterminating and nonrepeating. Examples are $\sqrt{15}$, π, and $-\sqrt{7}$. An irrational number cannot be represented in the form a/b, where a and b are integers.

- The set R of **real numbers**, whose elements consist of all rational and irrational numbers.

We can see how the different sets of numbers relate to one another from the diagram in Figure 1.1. Notice how all natural numbers also belong to the set of whole numbers; how all whole numbers also belong to the integers; how all integers are included in the rational numbers; and how all rational and irrational numbers are also real numbers.

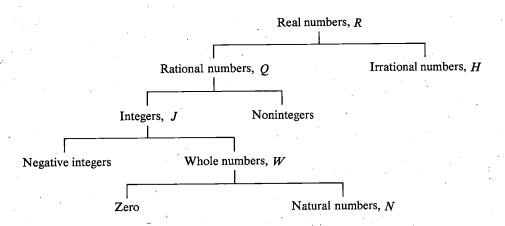

Figure 1.1

You may already be familiar with these sets and their properties from elementary algebra. Nevertheless, because the properties of real numbers are so important, we shall review many of them in the remainder of this chapter. Other properties will be considered in the following chapters.

EXERCISE 1.1

A ■ *Specify each set by listing the members.*

Examples   **a.** {integers between -5 and -1}   **b.** {whole numbers greater than 4}

Solutions   **a.** $\{-4, -3, -2\}$ **b.** $\{5, 6, \ldots\}$

    1. {natural numbers between 2 and 6}   **2.** {integers between -3 and 4}

    3. {first three whole numbers}   **4.** {first three natural numbers}

    5. {natural numbers greater than 4}   **6.** {integers less than 2}

    7. {odd natural numbers between 4 and 10}   **8.** {even integers between -7 and -1}

■ *Specify each set by a word phrase.* [*Note: There may be more than one way to specify the given set.*]

Examples   **a.** $\{3, 6, 9\}$ **b.** $\{1, 3, 5, \ldots\}$.

Solutions   **a.** {first three natural-number multiples of 3}

     b. {odd natural numbers}

    9. $\{4, 5, 6\}$ **10.** $\{-6, -5, -4\}$ **11.** $\{8, 10, 12\}$

   12. $\{-7, -5, -3\}$ **13.** $\{8, 9, 10, \ldots\}$ **14.** $\{\ldots, -4, -3, -2\}$

■ *Let $A = \{-5, -\sqrt{15}, -3.44, -\frac{2}{3}, 0, \frac{1}{5}, \frac{7}{3}, 6.1, 8\}$. List the members of the given set.*

Examples   **a.** $\{x | x \in A$ and x an integer$\}$ **b.** $\{x | x$ is a positive number in $A\}$

Solutions   **a.** All members of A that are also integers, $\{-5, 0, 8\}$.

     b. All positive members of A, $\{\frac{1}{5}, \frac{7}{3}, 6.1, 8\}$.

15. $\{x \mid x \in A \text{ and } x \text{ a whole number}\}$ **16.** $\{x \mid x \in A \text{ and } x \text{ a natural number}\}$

17. $\{x \mid x \in A \text{ and } x \text{ an irrational number}\}$ **18.** $\{x \mid x \in A \text{ and } x \text{ a rational number}\}$

19. $\{x \mid x \text{ is a negative real number in } A\}$ **20.** $\{x \mid x \text{ is a real number in } A\}$

■ *State whether the given set is finite or infinite.*

Example $\{$whole numbers greater than 10,000$\}$

Solution   Since there is no greatest whole number, the set is infinite.

21. $\{$whole numbers less than 10,000$\}$ **22.** $\{$integers less than 10,000$\}$

23. $\{$integers between -3 and 4$\}$ **24.** $\{$integers greater than $-4\}$

25. $\{$rational numbers between -1 and 0$\}$ **26.** $\{$rational numbers between 0 and 1$\}$

■ *Replace the question mark with either $\in$ or $\notin$ to form a true statement.*

Examples   **a.** -2 ? $\{$real numbers$\}$ **b.** ½ ? $\{$integers$\}$

Solutions   **a.** $-2 \in \{$real numbers$\}$ **b.** ½ $\notin \{$integers$\}$

27. 2 ? $\{$rational numbers$\}$ **28.** 0 ? $\{$natural numbers$\}$

29. ⅘ ? $\{$real numbers$\}$ **30.** -3 ? $\{$whole numbers$\}$

31. -4 ? $\{$irrational numbers$\}$ **32.** -6 ? $\{$integers$\}$

B **33.** If $A = B$ and $4 \notin B$, can 4 be an element of A?

34. If $A \neq B$ and $4 \in B$, must 4 be an element of A?

35. If $A \neq B$ and $B = C$, can $A = C$?

36. If $A \neq B$ and $B \neq C$, must $A \neq C$?

1.2

AXIOMS OF EQUALITY AND ORDER

In mathematics, formal assumptions about numbers or their properties are called **axioms** or **postulates**. Such assumptions are formal statements about properties which we propose to assume as always valid. While we are free to formulate axioms in any way we please, it is clearly desirable that any axioms we adopt lead to useful consequences. A set of useful axioms must not lead to contradictory conclusions.

The words *property*, *law*, and *principle* are sometimes used to denote assumptions, although these words may also be applied to certain consequences of axioms. In this book we use, in each situation, the word we believe to be the one most frequently encountered. The first such assumptions to be considered concern equality.

Properties
of equality

An **equality**, or an "is equal to" assertion, is a mathematical statement that *two symbols, or groups of symbols, are names for the same number*. A number has an infinite variety of names. Thus 3, $\%_2$, $4 - 1$, and $2 + 1$ are all names for the same number; hence, the equality

$$4 - 1 = 2 + 1$$

is a statement that "$4 - 1$" and "$2 + 1$" are different names for the same number.

We shall assume that the "is equal to" (=) relationship has the properties listed below.

E-1 $a = a$.   **Reflexive property**

E-2 *If* $a = b$, *then* $b = a$.   **Symmetric property**

E-3 *If* $a = b$ *and* $b = c$, *then* $a = c$.   **Transitive property**

E-4 *If* $a = b$, *then* b *may be replaced by* a *or*   **Substitution property**
         a *by* b *in any statement without altering*
         the truth or falsity of the statement.

Examples

a. Reflexive property: $15 = 15$.

b. Symmetric property: If $t = 8$, then $8 = t$.

c. Transitive property: If $x + 2 = y$ and $y = 3$, then $x + 2 = 3$.

d. Substitution property: If $x = 5$, and $x + y = 2$, then $5 + y = 2$.

The transitive property can be viewed as a special case of the substitution property. Thus, in Example c above, we can view the equality $x + 2 = 3$ as being obtained from $x + 2 = y$ by *substituting* 3 for y.

We refer to the symbol (or the number it names) to the left of an equals sign as the *left-hand member*, and that to the right as the *right-hand member* of the equality.

Number
lines

For each real number, there corresponds one and only one point on a line, and vice versa. Hence, a geometric line can be used to visualize relationships between real numbers. For example, to represent 1, 3, and 5 on a line, we scale a straight line in convenient units, with increasing positive direction indicated by an arrow, and mark the required points with dots on the line. This kind of line is called a **number line**, and this particular number line is shown in Figure 1.2. The real numbers corresponding to points on the number line are called the **coordinates** of the points, and the points are called the **graphs** of the numbers.

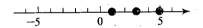

Figure 1.2

Positive numbers are associated with the points on the line to the right of a point called the **origin** (labeled 0); negative numbers are associated with the points on the line to the left of the origin. The number 0 is neither positive nor negative; it serves as a point of separation for the positive and negative numbers.

Properties of order

We now define the real number b to be *less than* the real number a if $a - b$ is positive. The relationship that establishes the order between b and a is called an **inequality**. In symbols, we write

$$b < a \quad \text{which is read} \quad \text{``}b \text{ is } less\ than\ a\text{,''}$$

or

$$a > b \quad \text{which is read} \quad \text{``}a \text{ is } greater\ than\ b\text{.''}$$

Furthermore, two conditions such as $a < b$ and $b < c$ can be written as the continued inequality $a < b < c$, which is read "a is less than b and b is less than c" or simply "b is between a and c."

It is also possible to combine the concepts of equality and inequality and write two statements at once:

$$\leq \quad \text{is read} \quad \text{``is less than or equal to,''}$$

and

$$\geq \quad \text{is read} \quad \text{``is greater than or equal to.''}$$

Examples

a. $y \leq 10$ is read "y is less than or equal to 10."

b. $x \geq 8$ is read "x is greater than or equal to 8."

c. $-2 < x \leq 5$ is read "-2 is less than x and x is less than or equal to 5" or "x is between -2 and 5, or x equals 5."

A real number that is not a negative number is called a **nonnegative** number. It can be zero or positive. In symbols, we write

$$a \geq 0$$

to indicate that a is nonnegative. Similarly, $a \leq 0$ means that a is a **nonpositive** number. It can be zero or negative.

Examples

a. 4 and 0 are both nonnegative numbers.

b. -4 and 0 are both nonpositive numbers.

Inequalities such as

$$1 < 2 \quad \text{and} \quad 3 < 5$$

are said to be of the *same sense*, because the left-hand member is less than the right-hand member in each case. Inequalities such as

$$1 < 2 \quad \text{and} \quad 5 > 3$$

are said to be of *opposite sense*, because in one case the left-hand member is less than the right-hand member and in the other case the left-hand member is greater than the right-hand member.

We assume the following property concerning the order of real numbers:

If $a < b$ and $b < c$, then $a < c$.   **Transitive property**

Examples

a. If $x < 7$ and $7 < y$, then $x < y$.

b. If $6 < x$ and $x < y + z$, then $6 < y + z$.

We have noted in the foregoing discussion that certain algebraic statements concerning the order of real numbers can be interpreted geometrically. We summarize some of the more common correspondences in Table 1.1, where in each case a, b, and c are real numbers.

Table 1.1

Algebraic statement	Geometric statement	
1. $a > 0$; a is positive	1. The graph of a lies to the right of the origin.	
2. $a < 0$; a is negative	2. The graph of a lies to the left of the origin.	
3. $a > b$	3. The graph of a lies to the right of the graph of b.	
4. $a < b$	4. The graph of a lies to the left of the graph of b.	
5. $a < c < b$	5. The graph of c is to the right of the graph of a and to the left of the graph of b.	

Graphs of
real numbers

Number lines can be used to display infinite sets of points as well as finite sets. For example, Figure 1.3a is the graph of the set of all integers greater than or equal to 2 *and* less than 5. The graph consists of three points. Figure 1.3b is the graph of the set of all real numbers greater than or equal to 2 and less than 5. The graph is of an *infinite set of points*. The closed dot on the left end of the shaded portion of the graph indicates that the endpoint is part of the graph, while the open dot on the right end indicates that the endpoint is not in the graph.

a.   b.

Figure 1.3

The set of numbers whose graph is shown in Figure 1.3a can be described in set-builder notation by

$$\{x|2 \leq x < 5 \text{ and } x \in J\}$$

(read "the set of all x such that x is greater than or equal to 2 *and* less than 5, and x is an element of the set of integers"). Similarly, the set of numbers whose graph is shown in Figure 1.3b can be described by

$$\{x|2 \leq x < 5 \text{ and } x \in R\}.$$

When no question exists relative to the replacement set for x, the notation $\{x|2 \leq x < 5, x \in R\}$ is abbreviated to $\{x|2 \leq x < 5\}$.

EXERCISE 1.2

A   ■ *Replace each question mark to make the given statement an application of the given property of equality or order.*

Example   If $z = 2$ and $2 = t$, then $? = t$; transitive property.

Solution   If $z = 2$ and $2 = t$, then $z = t$.

1. $3r = ?$; reflexive property.
2. If $n = t + 3$, then $? = n$; symmetric property.
3. If $r = 6$ and $r - 3 = t$, then $? - 3 = t$; substitution property.
4. If $4 < x$ and $x < y$, then $? < ?$; transitive property.
5. If $n < 3$ and $3 < t$, then $? < ?$; transitive property.

6. If $a = c$ and $c = 4$, then $\underline{?} = 4$; transitive property.

7. If $r = n$ and $n + 6 = 8$, then $\underline{?} + 6 = 8$; substitution property.

8. If $t = 4$ and $5 \cdot t = 6s$, then $5 \cdot \underline{?} = 6s$; substitution property.

9. $6 + x = \underline{?}$; reflexive property.

10. If $2 + x = y$, then $y = \underline{?}$; symmetric property.

■ *Express each relation using symbols.*

Examples   **a.** 6 is less than 10

b. x is greater than or equal to 0 and less than 3

Solutions   **a.** $6 < 10$ **b.** $0 \le x < 3$

11. 8 is greater than 5   **12.** -5 is greater than -8

13. -6 is less than -4 **14.** -3 is less than 4

15. $x + 1$ is negative   **16.** $x - 3$ is positive

17. $x - 4$ is nonpositive   **18.** $x + 2$ is nonnegative

19. y is between -2 and 3   **20.** y is between -4 and 0

21. x is greater than or equal to 1 and less than 7

22. $3t$ is greater than or equal to 0 and less than or equal to 4

■ *Replace each question mark with an appropriate order symbol to form a true statement.*

Examples   **a.** $-8 \; \underline{?} \; -2$ **b.** $4 \; \underline{?} \; 0 \; \underline{?} \; -1$

Solutions   **a.** $-8 < -2$ **b.** $4 > 0 > -1$

23. $-2 \; \underline{?} \; 8$ **24.** $3 \; \underline{?} \; 6$ **25.** $-7 \; \underline{?} \; -13$ **26.** $0 \; \underline{?} \; -5$

27. $-6 \; \underline{?} \; -3$ **28.** $\dfrac{-3}{2} \; \underline{?} \; \dfrac{-3}{4}$ **29.** $1\dfrac{1}{2} \; \underline{?} \; \dfrac{3}{2}$ **30.** $3 \; \underline{?} \; \dfrac{6}{2}$

31. $3 \; \underline{?} \; 5 \; \underline{?} \; 7$ **32.** $3 \; \underline{?} \; 0 \; \underline{?} \; -4$ **33.** $-7 \; \underline{?} \; 0 \; \underline{?} \; 2$ **34.** $-5 \; \underline{?} \; -2 \; \underline{?} \; 0$

■ *Graph each set of integers on a separate number line.*

Examples   **a.** $\{-4, -1, 3, 5\}$ **b.** $\{x \mid -2 < x \le 3\}$

Solutions   **a.**   **b.**

35. $\{-3, 0, 1, 4\}$

36. $\{-5, -2, 0, 3\}$

37. $\{x | x > 2\}$

38. $\{x | x < -3\}$

39. $\{x | x \geq -4\}$

40. $\{x | x \leq 3\}$

41. $\{x | -1 \leq x < 6\}$

42. $\{x | -5 < x \leq -1\}$

■ *Graph each set of real numbers on a separate number line.*

Examples   **a.** $\{x | x \geq -2\}$ **b.** $\{y | 30 \leq y < 40\}$

Solutions   **a.**

b.

43. $\{x | x \geq -5\}$

44. $\{y | 3 < y \leq 7\}$

45. $\{t | t < -3\}$

46. $\{s | s \leq 8\}$

47. $\{x | -5 \leq x \leq -3)$

48. $\{y | -5 \leq y < 10\}$

49. $\{z | 22 < z \leq 28\}$

50. $\{x | -25 \leq x < -18\}$

51. $\{r | 50 < r < 58\}$

52. $\{y | 23 \leq y \leq 28\}$

53. $\{z | -30 \leq z\}$

54. $\{x | x < 50\}$

1.3

SOME PROPERTIES
OF THE REAL NUMBERS

Properties for addition and multiplication

In addition to the axioms of equality and order, we assume the following properties for addition and multiplication of real numbers.

R-1	$a + b$ is a real number.	**Closure for addition**
R-2	$a + b = b + a.$	**Commutative property of addition**
R-3	$(a + b) + c = a + (b + c).$	**Associative property of addition**
R-4	ab is a real number.	**Closure for multiplication**
R-5	$ab = ba.$	**Commutative property of multiplication**
R-6	$(ab)c = a(bc).$	**Associative property of multiplication**
R-7	$a(b + c) = (ab) + (ac).$	**Distributive property**
R-8	*There exists a unique number 0 with the property* $$a + 0 = a \quad and \quad 0 + a = a.$$	**Identity element for addition**

Properties continued overleaf

R-9 *There exists a unique* **Identity element for multiplication**
number 1 *with the*
property

$$a \cdot 1 = a \quad and \quad 1 \cdot a = a.$$

R-10 *For each real number a,* **Additive-inverse property**
there exists a unique real
number $-a$ *(opposite of a)*
with the property

$$a + (-a) = 0 \quad and \quad (-a) + a = 0.$$

R-11 *For each real number a* **Multiplicative-inverse (reciprocal)**
except 0, *there exists a* **property**
unique real number 1/a
(**reciprocal** *of a) with the*
property

$$a\left(\frac{1}{a}\right) = 1 \quad and \quad \left(\frac{1}{a}\right)a = 1.$$

Parentheses are used in some of the above properties to indicate an order of operations; operations enclosed in parentheses are performed before any others. Note that R-7 is the only one of the properties that involves the operation of addition *and* the operation of multiplication.

Examples

	For addition	*For multiplication*
a. Closure property:	$3 + 4$ is a real number	$3 \cdot 4$ is a real number
b. Commutative property:	$3 + 4 = 4 + 3$	$3 \cdot 4 = 4 \cdot 3$
c. Associative property:	$(2 + 3) + 4 = 2 + (3 + 4)$	$(2 \cdot 3) \cdot 4 = 2 \cdot (3 \cdot 4)$
d. Identity element:	$3 + 0 = 3$	$3 \cdot 1 = 3$
e. Inverse property:	$3 + (-3) = 0$	$3\left(\frac{1}{3}\right) = 1$
f. Distributive property:	$3(4 + 5) = 3(4) + 3(5)$	

Some important theorems

Axioms R-1 through R-11 together with axioms E-1 through E-4 imply other properties of the real numbers. Such implications are generally stated as **theorems**, which are logical consequences of the axioms and other theorems.

Theorems generally consist of two parts: an "if" part, called the **hypothesis**, and a "then" part, called the **conclusion**. Proving a theorem consists of formally showing that if the hypothesis is true, then, because of the axioms, the conclusion must be true. We do not propose to emphasize proofs in this course, so whenever we do wish to argue the truth of a theorem, we shall do so informally.

Ordinarily, part of the hypothesis of a theorem involves a statement of the replacement sets of the variables involved. Because, except for a few special cases, the replacement set for variables used in this book is R, the set of real numbers, let us make the following simplifying assumption: *In any expression or equation involving variables in this book, unless otherwise stated, the replacement set for the variables will be the set R of real numbers.*

The following is an example of a theorem:

▶ *If $a = b$, then*

$$a + c = b + c \quad and \quad c + a = c + b. \qquad \textbf{Addition property of equality}$$

To see that this is so, we need only observe that by the reflexive property for equality E-1 (page 6), it is true that

$$a + c = a + c.$$

Since we are given that $a = b$, we can use the substitution property E-4 to replace a in the right-hand member of this equality to produce

$$a + c = b + c,$$

as was to be proven. From the commutative law of addition, we also have that

$$c + a = c + b.$$

Thus, if the hypothesis "$a = b$" is true, then the conclusion "$a + c = b + c$ and $c + a = c + b$" follows logically.

The following theorem is closely analogous to the addition property:

▶ *If $a = b$, then*

$$ac = bc \quad and \quad ca = cb. \qquad \textbf{Multiplication property of equality}$$

The proof exactly parallels that of the previous theorem.

Examples   **a.** If $x = y$, then $x + 3 = y + 3$ and $5x = 5y$.

b. If $x + y = z$, then $x + y + (-y) = z + (-y)$.

c. If $\dfrac{x + y}{4} = z$, then $4\left(\dfrac{x + y}{4}\right) = 4z$.

Two other important theorems detail the role of 0 in a product and assert a property of negatives.

▶ $a \cdot 0 = 0$ **Zero-factor property**

Examples   **a.** $3 \cdot 0 = 0$ **b.** $-8 \cdot 0 = 0$ **c.** $0 \cdot 0 = 0$

We have given names to the foregoing theorems because they are used frequently throughout the book. In general, the properties we obtain as the result of our arguments are not named.

Since the number $-a$ is assumed to be unique, it follows that if $a + b = 0$, then $b = -a$ and $a = -b$; that is, if the sum of two numbers is 0, one must be positive and the other negative. Thus, the opposite of a negative number is a positive number; that is,

$$(-3) + [-(-3)] = 0$$

implies that

$$-(-3) = 3.$$

This example suggests the following theorem where a can be positive or negative:

▶ $-(-a) = a$

This property expresses the fact that *the opposite of the opposite of a number is the number itself.*

Examples   **a.** $-(-4) = 4$ **b.** $-(-8) = 8$ **c.** $-[-(-6)] = -6$

Absolute value of a number

Sometimes we wish to consider only the nonnegative member of a pair of numbers a and $-a$. For example, since the graphs of $-a$ and a are each located the same distance from the origin, when we wish to refer simply to this distance and not to its direction to the left or right of 0, we can use the notation $|a|$ (read "the absolute value of a"). More formally:

▶
$$|a| = \begin{cases} a, & \text{if } a \geq 0, \\ -a, & \text{if } a < 0. \end{cases}$$

From this definition, $|a|$ is always nonnegative.

Examples   **a.** $|3| = 3$ **b.** $|0| = 0$

c. $|-3| = -(-3) = 3$ **d.** If $x < 0$, then $|x| = -x$.

EXERCISE 1.3

A ▪ *In Problems 1–18, replace each question mark to make the given statement an application of the given property.*

Examples   **a.** $2(3 + 1) = 2 \cdot 3 + \underline{?};$ distributive property

b. $5 \cdot \underline{?} = 0;$ zero-factor property

Solutions   **a.** $2(3 + 1) = 2 \cdot 3 + 2 \cdot 1$ **b.** $5 \cdot 0 = 0$

1. $12 + x = x + \underline{?};$ commutative property of addition
2. $(2m)n = 2(\underline{?});$ associative property of multiplication
3. $3(x + y) = 3x + \underline{?};$ distributive property
4. $(2 + z) + 3 = 2 + (\underline{?});$ associative property of addition
5. $4 \cdot t = \underline{?};$ commutative property of multiplication
6. $7 + \underline{?} = 0;$ additive-inverse property
7. $3 \cdot \dfrac{1}{3} = \underline{?};$ multiplicative-inverse property
8. $m + \underline{?} = m;$ identity element for addition
9. $r \cdot \underline{?} = r;$ identity element for multiplication
10. $r + (s + 2) = (s + 2) + \underline{?};$ commutative property of addition
11. $3(x + y) = \underline{?} + \underline{?};$ distributive property
12. $6 \cdot \dfrac{1}{6} = \underline{?};$ commutative property of multiplication
13. If $z = 4,$ then $z + 5 = 4 + \underline{?};$ additive property of equality
14. $(x + y) + z = x + \underline{?};$ associative property of addition
15. If $x = 6,$ then $-3x = \underline{?} \cdot 6;$ multiplicative property of equality
16. $15 \cdot \underline{?} = 0;$ zero-factor property
17. $-4(3x) = (-4 \cdot 3)\underline{?};$ associative property of multiplication
18. If $m = t,$ then $m + 3 = \underline{?} + \underline{?};$ additive property of equality
19. If $x < 0,$ does $-x$ represent a positive number or a negative number?
20. If $x < 0,$ does $-(-x)$ represent a positive number or a negative number?

▪ *Rewrite each expression without using absolute-value notation.*

Examples   **a.** $|-9|$ **b.** $|x + 5|$ **c.** $|-y|$

Solutions   **a.** 9   **b.** $x + 5,$ if $x + 5 \geq 0;$ **c.** $-y,$ if $-y \geq 0;$
                                    $-(x + 5),$ if $x + 5 < 0$ $y,$ if $-y < 0$

21. $|-3|$ 22. $|-7|$ 23. $|4|$ 24. $|6|$

25. $-|-2|$ 26. $-|-4|$ 27. $-|5|$ 28. $-|7|$

29. $|x|$ 30. $|-x|$ 31. $|x-2|$ 32. $|x+3|$

B   33. Is the set $\{0, 1\}$ closed with respect to addition? With respect to multiplication?

34. Is the set $\{-1, 0, 1\}$ closed with respect to addition? With respect to multiplication?

35. Is the set $\{1, 2\}$ closed with respect to addition? With respect to multiplication?

36. Which of the sets N, W, J, and Q (page 3) are closed with respect to addition? With respect to multiplication?

37. Must $|-x| = |x|$?   38. Can $|x| = -|x+3|$?

39. Can $|y| > |x| + |y|$?   40. Can $|y| < |x| + |y|$?

1.4
SUMS AND DIFFERENCES

Sums of real numbers

Addition is an operation that associates with each pair of real numbers a and b a third real number $a + b$, called the **sum** of a and b. We first observe that:

▶ *If $a > 0$ and $b > 0$, then*

$$a + b > 0.$$

The sum of two positive numbers is positive.

The properties of real numbers imply the familiar laws of signs for sums. For example, consider the sum $-2 + (-3)$. First note that, by the additive-inverse property,

$$2 + (-2) = 0 \quad \text{and} \quad 3 + (-3) = 0;$$

so

$$[2 + (-2)] + [3 + (-3)] = 0.$$

By means of the associative and commutative properties of addition, this can be rewritten as

$$(2 + 3) + [(-2) + (-3)] = 0.$$

Since by the additive-inverse property,

$$(2 + 3) + [-(2 + 3)] = 0,$$

it follows that

$$(-2) + (-3) = -(2 + 3).$$

The example on page 16 suggests the following:

▶ *The sum of the negative numbers −a and −b is the opposite of the sum of their absolute values.*

Examples **a.** $(-3) + (-5) = -(3 + 5)$
$$= -8$$

b. $(-18) + (-6) = -(18 + 6)$
$$= -24$$

Arguments similar to the one above concerning the sum of a positive number and a negative number lead to the following:

▶ *The sum of a positive number and a negative number is equal to the nonnegative difference of their absolute values preceded by the sign of the number with the greater absolute value.*

Examples **a.** $8 + (-6) = 8 - |-6|$
$$= 2$$

b. $-3 + 5 = 5 - |-3|$
$$= 5 - 3 = 2$$

c. $-12 + 10 = -(|-12| - 10)$
$$= -(12 - 10) = -2$$

d. $5 + (-7) = -(|7| - 5)$
$$= -(7 - 5) = -2$$

We can usually perform the addition of signed numbers mentally.

Differences of real numbers We can define the **difference** $a - b$ of two positive numbers a and b as follows:

If $a > b > 0$, then $a - b$ is the number d,

$$a - b = d,$$

such that

$$a = b + d.$$

For example,

$$7 - 2 = 5 \quad \text{because} \quad 7 = 2 + 5.$$

This definition for the difference of two *positive numbers a* and b, for $a > b$, can apply to *all numbers*. Thus,

$$2 - 7 = -5 \quad \text{because} \quad 2 = 7 + (-5).$$

We can also define the difference of two numbers in terms of a sum. Note that

$$7 - 2 = 5 \quad \text{and} \quad 7 + (-2) = 5,$$

and that

$$2 - 7 = -5 \quad \text{and} \quad 2 + (-7) = -5.$$

These examples suggest that the difference $a - b$ is equal to the sum of a and the additive inverse of b; that is:

▶

$$a - b = a + (-b).$$

The difference of b subtracted from a equals the sum of a and $-b$.

Examples   **a.** $8 - (-3) = 8 + (3)$ **b.** $(-7) - (4) = (-7) + (-4)$
                     $= 8 + 3 = 11$ $= -11$

   c. $(-5) - (-2) = (-5) + (2)$ **d.** $3 - 8 = 3 + (-8)$
                     $= -5 + 2 = -3$ $= -5$

Three uses of   Since we have seen that the difference $a - b$ is given by $a + (-b)$, we may consider the
the minus   symbols $a - b$ as representing either the difference of a and b or, preferably, the sum of
sign a and $(-b)$. Note that we have now used the sign $+$ in two ways and the sign $-$ in three
ways. We have used these signs to denote positive and negative numbers and as signs
of operation to indicate the sum or difference of two numbers; we also have used the
sign $-$ to indicate the "opposite" of a number.

Examples   **a.** -3;   the sign denotes a negative integer.

   b. $7 - 3$;   the sign can be viewed as the operation of subtraction where $+3$ is
           subtracted from $+7$.

   c. $-a$;   the sign denotes the "opposite" of a. If a is positive, then $-a$ is negative. If a
           is negative, then $-a$ is positive.

In discussing the results of operations, it is convenient to use the term **basic
numeral**. For example, while the numeral "$3 + 5$" names the sum of the real numbers
3 and 5, we shall refer to "8" as the basic numeral for this number. Similarly, the basic
numeral for "$2 - 8$" is "-6."

EXERCISE 1.4

A   ■ *Rewrite each difference as a sum.*

Examples   **a.** $12 - 3$ **b.** $6 - (-7)$ **c.** $-3 - 5$

Solutions   **a.** $12 - 3$ **b.** $6 - (-7)$ **c.** $-3 - 5$
                     $= 12 + (-3)$ $= 6 + [-(-7)]$ $= -3 + (-5)$
                                                         $= 6 + 7$

1. $5 - 9$ 2. $1 - 4$ 3. $-3 - 4$ 4. $-5 - 7$
5. $-2 - (-2)$ 6. $-4 - (-6)$ 7. $0 - 5$ 8. $0 - (-5)$

■ *Write each sum or difference using a basic numeral.*

Examples   a. $3 + (-7)$ b. $-3 - (-7)$ c. $-2 + (-3) - 4$

Solutions   a. $3 + (-7) = -4$ b. $-3 - (-7)$ c. $-2 + (-3) - 4$
                                    $= (-3) + [-(-7)]$ $= (-2) + (-3) + (-4)$
                                    $= (-3) + (+7)$ $= -9$
                                    $= 4$

9. $4 + 7$ 10. $6 + 9$ 11. $-2 + 8$ 12. $-5 + 7$
13. $6 + (-3)$ 14. $5 + (-10)$ 15. $-2 + (-5)$ 16. $-6 + (-6)$
17. $8 - 3$ 18. $6 - 1$ 19. $4 - 7$ 20. $8 - 14$
21. $-2 - 7$ 22. $-5 - 11$ 23. $4 - (-2)$ 24. $8 - (-4)$
25. $-2 - (-4)$ 26. $-8 - (-5)$ 27. $6 - 2 + 7$ 28. $8 + 3 - 5$
29. $2 - 5 - 3$ 30. $6 - 4 - 2$ 31. $8 - 7 - (-2)$ 32. $4 - (-3) + 2$

Examples   a. $5 - (8 - 12)$ b. $(4 - 2 + 7) - (3 - 8)$

Solutions   a. $5 - (8 - 12) = 5 - (-4)$ b. $(4 - 2 + 7) - (3 - 8) = 9 - (-5)$
                    $= 5 + [-(-4)]$ $= 9 + [-(-5)]$
                    $= 5 + 4 = 9$ $= 9 + 5 = 14$

33. $7 - (5 - 2)$ 34. $8 - (10 - 2)$ 35. $(6 - 5) - 11$
36. $(3 - 7) - 1$ 37. $(6 - 1 + 8) - 3$ 38. $4 - (6 + 2 - 11)$
39. $(7 - 2) + (-3 + 1)$ 40. $(5 - 9) + (4 - 2)$
41. $(3 - 5 + 4) - (8 - 13)$ 42. $(38 - 25) + (13 - 17 - 2)$
43. $(5 + 24 - 29) + (12 - 8 + 4)$ 44. $(-15 + 3 - 1) - (9 - 5 + 12)$
45. $(23 - 15 + 18) - (16 - 3 + 2)$ 46. $(-27 + 3 - 6) - (18 + 1 - 12)$

B 47. If subtraction were commutative, we would be able to write $a - b = b - a$. Find a numerical example to show that subtraction is *not* commutative.

48. If subtraction were associative, we would be able to write $(a - b) - c = a - (b - c)$. Find a numerical example to show that subtraction is *not* associative.

49. Which of the following sets are closed under subtraction?
    a. natural numbers   b. whole numbers   c. integers

50. For what values of a and b
    a. is the sum $a + b$ positive?
    b. is the difference $a - b$ positive?

1.5

PRODUCTS AND QUOTIENTS

*Products of
real numbers*

Multiplication is an operation that associates with each pair of numbers a and b a third number $a \cdot b$, also written $(a)(b)$ or ab. The number ab is called the **product** of the **factors** a and b. Let us begin by observing that:

▶ *If $a > 0$ and $b > 0$, then*

$$ab > 0.$$

The product of two positive numbers is positive.

Now, what can be said about the product of a positive number and a negative number? Let us consider the product $3(-4)$. First consider the equality

$$4 + (-4) = 0.$$

By the multiplication property of equality,

$$3[4 + (-4)] = 3 \cdot 0;$$

and the distributive and zero-factor properties permit us to write

$$3 \cdot 4 + 3 \cdot (-4) = 0.$$

This implies that $3(-4)$ is the additive inverse of $3 \cdot 4$, which is equal to $-(3 \cdot 4)$. It can be shown in a similar way that $(-3)(-4)$ is the additive inverse of $-(3 \cdot 4)$, which is equal to $3 \cdot 4$.

*Laws of signs
for products*

In general, as a consequence of properties of the real numbers, we can obtain the following familiar laws of signs for products:

▶

$$(a)(b) = ab,$$
$$a(-b) = -(ab) = -ab,$$
$$(-a)(b) = -(ab) = -ab,$$
$$(-a)(-b) = ab,$$

where the symbol $-ab$ denotes the additive inverse of ab.

Thus, the product of two numbers with unlike signs is negative and the product of two numbers with like signs is positive.

Examples   **a.** $(3)(-2) = -6$ **b.** $(-3)(-2) = 6$

c. $(-2)(3) = -6$ **d.** $(3)(2) = 6$

Note that since $-1 \cdot a = -(1 \cdot a)$ and $1 \cdot a = a$,

▶
$$-1 \cdot a = -a.$$

Since each *pair* of negative factors yields a positive product, we have the following guideline:

▶ *If a product contains*

1. *an even number of negative factors, then the product is positive;*

2. *an odd number of negative factors, then the product is negative.*

Examples   **a.** $(-3)(2)(-4)(5) = (-3)(-4)(2)(5)$ **b.** $(-1)(-5)(3)(-2)(6)$
$= 12 \cdot 10 = 120$ $= (-1)(-5)(-2)(3)(6)$
$= -10 \cdot 18 = -180$

Prime and composite numbers

If a natural number greater than 1 has no factors that are natural numbers other than itself and 1, it is said to be a **prime number**. Thus, 2, 3, 5, 7, 11, etc., are prime numbers. A natural number greater than 1 that is not a prime number is said to be a **composite number**. Thus, 4, 6, 8, 9, 10, etc., are composite numbers. When a composite number is exhibited as a product of prime factors only, it is said to be **completely factored**. For example, although 30 may be factored into

$$(5)(6), \quad (10)(3), \quad (15)(2), \quad \text{or} \quad (30)(1),$$

if we continue the factorization, we arrive at the set of prime factors 2, 3, and 5 in each case. It is a fact that, except for order, *each composite number has one and only one factorization*. This is known as the **fundamental theorem of arithmetic**.

The words *composite* and *prime* are used in reference to natural numbers only. Any negative integer can, however, be expressed as the product $(-1)a$, where a is a natural number. Hence, if we refer to the completely factored form of a negative integer, we refer to the product of (-1) and the prime factors of the associated natural number. The composite numbers in this section will be less than 100 and hence the factors can be determined by inspection.

Examples   **a.** $24 = (2)(2)(2)(3)$ **b.** $-33 = -1(3)(11)$ **c.** 37 is prime

Quotients of real numbers

We can define the *quotient* of two real numbers in terms of multiplication.

▶ *The* **quotient** *of two numbers a and b is the number q*,

$$\frac{a}{b} = q,$$

such that $bq = a$.

Examples   **a.** $\frac{6}{2} = 3$ because $2 \cdot 3 = 6$.   **b.** $\frac{-6}{2} = -3$ because $2(-3) = -6$.

c. $\frac{6}{-2} = -3$ because $(-2)(-3) = 6$.   **d.** $\frac{-6}{-2} = 3$ because $(-2)(3) = -6$.

Since the quotient of two numbers a/b is a number q such that $bq = a$, the sign of the quotient of two signed numbers must be consistent with the laws of signs for the product of two signed numbers. Therefore,

$$\frac{+a}{+b} = +q, \qquad \frac{-a}{-b} = +q, \qquad \frac{+a}{-b} = -q, \qquad \frac{-a}{+b} = -q.$$

When a fraction bar is used to indicate that one algebraic expression is to be divided by another, the dividend is the numerator and the divisor is the denominator. Note that the denominator of a/b is restricted to nonzero numbers, for, if b is 0 and a is not 0, then there exists no q such that

$$0 \cdot q = a.$$

Again, if b is 0 and a is 0, then, for *any* q,

$$0 \cdot q = 0,$$

and the quotient is not unique. Thus:

▶ *Division by zero is not defined.*

Examples   **a.** $\frac{-5}{0}$ is not defined.   **b.** $\frac{1}{x-2}$ is not defined for $x = 2$, because $(2) - 2 = 0$.

The following alternative definition of a quotient is consistent with the definition above (see Problem 77, Exercise 1.5):

▶
$$\frac{a}{b} = a\left(\frac{1}{b}\right) \qquad (b \neq 0).$$

Examples   **a.** $\dfrac{2}{3} = 2\left(\dfrac{1}{3}\right)$ **b.** $\dfrac{a}{3b} = a\left(\dfrac{1}{3b}\right)$ **c.** $\dfrac{x}{y-x} = x\left(\dfrac{1}{y-x}\right)$

$(b \neq 0)$ $(y \neq x)$

EXERCISE 1.5

A   ■ *Write each product as a basic numeral.*

Examples   **a.** $(-3)(2)$ **b.** $(-7)(-5)$ **c.** $2(-3 \cdot 5)$

Solutions   **a.** $(-3)(2) = -6$ **b.** $(-7)(-5) = 35$ **c.** $2(-3 \cdot 5) = 2(-15)$

$= -30$

Common error: Note that in Example c, $2(-3 \cdot 5) \neq 2(-3) \cdot 2(5)$.

1. $(4)(-3)$ **2.** $(-5)(3)$ **3.** $(-2)(-6)$ **4.** $(-8)(-5)$

5. $(3)(-2)(-4)$ **6.** $(-5)(2)(-3)$ **7.** $(5)(-1)(4)$ **8.** $(2)(-2)(6)$

9. $(2)(3)(-1)(-4)$ **10.** $(5)(-2)(-1)(6)$ **11.** $(4)(0)(-2)(3)$

12. $(-6)(0)(2)(3)$ **13.** $-2(3 \cdot 4)$ **14.** $-2(2 \cdot 6)$

15. $(-4 \cdot 3)(-2)$ **16.** $(-5 \cdot 2)(-3)$ **17.** $(2 \cdot 3)(-4 \cdot 1)$

18. $(-2 \cdot 3)(3 \cdot 4)$ **19.** $-2(-3 \cdot 1 \cdot 4)$ **20.** $-3(-2 \cdot 2 \cdot 4)$

■ *Express each integer in completely factored form. If the integer is a prime number, so state.*

Examples   **a.** 36   **b.** 29   **c.** -51

Solutions   **a.** $36 = (2)(2)(3)(3)$ **b.** Prime   **c.** $-51 = -1(3)(17)$

21. 8   **22.** 26   **23.** 49   **24.** 18

25. 17   **26.** -16 **27.** -12 **28.** 23

29. 56   **30.** 65   **31.** -48 **32.** -52

■ *Write each quotient as a basic numeral.*

Examples   **a.** $\dfrac{-20}{5}$ **b.** $\dfrac{0}{-4}$ **c.** $\dfrac{-3}{0}$

Solutions overleaf

Solutions   a. $\dfrac{-20}{5} = -4$ b. $\dfrac{0}{-4} = 0$ c. $\dfrac{-3}{0}$; not defined

33. $\dfrac{-16}{4}$ **34.** $\dfrac{-32}{8}$ **35.** $\dfrac{39}{-3}$ **36.** $\dfrac{45}{-15}$

37. $\dfrac{-27}{-9}$ **38.** $\dfrac{-54}{-6}$ **39.** $\dfrac{0}{-7}$ **40.** $\dfrac{0}{-12}$

41. $\dfrac{-5}{0}$ **42.** $\dfrac{-27}{0}$ **43.** $-\left(\dfrac{-8}{2}\right)$ **44.** $-\left(\dfrac{-12}{-3}\right)$

■ *From the definition of a quotient, rewrite each equation in the form $a = bq$.*

Examples   a. $\dfrac{18}{-3} = -6$ b. $\dfrac{-27}{-3} = 9$ c. $\dfrac{-12}{4} = -3$

Solutions   a. $18 = (-3)(-6)$ b. $-27 = (-3)(9)$ c. $-12 = 4(-3)$

45. $\dfrac{15}{-3} = -5$ **46.** $\dfrac{46}{-23} = -2$ **47.** $\dfrac{-38}{19} = -2$ **48.** $\dfrac{-14}{2} = -7$

49. $\dfrac{-52}{-4} = 13$ **50.** $\dfrac{-28}{-7} = 4$ **51.** $\dfrac{0}{-7} = 0$ **52.** $\dfrac{-12}{1} = -12$

■ *Rewrite each quotient as a product in which one factor is a natural number and the other factor is the reciprocal (multiplicative inverse) of a natural number.*

Examples   a. $\dfrac{3}{4}$ b. $\dfrac{16}{5}$ c. $\dfrac{7}{9}$

Solutions   a. $\dfrac{3}{4} = 3\left(\dfrac{1}{4}\right)$ b. $\dfrac{16}{5} = 16\left(\dfrac{1}{5}\right)$ c. $\dfrac{7}{9} = 7\left(\dfrac{1}{9}\right)$

53. $\dfrac{7}{8}$ **54.** $\dfrac{27}{7}$ **55.** $\dfrac{3}{8}$ **56.** $\dfrac{9}{17}$

57. $\dfrac{82}{11}$ **58.** $\dfrac{3}{13}$ **59.** $\dfrac{7}{100}$ **60.** $\dfrac{3}{1000}$

■ *Rewrite each product as a quotient.*

61. $3\left(\dfrac{1}{2}\right)$ **62.** $8\left(\dfrac{1}{3}\right)$ **63.** $2\left(\dfrac{1}{7}\right)$ **64.** $\left(\dfrac{1}{100}\right)3$

65. $\left(\dfrac{1}{8}\right)5$ **66.** $\left(\dfrac{1}{7}\right)6$ **67.** $9\left(\dfrac{1}{2}\right)$ **68.** $7\left(\dfrac{1}{10,000}\right)$

69. Find an example to show that $2(3y) \neq (2 \cdot 3)(2y)$ for all y.

70. Find an example to show that $-3(xy) \neq (-3x)(-3y)$ for all x and y.

B ∎ *Under what conditions is each statement in Problems 71–74 a true statement?*

71. $\dfrac{x}{y} = 0$ **72.** $\dfrac{x}{y} \neq 0$ **73.** $\dfrac{x}{y} > 0$ **74.** $\dfrac{x}{y} < 0$

75. If division were commutative, we would be able to write $a \div b = b \div a$. Find a numerical example to show that division is not commutative.

76. If division were associative, we would be able to write $(a \div b) \div c = a \div (b \div c)$. Find a numerical example to show that division is not associative.

77. Show that if $b \neq 0$, $\dfrac{a}{b} = a \cdot \dfrac{1}{b}$. [*Hint:* Start with the fact that $(a/b) = q$ implies that $bq = a$. Then multiply both members by $1/b$.]

1.6

ORDER OF OPERATIONS

On page 12, we observed that parentheses are frequently used to show the order of performing operations. Thus,

$$a(b + c)$$

represents the product of a and the sum of b and c, while

$$(ab) + (ac)$$

represents the sum of the products ab and ac. Of course, the distributive property assures us that these expressions name the same numbers. Frequently, however, we find such expressions as $(ab) + (ac)$ written without parentheses, simply as

$$ab + ac.$$

We therefore need an agreement about the meaning of such expressions.

Order of operations

1. First, any expression within a symbol of inclusion (parentheses, brackets, fraction bars, etc.) is simplified.

2. Next, multiplications and divisions are performed as encountered in order from left to right.

3. Last, additions and subtractions are performed in order from left to right.

Example $\dfrac{4 + 3 \cdot 6}{2} - \dfrac{9 - 2 \cdot 2}{5}$ Simplify expressions grouped by the fraction bar.
Multiply first; then add and subtract.

$$= \dfrac{4 + 18}{2} - \dfrac{9 - 4}{5}$$

$$= \dfrac{22}{2} - \dfrac{5}{5}$$

Divide

$$= 11 - 1$$

Subtract

$$= 10$$

Common error: Note that $4 + 3 \cdot 6 \neq (4 + 3) \cdot 6$ and that $9 - 2 \cdot 2 \neq (9 - 2) \cdot 2$.

EXERCISE 1.6

A   ■ *Write each expression as a basic numeral.*

Examples   a. $3 + 2 \cdot 5$ b. $3(2 + 7) - 3 \cdot (-4)$ c. $\dfrac{6 + (-15)}{-3} + \dfrac{5}{4 - (-1)}$

Solutions

a. $3 + 2 \cdot 5 = 3 + 10$
$= 13$

b. $3(2 + 7) - 3 \cdot (-4)$
$= 3(9) + 12$
$= 27 + 12$
$= 39$

c. $\dfrac{6 + (-15)}{-3} + \dfrac{5}{4 - (-1)}$
$= \dfrac{-9}{-3} + \dfrac{5}{5}$
$= 3 + 1 = 4$

Common error: Note that in Example a, $3 + 2 \cdot 5 \neq (3 + 2) \cdot 5$.

1. $4 + 4 \cdot 4$

2. $7 - 6 \cdot 2$

3. $4(-1) + 2 \cdot 2$

4. $6 \cdot 3 + (-2)(5)$

5. $-2 \cdot (4 + 6) + 3$

6. $5(6 - 12) + 7$

7. $2 - 3(6 - 1)$

8. $4 - 7(8 + 2)$

9. $\dfrac{7 + (-5)}{2} - 3$

10. $\dfrac{12}{8 - 2} - 5$

11. $\dfrac{3(6 - 8)}{-2} - \dfrac{6}{-2}$

12. $\dfrac{5(3 - 5)}{2} - \dfrac{18}{-3}$

13. $6[3 - 2(4 + 1)] - 2$

14. $6[5 - 3(1 - 4)] + 3$

15. $(4 - 3)[2 + 3(2 - 1)]$

16. $(8 - 6)[5 + 7(2 - 3)]$

17. $64 \div [8(4 - 2[3 + 1])]$

18. $27 \div (3[9 - 3(4 - 2)])$

19. $5[3 + (8 - 1)] \div (-25)$

20. $-3[-2 + (6 - 1)] \div [18 \div (-2)]$

21. $[-3(8 - 2) + 3] \cdot [24 \div (2 - 8)]$

22. $[-2 + 3(5 - 8)] \cdot [-15 \div (5 - 2)]$

■ *Evaluate each expression for the given values of the variables.*

Example $a + (n - 1)d;$ $a = 2,$ $n = 5,$ and $d = -3$

Solution
$$
\begin{aligned}
a + (n - 1)d &= (2) + [(5) - 1](-3) \\
&= 2 + 4(-3) \\
&= 2 + (-12) \\
&= -10
\end{aligned}
$$

23. $\dfrac{5(F - 32)}{9};$ $F = 212$

24. $\dfrac{R + r}{r};$ $R = 12$ and $r = 2$

25. $\dfrac{E - e}{R};$ $E = 18,$ $e = 2,$ and $R = 4$

26. $\dfrac{a - rs_n}{1 - r};$ $r = 2,$ $s_n = 12,$ and $a = 4$

27. $P + Prt;$ $P = 1000,$ $r = 0.04,$ and $t = 2$

28. $R_0(1 + at);$ $R_0 = 2.5,$ $a = 0.05,$ and $t = 20$

29. The perimeter of a rectangle of length ℓ and width w is given by $2\ell + 2w.$ Determine the perimeter of a rectangle of length 16 centimeters and width 12 centimeters.

30. The area of a trapezoid with bases b and c and height h is given by $\frac{1}{2}h(b + c).$ Determine the area of a trapezoid with bases of 10 and 12 centimeters and height of 14 centimeters.

31. If P dollars is invested at simple rate r, then the amount of money that is accumulated after t years is given by $P + Prt.$ Determine the amount accumulated if \$800 is invested at 9% $(r = 0.09)$ for 3 years.

32. The net resistance of an electric circuit is given by $\dfrac{rR}{r + R},$ where r and R are two resistors in the circuit. Determine the net resistance if r and R are 10 and 20 ohms, respectively.

33. Find an example to show that $3 + 4 \cdot x \neq (3 + 4)x$ for all x.

34. Find an example to show that $7 - 2 \cdot y \neq (7 - 2)y$ for all y.

B   ■ *Simplify.*

35. $\left[\dfrac{7 - (-3)}{5 - 3}\right]\left[\dfrac{4 + (-8)}{3 - 5}\right]$

36. $\left[\dfrac{12 + (-2)}{3 + (-8)}\right]\left[\dfrac{6 + (-15)}{8 - 5}\right]$

37. $\left(3 - 2\left[\dfrac{5 - (-4)}{2 + 1} - \dfrac{6}{3}\right]\right) + 1$

38. $\left(7 + 3\left[\dfrac{6 + (-18)}{4 + 2}\right] - 5\right) + 3$

39. $\dfrac{8 - 6\left(\dfrac{5 + 3}{4 - 8}\right) - 3}{-2 + 4\left(\dfrac{6 - 3}{1 - 4}\right) + 6}$

40. $\dfrac{12 + 3\left(\dfrac{12 - 20}{3 - 1}\right) - 1}{-8 + 6\left(\dfrac{12 - 30}{2 - 5}\right) + 1}$

41. $\dfrac{3(3+2)-3\cdot3+2}{-3\cdot2+2(2-1)}$

42. $\dfrac{6-2\left(\dfrac{4+6}{5}\right)+8}{3-3\cdot2+8}$

43. $\dfrac{3|-3|-|-5|}{|-2|}-\dfrac{|-6|}{3}$

44. $\dfrac{2|5-8|}{|-3|}-\dfrac{|-4|+2}{|-2|}$

CHAPTER SUMMARY

[1.1]   A **set** is a collection of objects. The items in the collection are the **members**, or **elements**, of the set. **Natural numbers, whole numbers, integers, rational numbers**, and **irrational numbers** are all **real numbers**.

A **variable** is a symbol used to represent an unspecified element of a given set called the **replacement set** of the variable. A symbol used to denote a single element is called a **constant**.

[1.2]   The **equality** (=) and **order** ($<$, $>$) relations in the set of real numbers R are governed by a set of assumptions called **axioms**. Order relationships between real numbers can be pictured on a **number line**.

[1.3]   The operations of addition and multiplication are governed by the following axioms: The set of real numbers is **closed** under the operations of addition and multiplication. The operations are **commutative** and **associative**, and each operation has an **identity element**. Each number a has an **additive inverse (negative)** $-a$, and each nonzero number a has a **multiplicative inverse (reciprocal)** $1/a$. Multiplication **distributes** over addition. These properties imply the following properties, called **theorems**:

$$\text{If } a=b, \text{ then } a+c=b+c \text{ and } c+a=c+b;$$

and

$$\text{if } a=b, \text{ then } ac=bc \text{ and } ca=cb.$$

Furthermore,

$$a\cdot0=0,$$

and

$$-(-a)=a.$$

The **absolute value** of a number a is defined by

$$|a|=\begin{cases} a, & \text{if } a\ge0, \\ -a, & \text{if } a<0. \end{cases}$$

[1.4] The sum of two positive numbers is positive; the sum of two negative numbers is negative; and the sum of a positive number and a negative number is equal to the nonnegative difference of their absolute values preceded by the sign of the number with the greater absolute value.

The difference $a - b$ of two numbers is equal to the sum of a and the additive inverse of b:

$$a - b = a + (-b).$$

[1.5] The product of two positive numbers or two negative numbers is positive; the product of a positive number and a negative number is negative. For all a, $a \cdot 0 = 0$.

A natural number greater than 1 that has no natural-number factors other than itself and 1 is a **prime number**. A natural number greater than 1 that is not a prime number is a **composite number**, and it has one and only one prime factorization.

The quotient of two numbers a/b, with $b \neq 0$, is equal to the number q, such that $b \cdot q = a$. Alternatively,

$$\frac{a}{b} = a\left(\frac{1}{b}\right).$$

Division by 0 is not defined.

[1.6] Operations are performed in the following order:

1. First, any expression within a symbol of inclusion (parentheses, brackets, fraction bars, etc.) is simplified.

2. Next, multiplications and divisions are performed as encountered in order from left to right.

3. Last, additions and subtractions are performed in order from left to right.

The symbols introduced in this chapter are listed on the inside of the front cover.

REVIEW EXERCISES

A

[1.1] ■ *Let* $A = \left\{-3, -2.55, -\sqrt{2}, -\dfrac{3}{5}, 0, 1, 5.55\ldots, \dfrac{13}{2}\right\}$.

1. List the members in set A that are integers.

2. List the members in set A that are whole numbers.

[1.2]   **3.** Use symbols to show the relationship between a and c if $a<b$ and $b<c$.

4. Use symbols to show that a is equal to or less than 7.

5. Use symbols to show that b is between a and c.

6. Express the relation "y is greater than or equal to 6" using symbols.

7. Graph $\{x|-4\le x<5,\ x$ an integer$\}$.

8. Graph $\{x|-4\le x<5,\ x$ a real number$\}$.

[1.3]   **9.** By the commutative property of addition, $x+2=\underline{?}+\underline{?}$.

10. By the addition property of equality, if $z=18,$ then $z+t=\underline{?}$.

11. By the multiplicative-inverse property, $3\cdot\underline{?}=1$.

12. By the identity property for multiplication, $7\cdot\underline{?}=7$.

[1.4]   ■ *Write each sum or difference as a basic numeral.*

13. a. $2-(-4)+8$ **b.** $5-(-6)-3$

14. a. $3-|6|-|-3|$ **b.** $2+|6-12|-|3|$

[1.5]   ■ *Write each product as a basic numeral.*

15. a. $3(-2)(6)$ **b.** $-3(-6)(0)$ **c.** $7(-3)(-1)$ **d.** $-1(-2)(-3)(-4)$

■ *Express each integer in completely factored form.*

16. a. 24 **b.** 28 **c.** -42 **d.** -72

■ *Write each quotient as a basic numeral.*

17. a. $\dfrac{-24}{-2}$ **b.** $\dfrac{0}{-3}$ **c.** $\dfrac{-16}{2}$ **d.** $\dfrac{32}{-4}$

■ *Express each quotient as a product.*

18. a. $\dfrac{-7}{5}$ **b.** $\dfrac{24}{7}$ **c.** $-\dfrac{4}{5}$ **d.** $\dfrac{8}{3}$

[1.6]   ■ *Write each expression as a basic numeral.*

19. a. $5-\dfrac{3+9}{6}$ **b.** $\dfrac{6+2\cdot3}{10-4\cdot2}$

20. a. $\dfrac{6\cdot8\div4-2}{7-5}$ **b.** $\dfrac{4\cdot3-|-2|}{|-5|}$

B   **21.** If $A \ne B$ and $2 \in A$, must 2 be an element of B?

22. If $A \ne B$ and $B \ne C$, can $A = C$?

23. Can $y < |y - 1|$?

24. For what values of x is $|x| = x$?

25. Is the set of integers closed for:
a. Subtraction?   b. Division?

26. For what values of x and y will the following inequalities be true statements?
a. $x - y > 0$ b. $x - y < 0$

■ *Simplify each expression.*

27. $\left[\dfrac{4 - (-2)}{-3} \right] - 2[8 - 3 \cdot 4]$

28. $2 \left[4 - \dfrac{2 - (-6)}{4} \right] - 5 \left[3 - \dfrac{6 \cdot 3}{-4 - 2} \right]$

29. $\dfrac{3|-2| - (-4)}{|-5|} - \dfrac{|-4| - 4}{|-3|}$

30. $\dfrac{2|4 - 6| - |6 - 4|}{|4 - 6|} + 2[|5 - 6| - 3]$

2. POLYNOMIALS

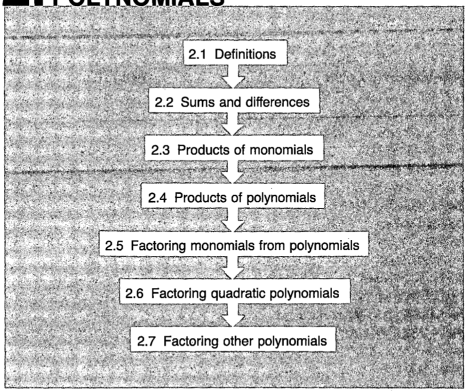

2.1 Definitions

2.2 Sums and differences

2.3 Products of monomials

2.4 Products of polynomials

2.5 Factoring monomials from polynomials

2.6 Factoring quadratic polynomials

2.7 Factoring other polynomials

2.1

DEFINITIONS

In Chapter 1 we saw that the sum of two numbers a and b is expressed by $a + b$ and their product is expressed by ab. Sometimes, to avoid any possible confusion, we used a centered dot to indicate multiplication; for example, $a \cdot b$, $x \cdot y$, $2 \cdot 3$. In other cases, we used parentheses around one or both of the symbols, as in $(2)(3)$, $2(3)$, and $(x)(x)$.

Powers

If the factors in a product are identical, the number of such factors is indicated by means of a natural-number superscript called an **exponent**. Thus, in the expression 3^4, which denotes $3 \cdot 3 \cdot 3 \cdot 3$, the exponent is 4. The expression 3^4 is called a **power**, and the number 3 is called the **base** of the power. In general:

▶ *If n is a natural number,*

$$a^n = a \cdot a \cdot a \cdots a \qquad (n \text{ factors}).$$

32

Examples   **a.** $x^2 = x \cdot x$ **b.** $y^3 = y \cdot y \cdot y$ **c.** $3^5 = 3 \cdot 3 \cdot 3 \cdot 3 \cdot 3$

If no exponent appears on a variable, as in y for example, the exponent 1 is assumed. Thus, $y = y^1$.

Algebraic expressions   Any meaningful collection of numerals, variables, and signs of operations is called an **algebraic expression** or simply an **expression**. In an expression of the form $A + B + C + \cdots$, A, B, and C are called **terms** of the expression. Thus, terms are separated by $+$, $-$, or $=$ and *not* by $\cdot$ or $\div$.

Examples   **a.** $3x + 4yz$ contains two terms. The first term, $3x$, contains two factors and the second term, $4yz$, contains three factors.

b. $3(x + 4y^2)$ contains only one term; however, the factor $x + 4y^2$ contains two terms.

Any factor or group of factors in a term is said to be the **coefficient** of the product of the remaining factors in the term. Thus, in the term $3xyz$, the product $3x$ is the coefficient of yz; y is the coefficient of $3xz$; 3 is the coefficient of xyz; and so on. Hereafter, the word "coefficient" will refer to a number unless otherwise indicated. For example, the coefficient of the term $4xy$ is 4; the coefficient of $-2a^2b$ is -2. The term **numerical coefficient** is sometimes used to emphasize this point. If no coefficient appears in a term, the coefficient is understood to be 1. Thus, x is viewed as $1x$.

Examples   **a.** In the expression $3x^2 - 5x$, the coefficient of x^2 is 3 and the coefficient of x is -5.

b. In the expression $x^2 - y$, the coefficient of x^2 is 1 and the coefficient of y is -1.

An algebraic expression of the form cx^n, where c is a constant and n is a *whole number*, or a product of such expressions, is called a **monomial**. A **polynomial** is an algebraic expression that contains only terms that are monomials. Thus,

$$x^2, \qquad \frac{1}{5}x - 2, \qquad 3x^2 - 2x + 1, \quad \text{and} \quad x^3 - 2x^2 + 1$$

are polynomials.

If the polynomial contains two or three terms, we refer to it as a **binomial** or **trinomial**, respectively. Polynomials with four or more terms do not have special names.

Examples   **a.** 3, $5x^3$, and x^2y are monomials.

b. $x + y$, $2x^2 - 3x$, and $4x + 3y$ are binomials.

c. $4x + 3y + 2z$, $x^2 + 2x + 1$, and $5xy + 2x - 3y$ are trinomials.

The degree of a monomial in one variable is given by the exponent on the variable. Thus, $3x^4$ is of fourth degree, and $7x^5$ is of fifth degree. The degree of a polynomial in one variable is the same as the degree of its term of largest degree.

Examples   **a.** $3x^2 + 2x + 1$ is a second-degree polynomial.

b. $x^5 - x - 1$ is a fifth-degree polynomial.

c. $2x - 3$ is a first-degree polynomial.

Order of
operations

Recall the agreement that was made in Section 1.6 concerning the order in which operations are to be performed. Because a power is a product,

> *simplify any powers of ungrouped bases in an expression before*
> *performing other multiplication or division operations.*

Examples   **a.** $5 + 3 \cdot 4^2$ **b.** $5 + (3 \cdot 4)^2$ **c.** $(5 + 3 \cdot 4)^2$

$\quad = 5 + 3 \cdot 16$ $= 5 + 12^2$ $= (5 + 12)^2$

$\quad = 5 + 48 = 53$ $= 5 + 144 = 149$ $= 17^2 = 289$

Particular care should be taken when evaluating powers that involve minus signs. For example, note that

$$(-2)^2 = (-2)(-2) = 4$$

and that

$$-2^2 = -(2)^2 = -2 \cdot 2 = -4.$$

Thus, $(-2)^2 \neq -2^2$; hence, $(-x)^2 \neq -x^2$ for all $x \neq 0$.

Names for
polynomials

Polynomials are frequently represented by symbols such as

$$P(x), \qquad D(y), \quad \text{and} \quad Q(z),$$

where the symbol in parentheses designates the variable. The symbols are read "P of x," "D of y," and "Q of z," respectively. For example, we might write

$$P(x) = x^2 - 2x + 1,$$

$$D(y) = y^6 - 2y^2 + 3y - 2,$$

$$Q(z) = 8z^4 + 3z^3 - 2z^2 + z - 1.$$

The notation $P(x)$ can be used to denote values of the polynomial for specific values of x. Thus, $P(2)$ represents the value of the polynomial $P(x)$ when x is replaced by 2.

Example   If $P(x) = x^2 - 2x + 1$,   then

$$P(2) = (2)^2 - 2(2) + 1 = 1,$$

$$P(3) = (3)^2 - 2(3) + 1 = 4,$$

$$P(-4) = (-4)^2 - 2(-4) + 1 = 25.$$

EXERCISE 2.1

A ■ *Identify each polynomial as a monomial, binomial, or trinomial and give the degree of the polynomial.*

Examples   **a.** $3n^2 - n$ **b.** $r^3 - 2r + r$

Solutions   **a.** Binomial; degree 2   **b.** Trinomial; degree 3

1. $2x^3 - x^2$ 2. $x^2 - 2x + 1$ 3. $5n^4$
4. $3n + 1$ 5. $3r^2 - r + 2$ 6. r^3
7. $y^3 - 2y^2 - y$ 8. $3y^2 + 1$

■ *Simplify.*

Examples   **a.** $(-4)^2 - 3^2$ **b.** $(3 + 1)^2 - \dfrac{4 \cdot 3^2}{6}$

Solutions   **a.** $(-4)^2 - 3^2 = 16 - 9$ **b.** $(3 + 1)^2 - \dfrac{4 \cdot 3^2}{6} = 4^2 - \dfrac{4 \cdot 9}{6}$
$$= 7$$
$$= 16 - \dfrac{36}{6}$$
$$= 16 - 6 = 10$$

Common errors: Note that in Example a, $-3^2 \neq (-3)(-3)$ or $+9$; in Example b, $4 \cdot 3^2 \neq (4 \cdot 3)^2$.

9. -5^2 10. $(-5)^2$ 11. $(-3)^2$
12. -3^2 13. $4^2 - 2 \cdot 3^2$ 14. $4^2 - (2 \cdot 3)^2$
15. $\dfrac{4 \cdot 2^3}{16} + 3 \cdot 4^2$ 16. $\dfrac{4 \cdot 3^2}{6} + (3 \cdot 4)^2$ 17. $\dfrac{3^2 - 5}{6 - 2^2} - \dfrac{6^2}{3^2}$
18. $\dfrac{3^2 \cdot 2^2}{4 - 1} + \dfrac{(-3)(2)^3}{6}$ 19. $\dfrac{(-5)^2 - 3^2}{4 - 6} + \dfrac{(-3)^2}{2 + 1}$ 20. $\dfrac{7^2 - 6^2}{10 + 3} - \dfrac{8^2 \cdot (-2)}{(-4)^2}$

■ *Given $x = 3$ and $y = -2$, evaluate each expression.*

Examples   **a.** $2xy - y^2$ **b.** $\dfrac{x^2 - y}{3x + 2} + x(-y)$

Solutions overleaf

Solutions   **a.** $2(3)(-2) - (-2)^2$ **b.** $\dfrac{(3)^2 - (-2)}{3(3) + 2} + 3[-(-2)]$

$\qquad\qquad = -12 - 4$ $= \dfrac{9 + 2}{9 + 2} + 3(2)$

$\qquad\qquad = -16$ $= \dfrac{11}{11} + 6 = 1 + 6 = 7$

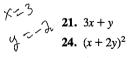

21. $3x + y$ **22.** $x - y^2$ **23.** $x^2 - 2y$

24. $(x + 2y)^2$ **25.** $x^2 - y^2$ **26.** $(3y)^2 - 3x$

27. $\dfrac{4x}{y} - xy$ **28.** $\dfrac{-xy^2}{6} + 2xy$ **29.** $\dfrac{(x - y)^2}{-5} + \dfrac{(xy)^2}{6}$

30. $(x + y)^2 + (x - y)^2$

■ *Evaluate each expression for the given values of the variables.*

31. $\dfrac{1}{2}gt^2;\quad g = 32$ and $t = 2$ **32.** $\dfrac{1}{2}gt^2 - 12t;\quad g = 32$ and $t = 3$

33. $\dfrac{Mv^2}{g};\quad M = 64,\ v = 2,$ and $g = 32$ **34.** $\dfrac{32(V - v)^2}{g};\quad V = 12,\ v = 4,$ and $g = 32$

35. $ar^{n-1};\quad a = 2,\ r = 3,$ and $n = 4$ **36.** $\dfrac{a - ar^n}{1 - r};\quad a = 4,\ r = 2,$ and $n = 3$

■ *Find the values of each expression for the specified values of the variable.*

Example   If $P(x) = 2x^2 - x + 3,$ find $P(-3)$ and $P(0)$.

Solution $P(-3) = 2(-3)^2 - (-3) + 3$ $P(0) = 2(0)^2 - (0) + 3$

$\qquad\qquad = 24$ $= 3$

37. If $P(x) = x^3 - 3x^2 + x + 1,$ find $P(2)$ and $P(-2)$.
38. If $P(x) = 2x^3 + x^2 - 3x + 4,$ find $P(3)$ and $P(-3)$.
39. If $D(x) = (2x - 4)^2 - x^2,$ find $D(2)$ and $D(0)$.
40. If $D(x) = (3x - 1)^2 + 2x^2,$ find $D(4)$ and $D(0)$.
41. If $P(x) = x^2 + 3x + 1$ and $Q(x) = x^3 - 1,$ find $P(3)$ and $Q(2)$.
42. If $P(x) = -3x^2 + 1$ and $Q(x) = 2x^2 - x + 1,$ find $P(0)$ and $Q(-1)$.
43. If $P(x) = x^5 + 3x - 1$ and $Q(x) = x^4 + 2x - 1,$ find $P(2)$ and $Q(-2)$.
44. If $P(x) = x^6 - x^5$ and $Q(x) = x^7 - x^6,$ find $P(-1)$ and $Q(-1)$.

B ■ *Given that* $P(x) = x + 1$, $Q(x) = x^2 - 1$, *and* $R(x) = x^2 + x - 1$, *find the value for each expression.*

Examples

a. $P(0) + Q(2)$ **b.** $P[Q(3)]$

Solutions

a.
$$P(0) = 0 + 1 = 1;$$
$$Q(2) = 2 - 1 = 4 - 1$$
$$= 3;$$
$$P(0) + Q(2) = 1 + 3 = 4$$

b.
$$Q(3) = (3)^2 - 1 = 9 - 1$$
$$= 8;$$
$$P[Q(3)] = P(8) = 8 + 1$$
$$= 9$$

45. $P(2) + R(3) - Q(1)$ **46.** $P(1) + R(-2) \cdot Q(-1)$

47. $P(-2) - R(1) \cdot Q(2)$ **48.** $P(3)[R(-1) + Q(1)]$

49. $P[Q(2)]$ **50.** $R[P(-1)]$

51. $Q[P(2)]$ **52.** $Q[R(-2)]$

53. Which axiom(s) from Chapter 1 justifies the assertion that if the variables in a monomial represent real numbers, the monomial represents a real number? That the same is true of any polynomial with real coefficients?

54. Argue that, for each real number $x > 0$, $(-x)^n = -x^n$ for n an odd natural number, and $(-x)^n = x^n$ for n an even natural number.
[*Hint:* Write $(-x)^n$ as $(-x)(-x) \ldots (-x) = (-1)(x)(-1)(x) \ldots (-1)(x).$]

2.2

SUMS AND DIFFERENCES

Rewriting sums

According to the symmetric property of equality, we can write the distributive property in the form

$$ab + ac = a(b + c);$$

and, using the commutative property, we can rewrite the above equation as

$$ba + ca = (b + c)a.$$

The distributive law in this form enables us to simplify certain polynomials.

Examples

a. $3x + 2x$
$$= (3 + 2)x$$
$$= 5x$$

b. $3y + 2y + 5y$
$$= (3 + 2 + 5)y$$
$$= 10y$$

c. $3x + 2y + x + 3y$
$$= 3x + x + 2y + 3y$$
$$= (3 + 1)x + (2 + 3)y$$
$$= 4x + 5y$$

Equivalent expressions Terms that differ only in their numerical coefficients are commonly called **like terms**. When we apply the distributive property to like terms, as in the examples above, we are *combining like terms*. The expression obtained from combining like terms represents the same number as the original expression for all real-number replacements of the variable or variables involved.

Expressions that represent (or name) *the same number for all replacements of the variables* are called **equivalent expressions**. Thus, in the examples above, we obtained equivalent expressions when we simplified each polynomial.

We can show that two expressions are *not* equivalent by simply finding replacement(s) for the variable(s) for which we obtain a false statement. For example, we can show that $-(xy)^2$ is not equivalent to $-xy^2$ by substituting numbers for x and y. Let us try 2 for x and 3 for y. Thus,

$$-(xy)^2 = -(2 \cdot 3)^2 = -6^2 = -36$$

and

$$-xy^2 = -2 \cdot 3^2 = -2 \cdot 9 = -18.$$

Hence, $-(xy)^2 \neq -xy^2$. We have used what is known as a **counterexample** to show that the two expressions are not equivalent.

Rewriting differences From the discussion on page 18 we know that $a - b$ is equal to $a + (-b)$. Thus:

▶ *We shall view the signs in any polynomial as indications of positive or negative coefficients, and the operation involved will be understood to be addition.*

Examples
a. $3x - 5x + 4x$
$\quad = (3x) + (-5x) + 4x$
$\quad = (3 - 5 + 4)x$
$\quad = 2x$

b. $(3x^2 + 2y^2) + (-x^2 + y^2)$
$\quad = 3x^2 + 2y^2 - x^2 + y^2$
$\quad = 3x^2 + (-x^2) + 2y^2 + y^2$
$\quad = (3 - 1)x^2 + (2 + 1)y^2$
$\quad = 2x^2 + 3y^2$

Sums can also be obtained in vertical form, with like terms vertically aligned.

Examples Express each sum as a single polynomial.

a. $(2x^2 - 3x + 1) + (-x + 3) + (x^2 + 5x - 2)$

b. $(3x^3 + 2x^2 - x + 1) + (3x^2 + 2x - 3) + (2x^3 - 4x^2 + x + 4)$

Solutions Write each sum in a vertical arrangement in which like terms are aligned.

a. $2x^2 - 3x + 1$
$\quad\quad\; - \;\; x + 3$
$\quad\;\; x^2 + 5x - 2$
$\quad\overline{\;\; 3x^2 + \;\; x + 2\;}$

b. $3x^3 + 2x^2 - \;\; x + 1$
$\quad\quad\quad\;\; 3x^2 + 2x - 3$
$\quad\; 2x^3 - 4x^2 + \;\; x + 4$
$\quad\overline{\;\; 5x^3 + \;\; x^2 + 2x + 2\;}$

We can write an expression such as $a + (b + c)$ without parentheses as $a + b + c$. However, for an expression such as

$$a - (b + c),$$

in which the parentheses are preceded by a negative sign, we first write

$$a - (b + c) = a + [-(b + c)].$$

Then, since

$$-(b + c) = -1(b + c) = -b - c,$$

we have

▶
$$a - (b + c) = a + [-b - c]$$
$$= a - b - c.$$

Thus,

an expression in parentheses preceded by a negative sign can be written equivalently without parentheses by replacing each term within the parentheses with its negative (additive inverse).

Examples
a. $(x^2 + 2x) - (2x^2 - 3x + 2)$
$= x^2 + 2x - 2x^2 + 3x - 2$
$= -x^2 + 5x - 2$

b. $(5x^2 - 2) - (x^2 - 2x) - (2x^2 - 4x + 3)$
$= 5x^2 - 2 - x^2 + 2x - 2x^2 + 4x - 3$
$= 2x^2 + 6x - 5$

In vertical form, we simply replace each term in the subtrahend with its negative, and then add.

Example

Subtract:

$$
\begin{array}{r}
3t^2 - 4t + 3 \\
(-) \quad -5t^2 + 2t - 2 \\
\hline
\end{array}
\quad \rightarrow \quad
\begin{array}{r}
3t^2 - 4t + 3 \\
(+) \quad 5t^2 - 2t + 2 \\
\hline
8t^2 - 6t + 5
\end{array}
$$

Simplifying expressions

In any expression where grouping devices occur within grouping devices, a great deal of difficulty can be avoided by removing the inner devices first and working outward.

Examples
a. $3x - [2 - (3x + 1)]$
$= 3x - [2 - 3x - 1]$
$= 3x - [1 - 3x]$
$= 3x - 1 + 3x$
$= 6x - 1$

b. $x^2 - [3x - (x^2 - 2)]$
$= x^2 - [3x - x^2 + 2]$
$= x^2 - 3x + x^2 - 2$
$= 2x^2 - 3x - 2$

It is usually helpful to simplify an expression inside a grouping device before removing the parentheses or brackets. Note that in Example a above, $2 - 3x - 1$ was written as $1 - 3x$ before removing the brackets.

The rewriting of a polynomial by combining like terms is a process called **simplifying** the polynomial, since the result is a polynomial with fewer terms than the original. If a polynomial contains no like terms or grouping symbols, we shall say that the polynomial is in **simple form**. Thus, the polynomial in simple form that is equivalent to $3x - [2 - (3x + 1)]$ in Example a above is $6x - 1$, which contains no grouping symbols and has all like terms combined.

EXERCISE 2.2

A   ■ *Write each expression in simplest form.*

Examples   **a.** $(x^2 + 3x) + (4 - 2x)$ **b.** $(3y^2 - 2y + 3) - (y^2 - 1)$

Solutions   **a.** $(x^2 + 3x) + (4 - 2x)$ **b.** $(3y^2 - 2y + 3) - (y^2 - 1)$
$\quad\quad = x^2 + 3x + 4 - 2x$ $\quad = 3y^2 - 2y + 3 - y^2 + 1$
$\quad\quad = x^2 + (3x - 2x) + 4$ $\quad = (3y^2 - y^2) + (-2y) + (3 + 1)$
$\quad\quad = x^2 + x + 4$ $\quad = 2y^2 - 2y + 4$

Common error: Note that in Example b, $-(y^2 - 1) \neq -y^2 - 1$.

1. $3x^2 + 4x^2$ **2.** $7x^3 - 3x^3$ **3.** $-6y + 3y$
4. $-5y - 6y$ **5.** $8z^2 - 8z^2 + z^2$ **6.** $-6z^3 + 6z^2 - z^2$
7. $3x^2y + 4x^2y - 2x$ **8.** $6xy^2 - 4xy^2 + 3y$ **9.** $3r^2 + (3r^2 + 4r)$
10. $r^2 - (2r^2 + r)$ **11.** $(s^2 - s) - 3s$ **12.** $(s^2 + s) - 3s^2$
13. $(2t^2 + 3t - 1) - (t^2 + t)$ **14.** $(t^2 - 4t + 1) - (2t^2 - 2)$
15. $(u^2 - 3u - 2) - (3u^2 - 2u + 1)$ **16.** $(2u^2 + 4u + 2) - (u^2 - 4u - 1)$

Example $(4x^2 - 3x + 2) + (x^2 - 4) + (3x^2 + 6x - 1)$

Solution $(4x^2 - 3x + 2) + (x^2 - 4) + (3x^2 + 6x - 1)$
$\quad\quad = 4x^2 - 3x + 2 + x^2 - 4 + 3x^2 + 6x - 1$
$\quad\quad = 8x^2 + 3x - 3$

17. $(4a^2 + 6a - 7) + (a^2 + 5a + 2) + (-2a^2 - 7a + 6)$
18. $(7c^2 - 10c + 8) + (8c + 11) + (-6c^2 - 3c - 2)$

19. $(4x^2y - 3xy + xy^2) + (-5x^2y + xy - 2xy^2) + (x^2y - 3xy)$

20. $(m^2n^2 - 2mn + 7) + (-2m^2n^2 + mn - 3) + (3m^2n^2 - 4mn + 2)$

Example $(4t^2 - 3t + 6) - (t^2 - 2t + 4)$

Solution

$$(4t^2 - 3t + 6) - (t^2 - 2t + 4) = 4t^2 - 3t + 6 - t^2 + 2t - 4$$
$$= 3t^2 - t + 2$$

21. $(2x^2 - 3x + 5) - (3x^2 + x - 2)$ **22.** $(4y^2 - 3y - 7) - (6y^2 - y + 2)$

23. $(4t^3 - 3t^2 + 2t - 1) - (5t^3 + t^2 + t - 2)$ **24.** $(4s^3 - 3s^2 + 2s - 1) - (s^3 - s^2 + 2s - 1)$

Example   Subtract $2t^2 - 3t + 2$ from the sum of $t^2 - t + 6$ and $3t^2 + 2t - 1$.

Solution

$$(t^2 - t + 6) + (3t^2 + 2t - 1) - (2t^2 - 3t + 2)$$
$$= t^2 - t + 6 + 3t^2 + 2t - 1 - 2t^2 + 3t - 2$$
$$= 2t^2 + 4t + 3$$

25. Subtract $4x^2 - 3x + 2$ from the sum of $x^2 - 2x + 3$ and $x^2 - 4$.

26. Subtract $2t^2 + 3t - 1$ from the sum of $2t^2 - 3t + 5$ and $t^2 + t + 2$.

27. Subtract the sum of $2b^2 - 3b + 2$ and $b^2 + b - 5$ from $4b^2 + b - 2$.

28. Subtract the sum of $7c^2 + 3c - 2$ and $3 - c - 5c^2$ from $2c^2 + 3c + 1$.

■ *Simplify.*

Examples   **a.** $x - [2x + (3 - x)]$ **b.** $[x^2 - (2x + 1)] - [2x^2 - (x - 3)]$

Solutions

a. $x - [2x + (3 - x)]$ **b.** $[x^2 - (2x + 1)] - [2x^2 - (x - 3)]$

     $= x - [2x + 3 - x]$ $= [x^2 - 2x - 1] - [2x^2 - x + 3]$

     $= x - [x + 3]$ $= x^2 - 2x - 1 - 2x^2 + x - 3$

     $= x - x - 3 = -3$ $= -x^2 - x - 4$

29. $y - [2y + (y + 1)]$ **30.** $3a + [2a - (a + 4)]$

31. $3 - [2x - (x + 1) + 2]$ **32.** $5 - [3y + (y - 4) - 1]$

33. $(3x + 2) - [x + (2 + x) + 1]$ **34.** $-(x - 3) + [2x - (3 + x) - 2]$

35. $[x^2 - (2x + 3)] - [2x^2 + (x - 2)]$ **36.** $[2y^2 - (4 - y)] + [y^2 - (2 + y)]$

37. Show by a counterexample that $-(x + 1)$ is not equivalent to $-x + 1$.

38. Show by a counterexample that $-(x - y)$ is not equivalent to $-x - y$.

B ■ *Simplify.*

39. $3y - (2x - y) - (y - [2x - (y - 2x)] + 3y)$

40. $[x - (y + x)] - (2x - [3x - (x - y)] + y)$

41. $[x - (3x + 2)] - (2x - [x - (4 + x)] - 1)$

42. $-(2y - [2y - 4y + (y - 2)] + 1) + [2y - (4 - y) + 1]$

■ *Given that* $P(x) = x - 1$, $Q(x) = x^2 + 1$, *and* $R(x) = x^2 - x + 1$, *write each expression in terms of* x.

43. $P(x) + Q(x) - R(x)$ **44.** $P(x) - Q(x) + R(x)$

45. $R(x) - [Q(x) - P(x)]$ **46.** $Q(x) - [R(x) + P(x)]$

2.3

PRODUCTS OF MONOMIALS

Consider the product

$$a^m a^n,$$

where m and n are natural numbers. Since

$$a^m = aaa \cdot \cdots \cdot a \quad (m \text{ factors}),$$

and

$$a^n = aaa \cdot \cdots \cdot a \quad (n \text{ factors}),$$

it follows that

$$a^m a^n = (\overbrace{aaa \cdot \cdots \cdot a}^{m \text{ factors}})(\overbrace{aaa \cdot \cdots \cdot a}^{n \text{ factors}})$$

$$= \overbrace{aaa \cdot \cdots \cdot a}^{m + n \text{ factors}}.$$

Hence, we have the following property:

▶ *For all natural numbers m and n,*

$$a^m a^n = a^{m+n}. \tag{1}$$

We shall refer to this property as the **first law of exponents**. Thus, we can simplify an expression for the product of two natural-number powers of the same base simply by adding the exponents and using the sum as an exponent on the same base.

Examples   **a.** $x^2 x^3 = x^{2+3} = x^5$ **b.** $xx^3 x^4 = x^{1+3+4} = x^8$ **c.** $y^3 y^4 y^2 = y^{3+4+2} = y^9$

We can use the commutative and associative properties of multiplication with the first law of exponents to multiply two or more monomials.

Examples   **a.** $(3x^2 y)(2xy^2)$ **b.** $(-2xy^4)(x^2 y)(4y^2)$

$\quad = 3 \cdot 2x^2 xyy^2$ $\quad = -2 \cdot 4xx^2 y^4 yy^2$

$\quad = 6x^3 y^3$ $\quad = -8x^3 y^7$

Now consider the powers $(a^m)^n$ and $(ab)^n$, where m and n are natural numbers:

$$(a^m)^n = a^m \cdot a^m \cdot a^m \cdot \cdots \cdot a^m \quad (n \text{ factors})$$

$$= a^{m+m+m+\cdots+m} \quad (n \text{ terms})$$

$$= a^{mn}$$

and

$$(ab)^n = (ab)(ab)(ab) \cdot \cdots \cdot (ab) \quad (n \text{ factors})$$

$$= \overbrace{(a \cdot a \cdot \cdots \cdot a)}^{n \text{ factors}} \overbrace{(b \cdot b \cdot \cdots \cdot b)}^{n \text{ factors}}$$

$$= a^n b^n.$$

Hence, we have the following properties:

▶ *For all natural numbers m and n,*

$$(a^m)^n = a^{mn} \tag{2}$$

and

$$(ab)^n = a^n b^n. \tag{3}$$

We shall refer to these properties as the **second** and **third laws of exponents**, respectively.

Examples   **a.** From (2),

$$(x^2)^3 = x^6 \quad \text{and} \quad (x^5)^2 = x^{10}.$$

b. From (3),

$$(xy)^3 = x^3 y^3 \quad \text{and} \quad (3x)^4 = 3^4 x^4 = 81x^4.$$

c. From (2) and (3),

$$(x^2 y^3)^3 = x^6 y^9 \quad \text{and} \quad (2x^3 y^2 z)^2 = 2^2 x^6 y^4 z^2 = 4x^6 y^4 z^2.$$

EXERCISE 2.3

A   ■ *Write each product as a polynomial in simplest form.*

Examples   **a.** $(7a^3)(4a^2b^4)$ **b.** $(-3c^2)(cd^2)(4c^3d)$

Solutions   **a.** $(7a^3)(4a^2b^4) = 7 \cdot 4 \cdot a^{3+2} \cdot b^4$ **b.** $(-3c^2)(cd^2)(4c^3d)$
$$= 28a^5b^4$$
$$\doteq -3 \cdot 4 \cdot c^{2+1+3}d^{2+1}$$
$$= -12c^6d^3$$

1. $(7t)(-2t^2)$ **2.** $(4c^3)(3c)$ **3.** $(4a^2b)(-10ab^2c)$

4. $(-6r^2s^2)(5rs^3)$ **5.** $(11x^2yz)(4xy^3z)$ **6.** $(-8abc)(-b^2c^3)$

7. $2(3x^2y)(x^3y^4)$ **8.** $-5(ab^3)(-3a^2bc)$ **9.** $(-r^3)(-r^2s^4)(-2rt^2)$

10. $(-5mn)(2m^2n)(-n^3)$ **11.** $(y^2z)(-3x^2z^2)(-y^4z)$ **12.** $(-3xy)(2xz^4)(3x^3y^2z)$

13. $(2rt)(-3r^2t)(-t^2)$ **14.** $-a^2(ab^2)(2a)(-3b^2)$ **15.** $-(2x)(x^2)(-3x)$

16. $(-2y^2)(y^2)(y)$ **17.** $z^3(-2z)(3z)(-1)$ **18.** $(-t)(2t^2)(-t)(0)$

■ *Write each expression as a polynomial in simplest form.*

Examples   **a.** $(2xy^3)^2$ **b.** $(xy^2z)^3 + (2x^2y)^4$

Solutions   **a.** $(2xy^3)^2 = 2^2x^2(y^3)^2$ **b.** $(xy^2z)^3 + (2x^2y)^4$
$$= 4x^2y^6$$
$$= x^3(y^2)^3z^3 + 2^4(x^2)^4y^4$$
$$= x^3y^6z^3 + 16x^8y^4$$

19. $(x^3)^2$ **20.** $(x^4)^3$ **21.** $(xy)^4$

22. $(xy)^2$ **23.** $(y^2z)^3$ **24.** $(yz^4)^2$

25. $(2xz^2)^3$ **26.** $(3x^2z)^2$ **27.** $(2xy^2z)^2$

28. $(3x^2yz^2)^3$ **29.** $(-2xy^3z)^3$ **30.** $(-3x^2yz^3)^3$

31. $(x^2y)^2 + (xy)^3$ **32.** $(2xy^3)^4 + (3xy)^2$

33. $(2xy^2z)^2 - (xyz^2)^2$ **34.** $(3x^2yz)^3 - (4x^2y^2z^2)^2$

B   ■ *Use the laws of exponents to simplify each expression. (Assume that variables in exponents represent natural numbers and that no denominator equals zero.)*

Examples   **a.** $x^n \cdot x^{n+3}$ **b.** $y^{2n+1} \cdot y^{3-n}$

Solutions   **a.** $x^n \cdot x^{n+3} = x^{n+n+3}$ **b.** $y^{2n+1} \cdot y^{3-n} = y^{2n+1+3-n}$
$$= x^{2n+3}$$
$$= y^{n+4}$$

35. $a^{2n} \cdot a^{n-3}$

36. $b^{-n} \cdot b^{2n+1}$

37. $x^{n^2-n} \cdot x^{2n-n^2}$

38. $y^{2n+6} \cdot y^{4-n}$

39. $a^{2n-2} \cdot a^{n+3}$

40. $b^{n+2} \cdot b^{2n-1}$

Examples

a. $(x^n y^m)^2$

b. $(x^{n+1} y^{n-1})^2$

Solutions

a. $(x^n y^m)^2 = x^{2n} y^{2m}$

b. $(x^{n+1} y^{n-1})^2 = x^{2(n+1)} y^{2(n-1)}$
$$= x^{2n+2} y^{2n-2}$$

41. $(x^{2n} y)^3$

42. $(x y^{3n})^2$

43. $(x^{n-2} y)^3$

44. $(x y^{n+2})^2$

45. $(x^{2n+1} y^{n-1})^3$

46. $(x^{n-2} y^{2n+1})^2$

2.4
PRODUCTS OF POLYNOMIALS

The associative property of addition can be used to extend the distributive property to cases where the right-hand factor contains more than two terms. Thus,

$$a[b + c + d] = a[(b + c) + d]$$
$$= a(b + c) + ad$$
$$= ab + ac + ad,$$
$$a[b + c + d + e] = a([(b + c) + d] + e)$$
$$= ab + ac + ad + ae, \quad \text{etc.}$$

We refer to this as the **generalized distributive property**.

Examples

a. $3x(x + y + z) = 3x(x) + 3x(y) + 3x(z)$
$$= 3x^2 + 3xy + 3xz$$

b. $-2ab^2(3a^2b - ab + 2ab^2) = (-2ab^2)(3a^2b) + (-2ab^2)(-ab) + (-2ab^2)(2ab^2)$
$$= -6a^3b^3 + 2a^2b^3 - 4a^2b^4$$

The distributive property can also be applied to simplify expressions for products of polynomials containing more than one term. For example,

$$(x - y)(3x + 2y) = 3x(x - y) + 2y(x - y) \tag{1}$$
$$= 3x^2 - 3xy + 2xy - 2y^2 \tag{2}$$
$$= 3x^2 - xy - 2y^2. \tag{3}$$

The four products in step (2) can be obtained more directly by writing the products in the form shown on page 46.

$$(3x + 2y)(x - y) = 3x^2 - 3xy + 2xy - 2y^2$$
$$= 3x^2 - xy - 2y^2.$$

This process is sometimes called the "FOIL" method, where "FOIL" represents:

the product of the First terms;
the product of the Outer terms;
the product of the Inner terms;
the product of the Last terms.

$(3x + 2y)(x - y)$

Examples

a. $(2x - 1)(x + 3)$
$= 2x^2 + 6x - x - 3$
$= 2x^2 + 5x - 3$

b. $(3x + 1)(2x - 1)$
$= 6x^2 - 3x + 2x - 1$
$= 6x^2 - x - 1$

Special products of binomials

The following products are special cases of the multiplication of binomials. Since they occur so frequently you should learn to recognize them on sight.

▶

I. $(x + a)^2 = (x + a)(x + a) = x^2 + 2ax + a^2$

II. $(x - a)^2 = (x - a)(x - a) = x^2 - 2ax + a^2$

III. $(x + a)(x - a) = x^2 - a^2$

Common error: Note that in I, $(x + a)^2 \neq x^2 + a^2$ and in II, $(x - a)^2 \neq x^2 - a^2$.

Examples

a. $(z - 3)^2$
$= (z - 3)(z - 3)$
$= z^2 - 2 \cdot 3z + 3^2$
$= z^2 - 6z + 9$

b. $(x + 4)^2$
$= (x + 4)(x + 4)$
$= x^2 + 2 \cdot 4x + 16$
$= x^2 + 8x + 16$

c. $(y + 5)(y - 5)$
$= y^2 - 5^2$
$= y^2 - 25$

More about products

The distributive property can also be used to simplify expressions involving products of polynomials containing more than two terms by multiplying each term of one factor by each of the terms in the other factor.

Example

$$(2x - 3)(x^2 + 2x - 1) = 2x(x^2 + 2x - 1) - 3(x^2 + 2x - 1)$$
$$= 2x^3 + 4x^2 - 2x - 3x^2 - 6x + 3$$
$$= 2x^3 + x^2 - 8x + 3$$

It is sometimes convenient to simplify products involving binomials or trinomials by using the familiar vertical form from arithmetic for computations. For example, the product $(2x - 3)(x^2 + 2x - 1)$ in the above example can be written as a polynomial in simple form as follows:

$$
\begin{array}{ll}
x^2 + 2x \ - 1 & \\
\quad\ \ 2x \ - 3 & \\
\hline
2x^3 + 4x^2 - 2x & [2x(x^2 + 2x - 1)] \\
\quad\ - 3x^2 - 6x + 3 & [-3(x^2 + 2x - 1)] \\
\hline
2x^3 + \ x^2 - 8x + 3 &
\end{array}
$$

Thus,

$$(2x - 3)(x^2 + 2x - 1) = 2x^3 + x^2 - 8x + 3.$$

The distributive property frequently plays a role in the simplification of expressions involving grouping devices. For example, we can begin to simplify

$$2[x - 3y + 3(y - x)] - 2(2x + y)$$

by applying the distributive property to $3(y - x)$, that is, to the *inner* set of parentheses, and then combining like terms.

Example

$$
\begin{aligned}
2[x - 3y + 3(y - x)] - 2(2x + y) &= 2[x - 3y + 3y - 3x] - 2(2x + y) \\
&= 2[-2x] - 2(2x + y) \\
&= -4x - 4x - 2y \\
&= -8x - 2y
\end{aligned}
$$

EXERCISE 2.4

A ■ *Write each expression as a polynomial in simple form.*

Examples

a. $-2(2x + y - z)$ **b.** $(x - 3)^2$ **c.** $(x - 4)(3x + 5)$

Solutions

a. $-2(2x + y - z)$
$= -4x - 2y + 2z$

b. $(x - 3)^2$
$= (x - 3)(x - 3)$
$= x^2 - 3x - 3x + 9$
$= x^2 - 6x + 9$

c. $(x - 4)(3x + 5)$
$= (x - 4)(3x + 5)$
$= 3x^2 + 5x - 12x - 20$
$= 3x^2 - 7x - 20$

1. $4y(x - 2y)$ 2. $3x(2x + y)$ 3. $-6x(2x^2 - x + 1)$

4. $-2y(y^2 - 3y + 2)$ 5. $-(x^2 - 4x - 1)$ 6. $-(2y^2 - y + 3)$

7. $(x + 3)^2$ 8. $(y - 4)^2$ 9. $(2y - 5)^2$

10. $(3x + 2)^2$ 11. $(x + 3)(x - 3)$ 12. $(x - 7)(x + 7)$

13. $(n + 2)(n + 8)$ 14. $(r - 1)(r - 6)$ 15. $(r + 5)(r - 2)$

16. $(y - 2)(y + 3)$ 17. $(y - 6)(y - 1)$ 18. $(z - 3)(z - 5)$

19. $(2z + 1)(z - 3)$ 20. $(3t - 1)(2t + 1)$ 21. $(4r + 3)(2r - 1)$

22. $(2z - 1)(3z + 5)$ 23. $(2x - a)(2x + a)$ 24. $(3t - 4s)(3t + 4s)$

Example

$(x - 4)(x^2 + 2x - 3)$

Solution

Method 1

$(x - 4)(x^2 + 2x - 3)$
$= x(x^2 + 2x - 3) - 4(x^2 + 2x - 3)$
$= x^3 + 2x^2 - 3x - 4x^2 - 8x + 12$
$= x^3 - 2x^2 - 11x + 12$

Method 2

$$
\begin{array}{r}
x^2 + 2x - 3 \\
x - 4 \\
\hline
x^3 + 2x^2 - 3x \\
-4x^2 - 8x + 12 \\
\hline
x^3 - 2x^2 - 11x + 12
\end{array}
$$

25. $(y + 2)(y^2 - 2y + 3)$ 26. $(t + 4)(t^2 - t - 1)$

27. $(x - 3)(x^2 + 5x - 6)$ 28. $(x - 7)(x^2 - 3x + 1)$

29. $(x - 2)(x - 1)(x + 3)$ 30. $(y + 2)(y - 2)(y + 4)$

31. $(z - 3)(z + 2)(z + 1)$ 32. $(z - 5)(z + 6)(z - 1)$

33. $(2x + 3)(3x^2 - 4x + 2)$ 34. $(3x - 2)(4x^2 + x - 2)$

35. $(2a^2 - 3a + 1)(3a^2 + 2a - 1)$ 36. $(b^2 - 3b + 5)(2b^2 - b + 1)$

Examples

a. $5 + 3[2a - 3(a - 2)]$ b. $-2(b - [2b + 3(b - 1)] + 3)$

Solutions

a. $5 + 3[2a - 3(a - 2)]$
$= 5 + 3[2a - 3a + 6]$
$= 5 + 3[-a + 6]$
$= 5 - 3a + 18$
$= 23 - 3a$

b. $-2(b - [2b + 3(b - 1)] + 3)$
$= -2(b - [2b + 3b - 3] + 3)$
$= -2(b - [5b - 3] + 3)$
$= -2(b - 5b + 3 + 3)$
$= -2(-4b + 6)$
$= 8b - 12$

37. $2[a - (a - 1) + 2]$ 38. $3[2a - (a + 1) + 3]$

39. $a[a - (2a + 3) - (a - 1)]$ 40. $-2a[3a + (a - 3) - (2a + 1)]$

41. $-[a - 3(a + 1) - (2a + 1)]$ **42.** $-[(a + 1) - 2(3a - 1) + 4]$

43. $2(a - [a - 2(a + 1) + 1] + 1)$ **44.** $-4(4 - [3 - 2(a - 1) + a] + a)$

45. $-x(x - 3[2x - 3(x + 1)] + 2)$ **46.** $x(4 - 2[3 - 4(x + 1)] - x)$

■ *Show that the left-hand member is equivalent to the right-hand member.*

Example $(ax - by)(cx - dy) = acx^2 - (ad + bc)xy + bdy^2$

Solution Rewriting the left-hand member yields

$$(ax - by)(cx - dy) = acx^2 - adxy - bcxy + bdy^2$$
$$= acx^2 - (ad + bc)xy + bdy^2.$$

Hence, the left-hand member of the original equation is equivalent to the right-hand member.

47. $(x + a)(x + b) = x^2 + (a + b)x + ab$ **48.** $(x + a)^2 = x^2 + 2ax + a^2$

49. $(x + a)(x - a) = x^2 - a^2$

50. $(ax + by)(cx + dy) = acx^2 + (ad + bc)xy + bdy^2$

51. $(x + a)(x^2 - ax + a^2) = x^3 + a^3$ **52.** $(x - a)(x^2 + ax + a^2) = x^3 - a^3$

53. Use a counterexample to show that $(x + y)^2$ is not equivalent to $x^2 + y^2$.

54. Use a counterexample to show that $(x - y)^2$ is not equivalent to $x^2 - y^2$.

B ■ *In Problems 55–66, assume that all variables in exponents denote natural numbers.*

Examples **a.** $2a^{2n}(3a^n - 2)$ **b.** $(2a^n + 1)(a^n - 2)$

Solutions
a. $2a^{2n}(3a^n - 2)$
  $= (2a^{2n})3a^n - (2a^{2n})2$
  $= 6a^{2n+n} - 4a^{2n}$
  $= 6a^{3n} - 4a^{2n}$

b. $(2a^n + 1)(a^n - 2)$
  $= 2a^n(a^n - 2) + 1(a^n - 2)$
  $= 2a^{2n} - 4a^n + a^n - 2$
  $= 2a^{2n} - 3a^n - 2$

55. $x^n(2x^n - 1)$ **56.** $3t^n(2t^n + 3)$ **57.** $a^{n+1}(a^n - 1)$

58. $b^{n-1}(b + b^n)$ **59.** $a^{2n+1}(a^n + a)$ **60.** $b^{2n+2}(b^{n-1} + b^n)$

61. $(1 + a^n)(1 - a^n)$ **62.** $(a^n - 3)(a^n + 2)$ **63.** $(a^{3n} + 2)(a^{3n} - 1)$

64. $(a^{3n} - 3)(a^{3n} + 3)$ **65.** $(2a^n - b^n)(a^n + 2b^n)$ **66.** $(a^{2n} - 2b^n)(a^{3n} + b^{2n})$

2.5

FACTORING MONOMIALS FROM POLYNOMIALS

In Section 1.5, we factored natural numbers that were not prime as the product of prime factors. Now we shall see how the distributive property in the form

$$ax + bx + cx + dx = (a + b + c + d)x$$

furnishes us a means of writing a polynomial as a single term comprised of two or more factors. By the distributive property,

$$3x^2 + 6x = 3x(x + 2).$$

Of course, we can also write

$$3x^2 + 6x = 3(x^2 + 2x)$$

or

$$3x^2 + 6x = 3x^2\left(1 + \frac{2}{x}\right) \qquad (x \neq 0)$$

or any other of an infinite number of such expressions. We are, however, primarily interested in factoring a polynomial into a unique form (except for signs and order of factors) referred to as the **completely factored form**. A polynomial with integral coefficients is in completely factored form if:

1. It is written as a product of polynomials with integral coefficients.

2. No polynomial—other than a monomial—can be further factored.

The restriction that the factors be polynomials means that all the variables involved have exponents from $\{1, 2, 3, \ldots\}$. Restricting the coefficients to integers prohibits such factorizations as

$$x + 3 = 3\left(\frac{1}{3}x + 1\right).$$

Because the choice of signs and order of factors is arbitrary, the factored form of an expression that seems most "natural" should be used, although this is admittedly not always easy to determine. For instance, the forms

$$a(1 - x - x^2) \quad \text{and} \quad -a(x^2 + x - 1)$$

are equivalent, but it is difficult to decide which is more "natural."

Common monomial factors can be factored from a polynomial by first identifying such common factors and then writing the resultant factored expression. For example, observe that the polynomial

$$6x^3 + 9x^2 - 3x$$

contains the monomial $3x$ as a factor of each term. We therefore write

$$6x^3 + 9x^2 - 3x = 3x(\qquad)$$

and insert within the parentheses the appropriate polynomial factor. This factor can be determined by inspection. We ask ourselves for the monomials that multiply $3x$ to yield $6x^3$, $9x^2$, and $-3x$. The final result appears as

$$6x^3 + 9x^2 - 3x = 3x(2x^2 + 3x - 1).$$

Examples

a. $6x^3 - 12x^2$
   $= 6x^2(x - 2)$

b. $6x^3 - 4x^2 + 2x$
   $= 2x(3x^2 - 2x + 1)$

Checking the result of factoring

We can always check the result of factoring an expression by multiplying the factors. In the examples above,

$$6x^2(x - 2) = 6x^3 - 12x^2$$

and

$$2x(3x^2 - 2x + 1) = 6x^3 - 4x^2 + 2x.$$

Notice that complete factorization of monomial factors is not required. Thus, in Example a above, it is not necessary that the form

$$6x^2(x - 2)$$

be written

$$2 \cdot 3 \cdot x \cdot x \cdot (x - 2)$$

in order for the expression to be considered completely factored.

Factoring $a - b$ as $-(b - a)$

One particularly useful factorization is of the form

$$a - b = (-1)(-a + b)$$
$$= (-1)(b - a)$$
$$= -(b - a).$$

Hence, we have the important relationship:

▶

$$a - b = -(b - a).$$

That is, $a - b$ and $b - a$ are negatives of each other.

Examples

a. $3x - y = -(y - 3x)$

b. $a - 2b = -(2b - a)$

EXERCISE 2.5

A ■ *Complete each factorization. Check by multiplying factors.*

Examples **a.** $x^5y^3 = x^2y(\underline{?})$ **b.** $12xy^3z^4 = 4y^2z^2(\underline{?})$

Solutions By inspection:

a. $x^5y^3 = x^2y(x^3y^2)$;   **b.** $12xy^3z^4 = 4y^2z^2(3xyz^2)$;
because $(x^2y)(x^3y^2) = x^5y^3$ because $4y^2z^2(3xyz^2) = 12xy^3z^4$

1. $x^3y^7 = xy^2(\underline{?})$ **2.** $x^6y^2 = xy(\underline{?})$ **3.** $-6x^4z^2 = 3xz^2(\underline{?})$
4. $-8x^3y^4 = 2x^2y^2(\underline{?})$ **5.** $4x^4y^4z^2 = 2xy^2(\underline{?})$ **6.** $12xy^5z^2 = 4y^2z(\underline{?})$

■ *Factor completely. Check by multiplying factors.*

Examples **a.** $6x - 18$ **b.** $18x^2y - 24xy^2$ **c.** $y(x - 2) + z(x - 2)$

Solutions **a.** $6x - 18$ **b.** $18x^2y - 24xy^2$ **c.** $y(x - 2) + z(x - 2)$
$\quad = 6(\underline{?} - \underline{?})$ $\quad = 6xy(\underline{?} - \underline{?})$ $\quad = (x - 2)(\underline{?} + \underline{?})$
$\quad = 6(x - 3)$; $\quad = 6xy(3x - 4y)$; $\quad = (x - 2)(y + z)$;
because   because   because
$\quad 6(x - 3) = 6x - 18$ $\quad 6xy(3x - 4y)$ $\quad (x - 2)(y + z)$
$\qquad\qquad = 18x^2y - 24xy^2$ $\qquad\qquad = y(x - 2) + z(x - 2)$

7. $2x + 6$ **8.** $3x - 9$ **9.** $4x^2 + 8x$
10. $3x^2y + 6xy$ **11.** $3x^2 - 3xy + 3x$ **12.** $x^3 - x^2 + x$
13. $24a^2 + 12a - 6$ **14.** $15r^2s + 18rs^2 - 3rs$ **15.** $2x^4 - 4x^2 + 8x$
16. $3n^4 - 6n^3 + 12n^2$ **17.** $12z^4 + 15z^3 - 9z^2$ **18.** $2x^2y^2 - 3xy + 5x^2$
19. $ay^2 + aby + ab$ **20.** $x^2y^2z^2 + 2xyz - xz$ **21.** $3m^2n - 6mn^2 + 12mn$
22. $6x^2y - 9xy^2 + 12x$ **23.** $15a^2c^2 - 12ac + 6ac^3$ **24.** $14xy + 21x^2y^2 - 28xyz$
25. $a(a + 3) + b(a + 3)$ **26.** $b(a - 2) + a(a - 2)$ **27.** $2x(x + 3) - y(x + 3)$
28. $y(y - 2) - 3x(y - 2)$ **29.** $2y(a + b) - x(a + b)$ **30.** $3x(2a - b) + 4y(2a - b)$

■ *Supply the missing factors or terms.*

Examples **a.** $-5x + 10 = -5(\underline{?})$ **b.** $x - 3 = -(\underline{?} - \underline{?})$

Solutions   a. $-5x + 10 = -5(x - 2)$ b. $x - 3 = -(-x + 3)$
$$= -(3 - x)$$

31. $7 - r = -(? - ?)$ **32.** $3m - 2n = -(? - ?)$ **33.** $2a - b = -(? - ?)$

34. $r^2 - s^2 t^2 = -(? - ?)$ **35.** $-2x + 2 = -2(?)$ **36.** $-6x - 9 = -3(?)$

37. $-ab - ac = ?(b + c)$ **38.** $-a^2 + ab = ?(a - b)$ **39.** $2x - 1 = -(?)$

40. $x^2 - 3 = -(?)$ **41.** $x - y + z = -(?)$ **42.** $3x + 3y - 2z = -(?)$

B ■ *Factor completely. (Assume that variables in the exponents denote natural numbers.)*

Examples   a. $x^{3n} + x^n$ b. $x^{n+2} - 2x^2$ c. $x^{2n} + x^{n+1} - x^n$

Solutions   a. $x^{3n} + x^n$ b. $x^{n+2} - 2x^2$ c. $x^{2n} + x^{n+1} - x^n$
$$= x^n(? + ?)$$ $$= x^n \cdot x^2 - 2x^2$$ $$= x^{2n} + x^n x - x^n$$
$$= x^n(x^{2n} + 1)$$ $$= x^2(? - ?)$$ $$= x^n(? + ? - ?)$$
$$= x^2(x^n - 2)$$ $$= x^n(x^n + x - 1)$$

43. $x^{2n} - x^n$ **44.** $x^{4n} + x^{2n}$ **45.** $x^{3n} - x^{2n} - x^n$

46. $y^{4n} + y^{3n} + y^{2n}$ **47.** $x^{n+2} + x^n$ **48.** $x^{n+2} + x^{n+1} + x^n$

49. $-x^{2n} - x^n = ?(x^n + 1)$ **50.** $-x^{5n} + x^{2n} = ?(x^{3n} - 1)$

51. $-x^{a+1} - x^a = -x^a(? + ?)$ **52.** $-y^{a+2} + y^2 = -y^2(? - ?)$

2.6

FACTORING QUADRATIC POLYNOMIALS

A common type of factoring involves one of the following quadratic (second-degree) binomials or trinomials.

▶
$$x^2 + (a + b)x + ab = (x + a)(x + b) \tag{1}$$
$$x^2 + 2ax + a^2 = (x + a)^2 \tag{2}$$
$$x^2 - a^2 = (x + a)(x - a) \tag{3}$$
$$acx^2 + (ad + bc)xy + bdy^2 = (ax + by)(cx + dy) \tag{4}$$
$$(ax)^2 - (by)^2 = (ax - by)(ax + by) \tag{5}$$

Again, we shall require integral coefficients and positive integral exponents on the variables when factoring these polynomials.

As an example of the application of form (1) on page 53, consider the trinomial

$$x^2 + 6x - 16.$$

We desire, if possible, to find two binomial factors,

$$(x + a)(x + b),$$

whose product is the given trinomial. We see from form (1) that a and b are two integers such that $a + b = 6$ and $ab = -16$; that is, their sum must be the coefficient of the linear term $6x$ and their product must be -16. By inspection, or by trial and error, we determine that the two numbers are 8 and -2, so that

$$x^2 + 6x - 16 = (x + 8)(x - 2).$$

Checking, we note that $(x + 8)(x - 2) = x^2 + 6x - 16$.

Examples   Factor:

a. $x^2 - 7x + 12$ **b.** $x^2 - x - 12$

Solutions   **a.** We want to find two numbers whose product is 12 and whose sum is -7. Since the product is positive and the sum is negative, the two numbers must both be negative. By inspection, or trial and error, the two numbers are -4 and -3. Hence,

$$x^2 - 7x + 12 = (x - 4)(x - 3).$$

b. We want to find two numbers whose product is -12 and whose sum is -1. Since the product is negative, the two numbers must be of opposite sign and their sum must be -1. By inspection, or trial and error, the two numbers are -4 and 3. Hence,

$$x^2 - x - 12 = (x - 4)(x + 3).$$

Although we have not specifically noted the check in the above examples, the check should be done mentally for each factorization.

Perfect-square trinomials   Form (2), the square of a binomial, is simply a special case of (1).

Examples   Factor:

a. $x^2 + 8x + 16$ **b.** $x^2 - 10x + 25$

Solutions   **a.** Two numbers whose product is 16 and whose sum is 8 are 4 and 4. Hence,

$$x^2 + 8x + 16 = (x + 4)(x + 4)$$
$$= (x + 4)^2.$$

b. Two numbers whose product is 25 and whose sum is -10 are -5 and -5. Hence,

$$x^2 - 10x + 25 = (x - 5)(x - 5)$$
$$= (x - 5)^2.$$

Equations of form (2) are sometimes called **perfect-square trinomials** because they are the square of a binomial.

Difference of two squares

Form (3) is another special case of (1), in which the coefficient of the first-degree term in x is 0. For example,

$$x^2 - 25 = x^2 + 0x - 25 = (x - 5)(x + 5).$$

In particular, form (3) states:

▶ *The difference of the squares of two numbers is equal to the product of the sum and the difference of the two numbers.*

The factors $x - 5$ and $x + 5$ in the above example are called **conjugates** of each other. In general, any binomials of the form $a - b$ and $a + b$ are called **conjugate pairs**.

Examples

Factor:

a. $x^2 - 81$　　　　　　　　　　　　**b.** $x^2 + 81$

Solutions

a. $x^2 - 81$ can be written as the difference of two squares, $(x)^2 - (9)^2$, which can be factored as conjugate pairs:

$$x^2 - 81 = x^2 - 9^2 = (x - 9)(x + 9).$$

b. $x^2 + 81$, equivalent to $x^2 + 0x + 81$, is *not* of form (3). It is *not* factorable, because no two real numbers have a product of 81 and a sum of 0.

Form (4) is a generalization of (1)—that is, in (4) we are confronted with a quadratic trinomial where the coefficient of the term of second degree in x is other than 1. The factoring of such a trinomial, for example $8x^2 - 9 - 21x$, is accomplished as follows:

1. Write in decreasing powers of x.

$$8x^2 - 21x - 9$$

2. Consider possible combinations of first-degree factors of the first term.

$$(8x \quad)(x \quad)$$
$$(4x \quad)(2x \quad)$$

Solution continued overleaf

3. Consider combinations of the factors ① of the last term.

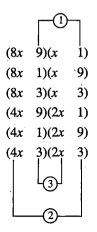

$$(8x \quad 9)(x \quad 1)$$
$$(8x \quad 1)(x \quad 9)$$
$$(8x \quad 3)(x \quad 3)$$
$$(4x \quad 9)(2x \quad 1)$$
$$(4x \quad 1)(2x \quad 9)$$
$$(4x \quad 3)(2x \quad 3)$$

4. Select the combination(s) of products ② and ③ whose sum(s) could be the second term ($-21x$).

$$(8x \quad 3)(x \quad 3)$$

5. Insert the proper signs.

$$(8x + 3)(x - 3)$$

Although this process can normally be done mentally, it is written in detail here for the purposes of illustration.

Signs on factors

Factoring trinomials of the form $Ax^2 + Bx + C$, where A is positive, can be facilitated by making use of the following considerations:

1. If both B and C are positive, both signs in the factored form are positive. For example, as a first step in factoring $6x^2 + 11x + 4$, we could write

$$(+)(+).$$

2. If B is negative and C is positive, both signs in the factored form are negative. Thus, as the first step in factoring $6x^2 - 11x + 4$, we could write

$$(-)(-).$$

3. If C is negative, the signs in the factored form are opposite. Thus, as a first step in factoring $6x^2 - 5x - 4$, we could write

$$(+)(-) \quad \text{or} \quad (-)(+).$$

Examples

a. $6x^2 + 5x + 1$
$= (+)(+)$
$= (3x + 1)(2x + 1)$

b. $6x^2 - 5x + 1$
$= (-)(-)$
$= (3x - 1)(2x - 1)$

c. $6x^2 - x - 1$
$= (+)(-)$
$= (3x + 1)(2x - 1)$

Difference of two squares Form (5) on page 53 is a special case of form (3), in which the coefficient of the xy term is 0.

Examples

a. $16y^2 - 1$
$$= (4y)^2 - (1)^2$$
$$= (4y - 1)(4y + 1)$$

b. $4x^2 - 9y^2$
$$= (2x)^2 - (3y)^2$$
$$= (2x - 3y)(2x + 3y)$$

If a polynomial of more than one term contains a common monomial factor in each of its terms, this monomial should be factored from the polynomial before seeking other factors.

Examples

a. $32x^2 - 84x - 36$
$$= 4(8x^2 - 21x - 9)$$
$$= 4(8x + 3)(x - 3)$$

b. $4x^2 - 100$
$$= 4(x^2 - 25)$$
$$= 4(x - 5)(x + 5)$$

Suggestions for factoring polynomials

1. Write a polynomial in one variable in descending powers of the variable.

2. Factor out any factors that are common to each term in the polynomial.

3. Factor any binomial that is the difference of two squares as the product of conjugate pairs.

4. Factor any trinomials that can be factored.

5. Check the result of the factoring by multiplying the factors.

EXERCISE 2.6

A ■ *Factor completely.*

Examples

a. $x^2 - 2x - 3$ b. $x^2 - 9y^2$ c. $5x^2 - 9x - 2$

Solutions

a. $x^2 - 2x - 3$
$$= (x - 3)(x + 1)$$

b. $x^2 - 9y^2$
$$= x^2 - (3y)^2$$
$$= (x - 3y)(x + 3y)$$

c. $5x^2 - 9x - 2$
$$= (5x + 1)(x - 2)$$

1. $x^2 + 5x + 6$ 2. $x^2 + 5x + 4$ 3. $y^2 - 7y + 12$

4. $y^2 - 7y + 10$ 5. $x^2 - x - 6$ 6. $x^2 - 2x - 15$

7. $y^2 - 3y - 10$ 8. $y^2 + 4y - 21$ 9. $x^2 - 25$

10. $x^2 - 36$ 11. $(xy)^2 - 1$ 12. $(xy)^2 - 4$

13. $y^4 - 9$ 14. $y^4 - 49$ 15. $x^2 - 4y^2$

16. $9x^2 - y^2$ 17. $4x^2 - 25y^2$ 18. $16x^2 - 9y^2$

19. $2x^2 + 3x - 2$ 20. $3x^2 - 7x + 2$ 21. $4x^2 + 7x - 2$

22. $6x^2 - 5x + 1$ 23. $3x^2 - 4x + 1$ 24. $4x^2 - 5x + 1$

25. $9x^2 - 21x - 8$ 26. $10x^2 - 3x - 18$ 27. $10x^2 - x - 3$

28. $8x^2 + 5x - 3$ 29. $4x^2 + 12x + 9$ 30. $4y^2 + 4y + 1$

31. $3x^2 - 7ax + 2a^2$ 32. $9x^2 + 9ax - 10a^2$ 33. $16x^2y^2 - 1$

34. $64x^2y^2 - 1$ 35. $9x^2y^2 + 6xy + 1$ 36. $4x^2y^2 + 12xy + 9$

Examples   a. $4a^3 - 5a^2 + a$ b. $8x^5 - 2x^3$

Solutions   a. $4a^3 - 5a^2 + a$ b. $8x^5 - 2x^3$

$\quad\quad\quad = a(4a^2 - 5a + 1)$ $\quad = 2x^3(4x^2 - 1)$

$\quad\quad\quad = a(4a - 1)(a - 1)$ $\quad = 2x^3(2x - 1)(2x + 1)$

37. $3x^2 + 12x + 12$ 38. $2x^2 + 6x - 20$ 39. $2a^3 - 8a^2 - 10a$

40. $2a^3 + 15a^2 + 7a$ 41. $4a^2 - 8ab + 4b^2$ 42. $20a^2 + 60ab + 45b^2$

43. $4x^2y - 36y$ 44. $x^2 - 4x^2y^2$ 45. $12x - x^2 - x^3$

46. $x^2 - 2x^3 + x^4$ 47. $x^4y^2 - x^2y^2$ 48. $x^3y - xy^3$

B   ■ *Factor completely.*

Examples   a. $x^4 + 2x^2 + 1$ b. $x^4 - 3x^2 - 4$

Solutions   a. $x^4 + 2x^2 + 1 = (x^2 + 1)(x^2 + 1)$ b. $x^4 - 3x^2 - 4$

$\quad\quad\quad\quad\quad = (x^2 + 1)^2$ $\quad = (x^2 - 4)(x^2 + 1)$

$\quad\quad\quad\quad\quad\quad\quad\quad\quad\quad\quad\quad = (x - 2)(x + 2)(x^2 + 1)$

49. $y^4 + 3y^2 + 2$ 50. $a^4 + 5a^2 + 6$ 51. $3x^4 + 7x^2 + 2$

52. $4x^4 - 11x^2 - 3$ 53. $x^4 + 3x^2 - 4$ 54. $x^4 - 6x^2 - 27$

55. $x^4 - 5x^2 + 4$ 56. $y^4 - 13y^2 + 36$ 57. $2a^4 - a^2 - 1$

58. $3x^4 - 11x^2 - 4$ 59. $x^4 + a^2x^2 - 2a^4$ 60. $4x^4 - 33a^2x^2 - 27a^4$

2.7

FACTORING OTHER POLYNOMIALS

A few other polynomials occur frequently enough to justify a study of their factorization.

Factoring by grouping

Sometimes a polynomial is factorable by grouping. For example, we can factor a from the first two terms of

$$ax + ay + bx + by$$

and b from the last two terms to obtain

$$a(x + y) + b(x + y).$$

Now, since the factor $x + y$ is a common factor of both terms, we can write the expression as

$$(a + b)(x + y).$$

Thus,

$$ax + ay + bx + by = (a + b)(x + y)$$

Example

Factor $3x^2y + 2y + 3xy^2 + 2x$.

Solution

We first rewrite the expression in the form

$$3x^2y + 2x + 3xy^2 + 2y.$$

We then factor the common monomial x from the first group of two terms and the common monomial y from the second group of two terms to obtain

$$x(3xy + 2) + y(3xy + 2).$$

If we now factor the common binomial $(3xy + 2)$ from each term, we have

$$(3xy + 2)(x + y).$$

Sum and difference of cubes

Two other factorizations are of special interest. These are the sum of two cubes, $x^3 + y^3$, and the difference of two cubes, $x^3 - y^3$. Our ability to factor these binomials is a result of observing that

$$(x + y)(x^2 - xy + y^2) = x^3 - x^2y + xy^2 + yx^2 - xy^2 + y^3$$
$$= x^3 + y^3$$

and that

$$(x - y)(x^2 + xy + y^2) = x^3 + x^2y + xy^2 - yx^2 - xy^2 - y^3$$
$$= x^3 - y^3.$$

Viewing the two equations from right to left, we have the following special factorizations:

▶

$$x^3 + y^3 = (x + y)(x^2 - xy + y^2)$$

and

$$x^3 - y^3 = (x - y)(x^2 + xy + y^2).$$

Examples

a. $8a^3 + b^3$
   $= (2a)^3 + b^3$
   $= (2a + b)[(2a)^2 - 2ab + b^2]$
   $= (2a + b)(4a^2 - 2ab + b^2)$

b. $x^3 - 27y^3$
   $= x^3 - (3y)^3$
   $= (x - 3y)[x^2 + 3xy + (3y)^2]$
   $= (x - 3y)(x^2 + 3xy + 9y^2)$

EXERCISE 2.7

A ■ *Factor.*

Examples

a. $yb - ya + xb - xa$

b. $x^2 + xb - ax - ab$

Solutions

a. $yb - ya + xb - xa$
   $= y(b - a) + x(b - a)$
   $= (b - a)(y + x)$

b. $x^2 + xb - ax - ab$
   $= x(x + b) - a(x + b)$
   $= (x + b)(x - a)$

1. $ax + a + bx + b$

2. $5a + ab + 5b + b^2$

3. $ax^2 + x + a^2x + a$

4. $a + ab + b + b^2$

5. $x^2 + ax + xy + ay$

6. $x^3 - x^2y + xy - y^2$

7. $3ab - cb - 3ad + cd$

8. $2ac - bc + 2ad - bd$

9. $3x + y - 6x^2 - 2xy$

10. $5xz - 5yz - x + y$

11. $a^3 + 2ab^2 - 2a^2b - 4b^3$

12. $6x^3 - 4x^2 + 3x - 2$

13. $x^2 - x + 2xy - 2y$

14. $2a^2 + 3a - 2ab - 3b$

15. $2a^2b + 6a^2 - b - 3$

16. $2ab^2 + 5a - 8b^2 - 20$

17. $x^3y^2 + x^3 - 3y^2 - 3$

18. $12 - 4y^3 - 3x^2 + x^2y^3$

Examples   **a.** $x^3 + 8$ **b.** $8x^3 - y^3$

Solutions   **a.** $x^3 + 8$ **b.** $8x^3 - y^3$

$\qquad = x^3 + (2)^3$ $\qquad = (2x)^3 - y^3$

$\qquad = (x + 2)(x^2 - 2x + 2^2)$ $\qquad = (2x - y)[(2x)^2 + 2xy + y^2]$

$\qquad = (x + 2)(x^2 - 2x + 4)$ $\qquad = (2x - y)(4x^2 + 2xy + y^2)$

19. $x^3 - 1$ **20.** $y^3 + 27$ **21.** $(2x)^3 + y^3$ **22.** $y^3 - (3x)^3$

23. $a^3 - 8b^3$ **24.** $27a^3 + b^3$ **25.** $(xy)^3 - 1$ **26.** $8 + x^3y^3$

27. $27a^3 + 64b^3$ **28.** $a^3 - 125b^3$

B   **29.** $x^3 + (x - y)^3$ **30.** $(x + y)^3 - z^3$ **31.** $(x + 1)^3 - 1$

     32. $x^6 + (x - 2y)^3$ **33.** $(x + 1)^3 - (x - 1)^3$ **34.** $(2y - 1)^3 + (y - 1)^3$

     35. Show that $ac - ad + bd - bc$ can be factored as $(a - b)(c - d)$ and as $(b - a)(d - c)$.

     36. Show that $a^2 - b^2 - c^2 + 2bc$ can be factored as $(a - b + c)(a + b - c)$.

CHAPTER SUMMARY

[2.1]   Expressions of the form a^n, where

$$a^n = aaa \cdot \cdots \cdot a \qquad (n \text{ factors}),$$

are called **powers**; a is the **base** and n is the **exponent** of the power.

Any meaningful collection of numerals, variables, and signs of operation is called an **expression**. In an expression of the form $A + B + C + \cdots$, A, B, and C are called **terms**. Any factor or group of factors in a term is the **coefficient** of the product of the remaining factors.

An algebraic expression of the form cx^n, where c is a constant and n is a whole number, or a product of such expressions, is called a **monomial**. A **polynomial** is an algebraic expression that contains only terms that are monomials. Polynomials of two and three terms are called **binomials** and **trinomials**, respectively.

The degree of a monomial in one variable is given by the exponents on the variable. The degree of a polynomial is the degree of its term of highest degree.

Polynomials are represented by symbols such as $P(x)$, $Q(z)$, etc., and the values of these polynomials for some specific value a are represented by $P(a)$, $Q(a)$, etc.

[2.2]   Terms that differ only in their numerical coefficients are called **like terms**. Two expressions that are equal for all real-number replacements of any variable or variables involved are **equivalent expressions**.

[2.3] The following laws of exponents are useful in rewriting powers:

$$a^m \cdot a^n = a^{m+n}, \qquad (a^m)^n = a^{mn}, \qquad (ab)^m = a^m b^m.$$

[2.4–2.7] The process of using the distributive property to rewrite a polynomial as a single term comprised of two or more factors is called **factoring**. If a polynomial of more than one term contains a common factor in each of its terms, this common factor should be factored from the polynomial first. Three special cases of factoring are:

$$x^2 - a^2 = (x - a)(x + a)$$
$$x^3 + a^3 = (x + a)(x^2 - ax + a^2)$$
$$x^3 - a^3 = (x - a)(x^2 + ax + a^2)$$

Some trinomials of the form $ax^2 + bx + c$ can be factored by trial and error. Some polynomials that consist of more than three terms can be factored by first grouping terms with like factors. Thus:

$$ax + ay + bx + by = a(x + y) + b(x + y)$$
$$= (a + b)(x + y).$$

The symbols introduced in this chapter are listed on the inside of the front cover.

REVIEW EXERCISES

A

[2.1] ■ *In Problems 1 and 2, identify each polynomial as a monomial, binomial, or trinomial. State the degree of the polynomial.*

1. a. $2y^3$ **b.** $3x^2 - 2x + 1$

2. a. $3x^2 - x^5$ **b.** $5y^4 - y^3 - y^2$

3. Find the value of $\dfrac{2x^2 - 3y}{x - y}$ for $x = -2$ and $y = 3$.

4. Find the value of $(x - y^2)^2 - xy$ for $x = 3$ and $y = -1$.

5. If $P(x) = 2x^2 - 3x - 1$, find   **a.** $P(3)$ **b.** $P(-2)$

6. If $Q(x) = x^3 - 2x^2 - x$, find   **a.** $Q(-1)$ **b.** $Q(-2)$

[2.2] ■ *Write each expression as a polynomial in simple form.*

7. a. $(2x - y) - (x - 2y + z)$ **b.** $(2x^2 - 3z^2) - (x - 2y) + (z^2 + y)$

■ *Simplify each expression.*

8. a. $2x - [x - (x - 3) + 2]$ **b.** $[x^2 - (x + 1)] - [2x^2 + (x - 1)]$

[2.3]   ■ *Simplify each expression.*

9. a. $(2x^2y)(-3xy^3)$ **b.** $(3xy)(2xz^2)(-y^2z)$
10. a. $(-2x^2y^3z)^3$ **b.** $(xy^2)^3 - (2x^3y)^2$

[2.4]   ■ *Write each expression as a polynomial in simple form.*

11. a. $2x(x^2 - 2x + 1)$ **b.** $2(2x - 1)(x + 3)$
12. a. $(y - 2)(y^2 - y + 1)$ **b.** $(z - 1)(z + 1)(z + 2)$
13. a. $3[x - 2(x + 1) - 3]$ **b.** $-x\{1 - 2[x + 3(x - 1)] + x\}$

[2.5]   ■ *Factor each polynomial.*

14. a. $12x^2 - 8x + 4$ **b.** $x^3 - 3x^2 - x$
15. a. $4y^3 - 8y^2$ **b.** $4x^3y^2 - 2x^2y^2 + 6xy^2$
16. a. $3x - y = -(?)$ **b.** $2x - y + z = -(?)$
17. a. $-x^2 + 2x = -x(?)$ **b.** $-6x^2y + 3xy - 3xy^2 = -3xy(?)$
18. a. $x(a + 2) - y(a + 2)$ **b.** $2a(x - y) + b(x - y)$

[2.6]   ■ *Factor each polynomial.*

19. a. $x^2 - 2x - 35$ **b.** $y^2 + 4y - 32$
20. a. $(xy)^2 - 36$ **b.** $a^2 - 49b^2$
21. a. $3y^2 + 11y - 4$ **b.** $x^3 + 3x^2 - 10x$
22. a. $9x^2 - 36$ **b.** $12x^2 - 3y^2$
23. a. $2x^2 + 3xy - 2y^2$ **b.** $6x^2 - xy - y^2$
24. a. $15a^2 + 28ab + 12b^2$ **b.** $12a^2 - 18ab + 6b^2$
25. a. $9x^2y^2 - 1$ **b.** $4x^2y^2 - x^2$
26. a. $2x^2 + 5ax - 3a^2$ **b.** $9x^2 - 12ax + 4a^2$

[2.7] ■ *Factor each polynomial.*

27. **a.** $2xy + 2x^2 + y + x$ **b.** $xy - 3x - y + 3$

28. **a.** $ax - 2bx + ay - 2by$ **b.** $2ax - 4ay + bx - 2by$

29. **a.** $(2x)^3 - y^3$ **b.** $x^3 + (4y)^3$

30. **a.** $27y^3 + z^3$ **b.** $x^3 - 8a^3$

B ■ *Given that* $P(x) = x^2 - 1$ *and* $Q(x) = x^2 + 3x - 1$, *find the value of each expression.*

31. $Q[P(2)]$ 32. $P[Q(-3)]$

■ *Given that* $P(x) = x + 1$, $Q(x) = x^2 - x$, *and* $R(x) = 2x^2 - 1$, *find the value of each expression.*

33. $P(2) - [Q(1) + R(0)]$ 34. $Q(-3) - [P(-2) - R(1)]$

■ *Write each expression without using parentheses. (Assume that variables in exponents denote natural numbers.)*

35. $(x^{n-1}y^n)^3$ 36. $(2x^n + 1)(x^n - 3)$

■ *Factor each expression completely. (Assume that variables in exponents denote natural numbers.)*

37. $x^{4n} + x^{2n}$ 38. $x^{n+2} - x^n$

39. $(x + y)^3 + (x - y)^3$ 40. $(x + y)^3 - (x - y)^3$

3. FRACTIONS

3.1

BASIC PROPERTIES

Recall that a fraction is an expression denoting a quotient. If the numerator (dividend) and the denominator (divisor) are polynomials, then the fraction is a **rational expression**. For example,

$$\frac{y}{y+1}, \qquad \frac{x^2 - 2x + 1}{x}, \qquad \frac{1}{x^2 + 1}, \quad \text{and} \quad \frac{x+1}{y}$$

are rational expressions. Any polynomial can be considered a rational expression, since it is the quotient of itself and 1. Thus,

$$x^2 + 2x, \quad 3y, \quad \text{and} \quad 5$$

are also rational expressions. For each replacement of the variable(s) for which the numerator and denominator of a fraction represent real numbers and for which the denominator is not zero, a rational expression represents a real number. Of course, for any value of the variable(s) for which the denominator vanishes (is equal to zero), the fraction does not represent a real number and is said to be undefined.

Fundamental principle of fractions

There are infinitely many fractions that correspond to a given quotient. Thus, for example,

$$\frac{1}{2}=\frac{2}{4}=\frac{3}{6}=\frac{4}{8}\cdots \quad \text{and} \quad \frac{3}{5}=\frac{6}{10}=\frac{9}{15}=\frac{12}{20}\cdots .$$

The properties of real numbers can be used to establish the following property which enables us to write such equivalent fractions.

► *An equivalent fraction is obtained if the numerator and the denominator of a fraction are each multiplied or divided by the same nonzero number.*

This property is commonly called **the fundamental principle of fractions** and is expressed in symbols as follows:

$$\frac{a}{b}=\frac{ac}{bc} \quad (b, c \neq 0).$$

Signs on fractions

There are three signs associated with a fraction: a sign for the numerator, a sign for the denominator, and a sign for the fraction itself. Although there are eight different possible symbols associated with the symbol "*a/b*" and the two signs "+" and "−," these symbols represent only two real numbers, *a/b* and its additive inverse −(*a/b*). The property below follows from the definition of a quotient and the fundamental principle of fractions.

►
$$\frac{a}{b}=\frac{-a}{-b}=-\frac{a}{-b}=-\frac{-a}{b} \quad (b \neq 0) \tag{1}$$

and

$$\frac{-a}{b}=\frac{a}{-b}=-\frac{a}{b}=-\frac{-a}{-b} \quad (b \neq 0). \tag{2}$$

Examples   a. $\dfrac{-2}{3}=\dfrac{2}{-3}=-\dfrac{2}{3}=-\dfrac{-2}{-3}$ b. $\dfrac{2}{3}=\dfrac{-2}{-3}=-\dfrac{2}{-3}=-\dfrac{-2}{3}$

We can use (1) and (2) to write a given fraction as an equivalent fraction by replacing any two of the fraction's three elements—the fraction itself, the numerator, and the denominator—with their negatives.

The forms *a/b* and −*a/b*, in which the sign of the fraction and the sign of the denominator are both positive, are generally the most convenient representations and will be referred to as **standard forms**. Thus,

$$\frac{-3}{5}, \quad \frac{3}{5}, \quad \text{and} \quad \frac{7}{10}$$

are in standard form, while

$$\frac{3}{-5}, \qquad -\frac{-3}{5}, \quad \text{and} \quad -\frac{7}{-10}$$

are not.

If the numerator or denominator of a fraction is an expression containing more than one term, there are alternative standard forms. For example, since

$$a - b = -(b - a),$$

we have

$$\frac{-b}{a - b} = \frac{-b}{-(b - a)} = \frac{b}{b - a},$$

and either

$$\frac{-b}{a - b} \quad \text{or} \quad \frac{b}{b - a}$$

may be taken as standard form, as convenience dictates. Observe that the quotient is not defined for $a - b = 0$ or $b - a = 0$; so we must make the restriction $a \neq b$.

Particular care should be taken when writing a fraction in standard form when the numerator contains more than one term. For example,

$$-\frac{a - b}{c} = \frac{-(a - b)}{c},$$

where the minus sign in the right member precedes *the entire numerator $a - b$*. This fraction can now be written as

$$\frac{-a + b}{c} \quad \text{or} \quad \frac{b - a}{c}.$$

Common error: In particular, note that $-\dfrac{a - b}{c} \neq \dfrac{-a - b}{c}$.

EXERCISE 3.1

A   ∎ *Write in standard form and specify any real values of the variables for which the fraction is undefined.*

Examples   **a.** $-\dfrac{5}{-y}$ **b.** $-\dfrac{a}{a - 2}$ **c.** $-\dfrac{x - 1}{3}$

Solutions overleaf

Solutions   **a.** $-\dfrac{5}{-y} = \dfrac{5}{y}$ $(y \neq 0)$ **b.** $-\dfrac{a}{a-2} = \dfrac{-a}{a-2}$ **c.** $-\dfrac{x-1}{3} = \dfrac{-(x-1)}{3}$

or $\dfrac{a}{2-a}$ $(a \neq 2)$ or $\dfrac{-x+1}{3}$

Common error: In Example c, note that $-\dfrac{x-1}{3} \neq \dfrac{-x-1}{3}$.

1. $\dfrac{1}{-4}$ **2.** $-\dfrac{1}{3}$ **3.** $-\dfrac{3}{-5}$ **4.** $-\dfrac{-3}{4}$

5. $\dfrac{-2}{-5}$ **6.** $\dfrac{-6}{-7}$ **7.** $-\dfrac{-3}{-7}$ **8.** $-\dfrac{-4}{-5}$

9. $-\dfrac{2x}{y}$ **10.** $\dfrac{x}{-3y}$ **11.** $-\dfrac{-3x}{4y}$ **12.** $-\dfrac{x}{-2y}$

13. $\dfrac{x+1}{-x}$ **14.** $\dfrac{x+3}{-x}$ **15.** $-\dfrac{x-y}{y+2}$ **16.** $-\dfrac{y-x}{y-1}$

■ *Write each fraction on the left as an equal fraction in standard form with the denominator shown on the right. (Assume that no denominator equals 0.)*

Examples   **a.** $\dfrac{-1}{2-x};\ \ \dfrac{}{x-2}$ **b.** $-\dfrac{a}{b-a};\ \ \dfrac{}{a-b}$ **c.** $\dfrac{3}{3x-2y};\ \ \dfrac{}{2y-3x}$

Solutions   **a.** $\dfrac{-1}{2-x}$ **b.** $-\dfrac{a}{b-a}$ **c.** $\dfrac{3}{3x-2y}$

$= \dfrac{-1}{-(x-2)}$ $= -\dfrac{a}{-(a-b)}$ $= \dfrac{3}{-(2y-3x)}$

$= \dfrac{1}{x-2}$ $= \dfrac{a}{a-b}$ $= \dfrac{-3}{2y-3x}$

17. $-\dfrac{4}{3-y};\ \ \dfrac{}{y-3}$ **18.** $\dfrac{-3}{2-x};\ \ \dfrac{}{x-2}$

19. $\dfrac{1}{x-y};\ \ \dfrac{}{y-x}$ **20.** $\dfrac{-6}{x-y};\ \ \dfrac{}{y-x}$

21. $\dfrac{x-2}{3-x};\ \ \dfrac{}{x-3}$ **22.** $\dfrac{2x-5}{3-y};\ \ \dfrac{}{y-3}$

23. Show by a counterexample that $-\dfrac{x+3}{2}$ is not equivalent to $\dfrac{-x+3}{2}$.

24. Show by a counterexample that $-\dfrac{2x-y}{3}$ is not equivalent to $\dfrac{-2x-y}{3}$.

B ■ *Write each fraction on the left as an equal fraction in standard form with the denominator shown on the right. (Assume that no denominator equals 0.)*

25. $\dfrac{x+1}{x-y}$; $\dfrac{\quad\quad}{y-x}$

26. $\dfrac{x+3}{y-x}$; $\dfrac{\quad\quad}{x-y}$

27. $-\dfrac{x-2}{x-y}$; $\dfrac{\quad\quad}{y-x}$

28. $-\dfrac{x-4}{x-2y}$; $\dfrac{\quad\quad}{2y-x}$

29. $\dfrac{-a+1}{-3a-b}$; $\dfrac{\quad\quad}{3a+b}$

30. $\dfrac{-a-1}{2b-3a}$; $\dfrac{\quad\quad}{3a-2b}$

3.2

REDUCING FRACTIONS

A fraction is said to be in lowest terms if the numerator and denominator do not contain common factors. The arithmetic fraction a/b, where a and b are integers and $b \neq 0$, is in lowest terms providing a and b are relatively prime—that is, providing they contain no common integral factor other than 1 or -1. If the numerator and denominator of a fraction are polynomials with integral coefficients, then the fraction is said to be in lowest terms if the numerator and denominator do not contain a common polynomial factor with integral coefficients other than 1 or -1.

To express a given fraction in lowest terms (to **reduce** the fraction), we can factor the numerator and denominator, and then apply the fundamental principle of fractions.

Examples

a. $\dfrac{yz^2}{y^3z} = \dfrac{z \cdot yz}{y^2 \cdot yz}$

$= \dfrac{z}{y^2} \quad (y, z \neq 0)$

b. $\dfrac{8x^3y}{6x^2y^3} = \dfrac{4x \cdot 2x^2y}{3y^2 \cdot 2x^2y}$

$= \dfrac{4x}{3y^2} \quad (x, y \neq 0)$

If binomial factors are involved, the process is the same.

Example

$\dfrac{12(x+y)(x-y)^2}{15(x+y)^2(x-y)} = \dfrac{4(x-y) \cdot [3(x+y)(x-y)]}{5(x+y) \cdot [3(x+y)(x-y)]}$

$= \dfrac{4(x-y)}{5(x+y)} \quad (x \neq y, -y).$

Slash lines are sometimes used to abbreviate the procedure in the above examples. For example, instead of writing

$$\frac{y}{y^2} = \frac{1 \cdot y}{y \cdot y} = \frac{1}{y} \quad (y \neq 0),$$

we can write

$$\frac{y}{y^2} = \frac{\overset{1}{\cancel{y}}}{\underset{y}{\cancel{y^2}}} = \frac{1}{y} \qquad (y \neq 0).$$

Reducing a fraction to lowest terms should be accomplished mentally whenever possible.

Note that the fundamental principle of fractions enables us to obtain an equivalent expression by dividing out any nonzero *factors* that are common to both numerator and denominator of a fraction. *The fundamental principle of fractions does not apply to common terms.* For example,

$$\frac{2xy}{3y} = \frac{2x}{3} \qquad (y \neq 0)$$

because y is a common factor in the numerator and denominator of the left member.

Common errors:

$$\frac{2x + y}{3 + y} \neq \frac{2x}{3} \tag{1}$$

because y *is not a common factor* of the numerator and the denominator. Furthermore,

$$\frac{5x + 3}{5y} \neq \frac{x + 3}{y} \tag{2}$$

because 5 *is not* a common factor of the *entire* numerator.

The examples in (1) and (2) above show two of the most common errors that are made in working with fractions.

More about reducing fractions   The division of a polynomial containing more than one term by a monomial may also be considered a special case of changing a fraction to lowest terms, providing the monomial is contained as a factor in each term of the polynomial.

Examples   a. $\dfrac{6y - 3}{3} = \dfrac{\cancel{3}(2y - 1)}{\cancel{3}}$

$\qquad\qquad = 2y - 1$

b. $\dfrac{9x^3 - 6x^2 + 3x}{3x}$

$\qquad = \dfrac{\cancel{3x}(3x^2 - 2x + 1)}{\cancel{3x}}$

$\qquad = 3x^2 - 2x + 1 \quad (x \neq 0)$

If the divisor is contained as a factor in the dividend, then the division of one polynomial by another polynomial, where each contains more than one term, can also be considered an example of reducing a fraction to lowest terms.

Examples

a. $\dfrac{y^2 - 5y + 4}{y - 1} = \dfrac{(y-4)(y-1)}{y-1}$

$= y - 4 \qquad (y \neq 1)$

b. $\dfrac{2x^2 + x - 15}{x + 3}$

$= \dfrac{(2x-5)(x+3)}{x+3}$

$= 2x - 5 \qquad (x \neq -3)$

EXERCISE 3.2

A ■ *Use the fundamental principle of fractions to reduce each fraction to lowest terms. (Assume no denominator is 0.)*

Examples

a. $\dfrac{x^3 y^4}{xy}$

b. $\dfrac{x^4 y^3}{xy^5}$

Solutions Using the fundamental principle of fractions:

a. $\dfrac{\overset{x^2 y^3}{\cancel{x^3 y^4}}}{\cancel{xy}} = x^2 y^3$

b. $\dfrac{\overset{x^3}{\cancel{x^4 y^3}}}{\underset{y^2}{\cancel{xy^5}}} = \dfrac{x^3}{y^2}$

1. $\dfrac{6x^3 y^2}{3xy}$

2. $\dfrac{12a^4 b^2}{4a^2 b^2}$

3. $\dfrac{14t^3 r^4}{7t^2 r^2}$

4. $\dfrac{22a^2 bc^3}{11ac^2}$

5. $\dfrac{14c^4 d^3}{-7c^2 d^3}$

6. $\dfrac{100m^2 n^3}{-5m^2 n^3}$

7. $\dfrac{a^5 b^7 c^6}{a^4 bc^3}$

8. $\dfrac{x^4 y^8 z^6}{xy^7 z^5}$

9. $\dfrac{6m^2 np^3}{-6m^2 np^3}$

10. $\dfrac{34a^6 b^2 c^4}{-17abc^3}$

11. $\dfrac{-12r^2 st^3}{-6rst^2}$

12. $\dfrac{-15xy^3 z^4}{-3y^2 z^4}$

Examples

a. $\dfrac{3y^2 - 12}{y - 2}$

b. $\dfrac{a - b}{b^2 - a^2}$

Solutions

a. $\dfrac{3y^2 - 12}{y - 2} = \dfrac{3(y^2 - 4)}{y - 2}$

$= \dfrac{3(y+2)(y-2)}{y-2}$

$= 3(y + 2)$

b. $\dfrac{a - b}{b^2 - a^2} = \dfrac{-1(b-a)}{(b+a)(b-a)}$

$= \dfrac{-1}{b + a}$

13. $\dfrac{5a - 10}{a - 2}$

14. $\dfrac{6x + 9}{10x + 15}$

15. $\dfrac{(a - b)^2}{a - b}$

16. $\dfrac{x^2 - 16}{x + 4}$

17. $\dfrac{6t^2 - 6}{t - 1}$

18. $\dfrac{4x^2 - 4}{x + 1}$

19. $\dfrac{2y^2 - 8}{y + 2}$ **20.** $\dfrac{5y^2 - 20}{y - 2}$ **21.** $\dfrac{3 - y}{y^2 - 9}$

22. $\dfrac{2 - y}{y^2 - 4}$ **23.** $\dfrac{y - x}{(x - y)^2}$ **24.** $\dfrac{2x - y}{y^2 - 4x^2}$

Examples **a.** $\dfrac{2x + 4}{4}$ **b.** $\dfrac{3x + 6}{3}$ **c.** $\dfrac{9x^2 + 3}{6x + 3}$

Solutions **a.** $\dfrac{2x + 4}{4} = \dfrac{\cancel{2}(x + 2)}{\cancel{2}(2)}$ **b.** $\dfrac{3x + 6}{3} = \dfrac{\cancel{3}(x + 2)}{\cancel{3}}$ **c.** $\dfrac{9x^2 + 3}{6x + 3} = \dfrac{\cancel{3}(3x^2 + 1)}{\cancel{3}(2x + 1)}$

$\qquad = \dfrac{x + 2}{2}$ $\qquad = x + 2$ $\qquad = \dfrac{3x^2 + 1}{2x + 1}$

Common errors: Note that, in Example a, $\dfrac{2x + 4}{4} \neq \dfrac{2x + \cancel{4}}{\cancel{4}}$; in Example b, $\dfrac{9x^2 + 3}{6x + 3} \neq \dfrac{9x^2 + \cancel{3}}{6x + \cancel{3}}$

and in Example c, $\dfrac{3x + 6}{3} \neq \dfrac{\cancel{3}x + 6}{\cancel{3}}$.

25. $\dfrac{4x + 6}{6}$ **26.** $\dfrac{2y - 8}{8}$ **27.** $\dfrac{9x - 3}{9}$

28. $\dfrac{5y - 10}{5}$ **29.** $\dfrac{ay - a}{a}$ **30.** $\dfrac{bx^2 + b}{bx + b}$

Examples **a.** $\dfrac{2y^3 - 6y^2 + 10y}{2y}$ **b.** $\dfrac{y^2 + y - 6}{y - 2}$ **c.** $\dfrac{x^3 + 8}{x^2 + 5x + 6}$

Solutions **a.** $\dfrac{2y^3 - 6y^2 + 10y}{2y}$ **b.** $\dfrac{y^2 + y - 6}{y - 2}$ **c.** $\dfrac{x^3 + 8}{x^2 + 5x + 6}$

$\qquad = \dfrac{\cancel{2y}(y^2 - 3y + 5)}{\cancel{2y}}$ $\qquad = \dfrac{(y + 3)\cancel{(y - 2)}}{1 \cdot \cancel{(y - 2)}}$ $\qquad = \dfrac{\cancel{(x + 2)}(x^2 - 2x + 4)}{\cancel{(x + 2)}(x + 3)}$

$\qquad = y^2 - 3y + 5$ $\qquad = y + 3$ $\qquad = \dfrac{x^2 - 2x + 4}{x + 3}$

31. $\dfrac{a^3 - 3a^2 + 2a}{a}$ **32.** $\dfrac{3x^3 - 6x^2 + 3x}{-3x}$ **33.** $\dfrac{y^2 + 5y - 14}{y - 2}$

34. $\dfrac{x^2 + 5x + 6}{x + 3}$ **35.** $\dfrac{x^2 - 5x + 4}{x^2 - 1}$ **36.** $\dfrac{y^2 - 2y - 3}{y^2 - 9}$

37. $\dfrac{2y^2 + y - 6}{y^2 + y - 2}$ **38.** $\dfrac{6x^2 - x - 1}{2x^2 + 9x - 5}$ **39.** $\dfrac{x^2 + xy - 2y^2}{x^2 - y^2}$

40. $\dfrac{4x^2 - 9y^2}{2x^2 + xy - 6y^2}$ **41.** $\dfrac{8y^3 - 27}{2y - 3}$ **42.** $\dfrac{8x^3 - 1}{2x - 1}$

43. Show by a counterexample that $\dfrac{2x+y}{y}$ is not equivalent to $2x$.

44. Show by a counterexample that $\dfrac{4x-y}{4}$ is not equivalent to $x-y$.

3.3

QUOTIENTS OF POLYNOMIALS

In Section 3.2, we used the fundamental principle of fractions to reduce fractions in which common factors occurred in numerators and denominators. However, we can use other methods to rewrite fractions if the polynomial in the numerator and the polynomial in the denominator do not have common factors.

Monomial denominators   If the divisor in a quotient of polynomials is a monomial, we can first rewrite the quotient as the sum of two or more fractions; that is,

$$\frac{a+b+c}{g} = \frac{a}{g} + \frac{b}{g} + \frac{c}{g}.$$

Examples   **a.** $\dfrac{6y^2 + 4y + 1}{2} = \dfrac{6y^2}{2} + \dfrac{4y}{2} + \dfrac{1}{2}$

$$= 3y^2 + 2y + \frac{1}{2}$$

b. $\dfrac{9x^3 - 6x^2 + 4}{3x}$

$$= \frac{9x^3}{3x} - \frac{6x^2}{3x} + \frac{4}{3x}$$

$$= 3x^2 - 2x + \frac{4}{3x} \quad (x \neq 0)$$

Long division   If the denominator in a quotient of polynomials is not a monomial, we can use a method similar to the long division process used in arithmetic to rewrite the quotient as the sum of a polynomial and a fraction. For example, consider the fraction

$$\frac{2x^2 + x - 7}{x + 3}.$$

We first write

$$x + 3 \overline{\smash{\big)}\, 2x^2 + x - 7}$$

and then divide $2x^2$ by x ($2x^2/x$) to obtain $2x$. Subtract the product of $2x$ and $x + 3$ from $2x^2 + x$, and "bring down" -7:

$$
\begin{array}{r}
2x \\
x + 3 \overline{\smash{\big)}\, 2x^2 + x - 7} \\
2x^2 + 6x \\
\hline
-5x - 7
\end{array}
$$

Change signs and add →

Solution continued overleaf

Then divide $-5x$ by x ($-5x/x$) to obtain -5. Subtract the product of -5 and $x + 3$ from $-5x - 7$.

$$
\begin{array}{r}
2x - 5 \\
x + 3 \overline{\smash{\big)}\, 2x^2 + x - 7} \\
\text{(Change signs and add)} \quad \underline{2x^2 + 6x } \\
-5x - 7 \\
\text{(Change signs and add)} \quad \underline{-5x - 15} \\
8
\end{array}
$$

Change signs and add $\rightarrow$

Change signs and add $\rightarrow$

Hence,

$$
\frac{2x^2 + x - 7}{x + 3} = 2x - 5 + \frac{8}{x + 3} \qquad (x \neq -3).
$$

When using this process of long division, it is necessary to write the dividend in descending powers of the variable. Furthermore, it is sometimes helpful to insert a term with a zero coefficient so that like terms will be aligned for convenient computation.

Example   Rewrite $\dfrac{3x - 1 + 4x^3}{2x - 1}$ using the long division process.

Solution   We first write $3x - 1 + 4x^3$ in descending powers as $4x^3 + 3x - 1$. Then we insert $0x^2$ between $4x^3$ and $3x$ and divide.

$$
\begin{array}{r}
2x^2 + x + 2 \\
2x - 1 \overline{\smash{\big)}\, 4x^3 + 0x^2 + 3x - 1} \\
\text{(Change signs and add)} \quad \underline{4x^3 - 2x^2 } \\
2x^2 + 3x \\
\text{(Change signs and add)} \quad \underline{2x^2 - x } \\
4x - 1 \\
\text{(Change signs and add)} \quad \underline{4x - 2} \\
1
\end{array}
$$

Change signs and add $\rightarrow$

Change signs and add $\rightarrow$

Change signs and add $\rightarrow$

Hence,

$$
\frac{3x - 1 + 4x^3}{2x - 1} = 2x^2 + x + 2 + \frac{1}{2x - 1} \qquad \left(x \neq \frac{1}{2} \right).
$$

We call an expression such as

$$
2x^2 + x + 2 + \frac{1}{2x - 1}
$$

a **mixed expression**, just as we call symbols such as $3\frac{1}{2}$ mixed numbers.

To avoid the necessity of always having to note restrictions on divisors (denominators), we shall assume in the remaining exercise sets in this chapter that no denominator is 0.

EXERCISE 3.3

A ■ *Divide.*

Examples

a. $\dfrac{2y^3 - 6y^2 + 4}{y}$

b. $\dfrac{27x^2y^2 + 18xy - 3}{9xy}$

Solutions

a. $\dfrac{2y^3 - 6y^2 + 4}{y}$

$= \dfrac{2y^3}{y} - \dfrac{6y^2}{y} + \dfrac{4}{y}$

$= 2y^2 - 6y + \dfrac{4}{y}$

b. $\dfrac{27x^2y^2 + 18xy - 3}{9xy}$

$= \dfrac{27x^2y^2}{9xy} + \dfrac{18xy}{9xy} - \dfrac{3}{9xy}$

$= 3xy + 2 - \dfrac{1}{3xy}$

1. $\dfrac{8a^2 + 4a + 1}{2}$

2. $\dfrac{15t^3 - 12t^2 + 5t}{3t^2}$

3. $\dfrac{7y^4 - 14y^2 + 3}{7y^2}$

4. $\dfrac{21n^4 + 14n^2 - 7}{7n^2}$

5. $\dfrac{18r^2s^2 - 15rs + 6}{3rs}$

6. $\dfrac{12x^3 - 8x^2 + 3x}{4x}$

7. $\dfrac{8a^2x^2 - 4ax^2 + ax}{2ax}$

8. $\dfrac{9a^2b^2 + 3ab^2 + 4a^2b}{ab^2}$

9. $\dfrac{25m^6 - 15m^3 + 7}{-5m^3}$

10. $\dfrac{36t^5 + 24t^3 - 12t}{-12t^2}$

11. $\dfrac{40m^4 - 25m^2 + 7m}{5m^2}$

12. $\dfrac{15s^{10} - 21s^5 + 6}{3s^2}$

Examples

a. $\dfrac{4y^2 - 4y - 5}{2y + 1}$

b. $\dfrac{16y^3 - 3y + 5}{y - 2}$

Solutions

a.

$$
\begin{array}{r}
2y \ - 3 \\
2y + 1 \overline{\big)\,4y^2 - 4y - 5} \\
\end{array}
$$

Change signs and add $\rightarrow$ $4y^2 + 2y$

$\quad\quad -6y - 5$

Change signs and add $\rightarrow$ $-6y - 3$

$\quad\quad\quad\quad -2$

$\dfrac{4y^2 - 4y - 5}{2y + 1} = 2y - 3 + \dfrac{-2}{2y + 1}.$

b.

$$
\begin{array}{r}
16y^2 + 32y \ + 61 \\
y - 2 \overline{\big)\,16y^3 +\ 0y^2 -\ 3y +\ \ 5} \\
\end{array}
$$

$\quad 16y^3 - 32y^2$

$\quad\quad 32y^2 -\ 3y$

$\quad\quad 32y^2 - 64y$

$\quad\quad\quad\quad 61y +\ \ 5$

$\quad\quad\quad\quad 61y - 122$

$\quad\quad\quad\quad\quad\quad 127$

$\dfrac{16y^3 - 3y + 5}{y - 2}$

$= 16y^2 + 32y + 61 + \dfrac{127}{y - 2}.$

13. $\dfrac{4y^2 + 12y + 7}{2y + 1}$

14. $\dfrac{2n^2 + 13n - 6}{2n - 1}$

15. $\dfrac{4t^2 - 4t - 5}{2t - 1}$

16. $\dfrac{2x^2 - 3x - 15}{2x + 5}$

17. $\dfrac{x^3 + 2x^2 + x + 1}{x - 2}$

18. $\dfrac{2x^3 - 3x^2 - 2x + 4}{x + 1}$

19. $\dfrac{2a - 3a^2 + a^4 - 1}{a + 3}$

20. $\dfrac{3 + 2b^2 + 2b^4}{b - 4}$

21. $\dfrac{4z^2 + 5z + 8z^4 + 3}{2z + 1}$

22. $\dfrac{7 - 3t^3 - 23t^2 + 10t^4}{2t + 3}$

23. $\dfrac{x^4 - 1}{x - 2}$

24. $\dfrac{y^5 + 1}{y - 1}$

B

Example

$$\dfrac{z^4 - 3z^3 + 2z^2 - 3z + 1}{z^2 + 2z - 1}$$

Solution

$$
\begin{array}{r}
z^2 - 5z\ \ + 13 \\
z^2 + 2z - 1\overline{\smash{\big)}\,z^4 - 3z^3 +\ \ 2z^2 -\ \ 3z +\ \ 1} \\
\underline{z^4 + 2z^3 -\ \ z^2} \\
-5z^3 +\ \ 3z^2 -\ 3z \\
\underline{-5z^3 - 10z^2 +\ \ 5z} \\
13z^2 -\ \ 8z +\ \ 1 \\
\underline{13z^2 + 26z - 13} \\
-34z + 14
\end{array}
$$

$$\dfrac{z^4 - 3z^3 + 2z^2 - 3z + 1}{z^2 + 2z - 1} = z^2 - 5z + 13 + \dfrac{-34z + 14}{z^2 + 2z - 1}.$$

25. $\dfrac{x^3 - 3x^2 + 2x + 5}{x^2 - 2x + 7}$

26. $\dfrac{2y^3 + 5y^2 - 3y + 2}{y^2 - y - 3}$

27. $\dfrac{4a^4 + 3a^3 - 2a + 1}{a^2 + 3a - 1}$

28. $\dfrac{2b^4 - 3b^2 + b + 2}{b^2 + b - 3}$

29. $\dfrac{t^4 - 3t^3 + 2t^2 - 2t + 1}{t^3 - 2t^2 + t + 2}$

30. $\dfrac{r^4 + r^3 - 2r^2 + r + 5}{r^3 + 2r + 3}$

31. Determine k so that the polynomial $x^3 - 3x + k$ has $x - 2$ as a factor.

32. Determine k so that the polynomial $x^3 + 2x^2 + k$ has $x + 3$ as a factor.

3.4

BUILDING FRACTIONS

Just as we changed fractions to equivalent fractions in lowest terms in Section 3.2 by applying the fundamental principle in the form

$$\frac{ac}{bc} = \frac{a}{b} \qquad (b, c \neq 0),$$

we can also change fractions to equivalent fractions in higher terms by applying the fundamental principle in the form

$$\frac{a}{b} = \frac{ac}{bc} \qquad (b, c \neq 0).$$

For example, $\frac{1}{2}$ can be changed to an equivalent fraction with a denominator of 8 by multiplying the numerator by 4 and the denominator by 4. Thus,

$$\frac{1}{2} = \frac{1 \cdot 4}{2 \cdot 4} = \frac{4}{8}.$$

This process is called **building** a fraction, and the number 4 is said to be a **building factor**. The fraction $\frac{4}{8}$ is said to be in **higher terms** than $\frac{1}{2}$.

In general, when building a/b to an equivalent fraction with bc as a denominator (i.e., $a/b = ?/bc$) we can usually determine the building factor c by inspection, and then multiply the numerator and the denominator of the original fraction by this building factor. If the building factor cannot be obtained by inspection, the desired denominator (bc) can be divided by the denominator of the given fraction (b) to determine the building factor (c).

Examples   Express each fraction as an equivalent fraction with the given denominator.

a. $\dfrac{5x}{3y} = \dfrac{?}{12y^2}$ **b.** $\dfrac{2}{y-1} = \dfrac{?}{y^2 - 1}$

Solutions   **a.** By inspection we note that the building factor is $4y$. Alternatively,

$$12y^2 \div 3y = 4y.$$

Hence,

$$\frac{5x(4y)}{3y(4y)} = \frac{20xy}{12y^2}.$$

b. We first factor $y^2 - 1$ as $(y-1)(y+1)$. Then by inspection we note that the building factor is $y + 1$. Alternatively,

$$(y^2 - 1) \div (y - 1) = y + 1.$$

Solution continued overleaf

Hence,

$$\frac{2(y+1)}{(y-1)(y+1)} = \frac{2y+2}{y^2-1}.$$

Least common denominator   Sometimes it is necessary to change two or more fractions with unlike denominators into fractions with a common denominator. In particular, the common denominator that is the most useful is the **least common multiple (LCM)** of the denominators, called the **least common denominator (LCD)**. The LCM of two or more natural numbers is the smallest natural number that is exactly divisible by each of the given numbers. Thus, 24 is the LCM of 6 and 8, because 24 is the smallest natural number each will divide into without a remainder.

In the example above it was easy to find the LCM of 6 and 8 (24) by inspection. Sometimes it is necessary to first factor each number in order to find the LCM.

To find the LCM of a set of natural numbers:

1. Express each number in completely factored form.

2. Write as factors of a product each *different* prime factor* occurring in any of the numbers, including each factor the greatest number of times it occurs in any one of the given numbers.

Example   Find the LCM of 12, 9, and 15.

Solution

The numbers: 12   9   15

Appear in prime factor form as: $2 \cdot 2 \cdot 3$ $3 \cdot 3$ $3 \cdot 5$

The LCM contains the factors: $2 \cdot 2 \cdot 3 \cdot 3 \cdot 5$

The LCM is: 180

The factors 2 and 3 are each used twice because 2 appears twice as a factor of 12 and 3 appears twice as a factor of 9.

We define the LCM of a set of polynomials as the polynomial of lowest degree yielding a polynomial quotient upon division by each of the given polynomials.

We can find the LCM of a set of polynomials with integral coefficients in a manner comparable to that used with a set of natural numbers.

Example   Find the LCM of $x^2 - 9$ and $x^2 - x - 6$.

---

*Prime factors were discussed in Section 1.5.

Solution

We first factor each polynomial.

$$x^2 - 9 \qquad x^2 - x - 6$$
$$(x - 3)(x + 3) \qquad (x - 3)(x + 2)$$

Hence, the LCM is $(x - 3)(x + 3)(x + 2)$.

We can now write two or more fractions with unlike denominators as equivalent fractions with a common denominator. In particular, with the least common denominator (LCD).

Example

Write the fractions $\frac{1}{12}, \frac{2}{9}$, and $\frac{4}{15}$ as equivalent fractions with the LCD of the fractions.

Solution

From the example above, the LCM of 12, 9, and 15 is 180. Hence, the LCD is 180 and we seek numerators such that

$$\frac{1}{12} = \frac{?}{180}, \qquad \frac{2}{9} = \frac{?}{180}, \quad \text{and} \quad \frac{4}{15} = \frac{?}{180}.$$

Since $180 \div 12 = 15$, $180 \div 9 = 20$, and $180 \div 15 = 12$, we have

$$\frac{1(15)}{12(15)} = \frac{15}{180}, \qquad \frac{2(20)}{9(20)} = \frac{40}{180}, \quad \text{and} \quad \frac{4(12)}{15(12)} = \frac{48}{180}.$$

A similar procedure is used if the fractions involve variables.

Example

Write the fractions $\frac{3}{x^2 - 9}$ and $\frac{4x}{x^2 - x - 6}$ as equivalent fractions with the LCD of the fractions.

Solution

From the example above, the LCM of $x^2 - 9$ and $x^2 - x - 6$ is $(x - 3)(x + 3)(x + 2)$. Hence, the LCD is $(x - 3)(x + 3)(x + 2)$ and we seek numerators such that

$$\frac{3}{x^2 - 9} = \frac{?}{(x - 3)(x + 3)(x + 2)}$$

and

$$\frac{4x}{x^2 - x - 6} = \frac{?}{(x - 3)(x + 3)(x + 2)}.$$

Since

$$(x - 3)(x + 3)(x + 2) \div (x - 3)(x + 3) = x + 2$$

and

$$(x - 3)(x + 3)(x + 2) \div (x - 3)(x + 2) = x + 3,$$

Solution continued overleaf

we have

$$\frac{3(x+2)}{(x^2-9)(x+2)} = \frac{3x+6}{(x-3)(x+3)(x+2)}$$

and

$$\frac{4x(x+3)}{(x^2-x-6)(x+3)} = \frac{4x^2+12x}{(x-3)(x+3)(x+2)}.$$

EXERCISE 3.4

A ■ *Express each fraction as an equivalent fraction with the given denominator.*

Examples **a.** $\dfrac{3}{4xy} = \dfrac{?}{8x^2y^2}$ **b.** $\dfrac{a+1}{3} = \dfrac{?}{6(a-3)}$

Solutions

a. Obtain the building factor by inspection, or note that

$$8x^2y^2 \div 4xy = 2xy$$

Multiply the numerator and the denominator of the given fraction by the building factor.

$$\frac{3}{4xy} = \frac{3(2xy)}{4xy(2xy)}$$

$$= \frac{6xy}{8x^2y^2}$$

b. Obtain the building factor by inspection, or note that

$$6(a-3) \div 3 = 2(a-3)$$

Multiply the numerator and the denominator of the given fraction by the building factor.

$$\frac{a+1}{3} = \frac{(a+1)(2)(a-3)}{3(2)(a-3)}$$

$$= \frac{2a^2-4a-6}{6(a-3)}$$

1. $\dfrac{2}{3} = \dfrac{?}{9}$ **2.** $\dfrac{3}{4} = \dfrac{?}{8}$ **3.** $\dfrac{-15}{7} = \dfrac{?}{14}$ **4.** $\dfrac{-12}{5} = \dfrac{?}{20}$

5. $4 = \dfrac{?}{5}$ **6.** $6 = \dfrac{?}{7}$ **7.** $\dfrac{2}{6x} = \dfrac{?}{18x}$ **8.** $\dfrac{5}{3y} = \dfrac{?}{21y}$

9. $\dfrac{-a^2}{b^2} = \dfrac{?}{b^3}$ **10.** $\dfrac{-a}{b} = \dfrac{?}{ab^2}$ **11.** $y = \dfrac{?}{xy}$ **12.** $x = \dfrac{?}{xy^3}$

Example $\dfrac{3}{2a-2b} = \dfrac{?}{4a^2-4b^2}$

Solution Factor denominators.

$$\frac{3}{2(a-b)} = \frac{?}{4(a-b)(a+b)}$$

Obtain the building factor by inspection, or note that

$$4(a - b)(a + b) \div 2(a - b) = 2(a + b).$$

Multiply the numerator and the denominator of the given fraction by the building factor $2(a + b)$.

$$\frac{3}{2a - 2b} = \frac{3 \cdot 2(a + b)}{2(a - b) \cdot 2(a + b)} = \frac{6(a + b)}{4a^2 - 4b^2}$$

13. $\dfrac{3}{a - b} = \dfrac{?}{a^2 - b^2}$ **14.** $\dfrac{5}{2a + b} = \dfrac{?}{4a^2 - b^2}$ **15.** $\dfrac{3x}{y + 2} = \dfrac{?}{y^2 - y - 6}$

16. $\dfrac{5x}{y + 3} = \dfrac{?}{y^2 + y - 6}$ **17.** $\dfrac{-2}{x + 1} = \dfrac{?}{x^2 + 3x + 2}$ **18.** $\dfrac{-3}{a + 2} = \dfrac{?}{a^2 + 3a + 2}$

■ *Find the LCM.*

Examples

  a. 24, 30, 20   **b.** $2a$, $4b$, $6ab^2$

Solutions

  a. $24 = 2 \cdot 2 \cdot 2 \cdot 3$,   **b.** $2a = 2 \cdot a$,

       $30 = 2 \cdot 3 \cdot 5$, $4b = 2 \cdot 2 \cdot b$,

       $20 = 2 \cdot 2 \cdot 5$ $6ab^2 = 2 \cdot 3 \cdot a \cdot b \cdot b$

       LCM: $2^3 \cdot 3 \cdot 5 = 120$ LCM: $2^2 \cdot 3ab^2 = 12ab^2$

19. 4, 6, 10   **20.** 3, 4, 5   **21.** 6, 8, 15

22. 4, 15, 18   **23.** 14, 21, 36   **24.** 4, 11, 22

25. $2ab$, $6b^2$ **26.** $12xy$, $24x^3y^2$ **27.** $6xy$, $8x^2$, $3xy^2$

28. $7x$, $8y$, $6z$ **29.** $(a - b)$, $a(a - b)^2$ **30.** $6(x + y)^2$, $4xy^2$

Examples

  a. $2a - 2$, $a - 1$ **b.** $x^2 - 1$, $2(x - 1)^2$

Solutions

  a. $2a - 2 = 2(a - 1)$,   **b.** $x^2 - 1 = (x - 1)(x + 1)$,

       $a - 1 = (a - 1)$ $2(x - 1)^2 = 2(x - 1)(x - 1)$

       LCM: $2(a - 1)$ LCM: $2(x - 1)(x - 1)(x + 1)$

31. $a^2 - b^2$, $a - b$ **32.** $x + 2$, $x^2 - 4$ **33.** $a^2 + 5a + 4$, $(a + 1)^2$

34. $x^2 - 3x + 2$, $(x - 1)^2$ **35.** $x^2 + 3x - 4$, $(x - 1)^2$ **36.** $x^2 - x - 2$, $(x - 2)^2$

37. $x^2 - x$, $(x - 1)^3$ **38.** $y^2 + 2y$, $(y + 2)^2$ **39.** $4a^2 - 4$, $(a - 1)^2$, 2

40. $3x^2 - 3$, $(x - 1)^2$, 4   **41.** x^3, $x^2 - x$, $(x - 1)^2$ **42.** y, $y^3 - y$, $(y - 1)^3$

■ *Write each fraction equivalently with the LCD of the given fractions.*

Example

$$\frac{3}{x^2 - 1} \text{ and } \frac{1}{2(x-1)^2}$$

Solution

From Example b above, the LCD of the fractions is $2(x-1)(x-1)(x+1)$. Now build each fraction to a fraction with this denominator:

$$\frac{3}{x^2 - 1} = \frac{3 \cdot 2(x-1)}{2(x-1)(x-1)(x+1)} = \frac{6x - 6}{2(x-1)(x-1)(x+1)}$$

and

$$\frac{1}{2(x-1)^2} = \frac{1 \cdot (x+1)}{2(x-1)(x-1)(x+1)} = \frac{x+1}{2(x-1)(x-1)(x+1)}.$$

43. $\dfrac{2}{3x}$ and $\dfrac{1}{6xy}$

44. $\dfrac{1}{4y}$ and $\dfrac{3}{8xy}$

45. $\dfrac{3}{4y^2}$ and $\dfrac{2}{3y}$

46. $\dfrac{1}{5x}$ and $\dfrac{3}{2x^2}$

47. $\dfrac{2}{a^2 - 1}$ and $\dfrac{3}{a - 1}$

48. $\dfrac{4}{(a-1)^2}$ and $\dfrac{2}{a - 1}$

49. $\dfrac{1}{a^2 - 5a + 4}$ and $\dfrac{2}{a^2 - 2a + 1}$

50. $\dfrac{3}{a^2 - 1}$ and $\dfrac{1}{a^2 + a - 2}$

51. $\dfrac{3y}{y^2 + 3y + 2}$ and $\dfrac{y}{y^2 + 4y + 4}$

52. $\dfrac{y}{y^2 + 5y + 4}$ and $\dfrac{y}{y^2 - 16}$

3.5

SUMS AND DIFFERENCES

Fractions with common denominators

Although we have defined fractions as symbols, they represent real numbers for permissible real-number replacements of any variables involved. Therefore, we can use the properties of real numbers to rewrite sums involving fractions in simpler form.

Since, for $c \neq 0$,

$$\frac{a}{c} = a\left(\frac{1}{c}\right) \quad \text{and} \quad \frac{b}{c} = b\left(\frac{1}{c}\right),$$

$$\frac{a}{c} + \frac{b}{c} = a\left(\frac{1}{c}\right) + b\left(\frac{1}{c}\right)$$

$$= (a + b)\frac{1}{c}$$

$$= \frac{a + b}{c}.$$

Furthermore, from the definition of a difference,

$$\frac{a}{c} - \frac{b}{c} = \frac{a}{c} + \left(\frac{-b}{c}\right)$$

$$= \frac{a-b}{c}.$$

Hence, we have the following properties for sums and differences:

▶
$$\frac{a}{c} + \frac{b}{c} = \frac{a+b}{c}$$

and

$$\frac{a}{c} - \frac{b}{c} = \frac{a-b}{c} \qquad (c \neq 0).$$

Examples   **a.** $\dfrac{2x}{9} + \dfrac{5x}{9} = \dfrac{2x+5x}{9}$ **b.** $\dfrac{3}{7y} - \dfrac{2}{7y} + \dfrac{5}{7y} = \dfrac{3-2+5}{7y}$

$\qquad\qquad\quad = \dfrac{7x}{9}$ $\qquad\qquad\qquad\qquad = \dfrac{6}{7y}$

Fractions
with unlike
denominators

If the fractions in a sum or difference have unlike denominators, we can build the fractions to equivalent fractions that have common denominators and then rewrite the sum as above.

Example $\dfrac{2y}{9} + \dfrac{5y}{6} - \dfrac{y}{4};$ the LCD is $3 \cdot 3 \cdot 2 \cdot 2 = 36$.

We build each fraction to a fraction with this LCD.

$$\frac{(4)2y}{(4)9} = \frac{8y}{36}; \qquad \frac{(6)5y}{(6)6} = \frac{30y}{36}; \qquad \frac{(9)y}{(9)4} = \frac{9y}{36}.$$

Then we obtain

$$\frac{8y}{36} + \frac{30y}{36} - \frac{9y}{36} = \frac{8y + 30y - 9y}{36} = \frac{29y}{36}.$$

A similar procedure is used if the fraction involves denominators that contain polynomials that must be factored.

Example $\dfrac{2}{x^2-9} - \dfrac{1}{x^2-x-6};$ the LCD is $(x-3)(x+3)(x+2)$.

We build each fraction to a fraction with this LCD.

$$\frac{2}{(x-3)(x+3)} = \frac{2(x+2)}{(x-3)(x+3)(x+2)}$$

$$\frac{1}{(x-3)(x+2)} = \frac{1(x+3)}{(x-3)(x+2)(x+3)}$$

Example continued overleaf

Then, we obtain

$$\frac{2(x+2)}{(x-3)(x+3)(x+2)} - \frac{(x+3)}{(x-3)(x+2)(x+3)} = \frac{2x+4-(x+3)}{(x-3)(x+3)(x+2)}$$
$$= \frac{x+1}{(x-3)(x+3)(x+2)}.$$

Since polynomials are easier to work with in factored form, it is usually advantageous to leave the LCM of a set of polynomials in factored form rather than carry out the indicated multiplication. Sometimes it is also convenient to leave numerators or denominators of fractions in factored form.

EXERCISE 3.5

A ■ *Write each sum or difference as a single fraction in lowest terms.*

Examples **a.** $\dfrac{x}{7} + \dfrac{2x}{7} - \dfrac{4}{7}$ **b.** $\dfrac{a+1}{b} - \dfrac{a-1}{b}$

Solutions **a.** $\dfrac{x}{7} + \dfrac{2x}{7} - \dfrac{4}{7} = \dfrac{x+2x-4}{7}$ **b.** $\dfrac{a+1}{b} - \dfrac{a-1}{b} = \dfrac{a+1-(a-1)}{b}$

$$= \frac{3x-4}{7} \qquad\qquad\qquad = \frac{a+1-a+1}{b} = \frac{2}{b}$$

Common error: Note that in Example b, $-(a-1) \neq -a-1$.

1. $\dfrac{x}{2} - \dfrac{3}{2}$ **2.** $\dfrac{y}{7} - \dfrac{5}{7}$ **3.** $\dfrac{a}{6} + \dfrac{b}{6} - \dfrac{c}{6}$

4. $\dfrac{x}{3} - \dfrac{2y}{3} + \dfrac{z}{3}$ **5.** $\dfrac{x-1}{2y} + \dfrac{x}{2y}$ **6.** $\dfrac{y+1}{b} + \dfrac{y-1}{b}$

7. $\dfrac{3}{x+2y} - \dfrac{x+3}{x+2y} - \dfrac{x-1}{x+2y}$ **8.** $\dfrac{2}{a-3b} - \dfrac{b-2}{a-3b} + \dfrac{b}{a-3b}$

9. $\dfrac{a+1}{a^2-2a+1} + \dfrac{5-3a}{a^2-2a+1}$ **10.** $\dfrac{x+4}{x^2-x+2} + \dfrac{2x-3}{x^2-x+2}$

Example $\dfrac{5}{a^2-9} - \dfrac{1}{a-3}$

Solution   Factor denominators and build each fraction to a fraction with the denominator $(a-3)(a+3)$.

$$\frac{5}{a^2-9}-\frac{1}{a-3}=\frac{5}{(a-3)(a+3)}-\frac{1(a+3)}{(a-3)(a+3)}$$

$$=\frac{5-(a+3)}{(a-3)(a+3)}=\frac{2-a}{(a-3)(a+3)}$$

The denominator $(a-3)(a+3)$ can also be expressed in the form a^2-9.

11. $\dfrac{2}{ax}-\dfrac{2}{x}$

12. $\dfrac{3}{by}-\dfrac{2}{b}$

13. $\dfrac{a-2}{6}-\dfrac{a+1}{3}$

14. $\dfrac{x-3}{4}+\dfrac{5-x}{10}$

15. $\dfrac{2x-y}{2y}+\dfrac{x+y}{x}$

16. $\dfrac{3a+2b}{3b}-\dfrac{a+2b}{6a}$

17. $\dfrac{5}{x-6}-\dfrac{3}{x+3}$

18. $\dfrac{1}{2x+1}-\dfrac{3}{x-2}$

19. $\dfrac{a}{3x+2}-\dfrac{a}{x-1}$

20. $\dfrac{a+1}{a+2}-\dfrac{a+2}{a+3}$

21. $\dfrac{7}{5x-10}-\dfrac{5}{3x-6}$

22. $\dfrac{2}{3y+6}-\dfrac{3}{2y+4}$

23. $\dfrac{5x-y}{3x+y}-\dfrac{6x-5y}{2x-y}$

24. $\dfrac{x+2y}{2x-y}-\dfrac{2x+y}{x-2y}$

25. $\dfrac{2}{x^2-x-2}-\dfrac{2}{x^2+2x+1}$

26. $\dfrac{1}{b^2-1}-\dfrac{1}{b^2+2b+1}$

27. $\dfrac{y}{y^2-16}-\dfrac{y+1}{y^2-5y+4}$

28. $\dfrac{8}{a^2-4b^2}-\dfrac{2}{a^2-5ab+6b^2}$

29. $\dfrac{y}{y^2-9}-\dfrac{1}{y^2+4y-21}$

30. $\dfrac{3a}{a^2+3a-10}-\dfrac{2a}{a^2+a-6}$

31. $\dfrac{1}{z^2-7z+12}+\dfrac{2}{z^2-5z+6}-\dfrac{3}{z^2-6z+8}$

32. $\dfrac{4}{a^2-4b^2}+\dfrac{2}{a^2+3ab+2b^2}+\dfrac{4}{a^2-ab-2b^2}$

33. $x+\dfrac{1}{x-1}-\dfrac{1}{(x-1)^2}$ $\left[\text{Hint: Write } x \text{ as } \dfrac{x}{1}.\right]$

34. $y-\dfrac{2y}{y^2-1}+\dfrac{3}{y+1}$ $\left[\text{Hint: Write } y \text{ as } \dfrac{y}{1}.\right]$

B ■ *Show that each pair of expressions is equivalent, given that* $c^2+s^2=1$. *Assume that the variables represent numbers for which each expression is defined. [Hint: Simplify the first expression.]*

35. $\dfrac{c}{s}+\dfrac{s}{c}$; $\dfrac{1}{sc}$

36. $\dfrac{1}{c^2}+\dfrac{1}{s^2}$; $\dfrac{1}{c^2s^2}$

37. $c+\dfrac{s^2}{c}$; $\dfrac{1}{c}$

38. $\left(\dfrac{c}{s}\right)^2+1$; $\dfrac{1}{s^2}$

39. $\dfrac{(c-1)(c+1)}{s^2}+1$;   0

40. $s+\dfrac{c(c-s)}{s}$; $\dfrac{1-cs}{s}$

41. $\dfrac{s-c}{s+c}+\dfrac{2sc}{s^2-c^2}$; $\dfrac{1}{s^2-c^2}$

42. $\dfrac{c^3+c^2s}{c^2-s^2}-s$; $\dfrac{1-cs}{c-s}$

3.6

PRODUCTS AND QUOTIENTS

Products of
fractions

In earlier mathematics courses, we learned that, for example,

$$\frac{2}{3}\cdot\frac{4}{5}=\frac{2\cdot4}{3\cdot5}=\frac{8}{15}.$$

In general,

►
$$\frac{a}{b}\cdot\frac{c}{d}=\frac{ac}{bd}\qquad (b,\,d\neq0).$$

Thus, the product of two fractions equals the product of their numerators divided by the product of their denominators. For example,

$$\frac{6x^2}{y}\cdot\frac{xy}{2}=\frac{6x^2\cdot xy}{y\cdot2}=\frac{6x^3y}{2y}$$

$$=\frac{3x^3(2y)}{1(2y)}=3x^3\qquad (y\neq0).$$

We can arrive at the same result for this example by using slash lines:

$$\frac{6x^2}{y}\cdot\frac{xy}{2}=\frac{\overset{3}{\cancel{6}x^2}}{\cancel{y}}\cdot\frac{\overset{1}{\cancel{xy}}}{\underset{1}{\cancel{2}}}=3x^3.$$

If any of the fractions, or any of the factors of the numerators or denominators of the fractions, have negative signs, it is advisable to proceed as if all the signs were positive and then attach the appropriate sign to the simplified product. If there is an even number of negative signs involved, the result has a positive sign; if there is an odd number of negative signs involved, the result has a negative sign.

Examples   **a.** $\dfrac{-x}{y}\cdot\dfrac{-2y^3}{x^2}=\dfrac{2y}{x}$ $(x,\,y\neq0)$

b. $\dfrac{-4x^2}{3y}\cdot\dfrac{-x}{2x}\cdot\dfrac{x}{5}=\dfrac{2x}{5}$

Quotients of fractions

When rewriting quotients

$$\frac{a}{b} \div \frac{c}{d} \qquad (b, c, d \neq 0),$$

we seek a quotient q such that

$$\left(\frac{c}{d}\right)q = \frac{a}{b} \qquad (b, c, d \neq 0).$$

To obtain q in terms of the other variables, we use the multiplication law of equality and multiply each member of this equality by d/c:

$$\left(\frac{d}{c}\right)\left(\frac{c}{d}\right)q = \left(\frac{d}{c}\right)\left(\frac{a}{b}\right),$$

from which

$$q = \left(\frac{d}{c}\right)\left(\frac{a}{b}\right) = \left(\frac{a}{b}\right)\left(\frac{d}{c}\right).$$

Therefore, it follows that:

$$\blacktriangleright \qquad \frac{a}{b} \div \frac{c}{d} = \frac{a}{b} \cdot \frac{d}{c} \qquad (b, c, d \neq 0). \tag{1}$$

Examples

a. $\dfrac{2x^3}{3y} \div \dfrac{4x}{5y^2}$

$$= \frac{\overset{x^2}{\cancel{2x^3}}}{\cancel{3y}} \cdot \frac{\overset{y}{\cancel{5y^2}}}{\underset{2}{\cancel{4x}}}$$

$$= \frac{5x^2y}{6} \qquad (x, y \neq 0)$$

b. $\dfrac{x^2 - 1}{x + 3} \div \dfrac{x^2 - x - 2}{x^2 + 5x + 6}$

$$= \frac{(x - 1)\cancel{(x+1)}}{\cancel{x+3}} \cdot \frac{\cancel{(x+3)}(x + 2)}{\cancel{(x+1)}(x - 2)}$$

$$= \frac{(x - 1)(x + 2)}{(x - 2)}$$

$$= \frac{x^2 + x - 2}{x - 2} \qquad (x \neq -3, -1, 2)$$

As special cases of (1), observe that

$$a \div \frac{c}{d} = \frac{a}{1} \cdot \frac{d}{c} = \frac{ad}{c}, \qquad \frac{a}{b} \div c = \frac{a}{b} \cdot \frac{1}{c} = \frac{a}{bc},$$

and

$$1 \div \frac{a}{b} = 1 \cdot \frac{b}{a} = \frac{b}{a}.$$

Examples

a. $2 \div \dfrac{3}{4} = \dfrac{2 \cdot 4}{3}$

$$= \frac{8}{3}$$

b. $\dfrac{5}{7} \div 3 = \dfrac{5}{7} \cdot \dfrac{1}{3}$

$$= \frac{5}{21}$$

c. $1 \div \dfrac{2}{3} = 1 \cdot \dfrac{3}{2}$

$$= \frac{3}{2}$$

EXERCISE 3.6

A ■ *Write each product as a single fraction in lowest terms.*

Examples   **a.** $\dfrac{-3c^2}{5ab} \cdot \dfrac{10a^2b}{9c}$ **b.** $\dfrac{4a^2-1}{a^2-4} \cdot \dfrac{a^2+2a}{4a+2}$

Solutions   **a.** $\dfrac{-\overset{c}{\cancel{3c^2}}}{\cancel{5ab}} \cdot \dfrac{\overset{2a}{\cancel{10a^2b}}}{\underset{3}{\cancel{9c}}} = \dfrac{-2ac}{3}$ **b.** $\dfrac{4a^2-1}{a^2-4} \cdot \dfrac{a^2+2a}{4a+2}$

$$= \frac{(2a-1)\cancel{(2a+1)}}{(a-2)\cancel{(a+2)}} \cdot \frac{a\cancel{(a+2)}}{2\cancel{(2a+1)}}$$

$$= \frac{a(2a-1)}{2(a-2)}$$

1. $\dfrac{16}{38} \cdot \dfrac{19}{12}$ **2.** $\dfrac{4}{15} \cdot \dfrac{3}{16}$ **3.** $\dfrac{21}{4} \cdot \dfrac{2}{15}$

4. $\dfrac{7}{8} \cdot \dfrac{48}{64}$ **5.** $\dfrac{24}{3} \cdot \dfrac{20}{36} \cdot \dfrac{3}{4}$ **6.** $\dfrac{3}{10} \cdot \dfrac{16}{27} \cdot \dfrac{30}{36}$

7. $\dfrac{7x}{12} \cdot \dfrac{3}{14x^2}$ **8.** $\dfrac{4a^2}{3} \cdot \dfrac{6b}{2a}$ **9.** $\dfrac{15n^2}{3p} \cdot \dfrac{5p^2}{n^3}$

10. $\dfrac{21t^2}{5s} \cdot \dfrac{15s^3}{7st}$ **11.** $\dfrac{-4}{3n} \cdot \dfrac{6n^2}{16}$ **12.** $\dfrac{14a^3b}{3b} \cdot \dfrac{-6}{7a^2}$

13. $\dfrac{-12a^2b}{5c} \cdot \dfrac{10b^2c}{24a^3b}$ **14.** $\dfrac{a^2}{xy} \cdot \dfrac{3x^3y}{4a}$ **15.** $\dfrac{-2ab}{7c} \cdot \dfrac{3c^2}{4a^3} \cdot \dfrac{-6a}{15b^2}$

16. $\dfrac{10x}{12y} \cdot \dfrac{3x^2z}{5x^3z} \cdot \dfrac{6y^2x}{3yz}$ **17.** $5a^2b^2 \cdot \dfrac{1}{a^3b^3}$ **18.** $15x^2y \cdot \dfrac{3}{45xy^2}$

19. $\dfrac{5x+25}{2x} \cdot \dfrac{4x}{2x+10}$ **20.** $\dfrac{3y}{4xy-6y^2} \cdot \dfrac{2x-3y}{12x}$

21. $\dfrac{4a^2-1}{a^2-16} \cdot \dfrac{a^2-4a}{2a+1}$ **22.** $\dfrac{9x^2-25}{2x-2} \cdot \dfrac{x^2-1}{6x-10}$

23. $\dfrac{x^2-x-20}{x^2+7x+12} \cdot \dfrac{(x+3)^2}{(x-5)^2}$ **24.** $\dfrac{4x^2+8x+3}{2x^2-5x+3} \cdot \dfrac{6x^2-9x}{1-4x^2}$

25. $\dfrac{x^2-6x+5}{x^2+2x-3} \cdot \dfrac{x^2-4x-21}{x^2-10x+25}$ **26.** $\dfrac{x^2-x-2}{x^2+4x+3} \cdot \dfrac{x^2-4x-5}{x^2-3x-10}$

27. $\dfrac{2x^2-x-6}{3x^2-4x+1} \cdot \dfrac{3x^2+7x+2}{2x^2+7x+6}$ **28.** $\dfrac{3x^2-7x-6}{2x^2-x-1} \cdot \dfrac{2x^2-9x-5}{3x^2-13x-10}$

29. $\dfrac{7a+14}{14a-28} \cdot \dfrac{4-2a}{a+2} \cdot \dfrac{a-3}{a+1}$ **30.** $\dfrac{5x^2-5x}{3} \cdot \dfrac{x^2-9x-10}{4x-40} \cdot \dfrac{y^2}{2-2x^2}$

■ *Write each quotient as a single fraction in lowest terms.*

Examples

a. $\dfrac{a^2b}{c} \div \dfrac{ab^3}{c^2}$

b. $\dfrac{3a^2 - 3}{2a + 2} \div \dfrac{3a - 3}{2}$

Solutions

a. $\dfrac{a^2b}{c} \div \dfrac{ab^3}{c^2} = \dfrac{\cancel{a^2 b}^{\,a}}{\cancel{c}} \cdot \dfrac{\cancel{c^2}^{\,c}}{\cancel{ab^3}_{b^2}}$

$= \dfrac{ac}{b^2}$

b. $\dfrac{3a^2 - 3}{2a + 2} \div \dfrac{3a - 3}{2}$

$= \dfrac{\cancel{3}\cancel{(a - 1)}\cancel{(a + 1)}}{\cancel{2}\cancel{(a + 1)}} \cdot \dfrac{\cancel{2}}{\cancel{3}\cancel{(a - 1)}}$

$= 1$

31. $\dfrac{3}{4} \div \dfrac{9}{16}$

32. $\dfrac{2}{3} \div \dfrac{9}{15}$

33. $\dfrac{xy}{a^2b} \div \dfrac{x^3y^2}{ab}$

34. $\dfrac{9ab^3}{x} \div \dfrac{3}{2x^3}$

35. $28x^2y^3 \div \dfrac{21x^2y^2}{5a}$

36. $24a^3b \div \dfrac{3a^2b}{7x}$

37. $\dfrac{4x - 8}{3y} \div \dfrac{6x - 12}{y}$

38. $\dfrac{6y - 27}{5x} \div \dfrac{4y - 18}{x}$

39. $\dfrac{a^2 - a - 6}{a^2 + 2a - 15} \div \dfrac{a^2 - 4}{a^2 + 6a + 5}$

40. $\dfrac{a^2 + 2a - 15}{a^2 + 3a - 10} \div \dfrac{a^2 - 9}{a^2 - 9a + 14}$

41. $\dfrac{x^2 + x - 2}{x^2 + 2x - 3} \div \dfrac{x^2 + 7x + 10}{x^2 - 2x - 15}$

42. $\dfrac{x^2 + 6x - 7}{x^2 + x - 2} \div \dfrac{x^2 + 5x - 14}{x^2 - 3x - 10}$

43. $\dfrac{10x^2 - 13x - 3}{2x^2 - x - 3} \div \dfrac{5x^2 - 9x - 2}{3x^2 + 2x - 1}$

44. $\dfrac{9x^2 + 3x - 2}{12x^2 + 5x - 2} \div \dfrac{9x^2 - 6x + 1}{8x^2 + 10x - 3}$

B   ■ *Write as a single fraction in lowest terms.*

45. $\dfrac{x^3 + y^3}{x} \div \dfrac{x + y}{3x}$

46. $\dfrac{8x^3 - y^3}{x + y} \div \dfrac{2x - y}{x^2 - y^2}$

47. $\dfrac{xy - 3x + y - 3}{x - 2} \div \dfrac{x + 1}{x^2 - 4}$

48. $\dfrac{2xy + 4x + 3y + 6}{2x + 3} \div \dfrac{y + 2}{y - 1}$

49. $\dfrac{a^2 - a}{a - 3} \cdot \dfrac{a + 1}{a^2 + 4a} \div \dfrac{a^2 - 3a - 4}{a^2 - 16}$

50. $\dfrac{y - 1}{y^2} \cdot \dfrac{y^2 + y}{y - 3} \div \dfrac{y^2 - 2y - 3}{y^2 - y - 6}$

■ *Let* $P(x) = \dfrac{x}{x-1}$, $Q(x) = \dfrac{x^2}{x^2-1}$, *and* $R(x) = \dfrac{x^3}{(x-1)^2}$. *Write each expression as a single fraction in lowest terms.*

51. $P(x) \div Q(x)$ **52.** $Q(x) \div R(x)$

53. $P(x) \cdot Q(x) \div R(x)$ **54.** $P(x) \cdot R(x) \div Q(x)$

3.7

COMPLEX FRACTIONS

A fraction that contains a fraction or fractions in either the numerator or the denominator or in both is called a **complex fraction**. In simplifying complex fractions, it is helpful to remember that fraction bars serve as grouping devices in the same sense as parentheses or brackets. Thus, the complex fraction

$$\frac{x + \dfrac{3}{4}}{x - \dfrac{1}{2}} \quad \text{means} \quad \left(x + \frac{3}{4}\right) \div \left(x - \frac{1}{2}\right).$$

Complex fractions can be simplified in either of two ways. In simple examples, it is easier to apply the fundamental principle of fractions and multiply the numerator and the denominator of the complex fraction by the LCD of all the fractions that appear. Thus, in the above example 4 is the LCD of ¾ and ½ and we have,

$$\frac{4\left(x + \dfrac{3}{4}\right)}{4\left(x - \dfrac{1}{2}\right)} = \frac{4(x) + 4\left(\dfrac{3}{4}\right)}{4(x) - 4\left(\dfrac{1}{2}\right)} = \frac{4x + 3}{4x - 2} \quad \left(x \neq \frac{1}{2}\right).$$

Alternatively, this fraction may be simplified by first representing the numerator as a single fraction and the denominator as a single fraction, and then writing the quotient as the product of the numerator and the reciprocal of the denominator:

$$\frac{x + \dfrac{3}{4}}{x - \dfrac{1}{2}} = \frac{\dfrac{4 \cdot x}{4 \cdot 1} + \dfrac{3}{4}}{\dfrac{2 \cdot x}{2 \cdot 1} - \dfrac{1}{2}} = \frac{\dfrac{4x + 3}{4}}{\dfrac{2x - 1}{2}}$$

$$= \frac{4x + 3}{4} \div \frac{2x - 1}{2} = \frac{4x + 3}{4} \cdot \frac{2}{2x - 1}$$

$$= \frac{4x + 3}{2(2x - 1)} = \frac{4x + 3}{4x - 2} \quad \left(x \neq \frac{1}{2}\right).$$

The first method is generally more convenient to use.

EXERCISE 3.7

A ■ *Write each complex fraction as a single fraction in lowest terms.*

Example

$$\frac{\dfrac{3}{a} - \dfrac{1}{2a}}{\dfrac{1}{3a} + \dfrac{5}{6a}}$$

Solution

The LCD for all fractions in the numerator and the denominator is $6a$. Multiply the numerator and the denominator by $6a$ and simplify.

$$\frac{6a\left(\dfrac{3}{a} - \dfrac{1}{2a}\right)}{6a\left(\dfrac{1}{3a} + \dfrac{5}{6a}\right)} = \frac{(6a)\dfrac{3}{a} - (6a)\dfrac{1}{2a}}{(6a)\dfrac{1}{3a} + (6a)\dfrac{5}{6a}} = \frac{18 - 3}{2 + 5} = \frac{15}{7}$$

1. $\dfrac{\dfrac{3}{7}}{\dfrac{2}{7}}$ **2.** $\dfrac{\dfrac{4}{5}}{\dfrac{7}{5}}$ **3.** $\dfrac{\dfrac{2}{9}}{\dfrac{7}{3}}$ **4.** $\dfrac{\dfrac{5}{2}}{\dfrac{21}{4}}$

5. $\dfrac{\dfrac{b}{c}}{\dfrac{b^2}{a}}$ **6.** $\dfrac{\dfrac{5x}{6y}}{\dfrac{4x}{5y}}$ **7.** $\dfrac{\dfrac{2x}{5y}}{\dfrac{3x}{10y^2}}$ **8.** $\dfrac{\dfrac{3ab}{4}}{\dfrac{3b}{8a^2}}$

9. $\dfrac{\dfrac{3}{4}}{4 - \dfrac{1}{4}}$ **10.** $\dfrac{\dfrac{1}{3}}{4 + \dfrac{2}{3}}$ **11.** $\dfrac{1 - \dfrac{2}{3}}{3 + \dfrac{1}{3}}$ **12.** $\dfrac{\dfrac{1}{2} + \dfrac{3}{4}}{\dfrac{1}{2} - \dfrac{3}{4}}$

13. $\dfrac{\dfrac{2}{a} + \dfrac{3}{2a}}{5 + \dfrac{1}{a}}$ **14.** $\dfrac{1 + \dfrac{2}{a}}{1 - \dfrac{4}{a^2}}$ **15.** $\dfrac{x + \dfrac{x}{y}}{1 + \dfrac{1}{y}}$ **16.** $\dfrac{1 + \dfrac{1}{x}}{1 - \dfrac{1}{x}}$

17. $\dfrac{1}{1 - \dfrac{1}{x}}$ **18.** $\dfrac{4}{\dfrac{2}{x} + 2}$ **19.** $\dfrac{y - 2}{y - \dfrac{4}{y}}$ **20.** $\dfrac{y + 3}{\dfrac{9}{y} - y}$

Example

$$\frac{a}{a + \dfrac{3}{3 + \dfrac{1}{2}}}$$

Solution overleaf

At top of page, handwritten notes:

$$\frac{a}{a+\dfrac{3}{3+\frac{1}{2}}} =$$

Solution First write the sum $3+\dfrac{1}{2}$ as $\dfrac{3\cdot 2}{1\cdot 2}+\dfrac{1}{2}=\dfrac{7}{2}$. Thus,

$$\frac{a}{a+\dfrac{3}{3+\dfrac{1}{2}}}=\frac{a}{a+\dfrac{3}{\dfrac{7}{2}}}.$$

Then write $\dfrac{3}{\dfrac{7}{2}}$ as $3\cdot\dfrac{2}{7}$ or $\dfrac{6}{7}$ and write the sum $a+\dfrac{6}{7}$ as $\dfrac{7\cdot a}{7\cdot 1}+\dfrac{6}{7}=\dfrac{7a+6}{7}$. Hence,

$$\frac{a}{a+\dfrac{3}{3+\dfrac{1}{2}}}=\frac{a}{\dfrac{7a+6}{7}}=a\cdot\frac{7}{7a+6}=\frac{7a}{7a+6}.$$

21. $2-\dfrac{2}{3+\dfrac{1}{2}}$

22. $1-\dfrac{1}{1-\dfrac{1}{3}}$

23. $a-\dfrac{a}{a+\dfrac{1}{4}}$

24. $x-\dfrac{x}{1-\dfrac{x}{1-x}}$

25. $1-\dfrac{1}{1-\dfrac{1}{y-2}}$

26. $2y+\dfrac{3}{3-\dfrac{2y}{y-1}}$

B

27. $\dfrac{1+\dfrac{1}{1-\dfrac{a}{b}}}{1-\dfrac{3}{1-\dfrac{a}{b}}}$

28. $\dfrac{1-\dfrac{1}{\dfrac{a}{b}+2}}{1+\dfrac{3}{\dfrac{a}{2b}+1}}$

29. $\dfrac{a+2-\dfrac{12}{a+3}}{a-5+\dfrac{16}{a+3}}$

30. $\dfrac{a+4-\dfrac{7}{a-2}}{a-1+\dfrac{2}{a-2}}$

31. $\dfrac{\dfrac{1}{ab}+\dfrac{2}{bc}+\dfrac{3}{ac}}{\dfrac{2a+3b+c}{abc}}$

32. $\dfrac{\dfrac{a}{bc}-\dfrac{b}{ac}+\dfrac{c}{ab}}{\dfrac{1}{a^2b^2}-\dfrac{1}{a^2c^2}+\dfrac{1}{b^2c^2}}$

■ *Show that each pair of expressions is equivalent, given that* $c^2+s^2=1$. *Assume that the variables represent numbers for which each expression is defined.* [*Hint: Simplify the first expression.*]

33. $\dfrac{\dfrac{(c+s)^2}{c-s}-\dfrac{1}{\dfrac{c-s}{2cs}}}{};\quad \dfrac{1}{c-s}$

34. $\dfrac{s}{\dfrac{s}{s+c}}-\dfrac{2sc}{s+c};\quad \dfrac{1}{s+c}$

35. $\dfrac{\dfrac{1}{s-c}}{\dfrac{s}{s^2-c^2}}+\dfrac{s-c}{s};\quad 2$

36. $\dfrac{\dfrac{1}{s^2}+\dfrac{1}{c^2}}{\dfrac{1}{s\cdot c}}-\dfrac{c}{s};\quad \dfrac{s}{c}$

CHAPTER SUMMARY

[3.1]   A **fraction** is an expression that denotes a quotient. If the numerator and denominator are polynomials, then the fraction is a **rational expression**.

The **fundamental principle of fractions** states:

If the numerator and the denominator of a fraction are multiplied or divided by the same nonzero number, the result is a fraction equivalent to the given fraction:

$$\frac{a}{b} = \frac{ac}{bc} \quad (b, c \neq 0).$$

For a given fraction, the replacement of any two of the three elements of the fraction—the fraction itself, the numerator, the denominator—by their negatives results in an equivalent fraction:

$$\frac{a}{b} = \frac{-a}{-b} = -\frac{a}{-b} = -\frac{-a}{b} \quad (b \neq 0),$$

$$\frac{-a}{b} = \frac{a}{-b} = -\frac{a}{b} = -\frac{-a}{-b} \quad (b \neq 0).$$

Standard forms for fractions are $\dfrac{a}{b}$ and $\dfrac{-a}{b}$.

[3.2]   A fraction is in **lowest terms** if the numerator and the denominator do not contain factors in common. We can reduce a fraction to lowest terms by using the fundamental principle of fractions to divide the numerator and the denominator by their common nonzero factors.

[3.3]   Some quotients of polynomials can be rewritten as equivalent **mixed expressions**. Sometimes a method similar to the long division process used in arithmetic can be used for this purpose.

[3.4]   The **least common denominator** (LCD) of a set of fractions with natural-number denominators is the smallest natural number that is exactly divisible by each of the denominators.

[3.5]   The operations of addition and subtraction of fractions are governed by the following properties:

$$\frac{a}{c} + \frac{b}{c} = \frac{a+b}{c} \quad \text{and} \quad \frac{a}{c} - \frac{b}{c} = \frac{a-b}{c} \quad (c \neq 0).$$

If the fractions in a sum or difference have unlike denominators, we can build each fraction to higher terms by using the fundamental principle of fractions in order to obtain the fractions with a common denominator.

[3.6] The operations of multiplication and division of fractions are governed by the following properties:

$$\frac{a}{b} \cdot \frac{c}{d} = \frac{ac}{bd} \quad (b, d \neq 0) \qquad \frac{a}{b} \div \frac{c}{d} = \frac{a}{b} \cdot \frac{d}{c} \quad (b, c, d \neq 0)$$

[3.7] A **complex fraction** is a fraction containing other fractions in its numerator or denominator or both. It is often convenient to use the fundamental principle to simplify such fractions.

REVIEW EXERCISES

A ■ *Assume that no denominator is* 0.

[3.1] **1. a.** Write six fractions equal to $\dfrac{1}{a-b}$ by changing the sign or signs of the numerator, the denominator, or the fraction itself.

b. What are the conditions on a and b for the fraction in part a to represent a positive number? A negative number?

2. Express $-\dfrac{1}{1-a}$ in standard form in two ways.

[3.2] ■ *Reduce each fraction to lowest terms.*

3. a. $\dfrac{4x^2y^3}{10xy^4}$ **b.** $\dfrac{2x^2-8}{2x+4}$ **4. a.** $\dfrac{4x^2-1}{1-2x}$ **b.** $\dfrac{8x^3+y^3}{4x^2-y^2}$

[3.3] ■ *Divide.*

5. a. $\dfrac{12x^2-6x+3}{2x}$ **b.** $\dfrac{6y^3+3y^2-y}{3y^2}$

6. a. $\dfrac{2y^2-3y+1}{2y+3}$ **b.** $\dfrac{6x^4-3x^3+2x+2}{x-1}$

[3.4] ■ *Express each given fraction as an equivalent fraction with the given denominator.*

7. a. $\dfrac{-3}{4} = \dfrac{?}{24}$ **b.** $\dfrac{x}{2y} = \dfrac{?}{2xy^2}$

8. a. $\dfrac{2}{x+3y} = \dfrac{?}{x^2-9y^2}$ **b.** $\dfrac{y}{3-y} = \dfrac{?}{y^2-4y+3}$

[3.5] ■ *Write each expression as a single fraction in lowest terms.*

9. a. $\dfrac{3x+y}{3} + \dfrac{x-y}{3}$ **b.** $\dfrac{x-2y}{4x} - \dfrac{2x-y}{4x}$

10. a. $\dfrac{2}{5x} - \dfrac{3}{4y} + \dfrac{7}{20xy}$ **b.** $\dfrac{y}{2y-6} + \dfrac{2}{3y-9}$

11. a. $\dfrac{5}{x-2} - \dfrac{3}{x+3}$ **b.** $\dfrac{1}{x+2y} + \dfrac{3}{x^2-4y^2}$

12. a. $\dfrac{2}{x^2-1} + \dfrac{3}{x^2-2x+1}$ **b.** $\dfrac{y}{y^2-16} - \dfrac{y+1}{y^2-5y+4}$

[3.6] ■ *Write each expression as a single fraction in lowest terms.*

13. a. $\dfrac{2x}{y^2} \cdot \dfrac{3y^3}{8x} \cdot \dfrac{4x^2}{9y}$ **b.** $\dfrac{4x}{2xy+8y^2} \cdot \dfrac{x+4y}{x}$

14. a. $\dfrac{x^2+2x-3}{x^2+6x+9} \cdot \dfrac{2x+6}{2x-2}$ **b.** $\dfrac{y^3-y}{2y+1} \cdot \dfrac{4y+2}{y^2+2y+1}$

15. a. $\dfrac{10x^2y^3}{9} \div \dfrac{2xy^2}{3}$ **b.** $\dfrac{2y-6}{y+2x} \div \dfrac{4y-12}{2y+4x}$

16. a. $\dfrac{y^2+4y+3}{y^2-y-2} \div \dfrac{y^2-4y-5}{y^2-3y-10}$ **b.** $\dfrac{x^2}{y-x} \div \dfrac{x^3-x^2}{x-y}$

[3.7] ■ *Write each expression as a single fraction in lowest terms.*

17. a. $\dfrac{\dfrac{3x}{2y}}{\dfrac{9x}{10y}}$ **b.** $\dfrac{\dfrac{2}{3}}{4-\dfrac{1}{3}}$ **18. a.** $\dfrac{2+\dfrac{3}{x}}{1-\dfrac{2}{3x}}$ **b.** $\dfrac{y-\dfrac{1}{y}}{y+\dfrac{1}{y}}$

19. a. $\dfrac{2}{x-\dfrac{1}{x}}$ **b.** $\dfrac{y}{\dfrac{y}{3}-\dfrac{y}{4}}$ **20. a.** $1+\dfrac{2}{1-\dfrac{1}{y+2}}$ **b.** $\dfrac{x+\dfrac{1}{x}}{1-\dfrac{1}{x-1}}$

B ■ *Divide.*

21. $\dfrac{x^3-2x^2+x+1}{x^2-x+2}$ **22.** $\dfrac{x^4-3x^2+4}{x^3-2x+1}$

23. $\dfrac{x^4+3x^3-x^2-6x-2}{x^2-2}$ **24.** $\dfrac{x^4-x^2-2}{x^2+4}$

■ *Simplify each expression.*

25. $1 + \dfrac{1 - \dfrac{1}{2}}{1 + \dfrac{1}{1 + \dfrac{1}{2}}}$

26. $\dfrac{3 - \dfrac{1}{\dfrac{x}{y} - 2}}{2 + \dfrac{3}{\dfrac{x}{2y} - 1}}$

27. $\dfrac{\dfrac{2}{ab} + \dfrac{1}{bc} - \dfrac{3}{ac}}{\dfrac{a - 3b + 2c}{abc}}$

28. $\dfrac{a - 3 - \dfrac{2}{a - 1}}{a + 2 - \dfrac{3}{a - 1}}$

29. $\dfrac{\dfrac{1}{a - b}}{\dfrac{a}{a^2 - b^2}} + \dfrac{ab - b^2}{ab}$

30. $\dfrac{a^2 - b^2}{\dfrac{a - b}{b}} - \dfrac{a - b}{\dfrac{1}{b}}$

4. FIRST-DEGREE EQUATIONS AND INEQUALITIES

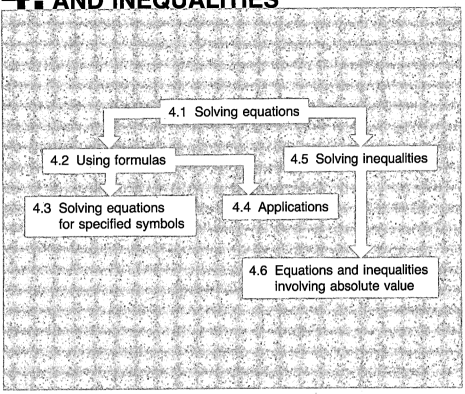

4.1 Solving equations

4.2 Using formulas

4.5 Solving inequalities

4.3 Solving equations for specified symbols

4.4 Applications

4.6 Equations and inequalities involving absolute value

4.1

SOLVING EQUATIONS

Solution of an equation

In mathematics, sentences involving equality relationships are called **equations**. The expressions (or sometimes the values of these expressions) on either side of the equals (=) sign are called the **members** of the equations. Values of the variable in an equation for which the equation becomes a true statement are called **solutions** of the equation and are said to *satisfy* the equation. For example, 2 is a solution of

$$x + 3 = 5,$$

because

$$2 + 3 = 5$$

is a true statement. The set of all solutions of an equation is called its **solution set** ({2} is the solution set of the above equation).

Equations that are satisfied by every element in the replacement set of the variable for which each member is defined are called **identities**. Equations that are not true for at least one element in the replacement set for which each member is defined are called **conditional equations**. For example,

$$x + 2 = x + 2$$

is an identity, and

$$x + 2 = 5$$

is a conditional equation (the equation is not true, for example, for $x = 10$). An equation such as

$$x = x + 2$$

that has *no solution* is a special case of a conditional equation.

One major type of equation is the polynomial equation in one variable. This is an equation that can be written in a form in which one member is 0 and the other member is a polynomial in simplest form.

Examples   **a.** $x^2 + 2x + 1 = 0$ is a second-degree equation in x.

b. $2y - 3 = 0$ is a first-degree equation in y.

c. $z^4 + z^3 - 1 = 0$ is a fourth-degree equation in z.

Equivalent equations   In this chapter we shall largely be concerned with conditional first-degree polynomial equations in one variable. The process of finding solution sets of such equations involves generating equations that have identical solution sets. Such equations are called **equivalent equations**. Thus,

$$2x + 1 = x + 4,$$
$$2x = x + 3,$$
$$x = 3$$

are equivalent equations, because {3} is the solution set of each. Equations whose solution sets contain at least one element assert that, for some or all values of the variable, the left-hand and right-hand members are names for the same number. Also, for each value of the variable, a polynomial represents a number. Thus, the equality axioms for the real numbers imply the following properties:

> *1. The addition of the same expression to each member of an equation produces an equivalent equation.*

> *2. The multiplication of each member of an equation by the same nonzero number produces an equivalent equation.*

The assertions above can be stated in symbols as follows:

► *If $P(x)$, $Q(x)$, and $R(x)$ are expressions, then for all values of x, the equation*

$$P(x) = Q(x)$$

is equivalent to

$$P(x) + R(x) = Q(x) + R(x), \tag{1}$$

$$P(x) \cdot R(x) = Q(x) \cdot R(x), \quad \textit{for} \quad R(x) \neq 0. \tag{2}$$

The application of these properties enables us to transform an equation whose solution may not be obvious through a series of equivalent equations until we reach an equation that has an obvious solution.

Any application of Property (1) or (2) is called an **elementary transformation**. An elementary transformation always results in an equivalent equation.

Example   Solve $3x + 5 = 11$.

Solution   We first add -5, the additive inverse of 5, to each member.

$$3x + 5 - 5 = 11 - 5$$
$$3x = 6$$

Now, by inspection, we can readily determine that 2 is the solution. However, if necessary, we can multiply each member by ⅓, the multiplicative inverse of 3, in order to obtain the variable x with a coefficient of 1.

$$\frac{1}{3}(3x) = \frac{1}{3}(6)$$
$$x = 2$$

In either case, the solution set is {2}.

Let us consider another example.

Example   Solve $3y - 7 = y + 5$.

Solution   We first add $-y + 7$ to each member, where $-y$ is the additive inverse of y and 7 is the additive inverse of -7.

$$3y - 7 + (-y + 7) = y + 5 + (-y + 7)$$
$$3y - y = 5 + 7.$$

Simplifying each member, we have

$$2y = 12.$$

Solution continued overleaf

We now have an equation in which one member includes only terms containing the variable and the other member includes only constants. By inspection, we can determine that 6 is the solution. Alternatively, we can multiply each member by ½, the multiplicative inverse of 2, to obtain the variable y with a coefficient of 1.

$$\frac{1}{2}(2y) = \frac{1}{2}(12),$$
$$y = 6$$

The solution set is $\{6\}$.

Multiplying by zero

Care must be exercised in the application of Property (2) above, for we have specifically excluded *multiplication by* 0. For example, the equation $x = 3$ with solution set $\{3\}$ is not equivalent to $0 \cdot x = 0 \cdot 3$, with solution set $\{x | x \in R\}$.

Example

Solve $\dfrac{x}{x-3} = \dfrac{3}{x-3} + 2.$ (3)

Solution

We might first multiply each member by $(x-3)$ to attempt to produce an equivalent equation that is free of fractions. We have

$$(x-3)\frac{x}{x-3} = (x-3)\frac{3}{x-3} + (x-3)2$$

or

$$x = 3 + 2x - 6,$$ (4)

from which

$$x = 3,$$

and 3 *appears* to be a solution of (3). But, upon substituting 3 for x in (3), we have

$$\frac{3}{0} = \frac{3}{0} + 2,$$

where neither member is defined. In obtaining Equation (4), each member of Equation (3) was multiplied by $(x-3)$, but if x is 3, then $(x-3)$ is 0, and our second property above is not applicable. Equation (4) is not equivalent to Equation (3) for $x = 3$, and Equation (3) has no solution.

We can always determine whether an apparent solution of an equation is really a solution by substituting the suggested solution in the original equation and verifying that the resulting statement is true. If each of the equations in a sequence is simply obtained by means of an application of Property (1) or Property (2), the sole purpose for such a check is to detect arithmetic errors. We shall dispense with checking solutions in the examples that follow *except in cases where we multiply or divide by an expression containing a variable.*

EXERCISE 4.1

A   ■ *Solve; that is, find the solution set.*

Examples

a. $z - 5 = 8$ **b.** $7t + 6 = 3t + 5$

Solutions

a. Add 5 to each member.   **b.** Add $-3t - 6$ to each member.

$$z = 13$$ $$4t = -1$$

The solution set is $\{13\}$.   Multiply each member by $\frac{1}{4}$

$$t = \frac{-1}{4}$$

The solution set is $\{-\frac{1}{4}\}$.

[*Note:* Of course, the solution set can be specified at any time the solution becomes evident by inspection.]

1. $x + 3 = 8$ **2.** $x - 5 = 7$ **3.** $6 = x - 4$

4. $9 = x + 2$ **5.** $3x = x + 4$ **6.** $3x - 4 = x$

7. $5x - 1 = 2x + 3$ **8.** $5x + 1 = 2x - 3$

9. $3(x + 2) = 14$ **10.** $2(x - 3) = 15$

11. $3x - (x - 4) = 12$ **12.** $5x - (x + 1) = 14$

Examples

a. $4 - (x - 1)(x + 2) = 8 - x^2$ **b.** $\frac{x}{3} + 4 = x - 2$

Solutions

a. Apply the distributive property.   **b.** Multiply each member by 3.

$$4 - (x^2 + x - 2) = 8 - x^2$$ $$3\left(\frac{x}{3} + 4\right) = 3(x - 2)$$
$$4 - x^2 - x + 2 = 8 - x^2$$ $$x + 12 = 3x - 6$$
$$-x^2 - x + 6 = 8 - x^2$$

Add $x^2 - 6$ to each member.   Add $-x + 6$ to each member.

$$-x = 2$$ $$18 = 2x$$

Multiply each member by -1.   Multiply each member by $\frac{1}{2}$.

$$x = -2$$ $$9 = x$$

The solution set is $\{-2\}$.   The solution set is $\{9\}$.

13. $2[x - (2x + 1)] = 6$ **14.** $4[2x + (3x - 1)] = 5$

15. $-2[x + 3(x - 1)] = 4 + x$ **16.** $-5[2x - 2(x + 1)] = 6 - x$

17. $(x - 1)(x + 3) = x^2 + 4$ **18.** $(2x - 3)(x + 2) = x + 4 + 2x^2$

19. $(x + 3)^2 = x^2 + 4x$ **20.** $(2x - 3)^2 = 4x^2 - 8$

21. $x - (x + 2)(x - 3) = 4 - x^2$ **22.** $3x - (x - 3)(x + 1) = 6x - x^2$

23. $\dfrac{3x}{4} = 6$ **24.** $\dfrac{2x}{3} = 8$ **25.** $\dfrac{3}{5}x = -12$

26. $\dfrac{2}{7}x = -8$ **27.** $7 + \dfrac{5x}{3} = x - 2$ **28.** $x + 4 = \dfrac{2}{5}x - 3$

29. $1 + \dfrac{x}{9} = \dfrac{4}{3}$ **30.** $4 + \dfrac{x}{5} = \dfrac{5}{3}$ **31.** $\dfrac{x}{5} - \dfrac{x}{2} = 9$

32. $\dfrac{x}{4} = 2 - \dfrac{x}{3}$ **33.** $\dfrac{2x - 1}{5} - \dfrac{x + 1}{2} = 0$ **34.** $\dfrac{2x}{3} - \dfrac{2x + 5}{6} = \dfrac{1}{2}$

Example $\dfrac{2}{3} = 6 - \dfrac{x + 10}{x - 3}$

Solution   Multiply each member by the LCD $3(x - 3)$.

$$3(x - 3)\dfrac{2}{3} = 3(x - 3)6 - 3(x - 3)\dfrac{x + 10}{x - 3}$$

$$2(x - 3) = 18(x - 3) - 3(x + 10)$$

Apply the distributive property.

$$2x - 6 = 18x - 54 - 3x - 30$$

Combine like terms and add $-15x + 6$ to each member.

$$-13x = -78$$

Multiply each member by $-\frac{1}{13}$.

$$x = 6$$

The solution set is $\{6\}$.

Check A check is required because each member of the equation was multiplied by an expression containing the variable. Equations (1) and (2) are equivalent for all values of x, except for $x - 3 = 0$ or $x = 3$. Hence, 6 is the solution.

35. $\dfrac{x}{x - 2} = \dfrac{2}{x - 2} + 7$ **36.** $\dfrac{2}{x - 9} = \dfrac{9}{x + 12}$

37. $\dfrac{2}{y + 1} + \dfrac{1}{3y + 3} = \dfrac{1}{6}$ **38.** $\dfrac{5}{x - 3} = \dfrac{x + 2}{x - 3} + 3$

39. $\dfrac{4}{2x - 3} + \dfrac{4x}{4x^2 - 9} = \dfrac{1}{2x + 3}$ **40.** $\dfrac{y}{y + 2} - \dfrac{3}{y - 2} = \dfrac{y^2 + 8}{y^2 - 4}$

B 41. Find a value for k in $3x - 1 = k$ so that the equation is equivalent to $2x + 5 = 1$.

42. For what value of k will the equation $2x - 3 = \dfrac{4 + x}{k}$ have as its solution set $\{-1\}$?

43. For what value of k will the equation $\dfrac{2y + k}{y - 5} + \dfrac{3}{4} = 8$ have as its solution set $\{9\}$?

44. For what value of k will the equation $\dfrac{y - 3}{3y + k} + 1 = \dfrac{12}{7}$ have as its solution set $\{-7\}$?

4.2

USING FORMULAS

Equations that express relationships between quantities of practical interest are called **formulas**. The methods of solving equations that we considered in Section 4.1 can be used to solve a formula for a specified variable if we are given values for the other variables in the formula.

Example The relation between Fahrenheit and Celsius temperatures is given by

$$F = \frac{9}{5}C + 32.$$

What is the Celsius temperature if the Fahrenheit temperature is 77°?

Solution We replace F with 77 in the given formula and solve for C. Thus,

$$77 = \frac{9}{5}C + 32$$

$$(5)77 = (5)\frac{9}{5}C + (5)32$$

$$385 = 9C + 160$$

$$225 = 9C$$

$$25 = C.$$

The Celsius temperature is 25°.

Example The number N of shirts sold by a clothing store chain is found to be related to the sale price p by the formula

$$N = \frac{80,000}{8 + p}, \qquad 8 < p < 16.$$

At what price p per shirt could the chain expect to sell shirts if it wished to sell 4000 shirts?

Solution overleaf

Solution   We replace N with 4000 in the given formula and solve for p. Thus,

$$4000 = \frac{80,000}{8 + p}$$

$$(8 + p)4000 = (8 + p)\left(\frac{80,000}{8 + p}\right)$$

$$32,000 + 4000p = 80,000$$

$$4000p = 48,000$$

$$p = 12.$$

The shirts should be priced at $12 each.

EXERCISE 4.2

A   1. Solve $s = \frac{1}{2}at^2$ for a, given that $s = 64$ and $t = 4$.

2. Solve $S = \frac{a}{1 - r}$ for a, given that $S = 12$ and $r = \frac{1}{2}$.

3. Solve $A = \frac{h}{2}(b + c)$ for b, given that $A = 64$, $h = 6$, and $c = 2$.

4. Solve $\ell = a + (n - 1)d$ for n, given that $\ell = 58$, $a = -2$, and $d = 12$.

5. Solve $A = 2\pi rh + 2\pi r^2$ for h, given that $A = 640$, $r = 4$, and $\pi \approx 3.14$.

6. Solve $\frac{1}{r} = \frac{1}{s} + \frac{1}{t}$ for s, given that $r = 3$ and $t = 4$.

7. A projectile launched upward from the ground at a velocity of v feet per second is located at a height s above the ground at time t, as given by

$$s = vt - 16t^2.$$

With what velocity must an object be launched to obtain a height of 200 feet in 8 seconds?

8. A business estimates that its total sales of S units is related to the amount A spent on advertising and the number C of competitors in its marketing area by

$$S = \frac{15A}{C}.$$

How much should the business allocate to advertising if it has three competitors and wishes to have total sales of 5000 units?

9. If P is invested at a simple rate r, then the amount A that is accumulated after t years is given by

$$A = P + Prt.$$

If $1000 were invested at 11% (0.11) interest, how many years would be required to accumulate $2320?

10. Use the formula in Problem 9 to find out how long it will take a loan of $3000 at 12.5% (0.125) annual interest to be worth $3625.

11. The cutting speed S (in feet per minute) of a circular saw blade is given by

$$S = \frac{C\omega D}{12d},$$

where C (in feet) is the circumference of the saw blade, ω is the angular velocity in revolutions per minute of the driving motor, D is the diameter of the driving pulley, and d is the diameter of the driven pulley. What angular velocity must the driving motor attain to achieve a cutting speed $S = 7000$ feet per minute if $C = 3$ feet, $D = 4$ inches, and $d = 3$ inches?

12. Consumers are usually charged for their use of electricity in units of kilowatt-hours (kwh), which are calculated using the formula

$$kwh = \frac{E \cdot I \cdot t}{1000},$$

where E is the voltage, I is the current in amperes, and t is the time in hours. Find the amperes, I, if a motor operates for 2 hours on a 120 volt line and uses 2.88 kilowatt-hours of electricity.

13. The normal weight w, in pounds, of a male between 60 and 70 inches tall is related to his height h, in inches, by

$$w = \frac{11}{2}h - 220.$$

What should be the height of a boy who weighs 143 pounds?

14. The estimated demand D for a television set is related to its price P (in dollars) by

$$D = -60P + 12{,}000, \qquad 80 < P < 140.$$

At what price should the set be sold to produce a demand for 4800 sets?

15. The pressure p, in pounds per square inch, is related to the depth d, in feet below the surface of an ocean, by

$$p = \frac{5}{11}d + 15.$$

At what depth is the pressure 75 pounds per square inch?

16. The horsepower (hp) generated by water flowing over a dam of height s, in feet, is given by

$$hp = \frac{62.4N \cdot s}{33{,}000},$$

where N is the number of cubic feet of water flowing over the dam per minute. Find the number of cubic feet of water that flowed over a 165-foot dam if 468 hp were generated.

17. The monthly benefit B paid on a certain retirement plan is determined by

$$B = \frac{2}{5}w\left(1 + \frac{2n}{100}\right),$$

where w is the average monthly wage of the worker and n is the number of years worked. How many years must a person work to receive a monthly benefit of $512 if his average monthly wage is $800?

18. The proceeds A of a loan with maturity value S at a simple discount rate of d per year is given by

$$A = S(1 - dn),$$

where n is the number of years of the loan. What is the maturity value of a loan payable in 6 months at 9% (0.09) simple discount if its proceeds now are $17,600?

19. The net resistance R_n of an electrical circuit is given by

$$\frac{1}{R_n} = \frac{1}{R_1} + \frac{1}{R_2},$$

where R_1 and R_2* are individual resistors in parallel. If one of the individual resistors is 40 ohms, what must the other be if the net resistance is 30 ohms?

20. The net resistance R_n of an electrical circuit containing resistors R_1, R_2, and R_3* in parallel is given by

$$\frac{1}{R_n} = \frac{1}{R_1} + \frac{1}{R_2} + \frac{1}{R_3}.$$

If two of the individual resistors are 20 ohms and 40 ohms, what must the third resistor be if the net resistance is 10 ohms?

21. A manufacturer plans to depreciate a large lathe that cost $12,000. The value V in t years is given by the formula

$$V = C\left(1 - \frac{t}{15}\right),$$

where C is the original cost. In how many years will the value be $5,000? In how many years will the lathe be completely depreciated?

22. The force F necessary to raise a weight w using a special pulley system is given by

$$F = \frac{w}{2}\left(\frac{D_1 - D_2}{D_1}\right).$$

Find the weight that can be lifted using a force of 45 pounds, where $D_1 = 25$ and $D_2 = 22$.

23. Three units are normally used as a measure of temperature, namely, Celsius (C), Fahrenheit (F), and Kelvin (K). These units are related by the formulas

$$C = \frac{5F - 160}{9} \quad \text{and} \quad K = C + 273.$$

What Fahrenheit temperature corresponds to a temperature of 290° Kelvin?

24. Use the formulas in Problem 23 to find the Kelvin temperature that corresponds to a temperature of 98.6° Fahrenheit.

■ **a.** *Translate each sentence into a formula.*

  b. *Solve for the specified quantity.*

---

*The subscripts 1, 2, and 3 used in R_1, R_2, and R_3 in the equations are read "R sub one," "R sub two," and "R sub three."

25. The price (P) of n items of equal value is the price (p) of one item times the number (n) of items. Determine the price of one pair of tennis shoes if 24 pairs cost $528.

26. The average rate (r) of a vehicle is the distance (d) traveled divided by the time (t). Determine the time it would take to travel 640 kilometers at an average rate of 48 kilometers per hour.

27. The area (A) of a triangle equals one-half the height (h) times the base (b). Find the height of a triangle if the area is 20 square centimeters and the base is 4 centimeters.

28. The perimeter (P) of a rectangle equals the sum of two times the length (ℓ) and two times the width (w). Find the length of a rectangle if the perimeter is 128 centimeters and the width is 21 centimeters.

29. The lateral surface area (S) of a right circular cylinder is the product π times twice the radius (r) of the base times the height (h). Find an approximation for the height of a cylinder if the radius of the circular base is 2 meters and the surface area is 50.24 square meters. (Use $\pi \approx 3.14$.)

30. Simple interest (I) on money invested is the product of the principal (P) invested times the rate (r) of interest times the number of years (t) the money is invested. How many years would be required for an investment of $600 at 12% (0.12) simple interest to yield $324?

4.3

SOLVING EQUATIONS FOR SPECIFIED SYMBOLS

An equation containing more than one variable, or containing symbols representing constants such as a, b, and c, can be solved for one of the symbols in terms of the remaining symbols by using the methods developed in Section 4.1. In general, we apply elementary transformations until the desired symbol stands alone as one member (ordinarily the left-hand member) of an equation.

Example   Solve $ax = 2r - cx$ for x.

Solution   Add cx to each member.

$$ax + cx = 2r - cx + cx$$

$$ax + cx = 2r$$

Factor x from the two terms in the left-hand member.

$$x(a + c) = 2r$$

Multiply each member by $\dfrac{1}{a + c}$ [or divide by $(a + c)$, the coefficient of x].

$$x = \frac{2r}{a + c} \qquad (a + c \neq 0)$$

The following suggestions, which were followed in the preceding example, may be helpful when rewriting equations.

To solve a first-degree equation for a specified variable:

1. Transform the equation to a form where all terms containing the specified variable are in one member and all terms not containing that variable are in the other member.

2. Combine like terms in each member.

3. If unlike terms containing the specified variable remain, factor that variable from the terms.

4. Divide each member by the coefficient of the specified variable.

Sometimes it is helpful first to rewrite an equation that contains fractions as an equivalent equation "free" of fractions.

Example   Solve $\dfrac{2}{a} - \dfrac{3}{y} = \dfrac{1}{b}$ for y.

Solution   Multiply each member of the equation by the least common denominator aby in order to write the equation without fractions.

$$aby\left(\frac{2}{a} - \frac{3}{y}\right) = aby\left(\frac{1}{b}\right)$$
$$2by - 3ab = ay$$

Add $3ab - ay$ to each member in order to obtain an equation in which all terms containing y are in one member and all terms not containing y are in the other member.

$$2by - 3ab + (3ab - ay) = ay + (3ab - ay)$$

Combine like terms.

$$2by - ay = 3ab$$

Factor y from the two terms in the left-hand member.

$$y(2b - a) = 3ab$$

Multiply each member by $1/(2b - a)$ [or divide by $(2b - a)$, the coefficient of y].

$$\frac{y(2b - a)}{2b - a} = \frac{3ab}{2b - a}$$

$$y = \frac{3ab}{2b - a} \qquad (2b - a \neq 0)$$

EXERCISE 4.3

A ■ *Solve for x or y. (Leave the results in the form of an equation equivalent to the given equation. Assume that no denominator is 0.)*

Example

$$a = \frac{2}{b}x - c$$

Solution

Multiply each member by b to obtain an equation that is free of fractions.

$$b(a) = b\left(\frac{2}{b}x - c\right)$$

$$ab = 2x - bc$$

Add bc to each member to obtain an equation in which all terms containing the specified variable are in one member and all terms not containing that variable are in the other member.

$$ab + bc = 2x - bc + bc$$

$$ab + bc = 2x$$

Multiply each member by ½ (or divide by 2).

$$\frac{ab + bc}{2} = x \quad \text{or} \quad x = \frac{ab + bc}{2}$$

1. $2x - a = b$ **2.** $3x + b = c$ **3.** $ay + c = b$

4. $by - a = c$ **5.** $2ax + b = c$ **6.** $3bx - c = a$

7. $\frac{a}{b}y + c = 0$ **8.** $b = \frac{1}{c}y + a$ **9.** $\frac{x + a}{b} = c$

10. $b = \frac{y - a}{c}$ **11.** $\frac{c - 2y}{b} = a$ **12.** $\frac{b - 3x}{a} = c$

Example

$$\frac{5}{2}by - a = ay$$

Solution

Multiply each member by 2.

$$5by - 2a = 2ay$$

Add $2a - 2ay$ to each member.

$$5by - 2ay = 2a$$

Factor the left-hand member.

$$y(5b - 2a) = 2a$$

Solution continued overleaf

Multiply each member by $1/(5b - 2a)$ [or divide by $(5b - 2a)$].

$$y = \frac{2a}{5b - 2a}$$

13. $ax = a - x$

14. $y = b + by$

15. $a(a - x) = b(b - x)$

16. $4y - 3(y - b) = 8b$

17. $(x - 4)(b + 2) = b$

18. $(y - 2)(a + 3) = 2a$

19. $\dfrac{2}{x} + \dfrac{2}{a} = 6$

20. $\dfrac{2}{y} - \dfrac{3}{b} = 4$

21. $\dfrac{a}{y} - \dfrac{1}{2} = \dfrac{a}{2y}$

22. $\dfrac{c}{3} + \dfrac{1}{x} = \dfrac{c}{6x}$

23. $\dfrac{1}{a} + \dfrac{1}{b} = \dfrac{1}{x}$

24. $\dfrac{1}{a} - \dfrac{1}{b} = \dfrac{1}{x}$

■ *Solve each formula for the specified variable.*

Example $v = k + gt,$ for g

Solution   Add $-k$ to each member.

$$v - k = gt$$

Multiply each member by $1/t$ (or divide by t).

$$\frac{v - k}{t} = g \quad \text{or} \quad g = \frac{v - k}{t}$$

25. $v = k + gt,$ for k

26. $E = mc^2,$ for m

27. $f = ma,$ for m

28. $I = prt,$ for p

29. $pv = K,$ for v

30. $E = IR,$ for R

31. $s = \dfrac{1}{2}at^2,$ for a

32. $p = 2\ell + 2w,$ for ℓ

33. $v = k + gt,$ for t

34. $y = \dfrac{k}{z},$ for z

35. $V = \ell wh,$ for h

36. $W = I^2R,$ for R

37. $180 = A + B + C,$ for B

38. $S = \dfrac{a}{1 - r},$ for r

39. $A = \dfrac{h}{2}(b + c),$ for c

40. $S = 2r(r + h),$ for h

41. $S = 3\pi d + 5\pi D,$ for d

42. $A = 2\pi rh + 2\pi r^2,$ for h

43. $\ell = a + (n - 1)d,$ for n

44. $\dfrac{1}{r} = \dfrac{1}{s} + \dfrac{2}{t},$ for r

45. $S = 2(\ell w + \ell h + wh),$ for ℓ

46. $S = \dfrac{n}{2}(a + s_n),$ for s_n

B ■ *The following formulas were used in exercises in Section* 4.2. *Solve each equation for the specified variable in terms of the other variables.*

47. $V = C\left(1 - \dfrac{t}{15}\right)$, for t

48. $\dfrac{1}{R_n} = \dfrac{1}{R_1} + \dfrac{1}{R_2} + \dfrac{1}{R_3}$, for R_3

49. $B = \dfrac{2}{5}w\left(1 + \dfrac{2n}{100}\right)$, for n

50. $C = \dfrac{5F - 160}{9}$ and $K = C + 273$, for F in terms of K

4.4

APPLICATIONS

Word problems state relationships between numbers. Problems may be explicitly concerned with numbers, or they may be concerned with numerical measures of physical quantities. In either event, we seek a number or numbers for which the stated relationships hold. To express the quantitative ideas symbolically in the form of an equation is the most difficult part of solving word problems. Unfortunately, there is no single means available to do this. However at this time we will work with equations in one variable and the following suggestions may be helpful:

Solving word problems

1. Represent the unknown quantities using word phrases.

2. Represent each unknown quantity in terms of symbols. Since at this time we are using one variable only, all relevant quantities should be represented in terms of this variable.

3. Where applicable, draw a sketch and label all known quantities.

4. Find a quantity that can be represented in two different ways and write these representations as an equation (mathematical model). Summarizing information in tabular form is sometimes helpful.

5. Solve the resulting equation.

6. Check the results to see that they are meaningful for the conditions of the problem.

As much or more attention should be given to the setting up of equations as is given to their solution. Although some of the problems in this book can be solved without recourse to algebra, the practice obtained in setting up the equations involved will be helpful when more difficult problems are encountered. Therefore, we suggest that you *read carefully each step of the solutions of the illustrative examples in Exercise 4.4.*

EXERCISE 4.4

A   ■ **a.** *Represent the unknown quantities using word phrases.*

b. *Set up an equation.*   **c.** *Solve.*

Example   A small business calculator sells for $28 less than a scientific model. If the cost to buy both is $158, how much does each cost?

Solution   **a.** Express the *two* quantities asked for in *two* simple phrases and represent these quantities using one variable.

$$\text{cost of scientific calculator:} \quad c$$
$$\text{cost of business calculator:} \quad c - 28$$

b. Write an equation relating the unknown quantities.

$$c + (c - 28) = 158$$

c. Solve the equation.

$$c + c - 28 = 158$$
$$2c - 28 = 158$$
$$2c = 186$$
$$c = 93$$
$$c - 28 = 65$$

The scientific calculator costs $93 and the business model costs $65.

1. One year the Dean of Admissions at City College selected ¾ of all applicants for the freshman class. How many students applied for entrance if the college selected 600 students?

2. A part-time secretary takes home ⅘ of her total salary. What is her total salary if she takes home $120 per week?

3. In a student body election, 584 students voted for one or the other of two candidates for president. If the winner received 122 more votes than the loser, how many votes were cast for each candidate?

4. The profit in selling an item is $15 less than the cost of the item. If the item sells for $85, what is its cost, and how much profit results from the sale?

5. A saleswoman earns a commission of 2% (0.02) on her sales. How much must she sell to have an income of $300 per week?

6. A company contracts to produce 1200 gears that meet certain specifications. If 3% (0.03) of the gears manufactured are defective, what is the total number of gears that must be manufactured?

Using Proportions

*A quotient of two numbers, a/b, is sometimes called a **ratio**. A statement of equality of two ratios, a/b = c/d, called a **proportion**, can be used as a model for a variety of problems. A table can be helpful in setting up a proportion.*

Example   A typist can type 200 words in 5 minutes. Typing at the same rate, how long would it take the typist to type 10 pages of a manuscript that averages 240 words per page?

Solution

a. Express the quantity asked for in a word phrase and represent the quantity using a symbol.

> time to type 10 pages of 240 words, a total of 2400 words: t

b. Set up a table as shown in (a). Then enter 2400 and t as the numerators in the appropriate columns and 200 and 5 as the denominators in the appropriate columns, as shown in (b).

Words		*Time*
———	=	———

(a)

Words		*Time*
$\dfrac{2400}{200}$	=	$\dfrac{t}{5}$

(b)

c. Solve the equation:

$$(200)\frac{2400}{200} = (200)\frac{t}{5}$$
$$2400 = 40t$$
$$60 = t.$$

Thus, it would take 60 minutes (1 hour) to type the 10 pages.

7. A mason can lay 450 bricks in 3 hours. At the same rate, how many bricks can he lay in a 40-hour week?

8. An automobile uses 6 gallons of gasoline to travel 102 miles. How many gallons are required to make a trip of 459 miles?

9. If 2.5 pounds of tin are required to make 12 pounds of a certain alloy, how many pounds of tin are needed to make 90 pounds of the alloy?

10. A man walks approximately 6.4 miles in 2 hours. At the same rate, how far can he walk in 5 hours?

Coin Problems

The basic idea of problems involving coins (or bills) is that the value of a number of coins (or bills) of the same denomination is equal to the product of the value of a single coin (or bill) and the total number of coins (or bills).

$$\begin{bmatrix} value\ of \\ n \\ coins \end{bmatrix} = \begin{bmatrix} value\ of \\ 1 \\ coin \end{bmatrix} \times \begin{bmatrix} number \\ of \\ coins \end{bmatrix}$$

It is helpful in most coin problems to think in terms of cents rather than dollars. Tables can also be helpful in solving coin problems.

Example   A collection of coins consisting of dimes and quarters has a value of \$11.60. How many dimes and quarters are in the collection if there are 32 more dimes than quarters?

Solution   **a.** Express the *two* quantities asked for in *two* simple phrases and represent the quantities using symbols.

$$\text{number of quarters: } x$$
$$\text{number of dimes: } x + 32$$

b. Set up a table.

Denomination	Value of 1 coin in cents	Number of coins	Value of coins in cents
Dimes	10	$x + 32$	$10(x + 32)$
Quarters	25	x	$25x$

Write an equation relating the value of the quarters and the value of the dimes to the value of the entire collection.

$$\begin{bmatrix} value\ of \\ quarters \\ in\ cents \end{bmatrix} + \begin{bmatrix} value\ of \\ dimes \\ in\ cents \end{bmatrix} = \begin{bmatrix} value\ of \\ collection \\ in\ cents \end{bmatrix}$$

$$25x \quad + 10(x + 32) = \quad 1160$$

c. Solve the equation:

$$25x + 10x + 320 = 1160$$
$$35x = 840$$
$$x = 24$$
$$x + 32 = 56$$

Therefore, there are 24 quarters and 56 dimes in the collection.

11. A woman who has 12 more quarters than dimes has a total of $12.45 in quarters and dimes. How many of each kind of coin does she have?

12. A savings bank containing $1.75 in dimes, quarters, and nickels contains 3 more dimes than quarters, and 3 times as many nickels as dimes. How many coins of each kind are in the bank?

13. On an airplane flight for which first-class fare is $80 and tourist fare is $64 there were 42 passengers. If receipts for the flight totaled $2880, how many first-class and how many tourist passengers were on the flight?

14. An ice cream vendor bought 50 ice cream bars—chocolate-covered bars at 18¢ each and sandwich bars at 15¢ each. If the total cost was $8.10, how many bars of each kind did the vendor purchase?

Interest Problems

Simple interest problems involve the fact that the interest earned during a single year is equal to the amount invested times the annual rate of interest. Thus, $I = Ar$. Tables are sometimes helpful in summarizing data for this type of problem.

Example A man has an annual income of $12,000 from two investments. He has $10,000 more invested at 8% than he has invested at 12%. How much does he have invested at each rate?

Solution a. Express the *two* quantities asked for in *two* simple phrases and represent these quantities using symbols:

$$\text{amount invested at } 12\%: \quad A$$
$$\text{amount invested at } 8\%: \quad A + 10{,}000$$

b. Set up a table.

Investment Rate		Amount	Interest
12%	0.12	A	$0.12A$
8%	0.08	$A + 10{,}000$	$0.08(A + 10{,}000)$

Write an equation relating the interest from each investment and the total interest received.

$$\begin{bmatrix} \text{interest from} \\ \text{12\% investment} \end{bmatrix} + \begin{bmatrix} \text{interest from} \\ \text{8\% investment} \end{bmatrix} = [\text{total interest}]$$

$$0.12A \qquad + 0.08(A + 10{,}000) = \qquad 12{,}000$$

Solution continued overleaf

c. Solve for A. First multiply each member by 100.

$$12A + 8(A + 10,000) = 1,200,000$$
$$12A + 8A + 80,000 = 1,200,000$$
$$20A = 1,120,000$$
$$A = 56,000$$
$$A + 10,000 = 66,000$$

Therefore, \$56,000 is invested at 12% and \$66,000 is invested at 8%.

15. A woman has invested \$8000, part in a bank at 10% and part in a savings and loan association at 12%. If her annual return is \$844, how much has she invested at each rate?

16. A sum of \$2400 is split between an investment in a mutual fund paying 14% and one in corporate bonds paying 11%. If the return on the 14% investment exceeds that on the 11% investment by \$111 per year, how much is invested at each rate?

17. If \$3000 is invested in bonds at 8%, how much additional money must be invested in stocks paying 13% to make the earnings on the total investment 10%?

18. If \$6000 is invested in bonds at 9%, how much additional money must be invested in stocks at 12% to earn a return of 11% on the total investment?

Geometry Problems

Solving problems involving geometric figures often requires the use of the formulas listed in the table on page 437.

Example   The vertex angle of an isosceles triangle measures 30° more then the sum of the measures of its other two angles. What is the measure of each angle in the triangle?

Solution   A sketch is helpful. Use the fact that in an isosceles triangle, the two angles opposite the two sides of equal length have equal measure.

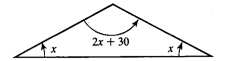

a. Express the *two* quantities asked for in *two* simple phrases and represent these quantities using symbols.

measure of each base angle: x

measure of the vertex angle: $2x + 30$

b. Write an equation using the fact that the sum of the measures of the angles in any triangle is 180°.

$$(x) + (x) + (2x + 30) = 180$$

c. Solve the equation:

$$4x + 30 = 180$$

$$4x = 150$$

$$x = 37\frac{1}{2}$$

$$2x + 30 = 105.$$

Therefore, the angles in the triangle measure $37\frac{1}{2}°$, $37\frac{1}{2}°$, and $105°$.

19. One angle of a triangle measures 2 times another angle, and the third angle measures $12°$ more than the sum of the measures of the other two. Find the measure of each angle.

20. The measure of the smallest angle of a triangle is $25°$ less than the measure of another angle, and $50°$ less than the measure of the third angle. Find the measure of each angle.

21. Each of the equal sides of an isosceles triangle is 3 centimeters longer than 2 times the length of the base. Find the length of each side of the triangle if its perimeter is 86 centimeters.

22. When the length of each side of a square is increased by 5 centimeters, the area is increased by 85 square centimeters. Find the length of a side of the original square.

Lever Problems

*A rigid bar rotating about a fixed point is called a **lever**. The fixed point is called the **fulcrum** of the lever. If a force F_1 is applied to one side of a lever at a distance d_1 from the fulcrum, and a force F_2 is applied on the other side of the lever at a distance d_2 from the fulcrum (as shown in the figure), then the lever will be in equilibrium if and only if*

$$F_1 d_1 = F_2 d_2.$$

*This relationship is called the **law of the lever** and has many practical applications.*

Example

What force must be applied to the end of a 10-centimeter beam to balance a weight of 6 grams hung on the other end if the beam rests on a fulcrum located $1\frac{1}{2}$ centimeters from the weight?

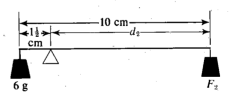

Solution

a. The force to be applied: F_2.

b. In the law of the lever, substitute

$$F_1 = 6, \qquad d_1 = \frac{3}{2}, \qquad d_2 = 10 - \frac{3}{2} = \frac{17}{2},$$

and write the resulting equation.

$$6 \cdot \frac{3}{2} = \frac{17}{2} \cdot F_2$$

Solution continued overleaf.

c. Solve the equation:

$$9 = \frac{17}{2} \cdot F_2$$

$$F_2 = \frac{18}{17}.$$

Therefore, the force to be applied is $1\frac{1}{17}$ grams.

23. A 24-gram weight is placed 2 centimeters further from the fulcrum on one side of a balanced lever than a 32-gram weight on the other side. How far is each weight from the fulcrum?

24. A 48-gram weight and a 36-gram weight are placed on opposite ends of a beam, each weight at a distance of 8 centimeters from the fulcrum. Where should a 24-gram weight be placed to balance the beam? [*Hint:* $F_1 d_1 = F_2 d_2 + F_3 d_3$.]

25. Where should the fulcrum of a 6-foot crowbar be placed so that a 200-pound man can just balance a 2200-pound weight?

26. Where should the fulcrum of a 9-foot crowbar be placed so that a 180-pound man can just balance a 900-pound weight?

B · Mixture Problems

The key to solving mixture problems is to recognize that the amount of a given substance in a mixture is obtained by multiplying the amount of the mixture by the percent (or rate) of the given substance in the mixture.

$$\begin{bmatrix} amount \\ of \\ substance \end{bmatrix} = \begin{bmatrix} percent\ (rate) \\ of\ substance \\ in\ mixture \end{bmatrix} \times \begin{bmatrix} amount \\ of \\ mixture \end{bmatrix}$$

Tables are especially helpful in mixture problems.

Example   How many liters of a 10% solution of acid should be added to 20 liters of a 60% solution of acid to obtain a 50% solution?

Solution   **a.** Express the quantity asked for in a simple phrase and represent the quantity using a symbol.

number of liters of 10% solution: n

b. Set up a table.

Percent acid in mixture		Amount of mixture	Amount of acid
10%	0.10	n	$0.10n$
60%	0.60	20	$0.60(20)$
50%	0.50	$n + 20$	$0.50(n + 20)$

A figure may be helpful. Here, the amount of pure acid in the solution is shown in red; the remainder is water.

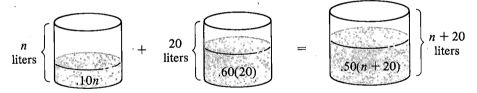

Write an equation relating the amount of *pure* acid.

$$\left[\begin{array}{c}\text{pure acid in}\\ \text{10\% solution}\end{array}\right] + \left[\begin{array}{c}\text{pure acid in}\\ \text{60\% solution}\end{array}\right] = \left[\begin{array}{c}\text{pure acid in}\\ \text{50\% solution}\end{array}\right]$$

$$0.10n \quad + \quad 0.60(20) \quad = \quad 0.50(n+20)$$

c. Solve the equation. First multiply each member by 100.

$$10n + 60(20) = 50(n+20)$$

$$10n + 1200 = 50n + 1000$$

$$-40n = -200$$

$$n = 5$$

Therefore, 5 liters of the 10% solution are needed.

27. How many liters of a 30% salt solution must be added to 40 liters of a 12% salt solution to obtain a 20% solution?

28. How many grams of an alloy containing 45% silver must be melted with an alloy containing 60% silver to obtain 40 grams of an alloy containing 48% silver?

29. How much pure alcohol should be added to 12 liters of a 45% solution to obtain a 60% solution?

30. How many liters of a 20% sugar solution should be added to 40 liters of a 32% solution to obtain a 28% solution?

31. A tank contains 20 gallons of a liquid fertilizer with a 16% nitrogen content. How many gallons of this solution should be drained from the tank and replaced with water so that the fertilizer will have a 12% nitrogen content?

32. An automobile radiator contains 32 quarts of a 40% antifreeze solution. How many quarts of this solution should be drained from the radiator and replaced with water if the resulting solution is to be 20% antifreeze?

Uniform-Motion Problems

Solving uniform-motion problems requires applying the fact that the distance traveled at a uniform rate is equal to the product of the rate and the time traveled. This relationship may be expressed by any one of the three equations

$$d = rt, \qquad r = \frac{d}{t}, \qquad t = \frac{d}{r}.$$

It is frequently helpful to use a table and a figure to solve these problems.

Example   An express train travels 150 miles in the same time that a freight train travels 100 miles. If the express goes 20 miles per hour faster than the freight, find the rate of each.

Solution   **a.** Express the *two* quantities asked for in *two* simple phrases and represent these quantities using symbols.

$$\text{rate for the freight train:} \quad r$$

$$\text{rate for the express train:} \quad r + 20$$

b. Set up a table.

	d	r	$t = d/r$
Freight	100	r	$\dfrac{100}{r}$
Express	150	$r + 20$	$\dfrac{150}{r + 20}$

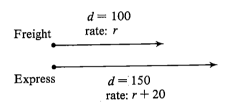

The fact that the times are equal is the significant equality in the problem.

$$[t \text{ freight}] = [t \text{ express}]$$

$$\frac{100}{r} = \frac{150}{r + 20}$$

c. Solve the equation:

$$r(r + 20)\left(\frac{100}{r}\right) = r(r + 20)\left(\frac{150}{r + 20}\right)$$

$$(r + 20)100 = (r)150$$

$$100r + 2000 = 150r$$

$$-50r = -2000$$

$$r = 40$$

$$r + 20 = 60.$$

Thus, the freight train's rate is 40 miles per hour and the express train's rate is 60 miles per hour.

33. An airplane travels 1260 miles in the same time that an automobile travels 420 miles. If the rate of the airplane is 120 miles per hour greater than the rate of the automobile, find the rate of each.

34. Two planes leave an airport at the same time and travel in opposite directions. If one plane averages 440 miles per hour over the ground and the other 560 miles per hour, in how long will they be 2500 miles apart?

35. A ship traveling at 20 knots is 5 nautical miles out from a harbor when another ship leaves the harbor at 30 knots sailing on the same course. How long does it take the second ship to catch up to the first?

36. A boat sails due west from a harbor at 36 knots. An hour later, another boat leaves the harbor on the same course at 45 knots. How far out at sea will the second boat overtake the first?

Work Problems

Some problems involve the accomplishment of a task in some fixed time when a steady rate of work is assumed. For example, if it takes 6 hours for a man to paint a room, then in one hour he can paint ⅙ of the room, in two hours he can paint (⅙)(2) of the room and in t hours he can paint (⅙)(t) of the room. In general in such problems, the part of the work done in t units of time equals the work done in one unit of time multiplied by t.

Example

One pipe can empty a tank in 6 hours, and a second pipe can empty the same tank in 9 hours. How long will it take both pipes to empty the tank?

Solution

a. Express the quantity asked for in a simple phrase and represent the quantity using a symbol.

number of hours for both pipes to empty the tank: t

b. A table can be helpful.

$$\begin{bmatrix} \text{part of tank} \\ \text{emptied in} \\ \text{1 hour} \end{bmatrix} \times \begin{bmatrix} \text{hours} \\ \text{together} \end{bmatrix} = \begin{bmatrix} \text{part of tank} \\ \text{emptied by} \\ \text{each pipe} \end{bmatrix}$$

	part of tank emptied in 1 hour	hours together	part of tank emptied by each pipe
Pipe 1	$\dfrac{1}{6}$	t	$\left(\dfrac{1}{6}\right)t$
Pipe 2	$\dfrac{1}{9}$	t	$\left(\dfrac{1}{9}\right)t$

Write an equation expressing the conditions of the problem.

$$\begin{bmatrix} \text{part of tank} \\ \text{emptied by pipe 1} \end{bmatrix} + \begin{bmatrix} \text{part of tank} \\ \text{emptied by pipe 2} \end{bmatrix} = \begin{bmatrix} \text{entire tank} \\ \text{emptied by both} \end{bmatrix}$$

$$\left(\frac{1}{6}\right)t \qquad + \qquad \left(\frac{1}{9}\right)t \qquad = \qquad 1$$

Solution continued overleaf

c. Solve the equation:

$$(18)\left(\frac{1}{6}\right)t + (18)\left(\frac{1}{9}\right)t = (18)1$$

$$3t + 2t = 18$$

$$5t = 18$$

$$t = \frac{18}{5}.$$

The tank will be emptied in $3\frac{3}{5}$ hours when both pipes are open.

37. It takes one pipe 30 hours to fill a tank, while a second pipe can fill the same tank in 45 hours. How long will it take both pipes running together to fill the tank?

38. One pipe can fill a tank in 4 hours and another can empty it in 6 hours. If both pipes are open, how long will it take to fill the empty tank?

39. A new billing machine can process a firm's monthly billings in 10 hours. By using an older machine together with the new machine, the billings can be completed in 6 hours. How long would it take the older machine to do the job alone?

40. A tractor plows $\frac{5}{9}$ of a field in 10 hours. By adding another tractor the job is finished in another 3 hours. How long would it take the second tractor to do the job alone?

4.5

SOLVING INEQUALITIES

Solution of an inequality

As we noted in Section 1.2, expressions such as

$$x + 3 \geq 10$$

and

$$\frac{-2y - 3}{3} < 5$$

are called **inequalities**. For appropriate values of the variable, one member of an inequality represents a real number that is less than ($<$), less than or equal to ($\leq$), greater than or equal to ($\geq$), or greater than ($>$) the real number represented by the other member.

Any element of the replacement set of the variable for which an inequality is true is called a **solution**, and the set of all solutions of an inequality is called the **solution set** of the inequality.

Equivalent inequalities

As in the case with equations, we shall solve a given inequality by generating a series of equivalent inequalities (inequalities that have the same solution set) until we arrive at one whose solution set is obvious. To do this we shall need some fundamental properties of inequalities. Notice that

$$2 < 3,$$

$$2 + 5 < 3 + 5,$$

and

$$2 - 5 < 3 - 5.$$

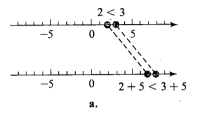
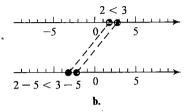

Figure 4.1

Parts a and b of Figure 4.1 demonstrate that the addition of 5 or -5 to each member of $2 < 3$ simply shifts the graphs of the members the same number of units to the right or left on the number line, with the order of the members left unchanged. This will be the case for the addition of any real number to each member of an inequality. Since for any real value of the variable for which an expression is defined the expression represents a real number, we generalize this idea and assert that:

1.   *The addition of the same expression representing a real number to each member of an inequality produces an equivalent inequality in the **same sense**.*

Next, if we multiply each member of

$$2 < 3$$

by 2, we have

$$4 < 6,$$

where the products form an inequality in the same sense. If, however, we multiply each member of

$$2 < 3$$

by -2, we have

$$-4 > -6,$$

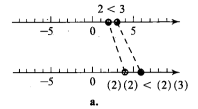

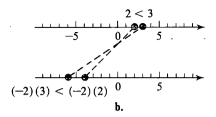

Figure 4.2

where the inequality is in the opposite sense. Figure 4.2 illustrates this. Multiplying each member of $2 < 3$ by 2 simply moves the graph of each member out twice as far in a positive direction (part a). Multiplying by -2 also doubles the absolute value of each member, but the products are negative and the sense of the inequality is reversed as shown in part b. In general, we assert that:

> 2. *If each member of an inequality is multiplied by the same expression representing a positive number, the result is an equivalent inequality in the **same sense**.*

> 3. *If each member of an inequality is multiplied by the same expression representing a negative number, the result is an equivalent inequality in the **opposite sense**.*

Statements 1, 2, and 3 above can be expressed in symbols as follows:

► *If $P(x)$, $Q(x)$, and $R(x)$ are expressions, then for all values of x for which these expressions represent real numbers,*

$$P(x) < Q(x)$$

is equivalent to

$$P(x) + R(x) < Q(x) + R(x), \tag{1}$$

$$P(x) \cdot R(x) < Q(x) \cdot R(x), \quad \textit{for} \quad R(x) > 0, \tag{2}$$

$$P(x) \cdot R(x) > Q(x) \cdot R(x), \quad \textit{for} \quad R(x) < 0. \tag{3}$$

These relationships are also true with $<$ replaced by $\leq$ and $>$ replaced by $\geq$.

Note that none of these assertions permits multiplying by 0 and that variables in expressions used as multipliers are restricted from values for which the expression vanishes or is not defined. The application of any of these properties is called an **elementary transformation**.

These properties can be applied to solve first-degree inequalities in the same way the equality properties are applied to solve first-degree equations. In the following examples and exercises in this section we assume that variables represent real numbers.

Example

Solve the inequality $\dfrac{x-3}{4} < \dfrac{2}{3}$.

Solution

By statement 2 above, we can multiply each member by 12 to obtain

$$3(x-3) < 8$$

or

$$3x - 9 < 8.$$

By statement 1, we can add 9 to each member, giving us

$$3x < 17.$$

And finally, by statement 2, we can multiply each member by $\frac{1}{3}$ to obtain

$$x < \frac{17}{3}.$$

The solution set is then $\{x \mid x < 17/3\}$.

Graphs of solution sets

The solution set of the inequality in the above example can be graphed as shown in Figure 4.3, where the red line represents the points whose coordinates are in the solution set.

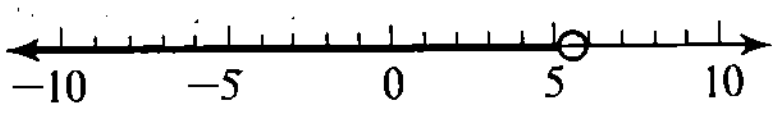

Figure 4.3

Intersection of two sets

Sometimes two conditions of inequality are placed on a variable. For example, we may want to consider $\{x \mid -2 < x \text{ and } x \le 5\}$, that is, the numbers in $\{x \mid -2 < x\}$ and also in $\{x \mid x \le 5\}$. We call the resulting set the **intersection** of the two given sets. In general:

> *If A and B are sets, the intersection of A and B consists of all those numbers that are in set A and also in set B. We designate this set by A ∩ B, read "the intersection of sets A and B."*

In the foregoing example, the two conditions on x are given by

$$\{x \mid -2 < x\} \cap \{x \mid x \le 5\},$$

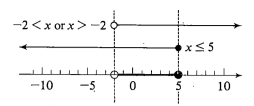

Figure 4.4

as illustrated in Figure 4.4. Since x must satisfy both conditions involved, the intersection is

$$\{x|-2<x\le 5\}.$$

A continued inequality of the form $a<x<b$, as in the example above, is an implied "and" statement. The two conditions $a<x$ *and* $x<b$ are being imposed on x.

Interval notation   A special notation is used sometimes to describe an interval of real numbers. The symbols (a, b) are used for the interval that includes all real numbers *between a* and *b*. The symbol "[" is used instead of "(" if a is included in the interval, and "]" is used instead of ")" if b is included in the interval.

Examples   **a.** $\{x|3<x<5\}=(3, 5)$ **b.** $\{y|6\le y\le 7\}=[6, 7]$

c. $\{z|1<z\le 3\}=(1, 3]$ **d.** $\{x|-2\le x<4\}=[-2, 4)$

In the event an interval includes all real numbers greater than a given real number or less than a given real number, we incorporate the symbols $+\infty$ and $-\infty$ in the notation. These symbols are read "positive infinity" and "negative infinity," respectively. They do not denote real numbers.

Examples   **a.** $\{x|x\ge 2\}=[2, +\infty)$ **b.** $\{y|y>-4\}=(-4, +\infty)$

c. $\{z|z\le 0\}=(-\infty, 0]$ **d.** $\{t|t<6\}=(-\infty, 6)$

Because an interval such as $[-5, 8]$ contains both of its endpoints, it is called a **closed interval**. For similar reasons, intervals such as $(3, 18]$ and $[-5, -2)$ are called **half-open intervals**, and an interval such as $(3, 8)$ or $(-\infty, 2)$ is called an **open interval**.

Disjoint sets   Sets that do not have any numbers in common are called **disjoint sets**. If A and B are sets, and

$$A \cap B = \varnothing,$$

then A and B are disjoint.

Examples

a. $\{2, 3\} \cap \{4, 5\} = \varnothing$

b. $\{x \,|\, x < 1\} \cap \{x \,|\, x > 3\} = \varnothing$

c. $(-5, -1) \cap (1, 5) = \varnothing$

d. $(-\infty, 3) \cap (3, +\infty) = \varnothing$

EXERCISE 4.5

A ■ *Solve and graph each solution set.*

Example $\dfrac{x+4}{3} \le 6 + x$

Solution Multiply each member by 3.

$$x + 4 \le 18 + 3x$$

Add $-4 - 3x$ to each member.

$$-2x \le 14$$

Multiply each member by $-\frac{1}{2}$ and reverse the sense of the inequality.

$$x \ge -7$$

The solution set is $\{x \,|\, x \ge -7\}$ or $[-7, +\infty)$.

1. $3x < 6$

2. $x + 7 > 8$

3. $x - 5 \le 7$

4. $2x - 3 < 4$

5. $3x - 2 > 1 + 2x$

6. $2x + 3 \le x - 1$

7. $\dfrac{2x - 6}{3} > 0$

8. $\dfrac{2x - 3}{2} \le 5$

9. $\dfrac{5x - 7x}{3} > 4$

10. $\dfrac{x - 3x}{5} \le 6$

11. $\dfrac{2x - 5x}{2} \le 7$

12. $\dfrac{x - 6x}{2} < -20$

13. $\dfrac{x}{2} + 1 < \dfrac{x}{3} - x$

14. $\dfrac{1}{2}(x + 2) \ge \dfrac{2x}{3}$

15. $2(x + 2) \le \dfrac{3}{4}x - 1$

16. $4(x - 3) \le 2x + 6$

17. $5(x + 2) > 3(x - 4)$

18. $3(2x + 1) \ge 5x - 6$

19. $4(2x + 1) - 3x \ge 4$

20. $0 < 2(x - 3) + 5x$

21. $x + 2 \le \dfrac{-2}{3}(x + 6)$

22. $\dfrac{2}{3}(x - 1) + \dfrac{3}{4}(x + 1) < 0$

23. $\dfrac{3}{4}(2x - 1) - \dfrac{1}{2}(4x + 3) \ge 0$

24. $\dfrac{3}{5}(3x + 2) - \dfrac{2}{3}(2x - 1) \le 2$

Example $4 < x + 4 \leq 6$

Solution   Add -4 to each member.

$$4 + (-4) < x + 4 + (-4) \leq 6 + (-4)$$
$$0 < x \leq 2$$

The solution set is $\{x | 0 < x \leq 2\}$ or $(0, 2]$.

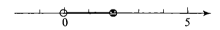

25. $4 < x - 2 < 8$	**26.** $0 \leq 2x \leq 12$	**27.** $-3 < 2x + 1 \leq 7$
28. $2 \leq 3x - 4 \leq 8$	**29.** $6 < 4 - x < 10$	**30.** $-3 < 3 - 2x < 9$

■ *Graph each set.*

Example $\{x | x > -4\} \cap \{x | x + 1 \leq 4\}$

Solution   Graph each inequality. Then locate the interval common to both number lines.

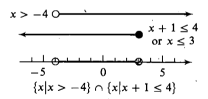

31. $\{x | x < 2\} \cap \{x | x > -2\}$	**32.** $\{x | x \leq 5\} \cap \{x | x \geq 1\}$
33. $\{x | x + 1 \leq 3\} \cap \{x | -(x + 1) \leq 3\}$	**34.** $\{x | 2x - 3 < 5\} \cap \{x | -(2x - 3) < 5\}$
35. $\{x | x - 7 < 2x\} \cap \{x | -3x \geq 4x + 28\}$	**36.** $\{x | 9 \leq 3x < 21\} \cap \{x | -10 < 5x \leq 15\}$

■ *Sentences describing inequalities are set up in the same way as those describing equalities, except that symbols such as $\leq$ or $>$ replace the symbol $=$.*

Example   A student must have an average of 80–90% on five tests in a course to receive a B. Her grades on the first four tests were 98%, 76%, 86%, and 92%. What grade on the fifth test would give her a B in the course?

Solution   **a.** Express the quantity asked for in a simple phrase and then represent the quantity symbolically.

grade (in percent) on the fifth test: x

b. Write an inequality expressing the word sentence.

$$80 \le \frac{98 + 76 + 86 + 92 + x}{5} < 90$$

c. Solve the inequality.

$$400 \le 352 + x < 450$$
$$48 \le \quad x \quad < 98$$

The solution set is $\{x \mid 48 \le x < 98\}$. Therefore, any grade equal to or greater than 48 and less than 98 would give the student a B in the course.

37. In the preceding example, what grade on the fifth test would give the student a B if her grades on the first four tests were 78%, 64%, 88%, and 76%?

38. In the preceding example, what grade on the final examination would give a student a B if her grades on the first four hourly tests were 72%, 68%, 84%, and 70%, and the final examination counted for two hourly tests?

39. The Fahrenheit and Celsius temperature scales are related by the formula

$$C = \frac{5}{9}(F - 32).$$

Within what range must the temperature be in degrees Fahrenheit for the temperature to lie between 30°C and 40°C?

40. Within what range must the temperature be in degrees Fahrenheit for the temperature to lie between −10°C and 20°C?

41. The Alpha Car Company rents a compact car for $24 per day. The Beta Car Company rents a similar car for $18 per day plus an initial fee of $90. For what rental period would it be cheaper to rent from Beta?

42. Employment Agency A charges a commission of $50 plus 15% of the first month's salary for its services. Agency B charges a commission of $80 plus 12% of the first month's salary. When is it cheaper to use Agency B?

43. A woman wishes to invest $10,000, part at 9% and part at 12%. What is the least amount she can invest at 12% if she wishes an annual return of at least $1008?

44. A man can sail upstream in a river at an average rate of 4 miles per hour, and downstream at an average rate of 6 miles per hour. What is the distance he can sail upstream if he starts at 8:00 AM and must be back no later than 6:00 PM?

B ■ *Graph the given intervals on the same line graph, and represent the numbers involved using set-builder notation.*

Examples   **a.** $(-\infty, -7)$; $[-3, 0]$ **b.** $[-4, 0]$; $(3, +\infty)$

Solutions overleaf

Solutions

a.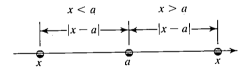

$\{x|x<-7\}; \quad \{x|-3\le x\le 0\}$

b.

$\{x|-4\le x\le 0\}; \quad \{x|x>3\}$

45. $[-8, 2]; \quad (3, 7]$

46. $(-4, 0]; \quad (2, +\infty)$

47. $[-7, -3]; \quad (0, 4]$

48. $(-\infty, 0); \quad (0, +\infty)$

49. $(-5, -3]; \quad (-2, 0]; \quad (1, 3)$

50. $(-\infty, -4]; \quad (-2, 0]; \quad (2, +\infty)$

4.6

EQUATIONS AND INEQUALITIES INVOLVING ABSOLUTE VALUE

Equations involving absolute value

In Section 1.4, we defined the absolute value of a real number by

$$|x| = \begin{cases} x, & \text{if } x \ge 0 \\ -x, & \text{if } x < 0 \end{cases}$$

and interpreted it in terms of distance on a number line. For example, $|-5| = 5$ by definition, but 5 also denotes the distance between the graph of -5 and the origin. More generally, we have

$$|x - a| = \begin{cases} x - a, & \text{if } x - a \ge 0, & \text{or equivalently, if } x \ge a \\ -(x - a), & \text{if } x - a < 0, & \text{or equivalently, if } x < a \end{cases}$$

and $|x - a|$ can be interpreted on a line graph as denoting the distance the graph of x is located from the graph of a, as shown in Figure 4.5.

$$x < a \qquad\qquad x > a$$

$$\xleftarrow{\hspace{0.5em}} |x - a| \xrightarrow{\hspace{0.5em}} \xleftarrow{\hspace{0.5em}} |x - a| \xrightarrow{\hspace{0.5em}}$$

$$\underset{x}{\ominus} \qquad\qquad \underset{a}{\ominus} \qquad\qquad \underset{x}{\ominus}$$

Figure 4.5

We can solve equations of the form

$$|x - a| = b$$

or, more generally, of the form

$$|ax - b| = c$$

by appealing to the definition of absolute value. Formally, we have:

▶ $|ax - b| = c$ *is equivalent to the joint statement*

$$ax - b = c \qquad or \qquad -(ax - b) = c. \qquad\qquad (1)$$

Example   Solve $|2x - 3| = 5$.

Solution   This equation implies that

$$2x - 3 = 5 \qquad \text{or} \qquad -(2x - 3) = 5$$
$$2x = 8 \qquad\qquad\qquad 2x - 3 = -5$$
$$x = 4 \qquad\qquad\qquad\quad 2x = -2$$
$$\qquad\qquad\qquad\qquad\qquad x = -1$$

Hence, the solution set is $\{4, -1\}$.

Union of two sets   Note that this set is really a combination of the member of $\{4\}$ and the member of $\{-1\}$. We call such a combination the **union** of the two sets.

> *If A and B are sets, the **union** of A and B consists of all those members that are either in set A or in set B or in both. We denote this set by A $\cup$ B, read "the union of sets A and B."*

In the above example, we have

$$\{4\} \cup \{-1\} = \{4, -1\}.$$

Any equation of the form $|ax - b| = c$ is not a polynomial equation, because the variable appears within the absolute-value symbols. Hence, such an equation cannot be assigned a degree. For example, although the equation

$$|2x - 3| = 5$$

contains the first-degree polynomial $2x - 3$, it has *two* solutions, 4 and -1. The equation is equivalent to *two* first-degree equations.

Inequalities involving absolute value   Consider the graph of the solution set of $|x| < 5$ in Figure 4.6. In this simple case, the fact that this is indeed the graph of the inequality can be verified by inspection. Note

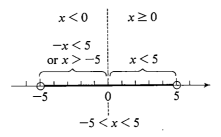

Figure 4.6

that if $x \geq 0$, then $x < 5$, and if $x < 0$, then $x > -5$. These two conditions can be expressed as the continued inequality $-5 < x < 5$.

Now consider the graph of the solution set of $|x| > 5$ in Figure 4.7, which can also be verified by inspection.

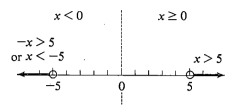

Figure 4.7

Note that in this case the two conditions $x < -5$ and $x > 5$ cannot be written as a continued inequality.

The above examples suggest the following properties which can be used to solve inequalities that involve absolute value.

▶ $|ax - b| < c$ *is equivalent to* $-c < ax - b < c,$ (2)

and

          $|ax - b| > c$ *is equivalent to* $ax - b > c$ *or* $-(ax - b) > c.$ (3)

Example   Solve $|x + 1| < 3$.

Solution   From statement 2 above, this inequality is equivalent to

$$-3 < x + 1 < 3,$$

from which we have

$$-4 < x < 2.$$

Hence, the solution set is $(-4, 2)$, with the graph as shown.

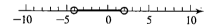

Example   Solve $|x + 1| \geq 3$.

Solution   From statement 3 above, this inequality is equivalent to the two inequalities

$$x + 1 \geq 3 \quad \text{or} \quad -(x + 1) \geq 3,$$

from which we have

$$x \geq 2 \quad \text{or} \qquad x \leq -4.$$

Hence, the solution set is $(-\infty, -4] \cup [2, +\infty)$, with the graph as shown.

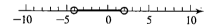

EXERCISE 4.6

A ■ *Solve.*

Example $|x + 5| = 8$

Solution Using statement 1 in the text, rewrite the equation as two first-degree equations.

$$x + 5 = 8 \qquad \text{or} \qquad -(x + 5) = 8$$

Solve each equation.

$$x = 3 \qquad \text{or} \qquad -x - 5 = 8$$
$$-x = 13$$
$$x = -13$$

The solution set is $\{3\} \cup \{-13\} = \{3, -13\}$.

1. $|x| = 5$ **2.** $|x| = 7$ **3.** $|x - 4| = 9$

4. $|x - 3| = 7$ **5.** $|2x + 1| = 13$ **6.** $|3x - 1| = 5$

7. $|4 - 3x| = 1$ **8.** $|6 - 5x| = 4$ **9.** $|4x + 3| = 0$

10. $|3x - 7| = 0$ **11.** $\left|x - \dfrac{3}{4}\right| = \dfrac{1}{4}$ **12.** $\left|x + \dfrac{2}{3}\right| = \dfrac{1}{3}$

13. $\left|1 + \dfrac{3}{2}x\right| = \dfrac{1}{4}$ **14.** $\left|1 - \dfrac{1}{2}x\right| = \dfrac{3}{4}$ **15.** $\left|\dfrac{1}{3} - 4x\right| = \dfrac{5}{6}$

16. $\left|\dfrac{2}{3} + 2x\right| = \dfrac{7}{12}$ **17.** $\left|\dfrac{2}{3}x - \dfrac{1}{2}\right| = 0$ **18.** $\left|\dfrac{3}{4}x + \dfrac{3}{8}\right| = 0$

■ *Solve and graph each solution set.*

Example $|1 - 2x| \le 7$

Solution Using statement 2 in the text, rewrite without the absolute-value symbol.

$$-7 \le 1 - 2x \le 7$$

Add -1 to each expression.

$$-8 \le -2x \le 6$$

Multiply each expression by $-\frac{1}{2}$ and reverse the sense of the inequality.

$$4 \ge x \ge -3$$

The solution set is $\{x \mid -3 \le x \le 4\}$ or $[-3, 4]$.

19. $|x| < 2$ **20.** $|x| < 5$ **21.** $|x + 3| \leq 4$ **22.** $|x + 1| \leq 8$

23. $|2x - 5| < 3$ **24.** $|2x + 4| < 6$ **25.** $|4 - x| \leq 8$ **26.** $|5 - 2x| \leq 15$

Example $|3x - 6| > 9$

Solution   Since the inequality is of the form $|x - a| > b$, the intervals involved are disjoint, and the two inequalities must be considered separately. Using statement 3 in the text, rewrite without the absolute-value symbol and solve.

$$3x - 6 > 9 \qquad \text{or} \qquad -(3x - 6) > 9$$
$$3x > 15 \qquad\qquad\qquad 3x - 6 < -9$$
$$x > 5 \qquad\qquad\qquad\qquad 3x < -3$$
$$x < -1$$

The solution set is $\{x | x < -1\} \cup \{x | x > 5\}$ or $(-\infty, -1) \cup (5, +\infty)$.

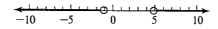

27. $|x| > 3$ **28.** $|x| \geq 5$ **29.** $|x - 2| > 5$

30. $|x + 5| > 2$ **31.** $|3 - 2x| \geq 7$ **32.** $|4 - 3x| > 10$

CHAPTER SUMMARY

[4.1]   A replacement for the variable in an equation or inequality that results in a true statement is called a **solution** of the equation or inequality; the set of all solutions is called the **solution set**.

A **conditional equation** has at least one solution, but is not true for at least one member in the replacement set of the variable.

Equations or inequalities that have identical solution sets are called **equivalent equations** or **equivalent inequalities**, respectively. A conditional equation can be solved by applying **elementary transformations** to generate a sequence of equivalent equations until an equation is obtained that can be solved by inspection.

The addition of the same expression representing a real number to each member of an equation or the multiplication of each member of an equation by the same expression representing a nonzero real number produces an equivalent equation.

If $P(x)$, $Q(x)$, and $R(x)$ are expressions, then for all values of x,

$$P(x) = Q(x)$$

is equivalent to

$$P(x) + R(x) = Q(x) + R(x),$$
$$P(x) \cdot R(x) = Q(x) \cdot R(x), \quad for \quad R(x) \neq 0.$$

[4.2] A formula can be solved for a specified variable given values for the other variables in the formula.

[4.3] An equation containing more than one variable can be rewritten in other forms by generating equivalent equations.

[4.4] Equations can be used as mathematical models to solve practical problems.

[4.5] A **conditional inequality** is not true for at least one member in the replacement set of the variable. Such an inequality can be solved by producing equivalent inequalities until one is obtained that can be solved by inspection.

If $P(x)$, $Q(x)$, and $R(x)$ are expressions, then for all values of x,

$$P(x) < Q(x)$$

is equivalent to

$$P(x) + R(x) < Q(x) + R(x),$$
$$P(x) \cdot R(x) < Q(x) \cdot R(x), \quad for \quad R(x) > 0,$$
$$P(x) \cdot R(x) > Q(x) \cdot R(x), \quad for \quad R(x) < 0.$$

The **intersection** of two sets, $A \cap B$, is the set containing all members that are in both set A and set B. If the intersection is $\varnothing$, then A and B are **disjoint sets**.

An **interval** of real numbers is an infinite set. An interval may be **closed** (two endpoints), **half-open** (one endpoint), or **open** (no endpoints).

[4.6] $|ax - b| = c$ is equivalent to $ax - b = c$ or $-(ax - b) = c$;

        $|ax - b| < c$ is equivalent to $-c < ax - b < c$;

        $|ax - b| > c$ is equivalent to $ax - b > c$ or $-(ax - b) > c$.

The **union** of two sets, $A \cup B$, is the set containing all members that are in either set A or set B or both.

The symbols introduced in this chapter are listed on the inside of the front cover.

REVIEW EXERCISES

A

[4.1] ■ *Solve.*

    1. a. $2[x - (2x + 1)] = 6$ **b.** $2 + \dfrac{x}{3} = \dfrac{5}{6}$

    2. a. $\dfrac{x}{x+1} + \dfrac{4}{5} = 6$ **b.** $1 - \dfrac{x-2}{x-3} = \dfrac{3}{x-1}$

[4.2]   **3.** Solve $N = \dfrac{5t}{3} - c$ for t, given that $N = 10$ and $c = 5$.

       4. Solve $C = 10 + 2(p - t)$ for p, given that $C = 38$ and $t = 3$.

       5. Solve $\dfrac{1}{R_n} = \dfrac{1}{R_1} + \dfrac{1}{R_2}$ for R_1, given that $R_2 = 30$ and $R_n = 10$.

       6. The volume V (in cubic centimeters) of a particular gas is related to the temperature T (in Celsius units) by the formula $V = (5/3)T + 455$. Find the temperature when the volume is 540 cubic centimeters.

[4.3]   **7.** Solve $\dfrac{x-y}{3} = \dfrac{x+y}{2}$ for x in terms of y.   **8.** Solve $\dfrac{x-y}{3} = \dfrac{x+y}{2}$ for y in terms of x.

       9. Solve $\ell = a + nd - d$ for d.   **10.** Solve $s = \dfrac{1}{2}at^2 + 2t$ for a.

[4.4]   ■ *Solve.*

       11. A part-time secretary has a take-home pay of $180 per week. What is her total salary if 20% has been deducted for income taxes?

       12. A typist works for $6.50 per hour. Ten percent of her total wages are deducted for income taxes. How many hours must she work in a month to take home $819?

       13. A man invested $7200, part at 12%, and part at 10%. How much did he invest at each rate if the yearly return from both investments amounted to $768?

       14. One angle of a triangle measures three times another angle, and the third angle measures 96° more than the sum of the other two. Find the measure of each angle.

[4.5]   ■ *Solve and graph each solution set. Represent the solution set in set-builder notation and interval notation.*

       15. $\dfrac{x-3}{4} \le 6$ **16.** $2(x - 1) \le \dfrac{2}{3}x$

       17. $1 < 2x - 3 \le 7$ **18.** $-5 < 4 - 3x < 1$

  ■ *Graph*

       19. $\{x \mid x > -5\} \cap \{x \mid x < 2\}$ **20.** $\{x \mid x + 2 \ge -1\} \cap \{x \mid 0 < x < 4\}$

[4.6]   ■ *Solve.*

       21. a. $|2x + 1| = 7$ **b.** $|3 - 4x| = 8$

■ *Solve and graph each solution set.*

22. a. $|x + 3| \geq 5$ **b.** $|x - 2| < 5$

B   **23.** For what value of k will the equation $x - 4 = \dfrac{2 - x}{k}$ have as its solution set $\{-2\}$?

24. Find a value for k in $4x - 1 = k + 2$ so that the equation is equivalent to $x - 2 = 3x + 1$.

■ *Solve.*

25. A manufacturer wishes to increase 10 gallons of an antifreeze solution that is 20% alcohol to 25% alcohol. How much pure alcohol must be added to the original solution?

26. Two cars start together and travel in the same direction, one going three times as fast as the other. How fast is each traveling if they are 100 miles apart at the end of one hour?

27. A passenger train and a freight train leave a station at the same time and travel in opposite directions. The passenger train averages 60 miles per hour and the freight train averages 20 miles per hour. How long after leaving the station will they be 200 miles apart?

28. One printing machine can run the necessary copies for the daily circulation of a newspaper in 5 hours. A second printer can run the same number of copies in 4 hours. How long does it take to run the copies when both machines are working?

29. How soon after 9 o'clock will the hour and minute hands of a clock be together?

30. The length of each side of a square is decreased by 6 inches. If the area is decreased by 84 square inches, find the length of the side of the original square.

5. EXPONENTS, ROOTS, AND RADICALS

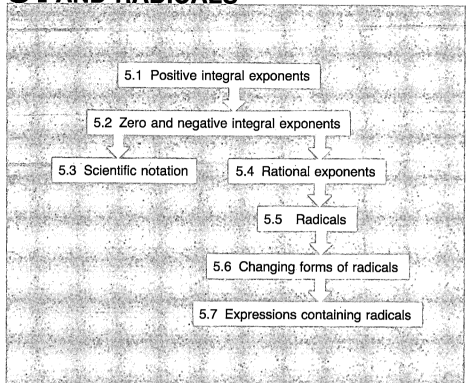

5.1 Positive integral exponents

5.2 Zero and negative integral exponents

5.3 Scientific notation

5.4 Rational exponents

5.5 Radicals

5.6 Changing forms of radicals

5.7 Expressions containing radicals

5.1

POSITIVE INTEGRAL EXPONENTS

In Section 2.1 the expression a^n, where n is a natural number, was defined by

▶ $$a^n = a \cdot a \cdot a \cdot \cdots \cdot a \qquad (n \text{ factors}).$$

The following three laws were developed from this definition in Section 2.3:

Laws of ▶
exponents
for products

    I. $a^m \cdot a^n = a^{m+n}$

   II. $(a^m)^n = a^{mn}$

  III. $(ab)^m = a^m b^m.$

Examples

  a. $x^5 \cdot x^2 = x^{5+2}$
           $= x^7$

  b. $(x^5)^2 = x^{5\cdot2}$
           $= x^{10}$

  c. $(x^2 y^3)^2 = (x^2)^2 (y^3)^2$
              $= x^4 y^6$

138

Laws of exponents for quotients   The following laws for quotients (Laws IV and V) are also useful in simplifying expressions involving positive integral exponents:

▶
$$\textbf{IV.} \quad \frac{a^m}{a^n} = a^{m-n} \quad (m > n, \ a \neq 0),$$

since

$$\frac{a^m}{a^n} = \frac{\overbrace{(a \cdot a \cdot a \cdots a)}^{(m-n) \text{ factors}} \overbrace{(a \cdot a \cdot a \cdots a)}^{n \text{ factors}}}{\underbrace{a \cdot a \cdot a \cdots a}_{n \text{ factors}}}$$

$$= a^{m-n}.$$

▶
$$\textbf{IVa.} \quad \frac{a^m}{a^n} = \frac{1}{a^{n-m}} \quad (m < n, \ a \neq 0),$$

since

$$\frac{a^m}{a^n} = \frac{\overbrace{a \cdot a \cdots a}^{m \text{ factors}}}{\underbrace{(a \cdot a \cdots a)}_{m \text{ factors}} \underbrace{(a \cdot a \cdots a)}_{n - m \text{ factors}}}$$

$$= \frac{1}{a^{n-m}}.$$

Examples   **a.** $\dfrac{x^4 y^6}{x^2 y} = x^{4-2} y^{6-1}$

$= x^2 y^5 \quad (x, y \neq 0)$

b. $\dfrac{x^2 y}{x^3 y^2} = \dfrac{1}{x^{3-2} y^{2-1}}$

$= \dfrac{1}{xy} \quad (x, y \neq 0)$

Note that the fourth law of exponents is consistent with the process of reducing fractions by using the fundamental principle of fractions. The same results would be obtained in the above examples by using the fundamental principle of fractions.

▶
$$\textbf{V.} \quad \left(\frac{a}{b}\right)^n = \frac{a^n}{b^n} \quad (b \neq 0),$$

since

$$\left(\frac{a}{b}\right)^n = \overbrace{\frac{a}{b} \cdot \frac{a}{b} \cdot \frac{a}{b} \cdots \left(\frac{a}{b}\right)}^{n \text{ factors}} = \frac{\overbrace{a \cdot a \cdot a \cdots a}^{n \text{ factors}}}{\underbrace{b \cdot b \cdot b \cdots b}_{n \text{ factors}}}$$

$$= \frac{a^n}{b^n}.$$

Examples   **a.** $\left(\dfrac{x}{y}\right)^3 = \dfrac{x^3}{y^3}$ $(y \neq 0)$ **b.** $\left(\dfrac{2x^2}{y}\right)^4 = \dfrac{(2x^2)^4}{y^4}$ $(y \neq 0)$

Ordinarily, two or more of the laws of exponents are required to simplify expressions containing exponents. In the last example, we can further simplify the result: From Law III,

$$\frac{(2x^2)^4}{y^4} = \frac{2^4(x^2)^4}{y^4} \qquad (y \neq 0),$$

and from Law II,

$$\frac{2^4(x^2)^4}{y^4} = \frac{16x^8}{y^4} \qquad (y \neq 0).$$

In the exercise sets in this chapter we shall assume that no denominator equals 0 unless otherwise stated.

EXERCISE 5.1

A   ■ *Using one or more of the laws of exponents, write each expression as a product or quotient in which each variable occurs only once, and all exponents are positive.*

Examples   **a.** $\dfrac{x^3 y^2}{xy^5}$ **b.** $\left(\dfrac{2x^3}{y}\right)^2$ **c.** $\dfrac{(-xy^2)^3}{(-x^2 y)^2}$

Solutions   **a.** $\dfrac{x^3 y^2}{xy^5} = \dfrac{x^{3-1}}{y^{5-2}}$ **b.** $\left(\dfrac{2x^3}{y}\right)^2 = \dfrac{2^2 \cdot x^6}{y^2}$ **c.** $\dfrac{(-xy^2)^3}{(-x^2 y)^2} = \dfrac{(-1)^3 x^3 y^6}{(-1)^2 x^4 y^2}$

$\qquad\qquad\quad = \dfrac{x^2}{y^3}$ $\qquad\qquad\qquad\qquad = \dfrac{4x^6}{y^2}$ $\qquad\qquad\qquad\qquad = \dfrac{-1 \cdot y^{6-2}}{x^{4-3}}$

$\qquad\qquad\qquad\qquad\qquad\qquad\qquad\qquad\qquad\qquad\qquad\qquad = \dfrac{-y^4}{x}$

1. $x^2 \cdot x^3$ **2.** $y \cdot y^4$ **3.** $a^3 \cdot a^5$ **4.** $b^5 \cdot b^4$

5. $(a^2)^3$ **6.** $(b^3)^4$ **7.** $(x^2)^3$ **8.** $(y^4)^2$

9. $(xy^2)^3$ **10.** $(x^2 y^3)^2$ **11.** $(abc^2)^4$ **12.** $(a^2 b^3 c)^2$

13. $(2x)(-2x)^3$ **14.** $(-3x^2)^2(-5x)$ **15.** $(ab^2)^3(-2a^2)^2$ **16.** $(a^2 b^2)^3(-ab^2)^3$

17. $\dfrac{x^5}{x^3}$ **18.** $\dfrac{y^2}{y^6}$ **19.** $\dfrac{x^2 y^4}{xy^2}$ **20.** $\dfrac{x^4 y^3}{x^2 y}$

21. $\left(\dfrac{x}{y^2}\right)^3$ **22.** $\left(\dfrac{y^2}{z^3}\right)^2$ **23.** $\left(\dfrac{2x}{y^2}\right)^3$ **24.** $\left(\dfrac{3y^2}{x}\right)^2$

25. $\left(\dfrac{-2x}{3y^2}\right)^3$ **26.** $\left(\dfrac{-x^2}{2y}\right)^4$ **27.** $\dfrac{(4x)^3}{(2x^2)^2}$ **28.** $\dfrac{(5x)^2}{(3x^2)^3}$

29. $\dfrac{(xy^2)^3}{(x^2y)^2}$

30. $\dfrac{(-xy^2)^2}{(x^2y)^3}$

31. $\dfrac{(xy)^2(x^2y)^3}{(x^2y^2)^2}$

32. $\dfrac{(-x)^2(-x^2)^4}{(x^2)^3}$

33. $\left(\dfrac{2x}{y^2}\right)^3\left(\dfrac{y^2}{3x}\right)^2$

34. $\left(\dfrac{x^2z}{2}\right)^2\left(-\dfrac{2}{x^2z}\right)^3$

35. $\left(\dfrac{-3}{y^2}\right)^2(2y^3)^2$

36. $\left(\dfrac{y}{x}\right)^2\left(-\dfrac{3}{4xy}\right)^3$

37. $\left[\left(\dfrac{r^2s^3t}{xy}\right)^3\left(\dfrac{x^2y}{r^3st^2}\right)^2\right]^2$

38. $\left[\left(\dfrac{a^3bc}{x^2y}\right)^4\left(\dfrac{x^2yz}{ab^2c^3}\right)^2\right]^2$

39. $\left(\dfrac{x^2}{a^2b}\right)^2\left(-\dfrac{ab}{x^3}\right)^3\left(\dfrac{x}{ab}\right)^2$

40. $\left(\dfrac{m^3n^2p}{r^2s}\right)^2\left(\dfrac{rs}{mn^2p^2}\right)^3\left(-\dfrac{mnp}{rs}\right)^2$

41. $\left(\dfrac{ab^2}{x}\right)^3\left(-\dfrac{x^2}{a^2b^3}\right)^2\left(\dfrac{ab^2}{x^3}\right)^3$

42. $\left(\dfrac{2x^2y}{3z}\right)^2\left(\dfrac{2z^2}{3xy^2}\right)^3\left(-\dfrac{4xz}{3y}\right)^2$

43. Use a counterexample to show that $(x^2 + y^2)^3$ is not equivalent to $x^6 + y^6$.

44. Rewrite $(x^2 + y^2)^3$ as an equivalent expression without using parentheses.

B

Examples

a. $\dfrac{x^n \cdot x^{n+1}}{x^{n-1}}$

b. $\dfrac{(y^{n-1})^2}{y^{n-2}}$

c. $(x^{n-1} \cdot x^{2n+3})^2$

Solutions

a. $\dfrac{x^n \cdot x^{n+1}}{x^{n-1}}$

$= x^{n+(n+1)-(n-1)}$

$= x^{n+2}$

b. $\dfrac{(y^{n-1})^2}{y^{n-2}}$

$= y^{(2n-2)-(n-2)}$

$= y^n$

c. $(x^{n-1} \cdot x^{2n+3})^2$

$= (x^{3n+2})^2$

$= x^{6n+4}$

45. $x^n \cdot x^n$

46. $\dfrac{x^{2n}x^n}{x^{n+1}}$

47. $\dfrac{(x^{n+1}x^{2n-1})^2}{x^{3n}}$

48. $\left(\dfrac{y^2 \cdot y^3}{y}\right)^{2n}$

49. $\left(\dfrac{x^{3n}x^{2n}}{x^{4n}}\right)^2$

50. $\dfrac{(y^{n+1})^n}{y^n}$

5.2

ZERO AND NEGATIVE INTEGRAL EXPONENTS

Zero exponent

Note that if the second law of exponents is to hold for the case where $m = n$, we have

$$\dfrac{a^n}{a^n} = a^{n-n} = a^0 \qquad (a \neq 0).$$

By the definition of a quotient,

$$\dfrac{a^n}{a^n} = 1 \qquad (a \neq 0).$$

Hence, for consistency, a^0 must be defined as 1:

$$a^0 = 1 \qquad (a \neq 0).$$

Examples   a. $3^0 = 1$ b. $(-4)^0 = 1$ c. $x^0 \neq 1 \quad \cdot (x = 0)$

Note that in Example c, if $x = 0$, then 0^0 is undefined. This follows from the fact that 0/0 is undefined.

Negative integer exponents

We would like the laws of exponents to hold also for negative exponents. We observe that for $a \neq 0$,

$$a^n \cdot a^{-n} = a^{n-n} = a^0 = 1.$$

Also, by the reciprocal property,

$$a^n \cdot \frac{1}{a^n} = 1 \qquad (a \neq 0).$$

Again, for consistency, we make the following definition:

$$a^{-n} = \frac{1}{a^n} \qquad (a \neq 0).$$

Therefore, because a^{-n} is the reciprocal of a^n, it follows that

$$\frac{1}{a^{-n}} = \frac{1}{\dfrac{1}{a^n}} = a^n \qquad (a \neq 0).$$

Examples   a. $3^{-2} = \dfrac{1}{3^2} = \dfrac{1}{9}$ b. $\dfrac{1}{2^{-3}} = 2^3 = 8$ c. $\dfrac{x^{-2}}{y^{-1}} = \dfrac{\dfrac{1}{x^2}}{\dfrac{1}{y}} = \dfrac{y}{x^2}$

It can be shown that all the laws of exponents given in Section 5.1 are valid for integral exponents (both positive, negative, and zero). These laws may be applied in any order.

Example   We can first apply Law IV to write

$$\left(\frac{x^3}{x^2}\right)^{-3} = (x)^{-3} = \frac{1}{x^3}.$$

Or we can first apply Law V and then Law II to write

$$\left(\frac{x^3}{x^2}\right)^{-3} = \frac{(x^3)^{-3}}{(x^2)^{-3}} = \frac{x^{-9}}{x^{-6}},$$

from which, by Law IV, we have

$$\frac{x^{-9}}{x^{-6}} = x^{-9-(-6)}$$

$$= x^{-9+6} = x^{-3} = \frac{1}{x^3}.$$

EXERCISE 5.2

A ■ *Write each expression as a basic numeral or fraction in lowest terms.*

Examples

a. $3 \cdot 5^{-2}$ b. $\dfrac{3}{2^{-3}}$ c. $4^2 + 4^{-2}$

Solutions

a. $3 \cdot 5^{-2} = 3 \cdot \dfrac{1}{5^2}$

$= 3 \cdot \dfrac{1}{25}$

$= \dfrac{3}{25}$

b. $\dfrac{3}{2^{-3}} = 3 \cdot \dfrac{1}{2^{-3}}$

$= 3 \cdot 2^3$

$= 3 \cdot 8$

$= 24$

c. $4^2 + 4^{-2} = 16 + \dfrac{1}{16}$

$= \dfrac{256}{16} + \dfrac{1}{16}$

$= \dfrac{257}{16}$

1. 2^{-1}

2. 3^{-2}

3. $\dfrac{1}{3^{-1}}$

4. $\dfrac{3}{4^{-2}}$

5. $(-2)^{-3}$

6. $\dfrac{1}{(-3)^{-2}}$

7. $\dfrac{5^{-1}}{3^0}$

8. $\dfrac{2^0}{3^{-2}}$

9. $\left(\dfrac{3}{5}\right)^{-1}$

10. $\left(\dfrac{1}{3}\right)^{-2}$

11. $\dfrac{5^{-1}}{3^{-2}}$

12. $\dfrac{3^{-3}}{6^{-2}}$

13. $3^{-2} + 3^2$

14. $5^{-1} + 25^0$

15. $4^{-1} - 4^{-2}$

16. $8^{-2} - 2^0$

■ *Write each expression as a product or quotient of powers in which each variable occurs only once, and all exponents are positive.*

Examples

a. $x^{-3} \cdot x^5$ b. $(x^2 y^{-3})^{-1}$ c. $\left(\dfrac{x^{-1} y^2 z^0}{x^3 y^{-4} z^2}\right)^{-1}$

Solutions overleaf

Solutions

a. $x^{-3} \cdot x^5$

$= x^{-3+5}$

$= x^2$

b. $(x^2 y^{-3})^{-1}$

$= x^{-2} y^3$

$= \dfrac{y^3}{x^2}$

c. $\left(\dfrac{x^{-1} y^2 z^0}{x^3 y^{-4} z^2}\right)^{-1}$

$= \dfrac{x y^{-2} z^0}{x^{-3} y^4 z^{-2}}$

$= \dfrac{x^4 z^2}{y^6}$

17. $x^2 y^{-3}$

18. $\dfrac{x^3}{y^{-2}}$

19. $(x^2 \cdot y)^{-3}$

20. $(xy^3)^{-2}$

21. $\left(\dfrac{x}{y^3}\right)^2$

22. $\left(\dfrac{2x}{y^2}\right)^3$

23. $\dfrac{(xy^2)^3}{(x^2 y)^2}$

24. $\left(\dfrac{3x}{y^2}\right)^2 \left(\dfrac{2y^3}{x}\right)^2$

25. $x^{-3} \cdot x^7$

26. $\dfrac{x^3}{x^{-2}}$

27. $(x^{-2} y^0)^3$

28. $(x^{-2} y^3)^0$

29. $\dfrac{x^{-1}}{y^{-1}}$

30. $\dfrac{x^{-3}}{y^{-2}}$

31. $\dfrac{8^{-1} x^0 y^{-3}}{(2xy)^{-5}}$

32. $\left(\dfrac{x^{-1} y^3}{2x^0 y^{-5}}\right)^{-2}$

33. $\left(\dfrac{x^0 y^2}{z^2}\right)^{-1}$

34. $\left(\dfrac{x^{-1} y z^0}{xy^{-1} z}\right)^{-1}$

35. $\left(\dfrac{x^2 y}{z^3}\right)\left(\dfrac{x}{z^2}\right)^{-1}$

36. $\left(\dfrac{2y^{-1}}{x^2}\right)^{-1} \cdot \dfrac{y^2}{x}$

37. $\left(\dfrac{x^{-2} y^2}{z^{-1}}\right)^{-1} \cdot \left(\dfrac{xy^0}{z}\right)^{-2}$

38. $\left(\dfrac{2y^2 x}{3z}\right)^2 \cdot \left(\dfrac{2x^2}{9z}\right)^{-2}$

■ *Write each expression as a single fraction involving positive exponents only.*

Examples

a. $x^{-1} + y^{-2}$

b. $(x^{-1} + x^{-2})^{-1}$

c. $x^{-1} + \dfrac{1}{x^{-1}}$

Solutions

a. $x^{-1} + y^{-2}$

$= \dfrac{1}{x} + \dfrac{1}{y^2}$

$= \dfrac{(y^2) 1}{(y^2) x} + \dfrac{1(x)}{y^2 (x)}$

$= \dfrac{y^2 + x}{xy^2}$

b. $(x^{-1} + x^{-2})^{-1}$

$= \left(\dfrac{1}{x} + \dfrac{1}{x^2}\right)^{-1}$

$= \left[\dfrac{(x) 1}{(x) x} + \dfrac{1}{x^2}\right]^{-1}$

$= \left(\dfrac{x + 1}{x^2}\right)^{-1}$

$= \dfrac{x^2}{x + 1}$

c. $x^{-1} + \dfrac{1}{x^{-1}}$

$= \dfrac{1}{x} + x$

$= \dfrac{1}{x} + \dfrac{x(x)}{(x)}$

$= \dfrac{1}{x} + \dfrac{x^2}{x}$

$= \dfrac{1 + x^2}{x}$

39. $x^{-2} + y^{-2}$

40. $x^{-1} - y^{-3}$

41. $\dfrac{x}{y^{-1}} + \dfrac{x^{-1}}{y}$

42. $\dfrac{x^{-1}}{y^{-1}} + \dfrac{y}{x}$

43. $(x - y)^{-2}$

44. $(x + y)^{-3}$

45. $x^{-1}y - xy^{-1}$ **46.** $xy^{-1} + x^{-1}y$ **47.** $\dfrac{x^{-1} - y}{x^{-1}}$

48. $\dfrac{x + y^{-1}}{y^{-1}}$ **49.** $\dfrac{x^{-1} + y^{-1}}{(xy)^{-1}}$ **50.** $\dfrac{x^{-2} - y^{-2}}{(xy)^{-1}}$

51. Use a counterexample to show that $(x + y)^{-2}$ is not equivalent to $\dfrac{1}{x^2 + y^2}$.

52. Use a counterexample to show that $(x + y)^{-2}$ is not equivalent to $\dfrac{1}{x^2} + \dfrac{1}{y^2}$.

B   ▪ *Write each expression as a product free of fractions in which each variable occurs only once.*

Examples   **a.** $x^{-2n} \cdot x^n$ **b.** $\left(\dfrac{x^{1-n}}{x^{2-n}}\right)^{-2}$ **c.** $\left(\dfrac{x^n y^{2n-1}}{y^n}\right)^2$

Solutions   **a.** $x^{-2n} \cdot x^n$ **b.** $\left(\dfrac{x^{1-n}}{x^{2-n}}\right)^{-2}$ **c.** $\left(\dfrac{x^n y^{2n-1}}{y^n}\right)^2$

$\qquad\qquad = x^{-2n+n} \qquad\qquad\quad = (x^{(1-n)-(2-n)})^{-2} \qquad = (x^n y^{2n-1-n})^2$

$\qquad\qquad = x^{-n} \qquad\qquad\qquad\quad = (x^{-1})^{-2} \qquad\qquad\quad = (x^n y^{n-1})^2$

$\qquad\qquad\qquad\qquad\qquad\qquad\qquad = x^2 \qquad\qquad\qquad\qquad = x^{2n} y^{2n-2}$

53. $a^{3-n}a^0$ **54.** $x^{-n}x^{n+1}$ **55.** $\left(\dfrac{a^{2n}}{a^{n+1}}\right)^{-2}$ **56.** $\dfrac{x^n y^{n+1}}{x^{2n-1} y^n}$

57. $\dfrac{b^n c^{2n-1}}{b^{n+1} c^{2n}}$ **58.** $\left(\dfrac{x^{n-1} y^n}{x^{-2} y^{-n}}\right)^2$ **59.** $\left(\dfrac{x^n}{x^{n-1}}\right)^{-1}$ **60.** $\left(\dfrac{a^{2n} b^{n-1}}{a^{n-1} b}\right)^2$

61. Show that $\left(\dfrac{a}{b}\right)^{-n} = \left(\dfrac{b}{a}\right)^n$.   **62.** Show that $\dfrac{a^m b^{-n}}{c^{-p} d^q} = \dfrac{a^m c^p}{b^n d^q}$.

5.3

SCIENTIFIC NOTATION

It is often very convenient to use an exponential form of notation called **scientific notation**. This form of notation is particularly useful in scientific applications of mathematics that involve very large or very small quantities. For example, the mass of the earth is approximately

$$5,980,000,000,000,000,000,000,000,000 = 5.98 \times 10^{27} \text{ grams,}$$

and the mass of a hydrogen atom is approximately

$$0.00000000000000000000000167 = 1.67 \times 10^{-24} \text{ gram.}$$

In each case, we have represented a number as the product of a number in the interval [1, 10) and a power of 10; that is, we have factored a power of 10 from each number. These examples suggest the following procedure.

To write a number in scientific notation:

1. Write the given number as the product of a number in the interval [1, 10) and a power of 10.

2. Count the number of places the decimal point was moved in going from the original number to the number between 1 and 10. This number is the absolute value of the exponent of 10. The exponent is positive if the original number is greater than 10 and negative if the original number is less than 1.

Examples   **a.** $478{,}000 = 4.78000 \times 10^5$ **b.** $0.00032 = 00003.2 \times 10^{-4}$

5 places   4 places

$= 4.78 \times 10^5$ $= 3.2 \times 10^{-4}$

A number written in scientific notation can be written in **standard form** by reversing the above procedure.

To go from scientific notation to decimal notation:

Move the decimal point the number of places indicated by the exponent on 10—to the right if the exponent is positive and to the left if it is negative.

Examples   **a.** $3.75 \times 10^4 = 3.7500 \times 10^4$ **b.** $2.03 \times 10^{-3} = .00203$

4 places   3 places

$= 37{,}500$ $= 0.00203$

Sometimes it is more convenient to express a number as a product of a power of 10 and a number that is not between 1 and 10. For example, under certain circumstances, any of the following forms may be a useful representation for 6280:

$$628 \times 10, \qquad 62.8 \times 10^2, \qquad 6.28 \times 10^3, \qquad 0.628 \times 10^4.$$

A factored form in which the factors do not contain decimals is generally easiest to use to simplify calculations.

Examples **a.** $\dfrac{248{,}000}{0.0124} = \dfrac{248 \times 10^3}{124 \times 10^{-4}}$ **b.** $\dfrac{0.0024 \times 0.0007}{0.000021}$

$= 2 \times 10^7$

$= 20{,}000{,}000$

$= \dfrac{24 \times 10^{-4} \times 7 \times 10^{-4}}{21 \times 10^{-6}}$

$= 8 \times 10^{-2} = 0.08$

Significant digits

When working with numbers, we sometimes use the term **significant digits** of a number to refer to the digits in the numbers that have meaning. In particular, the digits that are used to specify a number in scientific notation are significant digits.

Examples

a. 0.00321 has *three* significant digits because $0.00321 = 3.21 \times 10^{-3}$, and 3.21 has three digits.

b. 0.8005 has *four* significant digits because $0.8005 = 8.005 \times 10^{-1}$, and 8.005 has four digits.

Note that in Example **a** the zeros between the decimal point and the first nonzero digit 3 are not significant digits. However, in Example **b** the zeros between the nonzero digits 8 and 5 are significant.

Final zeros on the right of a decimal point are significant.

Examples

a. 4.300 has four significant digits,

b. 0.0530 has three significant digits. In scientific notation, $0.0530 = 5.30 \times 10^{-2}$.

Final zeros on a natural number are assumed to be not significant unless further information is available. For example, in the statement

"The height of a building is 4300 feet,"

without further information on how the measurement was made, we assume that 4300 has only two significant digits.

EXERCISE 5.3

A ▪ *Express each number using scientific notation.*

Examples **a.** 680,000 **b.** 0.000043 **c.** 0.00245

Solutions **a.** $680{,}000 = 6.8 \times 10^5$ **b.** $0.000043 = 4.3 \times 10^{-5}$ **c.** $0.00245 = 2.45 \times 10^{-3}$

1. 285	**2.** 3476	**3.** 21	**4.** 68,742
5. 8,372,000	**6.** 481,000	**7.** 0.024	**8.** 0.0063
9. 0.421	**10.** 0.000523	**11.** 0.000004	**12.** 0.0006

■ *Express each number using standard form.*

Examples   **a.** 1.01×10^3 **b.** 6.3×10^{-4} **c.** 4.317×10^{-2}

Solutions   **a.** $1.01 \times 10^3 = 1010$ **b.** $6.3 \times 10^{-4} = 0.00063$ **c.** $4.317 \times 10^{-2} = 0.04317$

13. 2.4×10^2 **14.** 4.8×10^3 **15.** 6.87×10^5 **16.** 8.31×10^4

17. 5.0×10^{-3} **18.** 8.0×10^{-1} **19.** 2.02×10^{-2} **20.** 4.31×10^{-3}

21. 12.27×10^3 **22.** 14.38×10^4 **23.** 23.5×10^{-4} **24.** 621.0×10^{-2}

Examples   **a.** $\dfrac{1}{4 \times 10^3}$ **b.** $\dfrac{1}{5 \times 10^{-1}}$

Solutions   **a.** $\dfrac{1}{4 \times 10^3} = \dfrac{1}{4} \times \dfrac{1}{10^3} = 0.25 \times 10^{-3}$ **b.** $\dfrac{1}{5 \times 10^{-1}} = \dfrac{1}{5} \times \dfrac{1}{10^{-1}} = 0.2 \times 10^1$

$\qquad\qquad\qquad\qquad\qquad = 0.00025$ $\qquad\qquad\qquad = 2$

25. $\dfrac{1}{2 \times 10^3}$ **26.** $\dfrac{1}{4 \times 10^4}$ **27.** $\dfrac{1}{8 \times 10^{-2}}$

28. $\dfrac{1}{5 \times 10^{-3}}$ **29.** $\dfrac{3}{5 \times 10^4}$ **30.** $\dfrac{5}{8 \times 10^2}$

■ *Compute.*

Examples   **a.** $\dfrac{10^6 \times 10^{-2} \times 10^{-1}}{10^4 \times 10^{-3}}$ **b.** $\dfrac{4 \times 10^3 \times 6 \times 10^{-5}}{8 \times 10^{-3}}$

Solutions   **a.** $\dfrac{10^6 \times 10^{-2} \times 10^{-1}}{10^4 \times 10^{-3}}$ **b.** $\dfrac{4 \times 10^3 \times 6 \times 10^{-5}}{8 \times 10^{-3}}$

$\qquad\qquad = \dfrac{10^3}{10}$ $\qquad = \dfrac{24 \times 10^{-2}}{8 \times 10^{-3}}$

$\qquad\qquad = 10^2 \quad \text{or} \quad 100$ $\qquad = 3 \times 10 \quad \text{or} \quad 30$

31. $\dfrac{10^3 \times 10^{-6}}{10^2}$ **32.** $\dfrac{10^3 \times 10^{-7} \times 10^2}{10^{-2} \times 10^4}$

33. $\dfrac{10^2 \times 10^5 \times 10^{-3}}{10^2 \times 10^2}$ **34.** $\dfrac{(4 \times 10^3)(6 \times 10^{-2})}{3 \times 10^{-7}}$

35. $\dfrac{(2 \times 10^2)^2(3 \times 10^{-3})}{2 \times 10^4}$ **36.** $\dfrac{(3 \times 10)^3(2 \times 10^{-1})}{2 \times 10^{-2}}$

37. $\dfrac{(2 \times 10^{-3})(6 \times 10^2)^2}{(2 \times 10^{-2})^2}$

38. $\dfrac{(8 \times 10^4)^2(3 \times 10)^3}{(6 \times 10^{-2})^2}$

■ *Specify the number of significant digits in each number.*

Examples **a.** 0.204 **b.** 0.000471 **c.** 0.9300

Solutions **a.** $0.204 = 2.04 \times 10^{-1}$; there are *three* significant digits.

b. $0.000471 = 4.71 \times 10^{-4}$; there are *three* significant digits.

c. $0.9300 = 9.300 \times 10^{-1}$; there are *four* significant digits.

39. 783.2 **40.** 29.4 **41.** 0.023 **42.** 0.00472

43. 0.430 **44.** 0.600 **45.** 0.0503 **46.** 0.02004

B ■ *Compute.*

Example $\dfrac{0.016 \times 3 \times 0.0028}{0.064}$

Solution First write each factor as the product of a natural number and a power of 10.

$$\frac{0.016 \times 3 \times 0.0028}{0.064} = \frac{16 \times 10^{-3} \times 3 \times 28 \times 10^{-4}}{64 \times 10^{-3}}$$

$$= \frac{16 \times 3 \times 28}{64} \times 10^{-4}$$

$$= 21 \times 10^{-4} \quad \text{or} \quad 0.0021$$

47. $\dfrac{0.6 \times 0.00084 \times 0.093}{0.00021 \times 0.00031}$

48. $\dfrac{0.065 \times 2.2 \times 50}{1.30 \times 0.011 \times 0.05}$

49. $\dfrac{28 \times 0.0006 \times 450}{1.5 \times 700 \times 0.018}$

50. $\dfrac{0.0054 \times 0.05 \times 300}{0.0015 \times 0.27 \times 80}$

51. $\dfrac{420 \times 0.0016 \times 800}{0.0028 \times 1200 \times 20}$

52. $\dfrac{0.0027 \times 0.004 \times 650}{260 \times 0.0001 \times 0.009}$

53. The speed of light is approximately 300,000,000 meters per second.

    a. Express this number in scientific notation.

    b. Express the speed of light in inches per second (1 inch equals 2.54 centimeters and 1 meter equals 100 centimeters).

54. One light-year is the number of miles traveled by light in 1 year (365 days), and the speed of light is approximately 186,000 miles per second. Express in scientific notation the number of miles in 1 light-year.

5.4

RATIONAL EXPONENTS

If the laws of exponents developed in Section 5.1 are to hold for rational exponents, meanings consistent with these laws must be assigned to powers with rational exponents. Let us examine exponents that are the reciprocals of natural numbers, that is, exponents of the form $1/n$, where n is a natural number.

$a^{1/n}$,
n a natural
number

We shall first define powers of the form $a^{1/n}$ to be consistent with Law II of exponents (page 138). We shall make this definition in two parts: for n an even natural number; and for n an odd natural number.

▶ *If n is an even natural number and $a > 0$, then $a^{1/n}$ is the positive number such that*
$$(a^{1/n})^n = a^{n/n} = a.$$

The number $a^{1/n}$ is called the **positive nth root of a.**

Examples   **a.** $16^{1/2} = 4$ because $4^2 = 16$

b. $-16^{1/2} = -(16)^{1/2} = -4$ because $-4^2 = -(4^2) = -16$

c. $(-16)^{1/2}$ is not defined in the set of real numbers because there is no real number a for which $a^2 = -16$

The restriction that $a > 0$ in the above definition for $a^{1/n}$ is not necessary if n is an odd natural number.

▶ *If n is an odd natural number, then $a^{1/n}$ is the number such that*
$$(a^{1/n})^n = a.$$

Examples   **a.** $(8)^{1/3} = 2$ because $2^3 = 8$

b. $(-8)^{1/3} = -2$ because $(-2)^3 = -8$

c. $(-64)^{1/3} = -4$ because $(-4)^3 = -64$

▶ *If n is an even or odd natural number,*
$$0^{1/n} = 0.$$

Examples   **a.** $0^{1/2} = 0$ **b.** $0^{1/3} = 0$

$a^{m/n}$,
n a natural
number

It can be shown that we can define powers with positive bases and positive rational exponents in two ways to be consistent with the laws of exponents:

▶ $$a^{m/n} = (a^{1/n})^m = (a^m)^{1/n} \quad (a^{1/n} \text{ a real number}).$$

Thus, we can look at $a^{m/n}$ in two ways, either as the mth power of the nth root of a, or as the nth root of the mth power of a.

Examples

a. $8^{2/3} = (8^{1/3})^2$
$= (2)^2 = 4$

b. $8^{2/3} = (8^2)^{1/3}$
$= (64)^{1/3} = 4$

Hereafter, we shall use whichever form is most convenient for the purpose at hand. In Example a above, the form $(8^{1/3})^2$ is preferred because it is easier to extract the root first than it is to recognize the root after the number is squared.

To extend meaning to powers with negative rational exponents, we define $a^{-m/n}$ as follows:

▶ *For m and n positive integers,*

$$a^{-m/n} = \frac{1}{a^{m/n}} \quad (a^{1/n} \text{ a real number, } a \neq 0).$$

Examples

a. $27^{-2/3} = \dfrac{1}{27^{2/3}} = \dfrac{1}{(27^{1/3})^2}$
$= \dfrac{1}{3^2} = \dfrac{1}{9}$

b. $(-8)^{-5/3} = \dfrac{1}{(-8)^{5/3}} = \dfrac{1}{[(-8)^{1/3}]^5}$
$= \dfrac{1}{(-2)^5} = -\dfrac{1}{32}$

With the definitions that we have made, it can be shown that powers with rational exponents—positive, negative, or 0—obey all the five laws of exponents set forth in Section 5.1.

Recall from Section 1.1 that any number that can be expressed as the quotient of two integers is called a rational number, and any real number that cannot be so expressed is called an irrational number. Some powers with rational-number exponents are rational numbers and some are irrational numbers. Any expression such as $a^{1/n}$ represents a rational number if and only if a is the nth power of a rational number.

Examples

a. $4^{1/2}$, $(-27/8)^{1/3}$, and $(81)^{1/4}$ are rational numbers equal to 2, $-3/2$, and 3, respectively.

b. $2^{1/2}$, $5^{1/3}$, and $7^{1/4}$ are irrational numbers, such that $(2^{1/2})^2 = 2$, $(5^{1/3})^3 = 5$, and $(7^{1/4})^4 = 7$.

In Section 5.5, we shall consider how to obtain rational approximations for some irrational numbers.

EXERCISE 5.4

A   ■ *Assume that all bases in this exercise are positive unless otherwise specified.*

■ *Write each expression using a basic numeral or fraction in lowest terms.*

Examples   **a.** $64^{1/2}$ **b.** $(-27)^{4/3}$ **c.** $8^{-2/3}$

Solutions   **a.** $64^{1/2} = 8$ **b.** $(-27)^{4/3}$ **c.** $8^{-2/3} = (8^{1/3})^{-2}$

$\qquad\qquad\qquad\qquad\qquad\qquad = [(-27)^{1/3}]^4 \qquad\qquad = (2)^{-2}$

$\qquad\qquad\qquad\qquad\qquad\qquad = [-3]^4 \qquad\qquad\qquad = \dfrac{1}{4}$

$\qquad\qquad\qquad\qquad\qquad\qquad = 81$

1. $9^{1/2}$ **2.** $25^{1/2}$ **3.** $32^{1/5}$ **4.** $27^{1/3}$

5. $(-8)^{1/3}$ **6.** $(-27)^{1/3}$ **7.** $27^{2/3}$ **8.** $32^{3/5}$

9. $81^{3/4}$ **10.** $125^{2/3}$ **11.** $(-8)^{4/3}$ **12.** $(-64)^{2/3}$

13. $16^{-1/2}$ **14.** $8^{-1/3}$ **15.** $16^{-3/4}$ **16.** $27^{-2/3}$

■ *Write each expression as a product or quotient of powers in which each variable occurs only once and all exponents are positive.*

Examples   **a.** $y^{3/4}y^{-1/2}$ **b.** $\dfrac{x^{5/6}}{x^{2/3}}$ **c.** $\dfrac{(x^{1/2}y^2)^2}{(x^{2/3}y)^3}$

Solutions   **a.** $y^{3/4}y^{-1/2}$ **b.** $\dfrac{x^{5/6}}{x^{2/3}} = \dfrac{x^{5/6}}{x^{4/6}}$ **c.** $\dfrac{(x^{1/2}y^2)^2}{(x^{2/3}y)^3} = \dfrac{xy^4}{x^2y^3}$

$\qquad\quad = y^{3/4+(-1/2)} \qquad\qquad\qquad = x^{1/6} \qquad\qquad\qquad\qquad = \dfrac{y}{x}$

$\qquad\quad = y^{1/4}$

17. $x^{1/3}x^{1/3}$ **18.** $y^{1/2}y^{3/2}$ **19.** $\dfrac{x^{2/3}}{x^{1/3}}$ **20.** $\dfrac{x^{3/4}}{x^{1/4}}$

21. $(a^{1/2})^3$ **22.** $(b^6)^{2/3}$ **23.** $x^{-3/4}x^{1/4}$ **24.** $y^{-2/3}y^{5/3}$

25. $(a^{2/3}b)^{1/2}$ **26.** $(a^{1/2}b^{1/3})^6$ **27.** $\left(\dfrac{a^6}{b^3}\right)^{2/3}$ **28.** $\left(\dfrac{a^{1/2}}{a^2}\right)^2$

29. $(r^{-2/3}t)^{-3}$ **30.** $(x^{1/4}y^{1/2})^8$ **31.** $\left(\dfrac{z^3}{t^6}\right)^{-1/3}$ **32.** $\left(\dfrac{a^{-1/2}}{b^{1/3}}\right)^6$

33. $\left(\dfrac{x^{-2}y^{-1/3}z}{x^{-5/3}y^{-2/3}z^{2/3}}\right)^3$ **34.** $\left(\dfrac{x^{1/4}y^{3/4}z^{-1}}{x^{-3/4}y^{1/4}z^0}\right)^2$

■ *Write each product so that each base of a power occurs at most once in each term.*

Examples   **a.** $y^{1/3}(y + y^{2/3})$ **b.** $x^{-3/4}(x^{1/4} + x^{3/4})$

Solutions   **a.** $y^{1/3}(y + y^{2/3})$ **b.** $x^{-3/4}(x^{1/4} + x^{3/4})$

$\qquad = y^{1/3+1} + y^{1/3+2/3}$ $\qquad\qquad = x^{-3/4+1/4} + x^{-3/4+3/4}$

$\qquad = y^{4/3} + y$ $\qquad\qquad\qquad = x^{-2/4} + x^0$

$\qquad\qquad\qquad\qquad\qquad\qquad\qquad = x^{-1/2} + 1$

35. $x^{1/2}(x + x^{1/2})$ **36.** $x^{1/5}(x^{2/5} + x^{4/5})$ **37.** $x^{1/3}(x^{2/3} - x^{1/3})$

38. $x^{3/8}(x^{1/4} - x^{1/2})$ **39.** $x^{-3/4}(x^{-1/4} + x^{3/4})$ **40.** $y^{-1/4}(y^{3/4} + y^{5/4})$

41. $t^{3/5}(t^{2/5} + t^{-3/5})$ **42.** $a^{-2/7}(a^{9/7} + a^{2/7})$ **43.** $b^{3/4}(b^{1/4} + b^{-1/2})$

44. $x^{5/6}(x^{-5/6} + x^{1/6})$

■ *Factor as indicated.*

Example $y^{3/4} = y^{1/4}(?)$

Solution   The missing factor is of the form y^n, where $1/4 + n = 3/4$. By inspection, $n = 2/4 = 1/2$.
Hence, $y^{3/4} = y^{1/4}(y^{1/2})$.

45. $x^{3/5} = x^{1/5}(?)$ **46.** $x^{7/8} = x^{3/8}(?)$ **47.** $x^{-1/3} = x^{-2/3}(?)$

48. $y^{-1/4} = y(?)$ **49.** $x^{1/3} = x(?)$ **50.** $y^{3/5} = y(?)$

51. Use a counterexample to show that $(a + b)^{1/2}$ is not equivalent to $a^{1/2} + b^{1/2}$.

52. Use a counterexample to show that $(a + b)^{-1/2}$ is not equivalent to $a^{-1/2} + b^{-1/2}$.

B   ■ *Simplify. Assume that $m, n > 0$.*

Examples   **a.** $\dfrac{(x^n)^{3/2}}{x^{n/2}}$ **b.** $(y^{2n} \cdot y^{n/2})^4$ **c.** $\left(\dfrac{a^{n+2} \cdot b^{n/2}}{a^n}\right)^2$

Solutions   **a.** $\dfrac{(x^n)^{3/2}}{x^{n/2}} = x^{3n/2-n/2}$ **b.** $(y^{2n} \cdot y^{n/2})^4 = y^{8n} \cdot y^{2n}$ **c.** $\left(\dfrac{a^{n+2} \cdot b^{n/2}}{a^n}\right)^2 = (a^2 \cdot b^{n/2})^2$

$\qquad\qquad = x^n$ $\qquad\qquad\qquad = y^{10n}$ $\qquad\qquad\qquad\qquad = a^4 b^n$

53. $x^n \cdot x^{n/2}$ **54.** $(a^2)^{n/2} \cdot (b^{2n})^{2/n}$ **55.** $\dfrac{x^{2n}}{x^{n/2}}$

56. $\left(\dfrac{a^n}{b}\right)^{1/2}\left(\dfrac{b}{a^{2n}}\right)^{3/2}$ **57.** $\dfrac{x^{3n}y^{2m+1}}{(x^ny^m)^{1/2}}$ **58.** $\left(\dfrac{m^a n^{2a}}{n^{4a}}\right)^{1/a}$

59. $\left(\dfrac{x^{2n} \cdot y^{3n}}{x^n}\right)^{1/3}$ **60.** $\left(\dfrac{x^{n+1} \cdot y^{n+2}}{xy^2}\right)^{1/n}$

■ *Factor as indicated.*

Example $y^{-1/2} + y^{1/2} = y^{-1/2}(? + ?)$

Solution   The first missing factor is 1 because $y^{-1/2} \cdot 1 = y^{-1/2}$. The second missing factor is of the form y^n, where $-\frac{1}{2} + n = \frac{1}{2}$. By inspection, $n = 1$. Hence,

$$y^{-1/2} + y^{1/2} = y^{-1/2}(1 + y).$$

61. $x^{3/2} + x = x(? + ?)$ **62.** $y^{5/2} + y = y(? + ?)$

63. $x - x^{2/3} = x^{1/3}(? - ?)$ **64.** $y^{2/3} - y = y^{1/3}(? - ?)$

65. Which is greater, $16^{1/4}$ or $16^{1/2}$? $(1/16)^{1/4}$ or $(1/16)^{1/2}$? Make a conjecture about the order of $a^{1/n}$ and $a^{1/m}$ when $n > m$ and $a > 1$, and when $n > m$ and $0 < a < 1$.

5.5

RADICALS

Radical notation

In Section 5.4 we agreed to refer to $a^{1/n}$ (when it exists) as the *n*th root of a. An alternative symbol is often used for $a^{1/n}$.

► *For all natural numbers $n \geq 2$,*

$$a^{1/n} = \sqrt[n]{a}.$$

In such a representation, the symbol $\sqrt{}$ is called a **radical**, a is called the **radicand**, n is called the **index**, and the expression is said to be a **radical of order n**. If no index is written, the index is understood to be 2, and the expression represents the nonnegative **square root** of the radicand.

We shall define $\sqrt[n]{a}$ to conform to our definition of $a^{1/n}$. It is convenient to do so in two parts: for n an even natural number; and for n an odd natural number.

► *For n an even natural number and $a \geq 0$, $\sqrt[n]{a}$ is the nonnegative number such that*

$$(\sqrt[n]{a})^n = a.$$

In particular, for $n = 2$,

$$(\sqrt{a})^2 = \sqrt{a}\sqrt{a} = a.$$

Examples   **a.** $\sqrt[4]{81} = 3$ because $3^4 = 81$ **b.** $\sqrt{16} = 4$ because $4^2 = 16$

c. $(\sqrt{5})^2 = \sqrt{5}\sqrt{5} = 5$ **d.** $(\sqrt[4]{5})^4 = 5$

Although each positive number a has two square roots, *the symbol $\sqrt{a}$ represents the positive value only*. Thus, the symbol $\sqrt{16}$ represents only the positive square root, 4, as in Example b above. The negative square root is denoted by $-\sqrt{16}$.

The restriction that $a > 0$ is not necessary if n is an odd natural number.

▶ *For n an odd natural number, $\sqrt[n]{a}$ is the number such that*

$$(\sqrt[n]{a})^n = a.$$

Examples

a. $\sqrt[3]{8} = 2$ because $2^3 = 8$

b. $\sqrt[3]{-8} = -2$ because $(-2)^3 = -8$

c. $(\sqrt[3]{4})^3 = \sqrt[3]{4}\sqrt[3]{4}\sqrt[3]{4}$

$\quad = 4$

d. $(\sqrt[3]{-2})^3 = \sqrt[3]{-2}\sqrt[3]{-2}\sqrt[3]{-2}$

$\quad = -2$

Since, from Section 5.4, for $a \geq 0$,

$$a^{m/n} = (a^m)^{1/n} = (a^{1/n})^m,$$

we may write a power with a rational exponent in radical form:

▶
$$a^{m/n} = \sqrt[n]{a^m} = (\sqrt[n]{a})^m \qquad (a \geq 0).$$

Note that the denominator of the exponent is the index of the radical, and the numerator of the exponent is either the exponent of the radicand or the exponent of the root.

Examples

a. $x^{2/3} = \sqrt[3]{x^2}$

$\quad = (\sqrt[3]{x})^2$

b. $8^{2/3} = \sqrt[3]{8^2} = \sqrt[3]{64} = 4$

or

$8^{2/3} = (\sqrt[3]{8})^2 = 2^2 = 4$

Since we have restricted the index of a radical to be a natural number, we must always express a fractional exponent in standard form (m/n or $-m/n$) before writing the power in radical form.

Examples

a. $x^{-(3/4)} = x^{(-3/4)}$

$\quad = \sqrt[4]{x^{-3}}$

b. $8^{-(2/3)} = 8^{-2/3} = \sqrt[3]{8^{-2}}$

$\quad = \sqrt[3]{\dfrac{1}{64}} = \dfrac{1}{4}$

Sometimes you may find it helpful to simplify radical expressions by first writing such expressions in exponential form.

Examples

a. $\sqrt[3]{8} = 8^{1/3}$

$\quad = 2$

b. $\sqrt[3]{-8} = (-8)^{1/3}$

$\quad = -2$

c. $\sqrt[3]{8^{-2}} = 8^{-2/3}$

$\quad = (8^{-2})^{1/3}$

$\quad = \left(\dfrac{1}{64}\right)^{1/3} = \dfrac{1}{4}$

$\sqrt[n]{a^n}$, *n* an even number

Because we have defined a radical with an even index to be a nonnegative number, we have the following special relationship for all values of a:

▶ $$\sqrt[n]{a^n} = |a| \qquad \textbf{(\textit{n} even)}.$$

Thus, the symbol $\sqrt[n]{a^n}$ represents only the nonnegative nth root of a^n when n is even. In particular,

$$\sqrt{a^2} = |a|.$$

In considering odd indices, there is no ambiguous interpretation possible. That is, for all values of a:

▶ $$\sqrt[n]{a^n} = a \qquad \textbf{(\textit{n} odd)}.$$

Examples

a. $\sqrt{2^2} = |2| = 2$ **b.** $\sqrt{(-2)^2} = |-2| = 2$

c. $\sqrt[3]{4^3} = 4$ **d.** $\sqrt[3]{(-4)^3} = -4$

Because, as observed in Section 5.4, $a^{1/n}$ represents a rational number if and only if a is the nth power of a rational number, the same fact is true of $\sqrt[n]{a}$. Thus,

$$\sqrt{4}, \qquad -\sqrt[3]{27}, \qquad \sqrt[4]{81/16}, \quad \text{and} \quad \sqrt[5]{-32}$$

are rational numbers equal to

$$2, \qquad -3, \qquad 3/2, \quad \text{and} \quad -2.$$

Approximations for irrational numbers

Although irrational numbers such as $\sqrt{5}$, $\sqrt[3]{9}$, $\sqrt[4]{15}$, and $\sqrt[5]{61}$ do not have terminating or repeating decimal representations, we can obtain decimal approximations correct to any desired degree of accuracy. The Table of Squares, Square Roots, and Prime Factors in Appendix D, or some calculators, can be used to obtain approximations for some irrational numbers. For example,

$$\sqrt{2} \approx 1.414,$$

where the symbol $\approx$ is read "is approximately equal to."

EXERCISE 5.5

A ■ *In Problems 1–64, assume each variable and each radicand represents a positive number.*

■ *Write each expression in radical form.*

Examples   **a.** $5^{1/2}$ **b.** $(xy)^{2/3}$ **c.** $(x - y^2)^{-1/2}$

Solutions   **a.** $5^{1/2} = \sqrt{5}$ **b.** $(xy)^{2/3} = \sqrt[3]{x^2y^2}$ **c.** $(x - y^2)^{-1/2}$

$$= \frac{1}{\sqrt{x - y^2}}$$

1. $3^{1/2}$	**2.** $2^{1/3}$	**3.** $x^{3/2}$	**4.** $y^{2/3}$
5. $3y^{1/4}$	**6.** $5x^{1/3}$	**7.** $xy^{1/3}$	**8.** $x^2y^{1/2}$
9. $(xy)^{1/3}$	**10.** $(x^2y)^{1/3}$	**11.** $-3x^{3/5}$	**12.** $-5y^{2/3}$
13. $(x + 2y)^{1/2}$	**14.** $(x - y)^{1/3}$	**15.** $(2x - y)^{2/3}$	**16.** $(3x + y)^{3/4}$
17. $4^{-1/3}$	**18.** $6^{-2/5}$	**19.** $x^{-2/3}$	**20.** $y^{-2/7}$

■ *Represent each expression with positive fractional exponents.*

Examples   **a.** $\sqrt{2^3}$ **b.** $\sqrt[3]{7a^2}$

Solutions   **a.** $\sqrt{2^3} = 2^{3/2}$ **b.** $\sqrt[3]{7a^2} = (7a^2)^{1/3}$

$$= 7^{1/3}a^{2/3}$$

21. $\sqrt{5}$	**22.** $\sqrt[3]{2}$	**23.** $\sqrt[3]{x^2}$	**24.** $\sqrt{y^3}$
25. $\sqrt{ab}$	**26.** $\sqrt[3]{ab^2}$	**27.** $x\sqrt{y}$	**28.** $y\sqrt[3]{x^2}$
29. $\sqrt[3]{2ab^2}$	**30.** $\sqrt[4]{a^3b}$	**31.** $\sqrt{a - b}$	**32.** $\sqrt[4]{a + 2b}$

Examples   **a.** $\sqrt[3]{a} - 3\sqrt{b}$ **b.** $\dfrac{3}{\sqrt{x - 1}}$

Solutions   **a.** $\sqrt[3]{a} - 3\sqrt{b} = a^{1/3} - 3b^{1/2}$ **b.** $\dfrac{3}{\sqrt{x - 1}} = \dfrac{3}{(x - 1)^{1/2}}$

33. $\sqrt{a} - 2\sqrt{b}$	**34.** $\sqrt[3]{b} + 2\sqrt{a}$	**35.** $\sqrt[3]{a}\sqrt{b}$	**36.** $\sqrt{a}\sqrt[3]{ab}$
37. $\dfrac{1}{\sqrt{x}}$	**38.** $\dfrac{2}{\sqrt[3]{y}}$	**39.** $\dfrac{2}{\sqrt{x + y}}$	**40.** $\dfrac{5}{\sqrt[3]{a - b}}$

■ *Find the root indicated.*

Examples   **a.** $\sqrt[3]{-8}$ **b.** $\sqrt[3]{x^6y^3}$ **c.** $-\sqrt[4]{81x^4}$

Solutions   **a.** $\sqrt[3]{-8} = -2$ **b.** $\sqrt[3]{x^6y^3} = x^2y$ **c.** $-\sqrt[4]{81x^4} = -3x$

Alternate solutions overleaf

Alternative solutions

a. $\sqrt[3]{-8}$
$\quad = (-8)^{1/3}$
$\quad = -2$

b. $\sqrt[3]{x^6 y^3}$
$\quad = (x^6 y^3)^{1/3}$
$\quad = x^2 y$

c. $-\sqrt[4]{81 x^4}$
$\quad = -(81 x^4)^{1/4}$
$\quad = -3x$

41. $\sqrt{16}$ **42.** $\sqrt{144}$ **43.** $-\sqrt{25}$ **44.** $-\sqrt{169}$

45. $\sqrt[3]{27}$ **46.** $\sqrt[3]{125}$ **47.** $\sqrt[3]{-64}$ **48.** $\sqrt[5]{-32}$

49. $\sqrt[3]{x^3}$ **50.** $\sqrt[5]{y^5}$ **51.** $\sqrt{x^4}$ **52.** $\sqrt{a^6}$

53. $\sqrt[3]{8y^6}$ **54.** $\sqrt[3]{27y^9}$ **55.** $-\sqrt{x^4 y^6}$ **56.** $-\sqrt{a^8 b^{10}}$

57. $\sqrt{\dfrac{4}{9} x^2 y^8}$ **58.** $\sqrt{\dfrac{9}{16} a^2 b^4}$ **59.** $\sqrt[3]{\dfrac{-8}{125} x^3}$ **60.** $\sqrt[3]{\dfrac{8}{27} a^3 b^6}$

61. $\sqrt[4]{16 x^4 y^8}$ **62.** $\sqrt[5]{-32 x^5 y^{10}}$ **63.** $-\sqrt[3]{-8 a^6 b^9}$ **64.** $-\sqrt[4]{81 a^8 b^{12}}$

■ *Graph each set of real numbers on a separate line graph. (Use a calculator or the Table of Squares, Square Roots, and Prime Factors in Appendix D to obtain rational-number approximations for irrational numbers.)*

Examples

a. $\sqrt{4}, \ -\sqrt{3}, \ \sqrt{17}, \ -\sqrt{9}$

b. $\sqrt{2}, \ -\sqrt{6}, \ 0, \ \sqrt{16}$

Solutions

a.

b.

65 $-\sqrt{7}, \ -\sqrt{1}, \ \sqrt{5}, \ \sqrt{9}$ **66.** $-\tfrac{2}{3}, \ 0, \ \sqrt{3}, \ -\sqrt{11}$

67. $-\sqrt{20}, \ -\sqrt{6}, \ \sqrt{1}, \ 6$ **68.** $\sqrt{41}, \ \sqrt{7}, \ -\sqrt{7}, \ \tfrac{3}{4}$

B ■ *In the foregoing problems, variables and radicands were restricted to represent positive numbers. In Problems 69–74, consider variables and radicands to represent elements of the set of real numbers and use absolute-value notation as needed.*

Examples

a. $\sqrt{16 x^2}$

b. $\sqrt{x^2 - 2xy + y^2}$

Solutions

a. $\sqrt{16 x^2} = 4|x|$

b. $\sqrt{x^2 - 2xy + y^2} = \sqrt{(x - y)^2}$
$\qquad\qquad\qquad\quad = |x - y|$

69. $\sqrt{4x^2}$ **70.** $\sqrt{9x^2 y^4}$ **71.** $\sqrt{x^2 + 2x + 1}$

72. $\sqrt{4x^2 - 4x + 1}$ **73.** $\dfrac{2}{\sqrt{x^2 + 2xy + y^2}}$ **74.** $\sqrt{x^4 + 2x^2 y^2 + y^4}$

5.6

CHANGING FORMS
OF RADICALS

From the definition of a radical and the laws of exponents, we can derive two important relationships.

Laws of ▶ *For a, b > 0 and n a natural number,*
radicals

$$\sqrt[n]{ab} = \sqrt[n]{a}\sqrt[n]{b}. \tag{1}$$

This follows from the fact that

$$\sqrt[n]{ab} = (ab)^{1/n} = a^{1/n}b^{1/n} = \sqrt[n]{a}\sqrt[n]{b}.$$

Relationship (1) can be used to write a radical in a form in which the radicand contains no prime factor or polynomial factor raised to a power greater than or equal to the index of the radical.

Examples   **a.** $\sqrt{18} = \sqrt{3^2}\sqrt{2}$ **b.** $\sqrt[3]{16x^3y^5} = \sqrt[3]{2^3x^3y^3}\sqrt[3]{2y^2}$
$$= 3\sqrt{2}$$ $$= 2xy\sqrt[3]{2y^2}$$

In each case the radicand was first factored into two factors, one of which consisted of factors raised to the same power as the index of the radical. This factor was then removed from the radicand.

The second important relationship involves quotients.

▶ *For a, b > 0 and n a natural number,*

$$\sqrt[n]{\frac{a}{b}} = \frac{\sqrt[n]{a}}{\sqrt[n]{b}}. \tag{2}$$

This follows from the fact that

$$\sqrt[n]{\frac{a}{b}} = \left(\frac{a}{b}\right)^{1/n} = \frac{a^{1/n}}{b^{1/n}} = \frac{\sqrt[n]{a}}{\sqrt[n]{b}}.$$

We can use relationship (2) to write a radical in a form in which the radicand contains no fraction.

Examples   **a.** $\sqrt{\frac{3}{4}} = \frac{\sqrt{3}}{\sqrt{4}} = \frac{\sqrt{3}}{2}$ **b.** $\sqrt[3]{\frac{5}{8}} = \frac{\sqrt[3]{5}}{\sqrt[3]{8}} = \frac{\sqrt[3]{5}}{2}$

If a radical expression in the denominator of a fraction cannot be written directly without radical notation, we can use the fundamental principle of fractions to obtain an equivalent form in which the denominator is free of radicals.

Examples   a. $\sqrt{\dfrac{1}{3}} = \dfrac{\sqrt{1}}{\sqrt{3}}$ b. $\sqrt{\dfrac{2}{5x}} = \dfrac{\sqrt{2}}{\sqrt{5x}}$

$\qquad\qquad\qquad = \dfrac{1\sqrt{3}}{\sqrt{3}\sqrt{3}}$ $= \dfrac{\sqrt{2}\sqrt{5x}}{\sqrt{5x}\sqrt{5x}}$

$\qquad\qquad\qquad = \dfrac{\sqrt{3}}{3}$ $= \dfrac{\sqrt{10x}}{5x}$

The foregoing process is called *rationalizing the denominator* of a fraction.

Example   Rationalize the denominator: $\dfrac{1}{\sqrt[3]{2x}}$.

Solution   Because we need a third power, $(2x)^3$, beneath the radical sign in the denominator in order to write it without a radical sign, we must multiply $2x$ by two additional factors of $2x$. Thus, using the fundamental principle of fractions we obtain

$$\dfrac{1}{\sqrt[3]{2x}} = \dfrac{1\sqrt[3]{2x}\sqrt[3]{2x}}{\sqrt[3]{2x}\sqrt[3]{2x}\sqrt[3]{2x}}$$

$$= \dfrac{\sqrt[3]{(2x)^2}}{\sqrt[3]{(2x)^3}}$$

$$= \dfrac{\sqrt[3]{4x^2}}{2x}.$$

Example   Rationalize the denominator: $\sqrt[5]{\dfrac{6}{16x^3}}$.

Solution   In this instance, we have

$$\sqrt[5]{\dfrac{6}{16x^3}} = \sqrt[5]{\dfrac{6}{2^4x^3}}.$$

Since we want a fifth power in the denominator, we need the factor $\sqrt[5]{2x^2}$. Thus,

$$\sqrt[5]{\dfrac{6}{16x^3}} = \dfrac{\sqrt[5]{6}\sqrt[5]{2x^2}}{\sqrt[5]{16x^3}\sqrt[5]{2x^2}}$$

$$= \dfrac{\sqrt[5]{12x^2}}{\sqrt[5]{32x^5}} = \dfrac{\sqrt[5]{12x^2}}{2x}.$$

Writing equivalent radical expressions

Application of relationships (1) and/or (2) on page 159 can be used to rewrite radical expressions in various ways, and, in particular, to write them in what is called **simplest form**.

A radical expression is in simplest form if:

1. The radicand contains no polynomial factor raised to a power equal to or greater than the index of the radical.

2. The radicand contains no fractions.

3. No radical expressions are contained in denominators of fractions.

Although we generally change the form of radicals to one of the forms implied above, there are times when such forms are not preferred. For example, in certain situations, $\sqrt{\frac{1}{2}}$ or $\frac{1}{\sqrt{2}}$ may be more useful than the equivalent form $\frac{\sqrt{2}}{2}$. In such cases, we may want to rationalize the numerator of a fraction.

Examples

Rationalize each numerator.

a. $\dfrac{\sqrt{2}}{2}$

b. $\dfrac{\sqrt{2x}}{x}$

Solutions

a. $\dfrac{\sqrt{2}}{2} = \dfrac{\sqrt{2} \cdot \sqrt{2}}{2 \cdot \sqrt{2}}$

$= \dfrac{2}{2\sqrt{2}} = \dfrac{1}{\sqrt{2}}$

b. $\dfrac{\sqrt{2x}}{x} = \dfrac{\sqrt{2x} \cdot \sqrt{2x}}{x \cdot \sqrt{2x}}$

$= \dfrac{2x}{x\sqrt{2x}} = \dfrac{2}{\sqrt{2x}}$

EXERCISE 5.6

A　■ *Assume that all variables in radicands in this exercise denote positive real numbers.*

　　■ *Change to simplest form.*

Examples

a. $\sqrt{300}$

b. $\sqrt[3]{2x^7 y^3}$

c. $\sqrt{2xy}\sqrt{8x}$

Solutions

a. $\sqrt{300} = \sqrt{100}\sqrt{3}$
$\qquad = 10\sqrt{3}$

b. $\sqrt[3]{2x^7 y^3} = \sqrt[3]{x^6 y^3}\cdot\sqrt[3]{2x}$
$\qquad = x^2 y\sqrt[3]{2x}$

c. $\sqrt{2xy}\sqrt{8x}$
$\qquad = \sqrt{16x^2 y}$
$\qquad = \sqrt{16x^2}\sqrt{y}$
$\qquad = 4x\sqrt{y}$

1. $\sqrt{18}$ 2. $\sqrt{50}$ 3. $\sqrt{20}$ 4. $\sqrt{72}$

5. $\sqrt{75}$ 6. $\sqrt{48}$ 7. $\sqrt{x^4}$ 8. $\sqrt{y^6}$

9. $\sqrt{x^3}$ 10. $\sqrt{y^{11}}$ 11. $\sqrt{9x^3}$ 12. $\sqrt{4y^5}$

13. $\sqrt{8x^6}$ 14. $\sqrt{18y^8}$ 15. $\sqrt[4]{x^5}$ 16. $\sqrt[3]{y^4}$

17. $\sqrt[5]{x^7y^9z^{11}}$ 18. $\sqrt[5]{a^{12}b^{15}}$ 19. $\sqrt[6]{a^7b^{12}c^{15}}$ 20. $\sqrt[6]{m^8n^7}$

21. $\sqrt[7]{3^7a^8b^9c^{10}}$ 22. $\sqrt[7]{4^8x^9y^{10}z^{14}}$ 23. $\sqrt{18}\sqrt{2}$ 24. $\sqrt{3}\sqrt{27}$

25. $\sqrt{xy}\sqrt{x^5y}$ 26. $\sqrt{a}\sqrt{ab^2}$ 27. $\sqrt[3]{2}\sqrt[3]{4}$ 28. $\sqrt[4]{3}\sqrt[4]{27}$

29. $\sqrt{3 \times 10^2}$ 30. $\sqrt{5 \times 10^3}$ 31. $\sqrt{60{,}000}$ 32. $\sqrt{800{,}000}$

■ *Rationalize denominators.*

Examples   a. $\sqrt{\dfrac{1}{3}}$ b. $\sqrt{\dfrac{2}{3x}}$

Solutions   a. $\sqrt{\dfrac{1}{3}} = \dfrac{\sqrt{1}\sqrt{3}}{\sqrt{3}\sqrt{3}}$ b. $\sqrt{\dfrac{2}{3x}} = \dfrac{\sqrt{2}\sqrt{3x}}{\sqrt{3x}\sqrt{3x}}$

$\qquad\qquad = \dfrac{\sqrt{3}}{3}$ $\qquad = \dfrac{\sqrt{6x}}{3x}$

33. $\sqrt{\dfrac{1}{5}}$ 34. $\sqrt{\dfrac{2}{3}}$ 35. $\dfrac{-1}{\sqrt{2}}$ 36. $\dfrac{-\sqrt{3}}{\sqrt{7}}$

37. $\sqrt{\dfrac{x}{2}}$ 38. $-\sqrt{\dfrac{y}{3}}$ 39. $-\sqrt{\dfrac{y}{x}}$ 40. $\sqrt{\dfrac{2a}{b}}$

41. $\dfrac{x}{\sqrt{x}}$ 42. $\dfrac{-x}{\sqrt{2y}}$ 43. $\sqrt{\dfrac{y}{2x}}$ 44. $\sqrt{\dfrac{y}{6x}}$

Examples   a. $\sqrt[3]{\dfrac{2}{y^2}}$ b. $\sqrt[5]{\dfrac{x}{4y^2}}$

Solutions   a. $\sqrt[3]{\dfrac{2}{y^2}} = \dfrac{\sqrt[3]{2}\,\sqrt[3]{y}}{\sqrt[3]{y^2}\,\sqrt[3]{y}}$ b. $\sqrt[5]{\dfrac{x}{4y^2}} = \dfrac{\sqrt[5]{x}\,\sqrt[5]{(2y)^3}}{\sqrt[5]{(2y)^2}\,\sqrt[5]{(2y)^3}}$

$\qquad\qquad = \dfrac{\sqrt[3]{2}\sqrt[3]{y}}{\sqrt[3]{y^3}} = \dfrac{\sqrt[3]{2y}}{y}$ $\qquad = \dfrac{\sqrt[5]{x(2y)^3}}{\sqrt[5]{(2y)^5}} = \dfrac{\sqrt[5]{8xy^3}}{2y}$

45. $\dfrac{1}{\sqrt[3]{x^2}}$ 46. $\dfrac{1}{\sqrt[4]{y^3}}$ 47. $\sqrt[3]{\dfrac{2}{3y}}$ 48. $\sqrt[4]{\dfrac{2}{3x}}$

49. $\sqrt[3]{\dfrac{x}{4y^2}}$ 50. $\sqrt[4]{\dfrac{x}{8y^3}}$ 51. $\sqrt[5]{\dfrac{3}{2x^3}}$ 52. $\sqrt[5]{\dfrac{2}{9y^2}}$

Examples

a. $\dfrac{\sqrt{a}\sqrt{ab^3}}{\sqrt{b}}$

b. $\dfrac{\sqrt[3]{16y^4}}{\sqrt[3]{y}}$

Solutions

a. $\dfrac{\sqrt{a}\sqrt{ab^3}}{\sqrt{b}} = \sqrt{\dfrac{a^2b^3}{b}}$

$= \sqrt{a^2b^2}$

$= ab$

b. $\dfrac{\sqrt[3]{16y^4}}{\sqrt[3]{y}} = \sqrt[3]{\dfrac{16y^4}{y}}$

$= \sqrt[3]{2^4y^3}$

$= 2y\sqrt[3]{2}$

53. $\dfrac{\sqrt{a^5b^3}}{\sqrt{ab}}$

54. $\dfrac{\sqrt{x}\sqrt{xy^3}}{\sqrt{y}}$

55. $\dfrac{\sqrt{98x^2y^3}}{\sqrt{xy}}$

56. $\dfrac{\sqrt{45x^3}\sqrt{y^3}}{\sqrt{5y}}$

57. $\dfrac{\sqrt[3]{8b^4}}{\sqrt[3]{a^6}}$

58. $\dfrac{\sqrt[3]{16r^4}}{\sqrt[3]{4t^3}}$

59. $\dfrac{\sqrt[5]{a}\sqrt[5]{b^2}}{\sqrt[5]{ab}}$

60. $\dfrac{\sqrt[5]{x^2}\sqrt[5]{y^3}}{\sqrt[5]{xy^2}}$

■ *Rationalize numerators.*

61. $\dfrac{\sqrt{3}}{3}$

62. $\dfrac{\sqrt{2}}{3}$

63. $\dfrac{\sqrt{x}}{\sqrt{y}}$

64. $\dfrac{\sqrt{xy}}{x}$

65. Use a counterexample to show that $(\sqrt{a} + \sqrt{b})^2$ is not equivalent to $a + b$.

66. Use a counterexample to show that $\sqrt{a + b}$ is not equivalent to $\sqrt{a} + \sqrt{b}$.

67. Use a counterexample to show that $\sqrt{a^2}$ is not equivalent to a.

68. Use a counterexample to show that $\sqrt{(a-1)^2}$ is not equivalent to $a - 1$.

B ■ *Reduce the order of each radical.*

Examples

a. $\sqrt[4]{5^2}$

b. $\sqrt[6]{9}$

c. $\sqrt[8]{x^2}$

Solutions

a. $\sqrt[4]{5^2} = 5^{2/4}$

$= 5^{1/2}$

$= \sqrt{5}$

b. $\sqrt[6]{9} = (3^2)^{1/6}$

$= 3^{1/3}$

$= \sqrt[3]{3}$

c. $\sqrt[8]{x^2} = x^{2/8}$

$= x^{1/4}$

$= \sqrt[4]{x}$

69. $\sqrt[4]{3^2}$

70. $\sqrt[6]{2^2}$

71. $\sqrt[6]{3^3}$

72. $\sqrt[8]{5^2}$

73. $\sqrt[6]{81}$

74. $\sqrt[10]{32}$

75. $\sqrt[6]{x^3}$

76. $\sqrt[9]{y^3}$

■ *Express as radicals of the same order and multiply.*

Examples

a. $\sqrt{3},\ \sqrt[3]{5}$

b. $\sqrt[3]{y},\ \sqrt[4]{y}$

Solutions overleaf

Solutions   **a.** Least common index is 6.   **b.** Least common index is 12.

$$\sqrt{3} = \sqrt[3\cdot2]{3^3} \qquad \sqrt[3]{5} = \sqrt[2\cdot3]{5^2}$$
$$= \sqrt[6]{27} \qquad\qquad = \sqrt[6]{25}$$
$$\sqrt[6]{27} \cdot \sqrt[6]{25} = \sqrt[6]{675}$$

$$\sqrt[3]{y} = \sqrt[4\cdot3]{y^4} \qquad \sqrt[4]{y} = \sqrt[3\cdot4]{y^3}$$
$$= \sqrt[12]{y^4} \qquad\qquad = \sqrt[12]{y^3}$$
$$\sqrt[12]{y^4} \cdot \sqrt[12]{y^3} = \sqrt[12]{y^7}$$

77. $\sqrt[3]{3}, \ \sqrt{2}$ **78.** $\sqrt[3]{2}, \ \sqrt[4]{2}$ **79.** $\sqrt[4]{5}, \ \sqrt{2}$ **80.** $\sqrt[3]{x}, \ \sqrt{x}$

5.7

EXPRESSIONS CONTAINING RADICALS

Sums and differences

The distributive property,

$$a(b + c) = ab + ac, \tag{1}$$

is assumed to hold for all real numbers. By the symmetric property of equality and the commutative property of multiplication, (1) can be written as

$$ba + ca = (b + c)a.$$

Since at this time all radical expressions have been defined so that they represent real numbers, the distributive property holds for radical expressions. Hence we may write sums or differences containing radicals of the same index and radicand as a single term.

Examples   **a.** $3\sqrt{3} + 4\sqrt{3} = (3 + 4)\sqrt{3}$ **b.** $7\sqrt{x} - 2\sqrt{x} = (7 - 2)\sqrt{x}$
$$= 7\sqrt{3} \qquad\qquad\qquad\qquad\qquad = 5\sqrt{x}$$

Products and factors

A direct application of (1) permits us to write certain products that contain parentheses as expressions without parentheses.

Examples   **a.** $x(\sqrt{2} + \sqrt{3}) = x\sqrt{2} + x\sqrt{3}$ **b.** $\sqrt{3}(\sqrt{2x} + \sqrt{6}) = \sqrt{6x} + \sqrt{18}$
$$= \sqrt{6x} + \sqrt{9}\sqrt{2}$$
$$= \sqrt{6x} + 3\sqrt{2}$$

In Chapter 2, we agreed to factor from each term of an expression only those common factors that are integers or positive integral powers of variables. However, we can (if we wish) consider other real numbers for factors. Thus, using the distributive property in the form

$$ab + ac = a(b + c),$$

radicals common to each term in an expression may be factored from the expression.

Examples   **a.** $a\sqrt{c} + b\sqrt{c} = \sqrt{c}(a + b)$ **b.** $\sqrt{a} + \sqrt{ab} = \sqrt{a} + \sqrt{a}\sqrt{b}$
$$= \sqrt{a}(1 + \sqrt{b})$$

Quotients

Recall from Section 5.6 that a monomial denominator of a fraction of the form $a/\sqrt{b}$ can be rationalized by multiplying the numerator and the denominator by $\sqrt{b}$.

Examples

a. $\dfrac{2}{\sqrt{3}} = \dfrac{2\sqrt{3}}{\sqrt{3}\sqrt{3}} = \dfrac{2\sqrt{3}}{3}$

b. $\dfrac{a}{\sqrt{b}} = \dfrac{a\sqrt{b}}{\sqrt{b}\sqrt{b}} = \dfrac{a\sqrt{b}}{b}$

The distributive property provides us with a means of rationalizing denominators of fractions in which radicals occur in one or both of the two terms of a binomial. To accomplish this, we first recall that

$$(a - b)(a + b) = a^2 - b^2,$$

where the product contains no first-degree term. Each of the two factors of a product exhibiting this property is said to be the **conjugate** of the other.

Now consider a fraction of the form

$$\frac{a}{b + \sqrt{c}} \qquad (b + \sqrt{c} \neq 0).$$

If we multiply the numerator and the denominator of this fraction by the conjugate of the denominator, the denominator of the resulting fraction will contain no term linear in $\sqrt{c}$; hence, it will be free of radicals. That is,

$$\frac{a(b - \sqrt{c})}{(b + \sqrt{c})(b - \sqrt{c})} = \frac{ab - a\sqrt{c}}{b^2 - c} \qquad (b^2 - c \neq 0),$$

where the denominator has been rationalized.

This process is equally applicable to fractions of the form

$$\frac{a}{\sqrt{b} + \sqrt{c}},$$

since

$$\frac{a(\sqrt{b} - \sqrt{c})}{(\sqrt{b} + \sqrt{c})(\sqrt{b} - \sqrt{c})} = \frac{a\sqrt{b} - a\sqrt{c}}{b - c} \qquad (b - c \neq 0).$$

Examples

Rationalize each denominator.

a. $\dfrac{2}{\sqrt{3} - 1}$

b. $\dfrac{x}{2 + \sqrt{x}}$

Solutions

a. $\dfrac{2}{\sqrt{3} - 1} = \dfrac{2(\sqrt{3} + 1)}{(\sqrt{3} - 1)(\sqrt{3} + 1)}$

$= \dfrac{2(\sqrt{3} + 1)}{3 - 1}$

$= \sqrt{3} + 1$

b. $\dfrac{x}{2 + \sqrt{x}} = \dfrac{x(2 - \sqrt{x})}{(2 + \sqrt{x})(2 - \sqrt{x})}$

$= \dfrac{x(2 - \sqrt{x})}{4 - x}$

$= \dfrac{2x - x\sqrt{x}}{4 - x}$

As noted in Section 5.6, we can rationalize the numerator of a fraction as well as the denominator.

Examples Rationalize each numerator.

a. $\dfrac{\sqrt{2}-1}{2}$

b. $\dfrac{\sqrt{x}+1}{x}$

Solutions

a. $\dfrac{\sqrt{2}-1}{2}=\dfrac{(\sqrt{2}-1)(\sqrt{2}+1)}{2(\sqrt{2}+1)}$

$=\dfrac{2-1}{2(\sqrt{2}+1)}$

$=\dfrac{1}{2\sqrt{2}+2}$

b. $\dfrac{\sqrt{x}+1}{x}=\dfrac{(\sqrt{x}+1)(\sqrt{x}-1)}{x(\sqrt{x}-1)}$

$=\dfrac{x-1}{x(\sqrt{x}-1)}$

$=\dfrac{x-1}{x\sqrt{x}-x}$

EXERCISE 5.7

A ■ *Assume that all radicands and variables in this exercise are positive real numbers.*

■ *Write each sum as a single term.*

Examples **a.** $3\sqrt{20}+\sqrt{45}$

b. $\sqrt{32x}+\sqrt{2x}-\sqrt{18x}$

Solutions

a. $3\sqrt{20}+\sqrt{45}$

$=3\cdot 2\sqrt{5}+3\sqrt{5}$

$=6\sqrt{5}+3\sqrt{5}$

$=9\sqrt{5}$

b. $\sqrt{32x}+\sqrt{2x}-\sqrt{18x}$

$=4\sqrt{2x}+\sqrt{2x}-3\sqrt{2x}$

$=2\sqrt{2x}$

1. $3\sqrt{7}+2\sqrt{7}$

2. $5\sqrt{2}-3\sqrt{2}$

3. $4\sqrt{3}-\sqrt{27}$

4. $\sqrt{75}+2\sqrt{3}$

5. $\sqrt{50x}+\sqrt{32x}$

6. $\sqrt{8y}-\sqrt{18y}$

7. $3\sqrt{4xy^2}-4\sqrt{9xy^2}$

8. $2\sqrt{8y^2z}+3\sqrt{32y^2z}$

9. $3\sqrt{8a}+2\sqrt{50a}-\sqrt{2a}$

10. $\sqrt{3b}-2\sqrt{12b}+3\sqrt{48b}$

11. $3\sqrt[3]{16}-\sqrt[3]{2}$

12. $\sqrt[3]{54}+2\sqrt[3]{128}$

■ *Write each expression without parentheses and all radicals in simple form.*

Examples **a.** $4(\sqrt{3}+1)$

b. $\sqrt{x}(\sqrt{2x}-\sqrt{x})$

c. $(\sqrt{x}-\sqrt{y})(\sqrt{x}+\sqrt{y})$

Solutions

a. $4(\sqrt{3} + 1)$
$= 4\sqrt{3} + 4$

b. $\sqrt{x}(\sqrt{2x} - \sqrt{x})$
$= x\sqrt{2} - x$

c. $(\sqrt{x} - \sqrt{y})(\sqrt{x} + \sqrt{y})$
$= x - y$

13. $2(3 - \sqrt{5})$

14. $5(2 - \sqrt{7})$

15. $\sqrt{2}(\sqrt{6} + \sqrt{10})$

16. $\sqrt{3}(\sqrt{12} - \sqrt{15})$

17. $(3 + \sqrt{5})(2 - \sqrt{5})$

18. $(1 - \sqrt{2})(2 + \sqrt{2})$

19. $(\sqrt{x} - 3)(\sqrt{x} + 3)$

20. $(2 + \sqrt{x})(2 - \sqrt{x})$

21. $(\sqrt{2} - \sqrt{3})(\sqrt{2} + 2\sqrt{3})$

22. $(\sqrt{3} - \sqrt{5})(2\sqrt{3} + \sqrt{5})$

23. $(\sqrt{5} - \sqrt{2})^2$

24. $(\sqrt{2} - 2\sqrt{3})^2$

■ *Change each expression to the form indicated.*

Examples

a. $3 + \sqrt{18} = 3(? + ?)$

b. $\sqrt{x} + \sqrt{xy} = \sqrt{x}(? + ?)$

Solutions

a. $3 + \sqrt{18} = 3 + 3\sqrt{2}$
$= 3(1 + \sqrt{2})$

b. $\sqrt{x} + \sqrt{xy} = \sqrt{x} + \sqrt{x}\sqrt{y}$
$= \sqrt{x}(1 + \sqrt{y})$

25. $2 + 2\sqrt{3} = 2(? + ?)$

26. $5 + 10\sqrt{2} = 5(? + ?)$

27. $2\sqrt{27} + 6 = 6(? + ?)$

28. $5\sqrt{5} - \sqrt{25} = 5(? - ?)$

29. $4 + \sqrt{16y} = 4(? + ?)$

30. $3 + \sqrt{18x} = 3(? + ?)$

31. $\sqrt{2} - \sqrt{6} = \sqrt{2}(? - ?)$

32. $\sqrt{12} - 2\sqrt{6} = 2\sqrt{3}(? - ?)$

■ *Reduce each fraction to lowest terms.*

Examples

a. $\dfrac{4 + 6\sqrt{3}}{2}$

b. $\dfrac{2x - \sqrt{8x^2}}{4x}$

c. $\dfrac{\sqrt{6} - \sqrt{8}}{\sqrt{2}}$

Solutions

a. $\dfrac{4 + 6\sqrt{3}}{2}$

$= \dfrac{2(2 + 3\sqrt{3})}{2}$

$= 2 + 3\sqrt{3}$

b. $\dfrac{2x - \sqrt{8x^2}}{4x}$

$= \dfrac{2x - 2x\sqrt{2}}{4x}$

$= \dfrac{2x(1 - \sqrt{2})}{2 \cdot 2x}$

$= \dfrac{1 - \sqrt{2}}{2}$

c. $\dfrac{\sqrt{6} - \sqrt{8}}{\sqrt{2}}$

$= \dfrac{\sqrt{2}\sqrt{3} - 2\sqrt{2}}{\sqrt{2}}$

$= \dfrac{\sqrt{2}(\sqrt{3} - 2)}{\sqrt{2}}$

$= \sqrt{3} - 2$

33. $\dfrac{2 + 2\sqrt{3}}{2}$

34. $\dfrac{6 + 2\sqrt{5}}{2}$

35. $\dfrac{6 + 2\sqrt{18}}{6}$

36. $\dfrac{8 - 2\sqrt{12}}{4}$

37. $\dfrac{x - \sqrt{x^3}}{x}$

38. $\dfrac{xy - x\sqrt{xy^2}}{xy}$

39. $\dfrac{x\sqrt{y} - \sqrt{y^3}}{\sqrt{y}}$

40. $\dfrac{\sqrt{x} - y\sqrt{x^3}}{\sqrt{x}}$

■ *Rationalize denominators.*

Examples   **a.** $\dfrac{3}{\sqrt{2}-1}$ **b.** $\dfrac{1}{\sqrt{x}-\sqrt{y}}$

Solutions   **a.** $\dfrac{3}{\sqrt{2}-1}=\dfrac{3(\sqrt{2}+1)}{(\sqrt{2}-1)(\sqrt{2}+1)}$ **b.** $\dfrac{1}{\sqrt{x}-\sqrt{y}}=\dfrac{1(\sqrt{x}+\sqrt{y})}{(\sqrt{x}-\sqrt{y})(\sqrt{x}+\sqrt{y})}$

$\qquad\qquad =\dfrac{3\sqrt{2}+3}{2-1}$ $\qquad\qquad =\dfrac{\sqrt{x}+\sqrt{y}}{x-y}$

$\qquad\qquad =3\sqrt{2}+3$

41. $\dfrac{4}{1+\sqrt{3}}$ **42.** $\dfrac{1}{2-\sqrt{2}}$ **43.** $\dfrac{2}{\sqrt{7}-2}$ **44.** $\dfrac{2}{4-\sqrt{5}}$

45. $\dfrac{x}{\sqrt{x}-3}$ **46.** $\dfrac{y}{\sqrt{3}-y}$ **47.** $\dfrac{\sqrt{6}-3}{2-\sqrt{6}}$ **48.** $\dfrac{\sqrt{x}+\sqrt{y}}{\sqrt{x}-\sqrt{y}}$

■ *Write each expression as a single fraction in which the denominator is rationalized.*

49. $\dfrac{1}{\sqrt{2}}+\dfrac{1}{\sqrt{3}}$ **50.** $\dfrac{3}{\sqrt{6}}-\dfrac{2}{\sqrt{3}}$ **51.** $\sqrt{3}-\dfrac{2}{\sqrt{3}}$

52. $\sqrt{5}+\dfrac{1}{\sqrt{5}}$ **53.** $\sqrt{x}+\dfrac{2x}{\sqrt{x}}$ **54.** $\dfrac{3}{\sqrt{2x}}-\dfrac{1}{\sqrt{x}}$

55. $\sqrt{x+1}-\dfrac{x}{\sqrt{x+1}}$ **56.** $\sqrt{x^2-2}-\dfrac{x^2+1}{\sqrt{x^2-2}}$

57. $\dfrac{x}{\sqrt{x^2+1}}-\dfrac{\sqrt{x^2+1}}{x}$ **58.** $\dfrac{x}{\sqrt{x^2-1}}+\dfrac{\sqrt{x^2-1}}{x}$

■ *Rationalize numerators.*

Examples   **a.** $\dfrac{\sqrt{3}-2}{2}$ **b.** $\dfrac{\sqrt{x}-2}{x}$

Solutions   **a.** $\dfrac{\sqrt{3}-2}{2}=\dfrac{(\sqrt{3}-2)(\sqrt{3}+2)}{2(\sqrt{3}+2)}$ **b.** $\dfrac{\sqrt{x}-2}{x}=\dfrac{(\sqrt{x}-2)(\sqrt{x}+2)}{x(\sqrt{x}+2)}$

$\qquad\qquad =\dfrac{3-4}{2\sqrt{3}+4}$ $\qquad\qquad =\dfrac{x-4}{x\sqrt{x}+2x}$

$\qquad\qquad =\dfrac{-1}{2\sqrt{3}+4}$

59. $\dfrac{1-\sqrt{2}}{2}$ **60.** $\dfrac{\sqrt{3}+\sqrt{2}}{\sqrt{3}}$ **61.** $\dfrac{\sqrt{x}-1}{3}$

62. $\dfrac{4-\sqrt{2y}}{2}$ **63.** $\dfrac{\sqrt{x}-\sqrt{y}}{x}$ **64.** $\dfrac{2\sqrt{x}+\sqrt{y}}{\sqrt{xy}}$

CHAPTER SUMMARY

[5.1–5.2] For n a natural number,

$$a^n = a \cdot a \cdot a \cdot \cdots \cdot a \qquad (n \text{ factors});$$

for n an integer,

$$a^{-n} = \frac{1}{a^n} \qquad (a \neq 0);$$

also,

$$a^0 = 1 \qquad (a \neq 0).$$

The laws of exponents are determined by the definitions adopted for powers:

$$\text{I.} \quad a^m \cdot a^n = a^{m+n}$$

$$\text{II.} \quad (a^m)^n = a^{mn}$$

$$\text{III.} \quad (ab)^n = a^n b^n$$

$$\text{IV.} \quad \frac{a^m}{a^n} = a^{m-n} \qquad (a \neq 0)$$

$$\text{V.} \quad \left(\frac{a}{b}\right)^n = \frac{a^n}{b^n} \qquad (b \neq 0)$$

[5.3] A number is expressed in **scientific notation** when it is expressed as a product of a number between 1 and 10 and a power of 10.

[5.4] For each natural number n, the number $a^{1/n}$, when it exists, is called an **nth root of a**.

If n is an even natural number and $a \geq 0$, then $a^{1/n}$ is the nonnegative number such that

$$(a^{1/n})^n = a.$$

If n is an odd natural number, then $a^{1/n}$ is the number such that

$$(a^{1/n})^n = a.$$

For $a^{1/n}$ a real number,

$$a^{m/n} = (a^{1/n})^m = (a^m)^{1/n}.$$

[5.5] An nth root of a is also designated by a **radical expression** $\sqrt[n]{a}$, where n is the **index** and a is the **radicand**. For all natural numbers $n \geq 2$,

$$\sqrt[n]{a} = a^{1/n}.$$

For n an even natural number:

If $a > 0$, $\sqrt[n]{a}$ is the positive number such that

$$(\sqrt[n]{a})^n = a.$$

For $n = 2$,

$$(\sqrt{a})^2 = \sqrt{a}\sqrt{a} = a.$$

If $a < 0$,

$$\sqrt[n]{a^n} = |a|.$$

For n an odd natural number,

$$(\sqrt[n]{a})^n = a.$$

For $\sqrt[n]{a}$ a real number,

$$a^{m/n} = \sqrt[n]{a^m} = (\sqrt[n]{a})^m.$$

[5.6] Two laws of radicals follow from corresponding laws for exponents:

$$\sqrt[n]{ab} = \sqrt[n]{a}\sqrt[n]{b} \quad \text{and} \quad \sqrt[n]{\frac{a}{b}} = \frac{\sqrt[n]{a}}{\sqrt[n]{b}}.$$

A radical expression is in **simplest form** if:

1. The radicand contains no polynomial factor raised to a power equal to or greater than the index of the radical.

2. The radicand contains no fractions.

3. No radical expressions are contained in denominators of fractions.

[5.7] Expressions containing radicals can be rewritten using properties of real numbers. In particular, the fundamental principle of fractions can be used to rationalize the denominator (or the numerator) of a fraction containing radicals.

The symbols introduced in this chapter are listed on the inside of the front cover.

REVIEW EXERCISES

A ■ *Assume that all variables denote positive numbers.*

[5.1–5.2] ■ *Simplify. In Problems 1–18, assume that no denominator equals 0.*

1. a. $\dfrac{x^4 y^3}{xy}$ **b.** $(3x^2 y)^3$

2. a. $\dfrac{x^{-2} \cdot y^{-1}}{x}$ **b.** $\left(\dfrac{x^2 y^{-2}}{x^{-2} y}\right)^{-2}$

3. a. $3^{-2} + 2^{-1}$ **b.** $xy^{-1} + x^{-1} y$

4. a. $\dfrac{(2x+1)^{-1}}{(2x)^{-1} + 1^{-1}}$ **b.** $\dfrac{x^{-2} - y^{-2}}{(x-y)^{-2}}$

[5.3] ■ *Write each number in scientific notation.*

   5. a. 0.000000000023   **b.** 307,000,000,000

■ *Specify the number of significant digits in each number.*

   6. a. 0.0702   **b.** 0.4030

■ *Simplify.*

   7. a. $\dfrac{(2 \times 10^{-3})(3 \times 10^{4})}{6 \times 10^{-1}}$ **b.** $\dfrac{(4 \times 10^{3})(6 \times 10^{-4})}{3 \times 10^{-2}}$

[5.4] ■ *Simplify.*

   8. a. $(-27)^{2/3}$ **b.** $4^{-1/2}$

   9. a. $x^{3/2} \cdot x^{1/2}$ **b.** $\left(\dfrac{x^{2/3} y^{1/2}}{x^{1/3}}\right)^{6}$

■ *Multiply.*

   10. a. $x^{2/3}(x^{1/3} - x)$ **b.** $y^{-1/4}(y^{5/4} - y^{1/4})$

■ *Factor as indicated.*

   11. a. $x^{4/5} = x^{1/5}(?)$ **b.** $y^{-3/4} = y^{-1/2}(?)$

[5.5] ■ *Express in radical notation.*

   12. a. $(1 - x^2)^{2/3}$ **b.** $(1 - x^2)^{-2/3}$

■ *Express in exponential notation.*

13. a. $\sqrt[3]{x^2y}$ **b.** $\dfrac{1}{\sqrt[3]{(a+b)^2}}$

■ *Find each root indicated.*

14. a. $\sqrt{4y^2}$ **b.** $\sqrt[3]{-8x^3y^6}$

[5.6]   ■ *Simplify.*

15. a. $\sqrt{180}$ **b.** $\sqrt[4]{32x^4y^5}$

16. a. $\dfrac{x}{\sqrt{xy}}$ **b.** $\dfrac{\sqrt{xy}\sqrt{6x^3y}}{\sqrt{2xy}}$

17. a. $\sqrt[3]{\dfrac{1}{2}}$ **b.** $\sqrt[4]{\dfrac{2}{3x}}$

18. a. $\sqrt[3]{\dfrac{x^7y^5}{x^2y}}$ **b.** $\dfrac{\sqrt[3]{x^5y^4}}{\sqrt[3]{x^2y}}$

[5.7]   ■ *Simplify.*

19. a. $4\sqrt{12}+2\sqrt{75}$ **b.** $3\sqrt{2x}-\sqrt{32x}+4\sqrt{50x}$
20. a. $9\sqrt{2xy^2}-3y\sqrt{8x}$ **b.** $\sqrt{75x^3}-x\sqrt{3x}$

■ *Write each expression without parentheses and then write all radicals in simple form.*

21. a. $\sqrt{3}(\sqrt{6}-2)$ **b.** $\sqrt{5}(\sqrt{10}-\sqrt{5})$
22. a. $(2-\sqrt{3})(3-2\sqrt{3})$ **b.** $(\sqrt{x}+2)(\sqrt{x}-2)$

■ *Rationalize each denominator.*

23. a. $\dfrac{4}{2-\sqrt{3}}$ **b.** $\dfrac{2}{\sqrt{5}+2}$

24. a. $\dfrac{y}{\sqrt{y}-3}$ **b.** $\dfrac{x-y}{\sqrt{x}+\sqrt{y}}$

B   ■ *Simplify.*

25. $\left(\dfrac{x^{2n}y^{2n-1}}{y^n}\right)^2$

26. $\left(\dfrac{x^{3n}y^{1-n}}{x^{4n-1}}\right)^{-1}$

27. Compute: $\dfrac{0.0016 \times 0.00012 \times 270}{0.08 \times 0.00004 \times 81}$.

28. Factor as indicated: $x^{3/4} - x^{1/4} = x(? - ?)$.

29. Write $\sqrt{4x^2 - 8x + 4}$ $(x \in R)$ without radical notation.

30. Simplify $\sqrt[6]{16}$.

6. SECOND-DEGREE EQUATIONS AND INEQUALITIES

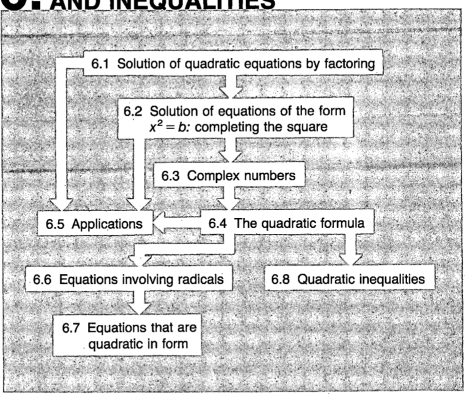

6.1 Solution of quadratic equations by factoring

6.2 Solution of equations of the form $x^2 = b$: completing the square

6.3 Complex numbers

6.5 Applications

6.4 The quadratic formula

6.6 Equations involving radicals

6.8 Quadratic inequalities

6.7 Equations that are quadratic in form

6.1

SOLUTION OF QUADRATIC EQUATIONS BY FACTORING

Standard form

A second-degree equation in one variable is called a **quadratic equation** in that variable. We shall designate as **standard form** for such equations

$$ax^2 + bx + c = 0,$$

where a, b, and c are constants representing real numbers and $a \neq 0$. If $b = 0$ or $c = 0$, then the equations are of the form

$$ax^2 + c = 0 \quad \text{or} \quad ax^2 + bx = 0$$

and are called **incomplete quadratic equations**.

Writing equivalent equations

In Chapter 4, we solved first-degree equations by performing certain elementary transformations. These transformations are equally applicable to equations of higher degree and, in particular, to quadratic equations. For example, adding $8x^2 - 6$ to each member of

$$2x = 6 - 8x^2,$$

we obtain

$$8x^2 + 2x - 6 = 0.$$

Then, dividing each member by 2, we have

$$4x^2 + x - 3 = 0.$$

The three equations are equivalent.

Solution by factoring

If the left-hand member of a quadratic equation in standard form is factorable, we may solve the equation by making use of the following principle:

▶ *The product of two factors equals 0 if and only if one or both of the factors equals 0.*

Thus:

$$ab = 0 \quad \textit{if and only if} \quad a = 0 \quad \textit{or} \quad b = 0.$$

Example

$(x - 1)(x + 2) = 0$ if and only if

$$x - 1 = 0 \qquad (x = 1)$$

or

$$x + 2 = 0 \qquad (x = -2).$$

The word "or" in the above statement and in the example is used in an inclusive sense to mean either one *or* the other *or* both.

The above principle enables us to solve quadratic equations if the left-hand member of an equation in standard form is factorable.

Example

Solve $x^2 + 2x - 15 = 0$. 　　　　　　　　　　　　　　　　　　　　 (1)

Solution

We first factor the left-hand member to obtain

$$(x + 5)(x - 3) = 0, \tag{2}$$

which will be true if and only if

$$x + 5 = 0 \qquad \text{or} \qquad x - 3 = 0$$
$$x = -5 \qquad\qquad\qquad x = 3.$$

Solution continued overleaf

Thus, either -5 or 3, when substituted for x in (2) or (1), will make the left-hand member 0. The solution set is $\{-5, 3\}$.

In general, the solution set of a quadratic equation can be expected to contain two elements. However, if the left-hand member of a quadratic equation in standard form is the square of a binomial, we find that the solution set contains only one member.

Example   Solve $x^2 - 2x + 1 = 0$.

Solution   Factoring the left-hand member, we have

$$(x - 1)(x - 1) = 0,$$

and the solution set is $\{1\}$, which contains only one element.

For reasons of convenience and consistency in more advanced work, the unique solution obtained in the above example is said to be of **multiplicity two**.

Writing equations from given solutions

Notice that the solution set of the quadratic equation

$$(x - r_1)(x - r_2) = 0 \tag{3}$$

is $\{r_1, r_2\}$*. Therefore, if r_1 and r_2 are given as solutions of a quadratic equation, the equation can be written directly as (3). By completing the indicated multiplication, the equation can be transformed to standard form.

Examples   **a.** If 2 and -3 are given as solutions of a quadratic equation, then

$$[x - (2)][x - (-3)] = 0.$$

That is,

$$(x - 2)(x + 3) = 0,$$

from which

$$x^2 + x - 6 = 0.$$

b. If ¼ and 3/2 are given as solutions of a quadratic equation, then

$$\left(x - \frac{1}{4}\right)\left(x - \frac{3}{2}\right) = 0,$$

from which

$$x^2 - \frac{1}{4}x - \frac{3}{2}x + \frac{3}{8} = 0$$

or

$$x^2 - \frac{7}{4}x + \frac{3}{8} = 0.$$

*The subscripts 1 and 2 used in r_1 and r_2 in this equation are used to identify constants. The symbols are read "r sub one" and "r sub two" or simply "r one" and "r two."

This equation can be transformed to one with integral coefficients by multiplying each member by 8 to obtain the equivalent equation

$$8x^2 - 14x + 3 = 0.$$

EXERCISE 6.1

A ■ *Solve. Determine your answer by inspection or follow the procedure in the examples.*

Example $x(x + 3) = 0$

Solution   Set each factor equal to 0 and solve the equations.

$$x = 0 \quad \text{or} \quad x + 3 = 0$$
$$x = -3$$

The solution set is $\{0, -3\}$.

1. $(x + 2)(x - 5) = 0$ 2. $(x + 3)(x - 4) = 0$ 3. $(2x + 5)(x - 2) = 0$
4. $(x + 1)(3x - 1) = 0$ 5. $x(2x + 1) = 0$ 6. $x(3x - 7) = 0$
7. $4(x - 6)(2x + 3) = 0$ 8. $5(2x - 7)(x + 1) = 0$ 9. $3(x - 2)(2x + 1) = 0$
10. $7(x + 5)(3x - 1) = 0$ 11. $4(2x - 5)(3x + 2) = 0$ 12. $2(3x - 4)(2x + 5) = 0$

Example $x^2 + x = 30$

Solution   Write in standard form (the right-hand member should be 0).

$$x^2 + x - 30 = 0$$

Factor the left-hand member.

$$(x + 6)(x - 5) = 0$$

Set each factor equal to 0 and solve the equations.

$$x + 6 = 0 \quad \text{or} \quad x - 5 = 0$$
$$x = -6 \qquad x = 5$$

The solution set is $\{-6, 5\}$.

13. $x^2 - 3x = 0$ 14. $x^2 + 5x = 0$ 15. $2x^2 = 6x$
16. $3x^2 = 3x$ 17. $x^2 - 9 = 0$ 18. $x^2 - 4 = 0$

19. $2x^2 - 18 = 0$ **20.** $3x^2 - 3 = 0$ **21.** $3x - \dfrac{4}{3x} = 0$

22. $5x - \dfrac{4}{5x} = 0$ **23.** $\dfrac{4}{9}x^2 - 1 = 0$ **24.** $\dfrac{9x^2}{25} - 1 = 0$

25. $x^2 - 5x + 4 = 0$ **26.** $x^2 + 5x + 6 = 0$ **27.** $x^2 - 5x - 14 = 0$

28. $x^2 - x - 42 = 0$ **29.** $3x^2 - 6x = -3$ **30.** $12x^2 = 8x + 15$

Example $3x(x + 1) = 2x + 2$

Solution   Write in standard form.

$$3x^2 + x - 2 = 0$$

Factor the left-hand member.

$$(3x - 2)(x + 1) = 0$$

Set each factor equal to 0 and solve the equations.

$$3x - 2 = 0 \quad \text{or} \quad x + 1 = 0$$

$$x = \frac{2}{3} \qquad\qquad x = -1$$

The solution set is $\{\tfrac{2}{3}, -1\}$.

31. $x(2x - 3) = -1$ **32.** $2x(x - 2) = x + 3$ **33.** $(x - 2)(x + 1) = 4$

34. $x(3x + 2) = (x + 2)^2$ **35.** $(x - 1)^2 = 2x^2 + 3x - 5$ **36.** $x(x + 1) = 4 - (x + 2)^2$

37. $t(t + 4) - 1 = 4$ **38.** $z(z + 5) + 18 = 4(1 - z)$ **39.** $2z(z + 3) = 3 + z$

40. $(n - 3)(n + 2) = 6$ **41.** $(t + 2)(t - 5) = 8$ **42.** $(x + 1)(2x - 3) = 3$

Example $x^2 - \dfrac{17}{3}x = 2$

Solution   Write in standard form.

$$x^2 - \frac{17}{3}x - 2 = 0$$

Multiply each member by 3 and simplify.

$$3(x^2) - 3\left(\frac{17}{3}x\right) - 3(2) = 3(0)$$
$$3x^2 - 17x - 6 = 0$$

Factor the left-hand member.

$$(3x + 1)(x - 6) = 0$$

Set each factor equal to 0 and solve the equations.

$$3x + 1 = 0 \quad \text{or} \quad x - 6 = 0$$

$$x = -\frac{1}{3} \qquad x = 6$$

The solution set is $\{-\frac{1}{3}, 6\}$.

43. $\dfrac{2x^2}{3} + \dfrac{x}{3} - 2 = 0$

44. $2x - \dfrac{5}{3} = \dfrac{x^2}{3}$

45. $\dfrac{x^2}{6} + \dfrac{x}{3} = \dfrac{1}{2}$

46. $\dfrac{x}{4} - \dfrac{3}{4} = \dfrac{1}{x}$

47. $3 = \dfrac{10}{x^2} - \dfrac{7}{x}$

48. $\dfrac{4}{3x} + \dfrac{3}{3x + 1} + 2 = 0$

■ *Given the solutions of a quadratic equation, r_1 and r_2, write the equation in standard form with integral coefficients.*

Example   ¾ and -2

Solution   Write in the form $(x - r_1)(x - r_2) = 0$.

$$\left(x - \frac{3}{4}\right)[x - (-2)] = 0$$

Write in standard form.

$$\left(x - \frac{3}{4}\right)(x + 2) = 0$$

$$x^2 + \frac{5}{4}x - \frac{3}{2} = 0$$

$$4x^2 + 5x - 6 = 0$$

49. -2 and 1 **50.** -4 and 3 **51.** 0 and -5 **52.** 0 and 5

53. 4 and $-\frac{3}{4}$ **54.** $-\frac{2}{3}$ and 3 **55.** $\frac{1}{2}$ and $\frac{3}{5}$ **56.** $\frac{2}{3}$ and $\frac{1}{5}$

6.2

SOLUTION OF EQUATIONS OF THE FORM $x^2 = b$; COMPLETING THE SQUARE

Quadratic equations of the form

$$x^2 = b \qquad (b \geq 0)$$

may be solved by a method often termed the **extraction of roots**. If the equation has a solution, then from the definition of a square root, x must be a square root of b. Since each positive number b has two square roots, we have two solutions if $b > 0$. These are given by

$$x = \sqrt{b} \quad \text{and} \quad x = -\sqrt{b};$$

the solution set is $\{\sqrt{b}, -\sqrt{b}\}$. If $b = 0$, we have one number, 0, that satisfies the equation, and the solution set of $x^2 = 0$ is $\{0\}$.

Examples   Solve

a. $2x^2 - 6 = 0$ **b.** $5x^2 = 0$

Solutions

a. $2x^2 - 6 = 0$ **b.** $5x^2 = 0$
$\qquad 2x^2 = 6$ $\qquad x^2 = 0$
$\qquad\quad x^2 = 3$ $\qquad\quad x = 0$
$\quad x = \sqrt{3} \quad \text{or} \quad x = -\sqrt{3}$ The solution set is $\{0\}$.

The solution set is $\{\sqrt{3}, -\sqrt{3}\}$.

Equations of the form

$$(x - p)^2 = q \qquad (q \geq 0)$$

may also be solved by the method of extraction of roots.

Example   Solve $(x - 2)^2 = 16$.

Solution   The equation

$$(x - 2)^2 = 16$$

implies that $(x - 2)$ is a number whose square is 16. Hence,

$$x - 2 = 4 \quad \text{or} \quad x - 2 = -4,$$

from which we have

$$x = 6 \quad \text{or} \qquad x = -2.$$

The solution set is $\{6, -2\}$.

Solution by completing the square   The method of extraction of roots can be used to find the solution set of any quadratic equation. Let us first consider a specific example,

$$x^2 - 4x - 12 = 0,$$

which can be written

$$x^2 - 4x \quad\;\; = 12.$$

If the square of one-half of the coefficient of the first-degree term,

$$\left[\frac{1}{2}(-4)\right]^2,$$

equal to 4 is added to each member, we obtain

$$x^2 - 4x + 4 = 12 + 4,$$

in which the left-hand member is the square of $(x - 2)$. Therefore, the equation can be written

$$(x - 2)^2 = 16,$$

and the solution set is obtained as above.

Now consider the general quadratic equation in standard form,

$$ax^2 + bx + c = 0,$$

for the special case where $a = 1$; that is,

$$x^2 + bx + c = 0. \qquad (1)$$

If we can factor the left-hand member of (1), we can solve the equation by factoring; if not, we can write it in the form

$$(x - p)^2 = q \qquad (q \geq 0),$$

which we can solve by the extraction of roots. We begin the latter process by adding $-c$ to each member of (1), which yields

$$x^2 + bx \qquad = -c. \qquad (2)$$

If we then add the square of one-half of the coefficient of x, $(b/2)^2$, to each member of (2), the result is

$$x^2 + bx + \left(\frac{b}{2}\right)^2 = -c + \left(\frac{b}{2}\right)^2, \qquad (3)$$

where the left-hand member is equivalent to $(x + b/2)^2$, and we have

$$\left(x + \frac{b}{2}\right)^2 = -c + \frac{b^2}{4}. \qquad (4)$$

Since we have performed only elementary transformations, (4) is equivalent to (2) and we can solve (4) by the method used in the example above.

The technique used to obtain equations (3) and (4) is called **completing the square**. We can determine the term necessary to complete the square in (2) by dividing the coefficient b of the linear term by 2 and squaring the result. The expression obtained,

$$x^2 + bx + \left(\frac{b}{2}\right)^2,$$

can then be written in the form $(x + b/2)^2$.

Example   Solve $x^2 - 3x - 1 = 0$.

Solution   First rewrite the equation with the constant term as the right-hand member.

$$x^2 - 3x \qquad = 1$$

Add$(-\frac{3}{2})^2$, the square of one-half of the coefficient of the first-degree term, to each member.

$$x^2 - 3x + \left(-\frac{3}{2}\right)^2 = 1 + \left(-\frac{3}{2}\right)^2$$

Rewrite the left-hand member as the square of a binomial and simplify the right-hand member.

$$\left(x - \frac{3}{2}\right)^2 = \frac{13}{4}$$

Set $(x - \frac{3}{2})$ equal to each square root of $\frac{13}{4}$.

$$x - \frac{3}{2} = \sqrt{\frac{13}{4}} \qquad \text{or} \qquad x - \frac{3}{2} = -\sqrt{\frac{13}{4}}$$

$$x = \frac{3}{2} + \frac{\sqrt{13}}{2} \qquad\qquad\qquad x = \frac{3}{2} - \frac{\sqrt{13}}{2}$$

The solution set is $\left\{ \dfrac{3 + \sqrt{13}}{2}, \dfrac{3 - \sqrt{13}}{2} \right\}$.

We began with the special case,

$$x^2 + bx + c = 0, \tag{5}$$

rather than the general form,

$$ax^2 + bx + c = 0,$$

because the term necessary to complete the square is more obvious when $a = 1$. However, a quadratic equation in standard form can always be written in the form (5) by multiplying each member of

$$ax^2 + bx + c = 0$$

by $\frac{1}{a}$ $(a \neq 0)$ and obtaining

$$x^2 + \frac{b}{a}x + \frac{c}{a} = 0.$$

EXERCISE 6.2

A ■ *Solve for x by the extraction of roots.*

Example $7x^2 - 63 = 0$

Solution Obtain an equivalent equation with x^2 as the only term in the left-hand member.
$$x^2 = 9$$

Set x equal to each square root of 9.
$$x = 3 \quad \text{or} \quad x = -3$$

The solution set is $\{3, -3\}$.

1. $x^2 = 100$ **2.** $x^2 = 16$ **3.** $9x^2 = 25$ **4.** $4x^2 = 9$

5. $2x^2 = 14$ **6.** $3x^2 = 15$ **7.** $4x^2 - 24 = 0$ **8.** $3x^2 - 9 = 0$

9. $\dfrac{2x^2}{3} = 4$ **10.** $\dfrac{3x^2}{5} = 3$ **11.** $\dfrac{4x^2}{3} = 27$ **12.** $\dfrac{9x^2}{2} = 50$

Example $(x + 3)^2 = 7$

Solution Set $x + 3$ equal to each square root of 7.
$$x + 3 = \sqrt{7} \qquad \text{or} \quad x + 3 = -\sqrt{7}$$
$$x = -3 + \sqrt{7} \qquad\qquad x = -3 - \sqrt{7}$$
The solution set is $\{-3 + \sqrt{7}, -3 - \sqrt{7}\}$.

13. $(x - 2)^2 = 9$ **14.** $(x + 3)^2 = 4$ **15.** $(2x - 1)^2 = 16$

16. $(3x + 1)^2 = 25$ **17.** $(x + 2)^2 = 3$ **18.** $(x - 5)^2 = 7$

19. $(x - 2)^2 = 12$ **20.** $(x + 3)^2 = 18$ **21.** $(x - 7)^2 = 4$

22. $(x + 3)^2 = 36$ **23.** $(2x - 5)^2 = 9$ **24.** $(3x + 4)^2 = 16$

25. $(3x + 5)^2 = 9$ **26.** $(2x + 3)^2 = 36$ **27.** $(7x - 1)^2 = 15$

28. $(5x + 3)^2 = 7$ **29.** $(8x - 7)^2 = 8$ **30.** $(5x - 12)^2 = 24$

■ *Solve by completing the square.*

Example $2x^2 + x - 1 = 0$

Solution overleaf

Solution   Rewrite the equation with the constant term as the right-hand member and the coefficient of x^2 equal to 1.

$$x^2 + \frac{1}{2}x \qquad = \frac{1}{2}$$

Add $[\frac{1}{2}(\frac{1}{2})]^2$, the square of one-half of the coefficient of the first-degree term, to each member.

$$x^2 + \frac{1}{2}x + \frac{1}{16} = \frac{1}{2} + \frac{1}{16}$$

Rewrite the left-hand member as the square of a binomial.

$$\left(x + \frac{1}{4}\right)^2 = \frac{9}{16}$$

Set $(x + \frac{1}{4})$ equal to each square root of $\frac{9}{16}$.

$$x + \frac{1}{4} = \frac{3}{4} \quad \text{or} \quad x + \frac{1}{4} = -\frac{3}{4}$$

$$x = \frac{1}{2} \qquad\qquad x = -1$$

The solution set is $\{\frac{1}{2}, -1\}$.

31. $x^2 + 4x - 12 = 0$	**32.** $x^2 - x - 6 = 0$	**33.** $x^2 - 2x + 1 = 0$
34. $x^2 + 4x + 4 = 0$	**35.** $x^2 + 9x + 20 = 0$	**36.** $x^2 - x - 20 = 0$
37. $x^2 - 2x - 1 = 0$.	**38.** $x^2 + 3x - 1 = 0$	**39.** $x^2 = 3 - 3x$
40. $x^2 = 5 - 5x$	**41.** $2x^2 + 4x - 3 = 0$	**42.** $3x^2 + x - 4 = 0$
43. $2x^2 - 5 = 3x$	**44.** $4x^2 - 3 = 2x$	**45.** $6x^2 - x - 3 = 0$
46. $2x^2 - 5x = 8$	**47.** $5x^2 - 3 = 2x$	**48.** $3x^2 - 7x = 3$

B   ■ *Solve for x in terms of a, b, and c.*

49. $x^2 - a = 0$	**50.** $x^2 - 2a = 0$	**51.** $\dfrac{ax^2}{b} = c$	**52.** $\dfrac{bx^2}{c} - a = 0$
53. $(x - a)^2 = 16$	**54.** $(x + a)^2 = 36$	**55.** $(ax + b)^2 = 9$	**56.** $(ax - b)^2 = 25$

57. Solve $a^2 + b^2 = c^2$ for b in terms of a and c.

58. Solve $S = h - r^2$ for r in terms of S and h.

59. Solve $A = P(1 + r)^2$ for r in terms of A and P.

60. Solve $s = \frac{1}{2}gt^2 + c$ for t in terms of s, g, and c.

61. Solve $x^2 + 9y^2 = 9$ for y in terms of x.

62. Solve $9x^2 + 4y^2 = 36$ for y in terms of x.

63. Solve $4x^2 - 9y^2 = 36$ for y in terms of x.

64. Solve $9x^2 - 25y^2 = 0$ for y in terms of x.

65. Solve $ax^2 + bx + c = 0$ for x in terms of a, b, and c by using the method of completing the square.

6.3

COMPLEX NUMBERS

In Section 6.2 we considered only quadratic equations with solutions in the set of real numbers. Some quadratic equations do not have real solutions. For example, if $b > 0$, then $x^2 = -b$ has no real-number solution, because there is no real number whose square is negative. For this reason, the expression $\sqrt{-b}$, for $b \in R, b > 0$, is undefined in the set of real numbers. In this section, we wish to consider a set of numbers that contains members whose squares are negative real numbers and that also contains members that are real numbers. We shall see that this new set of numbers, called the set C of **complex numbers**, provides solutions for all quadratic equations in one variable with real coefficients.

Imaginary numbers

Let us first assume a set of numbers among whose members are square roots of negative real numbers. We define $\sqrt{-b}$ and $-\sqrt{-b}$, where b is a positive real number, to be numbers whose squares are equal to $-b$. Thus:

▶ *For $b > 0$,*

$$(\sqrt{-b})^2 = -b \quad and \quad (-\sqrt{-b})^2 = -b.$$

In particular,

$$(\sqrt{-1})^2 = -1 \quad and \quad (-\sqrt{-1})^2 = -1.$$

It is customary to use the symbol i for $\sqrt{-1}$. With this convention, and assuming that $-i = -1 \cdot i$,

$$i = \sqrt{-1}, \quad -i = -\sqrt{-1}, \quad i^2 = -1, \quad and \quad (-i)^2 = -1.$$

Thus, in this new set of numbers, -1 has two square roots, i and $-i$. Next, assuming that $-\sqrt{-b} = -1 \cdot \sqrt{-b}$ and that

$$(i\sqrt{b})(i\sqrt{b}) = i^2 b = -1 \cdot b = -b,$$

it follows from the definition of $\sqrt{-b}$ that:

▶ $$\sqrt{-b} = \sqrt{-1}\sqrt{b} = i\sqrt{b} \quad and \quad -\sqrt{-b} = -1 \cdot \sqrt{-1}\sqrt{b} = -i\sqrt{b}.$$

Hence, a square root of any negative real number can be represented as the product of a real number and the number $\sqrt{-1}$ or i.

Examples   **a.** $\sqrt{-4} = \sqrt{-1}\sqrt{4}$ **b.** $-\sqrt{-3} = -\sqrt{-1}\sqrt{3}$
$= i\sqrt{4} = 2i$ $= -i\sqrt{3}$

The numbers represented by the symbols $\sqrt{-b}$ and $-\sqrt{-b}$, where b is a real number greater than zero, are called **pure imaginary numbers**.

Now, consider all possible expressions of the form $a + bi$, where a, $b \in R$ and $i = \sqrt{-1}$, which are the sums of all real numbers and all pure imaginary numbers. Such an expression names a complex number, that is, a number in the set C. If $b = 0$, then $a + bi = a$, and it is evident that the set R of real numbers is contained in the set C of complex numbers. If $b \neq 0$, then $a + bi$ is called an **imaginary number** (see Figure 6.1), where a is the **real part** of the number and b is the **imaginary part**. For example, the numbers -7, $3 + 2i$, and $4i$ are all complex numbers. However, -7 is also a real number, and $3 + 2i$ and $4i$ are also imaginary numbers. Furthermore, $4i$ is a pure imaginary number.

Complex numbers: $C = \{a + bi \,|\, a, b \in R\}$

$(b = 0)$ $(b \neq 0)$
Real numbers: $a + bi = a$*   Imaginary numbers: $a + bi$

$(a = 0)$
Pure imaginary numbers: $a + bi = bi$

Figure 6.1

Sums and   To add or subtract complex numbers, we simply add or subtract their real parts and their
differences   imaginary parts.

Examples   **a.** $(2 + 3i) + (5 - 4i)$ **b.** $(2 + 3i) - (5 - 4i)$
$= (2 + 5) + (3 - 4)i$ $= (2 - 5) + [3 - (-4)]i$
$= 7 - i$ $= -3 + 7i$

In general:

▶

$$(a + bi) + (c + di) = (a + c) + (b + d)i$$

and

$$(a + bi) - (c + di) = (a - c) + (b - d)i.$$

*See Figure 1.1, page 3, for sets of numbers that are contained in the set of real numbers.

Products

To multiply complex numbers, we treat them as though they were binomials and replace i^2 with -1.

Examples

a. $(2-i)(1+3i)$

$= 2 + 6i - i - 3i^2$

$= 2 + 6i - i - 3(-1)$

$= 2 + 6i - i + 3$

$= 5 + 5i$

b. $(3-i)^2 = (3-i)(3-i)$

$= 9 - 3i - 3i + i^2$

$= 9 - 6i + (-1)$

$= 8 - 6i$

More formally, we express $(a+bi)(c+di)$ as follows:

$$(a+bi)(c+di) = (ac-bd) + (ad+bc)i.$$

Ordinarily, we compute such products by multiplying, as we did in the above examples, rather than from the definition.

Radical notation

The symbol $\sqrt{-b}$ $(b>0)$ should be used with care, since certain relationships involving the square root symbol that are valid for real numbers are not valid when the symbol does not represent a real number. For instance,

$$\sqrt{-2}\sqrt{-3} = (i\sqrt{2})(i\sqrt{3}) = i^2\sqrt{6} = -\sqrt{6};$$
$$\sqrt{-2}\sqrt{-3} \neq \sqrt{(-2)(-3)} = \sqrt{6}.$$

To avoid difficulty with this point:

Rewrite all expressions of the form $\sqrt{-b}$ $(b>0)$ in the form $i\sqrt{b}$ before performing any computations.

Examples

a. $\sqrt{-2}(3-\sqrt{-5})$

$= i\sqrt{2}(3-i\sqrt{5})$

$= 3i\sqrt{2} - i^2\sqrt{10}$

$= 3i\sqrt{2} - (-1)\sqrt{10}$

$= \sqrt{10} + 3i\sqrt{2}$

b. $(2+\sqrt{-3})(2-\sqrt{-3})$

$= (2+i\sqrt{3})(2-i\sqrt{3})$

$= 4 - 3i^2$

$= 4 - 3(-1)$

$= 7$

Quotients

The quotient of two complex numbers can be found by using the following property, which is analogous to the *fundamental principle of fractions* in the set of real numbers. First recall from Section 5.7 that, for $b>0$, the *conjugate* of $a+\sqrt{b}$ is $a-\sqrt{b}$. Similarly, the conjugate of $a+\sqrt{-b}$ is $a-\sqrt{-b}$.

Examples

a. The conjugate of $2+3i$ is $2-3i$.

b. The conjugate of $-3-i$ is $-3+i$.

c. The conjugate of $2i$ is $-2i$.

d. The conjugate of $-4+i$ is $-4-i$.

The quotient

$$\frac{a + bi}{c + di}$$

of two complex numbers can be simplified by multiplying the numerator and the denominator by $c - di$, the conjugate of the denominator. That is,

$$\frac{a + bi}{c + di} = \frac{(a + bi)(c - di)}{(c + di)(c - di)}.$$

Examples

a. $\dfrac{4 + i}{2 + 3i} = \dfrac{(4 + i)(2 - 3i)}{(2 + 3i)(2 - 3i)}$

$= \dfrac{8 - 10i - 3i^2}{4 - 9i^2}$

$= \dfrac{8 - 10i + 3}{4 + 9}$

$= \dfrac{11}{13} - \dfrac{10}{13}i$

b. $\dfrac{3}{1 + \sqrt{-2}} = \dfrac{3}{1 + i\sqrt{2}}$

$= \dfrac{3(1 - i\sqrt{2})}{(1 + i\sqrt{2})(1 - i\sqrt{2})}$

$= \dfrac{3(1 - i\sqrt{2})}{1 - i^2(2)}$

$= \dfrac{3(1 - i\sqrt{2})}{3} = 1 - i\sqrt{2}$

Imaginary solutions

As noted at the beginning of this section, the set of complex numbers provides solutions for all quadratic equations in one variable with real coefficients.

Examples

a. $x^2 + 4 = 0$

b. $x^2 + 4x + 6 = 0$

Solutions

a. $x^2 = -4$

$x = \sqrt{-4}$ or $x = -\sqrt{-4}$

$= 2i$ $\qquad = -2i$

Hence, the solution set is $\{2i, -2i\}$.

b. $x^2 + 4x \quad = -6$

$x^2 + 4x + 4 = -6 + 4$

$(x + 2)^2 = -2$

$x + 2 = \sqrt{-2}$ or $x + 2 = -\sqrt{-2}$

$x = -2 + i\sqrt{2}$ or $x = -2 - i\sqrt{2}$

Hence, the solution set is $\{-2 + i\sqrt{2}, -2 - i\sqrt{2}\}$.

EXERCISE 6.3

A ■ *Write each expression in the form $a + bi$ or $a + ib$.*

Examples

a. $3\sqrt{-18}$

b. $2 - 3\sqrt{-16}$

Solutions

a. $3\sqrt{-18} = 3\sqrt{-1 \cdot 9 \cdot 2}$
$\quad\quad\quad = 3\sqrt{-1}\sqrt{9}\sqrt{2}$
$\quad\quad\quad = 3i(3)\sqrt{2}$
$\quad\quad\quad = 9i\sqrt{2}$

b. $2 - 3\sqrt{-16} = 2 - 3\sqrt{-1 \cdot 16}$
$\quad\quad\quad\quad = 2 - 3\sqrt{-1}\sqrt{16}$
$\quad\quad\quad\quad = 2 - 3i(4)$
$\quad\quad\quad\quad = 2 - 12i$

1. $\sqrt{-4}$ 2. $\sqrt{-9}$ 3. $\sqrt{-32}$
4. $\sqrt{-50}$ 5. $3\sqrt{-8}$ 6. $4\sqrt{-18}$
7. $3\sqrt{-24}$ 8. $2\sqrt{-40}$ 9. $5\sqrt{-64}$
10. $7\sqrt{-81}$ 11. $-2\sqrt{-12}$ 12. $-3\sqrt{-75}$
13. $4 + 2\sqrt{-1}$ 14. $5 - 3\sqrt{-1}$ 15. $3\sqrt{-50} + 2$
16. $5\sqrt{-12} - 1$ 17. $\sqrt{4} + \sqrt{-4}$ 18. $\sqrt{20} - \sqrt{-20}$

Examples

a. $(3 - 2i) + (5 + i)$

b. $(2 + 3i) - (4 - i)$

Solutions

a. $(3 - 2i) + (5 + i)$
$\quad = (3 + 5) + (-2 + 1)i$
$\quad = 8 - i$

b. $2 + 3i + (-4 + i)$
$\quad = [2 + (-4)] + (3 + 1)i$
$\quad = -2 + 4i$

19. $(2 + 4i) + (3 + i)$ 20. $(2 - i) + (3 - 2i)$ 21. $(4 - i) - (6 - 2i)$
22. $(2 + i) - (4 - 2i)$ 23. $3 - (4 + 2i)$ 24. $(2 - 6i) - 3$

Examples

a. $(4 + 2i)(2 - 3i)$

b. $\dfrac{3}{3 + i}$

Solutions

a. $(4 + 2i)(2 - 3i)$
$\quad = 8 - 12i + 4i - 6i^2$
$\quad = 8 - 8i - 6(-1)$
$\quad = 14 - 8i$

b. $\dfrac{3}{3 + i} = \dfrac{3(3 - i)}{(3 + i)(3 - i)}$
$\quad\quad = \dfrac{9 - 3i}{9 - i^2} = \dfrac{9 - 3i}{9 + 1}$
$\quad\quad = \dfrac{9}{10} - \dfrac{3}{10}i$

25. $(2 - i)(3 + 2i)$ 26. $(1 - 3i)(4 - 5i)$
27. $(3 + 2i)(5 + i)$ 28. $(-3 - i)(2 - 3i)$
29. $(6 - 3i)(4 - i)$ 30. $(7 + 3i)(-2 - 3i)$
31. $(2 - i)^2$ 32. $(2 + 3i)^2$
33. $(2 - i)(2 + i)$ 34. $(1 - 2i)(1 + 2i)$

35. $\dfrac{1}{3i}$ **36.** $\dfrac{-2}{5i}$ **37.** $\dfrac{3-i}{5i}$ **38.** $\dfrac{4+2i}{3i}$

39. $\dfrac{2}{1-i}$ **40.** $\dfrac{-3}{2+i}$ **41.** $\dfrac{2+i}{1+3i}$ **42.** $\dfrac{3-i}{1+i}$

43. $\dfrac{2-3i}{3-2i}$ **44.** $\dfrac{6+i}{2-5i}$ **45.** $\dfrac{3+2i}{5-3i}$ **46.** $\dfrac{-4-3i}{2+7i}$

■ *Write the given product or quotient in the form $a + bi$ or $a + ib$.*

Examples   **a.** $(1+\sqrt{-2})(3-\sqrt{-2})$ **b.** $\dfrac{1}{3-\sqrt{-4}}$

Solutions   **a.** $(1+\sqrt{-2})(3-\sqrt{-2})$ **b.** $\dfrac{1}{3-\sqrt{-4}} = \dfrac{1(3+2i)}{(3-2i)(3+2i)}$

$\qquad = (1+i\sqrt{2})(3-i\sqrt{2})$ $\qquad\qquad\qquad = \dfrac{3+2i}{9-4i^2}$

$\qquad = 3 - i\sqrt{2} + 3i\sqrt{2} - i^2 \cdot 2$ $\qquad\qquad\qquad = \dfrac{3}{13} + \dfrac{2}{13}i$

$\qquad = 3 + 2i\sqrt{2} + 1 \cdot 2$

$\qquad = 5 + 2i\sqrt{2}$

47. $\sqrt{-4}(1-\sqrt{-4})$ **48.** $\sqrt{-9}(3+\sqrt{-16})$

49. $(2+\sqrt{-9})(3-\sqrt{-9})$ **50.** $(4-\sqrt{-2})(3+\sqrt{-2})$

51. $\dfrac{3}{\sqrt{-4}}$ **52.** $\dfrac{-1}{\sqrt{-25}}$ **53.** $\dfrac{2-\sqrt{-1}}{2+\sqrt{-1}}$ **54.** $\dfrac{1+\sqrt{-2}}{3-\sqrt{-2}}$

■ *Solve.*

Examples   **a.** $x^2 + 8 = 0$ **b.** $x^2 + 2x + 2 = 0$

Solutions   **a.** $x^2 + 8 = 0$ **b.** $x^2 + 2x + 2 = 0$

$\qquad x^2 = -8$ $\qquad\qquad\qquad x^2 + 2x \quad\;\; = -2$

$\qquad x = \sqrt{-8} \;\; \text{or} \;\; x = -\sqrt{-8}$ $\qquad\qquad x^2 + 2x + 1 = -2 + 1$

$\qquad x = 2i\sqrt{2} \;\; \text{or} \;\; x = -2i\sqrt{2}$ $\qquad\qquad\quad (x+1)^2 = -1$

$\qquad \{2i\sqrt{2},\, -2i\sqrt{2}\}$ $\qquad\qquad x+1 = \sqrt{-1} \;\;\text{or}\;\; x+1 = -\sqrt{-1}$

$\qquad\qquad\qquad\qquad\qquad\qquad\quad x = -1 + i \qquad\;\; \text{or} \;\; x = -1 - i$

$\qquad\qquad\qquad\qquad\qquad\qquad\quad \{-1+i,\, -1-i\}$

55. $x^2 + 9 = 0$ **56.** $x^2 + 25 = 0$ **57.** $x^2 + 18 = 0$

58. $x^2 + 24 = 0$ **59.** $x^2 - x + 2 = 0$ **60.** $x^2 + x + 1 = 0$

61. $x^2 + 2x + 3 = 0$ **62.** $x^2 - 4x + 5 = 0$

B 63. Simplify. [*Hint:* $i^2 = -1$ and $i^4 = 1$.]

   a. i^6 **b.** i^{12} **c.** i^{15} **d.** i^{102}

64. Express with a positive exponent and simplify.

   a. i^{-1} **b.** i^{-2} **c.** i^{-3} **d.** i^{-6}

65. Evaluate $x^2 + 2x + 3$ for $x = 1 + i$.

66. Evaluate $2y^2 - y + 2$ for $y = 2 - i$.

67. For what values of x will $\sqrt{x - 5}$ be real? Imaginary?

68. For what values of x will $\sqrt{x + 3}$ be real? Imaginary?

6.4

THE QUADRATIC FORMULA

In Section 6.1 we considered quadratic equations that we were able to solve by factoring over the integers. Let us now develop a formula that can be used to solve all quadratic equations, including those having solutions in the set of complex numbers.

When a quadratic equation is written in its standard form,

$$ax^2 + bx + c = 0, \tag{1}$$

we can solve it by completing the square as follows:

$$x^2 + \frac{b}{a}x + \frac{c}{a} = 0$$

$$x^2 + \frac{b}{a}x + \left(\frac{b}{2a}\right)^2 = -\frac{c}{a} + \left(\frac{b}{2a}\right)^2$$

$$\left(x + \frac{b}{2a}\right)^2 = \frac{b^2}{4a^2} - \frac{c}{a}$$

$$\left(x + \frac{b}{2a}\right)^2 = \frac{b^2 - 4ac}{4a^2}$$

$$x + \frac{b}{2a} = \pm\sqrt{\frac{b^2 - 4ac}{4a^2}}$$

$$x = -\frac{b}{2a} \pm \frac{\sqrt{b^2 - 4ac}}{2a}.$$

When the right-hand member is written as a single fraction, the resulting equation,

$$x = \frac{-b \pm \sqrt{b^2 - 4ac}}{2a},$$

is called the **quadratic formula**. This is a formula for the solutions of a quadratic equation expressed in terms of the coefficients. The symbol $\pm$ is used to condense the two equations

▶
$$x = \frac{-b + \sqrt{b^2 - 4ac}}{2a} \qquad \text{or} \qquad x = \frac{-b - \sqrt{b^2 - 4ac}}{2a}$$

into a single equation. We need only substitute the coefficients a, b, and c of a given quadratic equation in the formula to find the solution set for the equation.

Although the quadratic formula can be used to solve any quadratic equation, it is particularly useful when the left-hand member of a quadratic equation in standard form cannot be factored easily.

Example   Solve $2x^2 - x - 2 = 0$.

Solution   Substitute 2 for a, -1 for b, and -2 for c in the quadratic formula.
$$x = \frac{-(-1) \pm \sqrt{(-1)^2 - 4(2)(-2)}}{2(2)}$$
$$= \frac{1 \pm \sqrt{1 + 16}}{4}$$

The solution set is $\left\{ \dfrac{1 + \sqrt{17}}{4}, \dfrac{1 - \sqrt{17}}{4} \right\}$.

Notice in the above example that we could have expressed the solutions as $\dfrac{1}{4} + \dfrac{\sqrt{17}}{4}$ and $\dfrac{1}{4} - \dfrac{\sqrt{17}}{4}$. In fact, it is clear from the form

$$x = \frac{-b}{2a} \pm \frac{\sqrt{b^2 - 4ac}}{2a}$$

that, if the solutions of a quadratic equation are irrational or imaginary, they are conjugates:

$$\frac{-b}{2a} + \frac{\sqrt{b^2 - 4ac}}{2a} \qquad \text{or} \qquad \frac{-b}{2a} - \frac{\sqrt{b^2 - 4ac}}{2a}.$$

Discriminant   In the quadratic formula, the number represented by $b^2 - 4ac$ is called the **discriminant** of the equation. If a, b, and c are real numbers, the discriminant affects the solution set in the following ways:

1. If $b^2 - 4ac = 0$, there is one real solution (of multiplicity two).

2. If $b^2 - 4ac > 0$, there are two unequal real solutions.

3. If $b^2 - 4ac < 0$, there are two unequal imaginary solutions.

EXERCISE 6.4

A   ■ *In Problems* 1–24, *solve for x, y, or z using the quadratic formula.*

Example $\dfrac{x^2}{4} + x = \dfrac{5}{4}$

Solution   Write in standard form.

$$4\left(\frac{x^2}{4} + x\right) = \left(\frac{5}{4}\right)(4)$$

$$x^2 + 4x = 5$$

$$x^2 + 4x - 5 = 0$$

Substitute 1 for a, 4 for b, and -5 for c in the quadratic formula.*

$$x = \frac{-(4) \pm \sqrt{(4)^2 - 4(1)(-5)}}{2(1)}$$

$$= \frac{-4 \pm \sqrt{16 + 20}}{2} = \frac{-4 \pm 6}{2}$$

The solution set is $\{-5, 1\}$.

1. $x^2 - 5x + 4 = 0$ 2. $x^2 - 4x + 4 = 0$ 3. $y^2 + 3y = 4$

4. $y^2 - 5y = 6$ 5. $z^2 = 3z - 1$ 6. $2z^2 = 7z - 6$

7. $0 = x^2 - \dfrac{5}{3}x + \dfrac{1}{3}$ 8. $0 = x^2 - \dfrac{1}{2}x + \dfrac{1}{2}$ 9. $5z + 6 = 6z^2$

10. $13z + 5 = 6z^2$ 11. $x^2 - 5x = 0$ 12. $y^2 + 3y = 0$

Example $x = \dfrac{-2}{2x - 1}$

Solution   Write in standard form.

$$(2x - 1)x = \frac{-2}{2x - 1}\,(2x - 1)$$

$$x(2x - 1) = -2$$

$$2x^2 - x + 2 = 0$$

Solution continued overleaf

---

*Note that the equation could be solved by the factoring method, as could many of the following exercises. However, the intent in this section is to practice using the quadratic formula.

Substitute 2 for a, -1 for b, and 2 for c in the quadratic formula.

$$x = \frac{-(-1) \pm \sqrt{(-1)^2 - 4(2)(2)}}{2(2)}$$

$$= \frac{1 \pm \sqrt{1 - 16}}{4} = \frac{1 \pm \sqrt{-15}}{4} = \frac{1 \pm i\sqrt{15}}{4}$$

The solution set is $\left\{ \dfrac{1 - i\sqrt{15}}{4}, \dfrac{1 + i\sqrt{15}}{4} \right\}$.

13. $4y^2 + 8 = 0$　　　　　　**14.** $2z^2 + 1 = 0$　　　　　　**15.** $2y^2 = y - 1$

16. $x^2 + 2x = -5$　　　　　**17.** $y = \dfrac{1}{y - 3}$　　　　　**18.** $2z = \dfrac{3}{z - 2}$

19. $3y = \dfrac{1 + y}{y - 1}$　　　　**20.** $2x = \dfrac{x + 1}{x - 1}$　　　　**21.** $2y^2 = y - 2 - y^2$

22. $3z^2 + 2z + 2 = z$　　　　**23.** $2x = \dfrac{-1}{2x - 1}$　　　　**24.** $y = \dfrac{-1}{y - 1}$

■ *In Problems 25–30, find only the discriminant and determine whether the solution(s) are:*
a. one real;　　　　*b. real and unequal;*　　　　*c. imaginary and unequal.*

Example　　　$x^2 - x - 3 = 0$

Solution　　　Substitute 1 for a, -1 for b, and -3 for c in the discriminant.

$$b^2 - 4ac = (-1)^2 - 4(1)(-3)$$
$$= 1 + 12 = 13$$

Because $13 > 0$, the solutions are real and unequal.

25. $x^2 - 7x + 12 = 0$　　　　　　　　**26.** $y^2 - 2y - 3 = 0$
27. $5x^2 + 2x + 1 = 0$　　　　　　　　**28.** $2y^2 + 3y + 7 = 0$
29. $9x^2 - 6x + 1 = 0$　　　　　　　　**30.** $4y^2 - 12y + 9 = 0$

B　■ *Solve for x in terms of the other variables or constants.*

Example　　　$x^2 - xy + y = 2$

Solution　　　Write the equation in standard form.

$$x^2 - yx + (y - 2) = 0$$

Substitute 1 for a, $-y$ for b, and $y - 2$ for c in the quadratic formula.

$$x = \frac{-(-y) \pm \sqrt{(-y)^2 - 4(1)(y-2)}}{2(1)}$$

$$= \frac{y \pm \sqrt{y^2 - 4y + 8}}{2}$$

31. $x^2 - kx - 2k^2 = 0$ $k > 0$

32. $2x^2 - kx + 3 = 0$

33. $ax^2 - x + c = 0$

34. $x^2 + 2x + k + 3 = 0$

35. $x^2 + 2x - y = 0$

36. $2x^2 - 3x + 2y = 0$

37. $3x^2 + xy + y^2 = 2$

38. $x^2 - 3xy + y^2 = 3$

39. In Problem 37 solve for y in terms of x.

40. In Problem 38 solve for y in terms of x.

Example

Determine k so that the solutions of $x^2 - 3x + k - 1 = 0$ are imaginary.

Solution

The solutions are imaginary if the discriminant $b^2 - 4ac < 0$. Substituting -3 for b, 1 for a, and $k - 1$ for c in $b^2 - 4ac$, we have

$$(-3)^2 - 4(1)(k - 1) < 0$$
$$9 - 4k + 4 < 0$$
$$-4k < -13$$
$$k > \frac{13}{4}.$$

41. Determine $k \neq 0$ so that $kx^2 + 4x + 1 = 0$ has one solution.

42. Determine k so that $x^2 - kx + 9 = 0$ has one solution.

43. Determine k so that the solutions of $x^2 + 2x + k + 3 = 0$ are real numbers.

44. Determine k so that the solutions of $x^2 - x + k = 2$ are not real numbers.

45. If r_1 and r_2 are solutions of a quadratic equation, show that

$$r_1 + r_2 = -\frac{b}{a} \qquad \text{and} \qquad r_1 \cdot r_2 = \frac{c}{a}.$$

$$\left[\textit{Hint:} \text{ Let } r_1 = \frac{-b + \sqrt{b^2 - 4ac}}{2a} \text{ and } r_2 = \frac{-b - \sqrt{b^2 - 4ac}}{2a}. \right]$$

6.5

APPLICATIONS

Some of the procedures suggested in Chapter 4 (page 111) for writing equations for word problems should be reviewed at this time. In some cases, the mathematical model we obtain for a physical situation is a quadratic equation and hence may have two solutions. It may be that one but not both of the solutions of the equation fits the physical situation. For example, if we were asked to find two consecutive *natural numbers* whose product is 72, we would write the equation

$$x(x+1) = 72$$

as our model. Solving this equation, we have

$$x^2 + x - 72 = 0$$
$$(x+9)(x-8) = 0,$$

where the solution set is $\{8, -9\}$. Since -9 is not a natural number, we must reject it as a possible answer to the original question; however, the solution 8 leads to the consecutive natural numbers 8 and 9. As additional examples, observe that we would not accept -6 feet as the height of a man or $2\frac{3}{4}$ for the number of people in a room.

A quadratic equation used as a model for a physical situation may have two, one, or no meaningful solutions—meaningful, that is, in a physical sense. Answers to word problems should always be checked in the original problem.

EXERCISE 6.5

A ■ *Solve.*

Example   The sum of an integer and two times its reciprocal is $11\frac{1}{3}$. What is the number?

Solution   An integer: x

Now write a model for the conditions on the variable:

$$x + 2 \cdot \frac{1}{x} = \frac{11}{3}.$$

Multiply each member by $3x$.

$$3x\left(x + \frac{2}{x}\right) = 3x\left(\frac{11}{3}\right)$$
$$3x^2 + 6 = 11x$$

Write in standard form and solve by the factoring method.

$$3x^2 - 11x + 6 = 0$$

$$(3x - 2)(x - 3) = 0$$

$$3x - 2 = 0 \quad \text{or} \quad x - 3 = 0$$

$$x = \frac{2}{3} \qquad\qquad x = 3$$

Since ⅔ is not an integer, the number 3 is the solution: $(3) + 2 \cdot \dfrac{1}{(3)} = \dfrac{11}{3}$.

1. Find two positive numbers that differ by 3 and have a product of 40.

2. Find two consecutive positive integers whose product is 56.

3. Find two consecutive negative integers whose product is 72.

4. Find two consecutive positive integers such that the sum of their squares is 61.

5. The positive numerator of a fraction is 1 less than the denominator. The sum of the fraction and 2 times its reciprocal is $4\frac{1}{12}$. Find the numerator and the denominator.

6. The positive numerator of a fraction is 2 less than the denominator. The sum of the fraction and 3 times its reciprocal is $2\frac{8}{5}$. Find the numerator and the denominator.

7. The sum of a number and its reciprocal is $1\frac{1}{4}$. What is the number?

8. The difference of a number and 2 times its reciprocal is $1\frac{1}{3}$. What is the number?

Example

The length of a rectangle is 4 centimeters greater than the width, and the area is 77 square centimeters. Find the dimensions of the rectangle.

Solution

Width: x

Length: $x + 4$

Write a model for the conditions on the variable. Since the area (77 square centimeters) of a rectangle is the product of its width and length, the model is

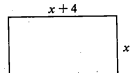

$$x(x + 4) = 77,$$

from which

$$x^2 + 4x - 77 = 0,$$

$$(x + 11)(x - 7) = 0.$$

The solution set is $\{-11, 7\}$. Since a dimension cannot be negative, the only acceptable value for the width x is 7. If $x = 7$, then $x + 4 = 11$, and the dimensions are 7 centimeters and 11 centimeters.

9. A rectangular lawn is 2 meters longer than it is wide. If its area is 63 square meters, what are the dimensions of the lawn?

10. The length of a rectangular steel plate is 2 centimeters greater than 2 times its width. If the area of the plate is 40 square centimeters, find its dimensions.

11. The length of a rectangle is 1 meter less than 4 times the width. Its area is 33 square meters. Find the dimensions.

12. The area of a triangle is 70 square centimeters. Find the lengths of the base and the altitude if the base is 13 centimeters longer than the altitude.

Example

A ball thrown vertically upward reaches a height h in feet, given by the equation $h = 64t - 16t^2$, where t is the time in seconds after the throw. How long will it take the ball to reach a height of 48 feet on its way up?

Solution

Time to reach 48 feet on way up: t

In this case, the model is given:

$$h = 64t - 16t^2.$$

Substitute 48 for h and solve for t.

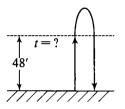

$$48 = 64t - 16t^2$$
$$16t^2 - 64t + 48 = 0$$
$$16(t^2 - 4t + 3) = 0$$
$$16(t - 1)(t - 3) = 0$$
$$t = 1 \quad \text{or} \quad t = 3$$

Use the figure to interpret the two solutions: it takes 1 second to reach 48 feet *on the way up*.

13. Using the information in the example above, determine how long after the ball was thrown it will take for the ball to reach a height of 28 feet on the way down.

14. Using the information in the example above, determine how long after the ball was thrown it will return to the ground.

15. The distance s a body falls in a vacuum is given by

$$s = v_0 t + \left(\frac{1}{2}\right) g t^2,$$

where s is in feet, t is in seconds, v_0 is the initial velocity in feet per second, and g is the constant of acceleration due to gravity (approximately 32 feet per second per second). How long will it take a body to fall 150 feet if v_0 is 20 feet per second?

16. Using the information in Problem 15, determine how long it will take a body to fall 150 feet if the body starts from rest $(v_0 = 0)$.

17. The sum of the squares of three consecutive negative integers is 434. Find the integers.

18. Find two consecutive integers whose cubes differ by 919.

19. The number of diagonals, D, of a polygon of n sides is given by

$$D = \frac{n}{2}(n - 3).$$

How many sides does a polygon with 90 diagonals have?

20. The formula

$$s = \frac{n}{2}(n + 1)$$

gives the sum of the first n natural numbers 1, 2, 3, ... How many consecutive natural numbers starting with 1 will give a sum of 406?

B   21. A book designer decides to use 48 square inches of type on a page and to make the height of the page twice the width. If the margin around the type is to be 2 inches uniformly, what are the dimensions of the page?

22. A garden measuring 12 by 18 meters has its area increased by 216 square meters by a border of uniform width on all sides. Find the width of the border.

23. A garden measuring 25 by 50 meters has its area increased by 318 square meters by a border of uniform width along both shorter sides and one longer side. Find the width of the border.

24. A tray is formed from a rectangular piece of metal whose length is 2 centimeters greater than its width by cutting a square with sides 2 centimeters in length from each corner, and then bending up the sides. Find the dimensions of the tray if the volume is 160 cubic centimeters.

25. A riverboat that travels at 18 miles per hour in still water can go 30 miles up a river in 1 hour less time than it can go 63 miles down the river. What is the speed of the current in the river?

26. If a boat travels 20 miles per hour in still water and takes 3 hours longer to go 90 miles up a river than it does to go 75 miles down the river, what is the speed of the current in the river?

27. A commuter takes a train 10 miles to his job in a city. The train returns him home at a rate 10 miles per hour greater than the rate of the train that takes him to work. If he spends a total of 50 minutes a day commuting, what is the rate of each train?

28. On a 50 mile trip, a woman traveled 10 miles in heavy traffic and then 40 miles in less congested traffic. If her average rate in heavy traffic was 20 miles per hour less than her average rate in light traffic, what was each rate if the trip took 1 hour and 30 minutes?

29. A student store sold $96 worth of notebooks in one month. The store could sell 20 fewer notebooks to get the same income if the price on each was raised $0.40. How much did each notebook sell for originally?

30. A theater now has an income of $1200 for each performance that is sold out. The management could obtain the same income by adding 60 seats and lowering the ticket price $1.00. How many sets does the theater have now?

6.6

EQUATIONS INVOLVING RADICALS

To solve equations containing radicals, we shall make use of the following fact:

▶ *If each member of an equation is raised to the same natural-number power, the solutions of the original equation are contained in the solution set of the resulting equation.*

In symbols, we have:

For n a natural number, the solution set of

$$[P(x)]^n = [Q(x)]^n \tag{1}$$

contains all the solutions of

$$P(x) = Q(x). \tag{2}$$

Extraneous solutions

Note that an application of this property does not always result in an equivalent equation. Equation (1) actually may have additional solutions that are not solutions of (2). With respect to (2), these are called **extraneous solutions**. For example, the solution set of the equation

$$x^2 = 9,$$

obtained from

$$x = 3$$

by squaring each member, is $\{3, -3\}$. This set contains -3 as an extraneous solution, since -3 does not satisfy the equation $x = 3$.

Because the result of applying the foregoing principle is not always an equivalent equation, each solution obtained through its use *must* be checked in the original equation to verify its validity. The check is part of the solution process.

Example

Solve $\sqrt{x-3} = 2$.

Solution

We first square each member to remove the radical, and then solve for x.

$$(\sqrt{x-3})^2 = 2^2$$
$$x - 3 = 4$$
$$x = 7$$

Check Does $\sqrt{7-3} = 2$? Yes. The solution set is $\{7\}$.

Sometimes it is necessary to square each member of an equation more than once in order to obtain an equation free of radicals.

Example   Solve $\sqrt{x-7} + \sqrt{x} = 7$.

Solution   We first add $-\sqrt{x}$ to each member to obtain

$$\sqrt{x-7} = 7 - \sqrt{x},$$

which contains only one term with a radical in the left-hand member. We now square each member to remove one radical.

$$(\sqrt{x-7})^2 = (7 - \sqrt{x})^2$$
$$x - 7 = 49 - 14\sqrt{x} + x$$
$$-56 = -14\sqrt{x}$$
$$4 = \sqrt{x}$$

Now we square each member again to obtain

$$(4)^2 = (\sqrt{x})^2,$$
$$16 = x.$$

Check Does $\sqrt{16-7} + \sqrt{16} = 7$? Yes. The solution set is $\{16\}$.

EXERCISE 6.6

A   ■ *Solve and check. If there is no solution, so state. Assume that all variables represent real numbers.*

Example $\sqrt{x+2} + 4 = x$

Solution   Obtain $\sqrt{x+2}$ as the only term in one member.

$$\sqrt{x+2} = x - 4$$

Square each member.

$$(\sqrt{x+2})^2 = (x-4)^2$$
$$x + 2 = x^2 - 8x + 16$$

Solve the quadratic equation.

$$x^2 - 9x + 14 = 0$$
$$(x-7)(x-2) = 0$$
$$x = 7 \qquad x = 2$$

Check Does $\sqrt{7+2} + 4 = 7$?   Yes.   Does $\sqrt{2+2} + 4 = 2$?   No.
2 is not a solution. The solution set is $\{7\}$.

1. $\sqrt{x} - 5 = 3$ 2. $\sqrt{x} - 4 = 1$ 3. $\sqrt{y+6} = 2$

4. $\sqrt{y-3} = 5$ 5. $3z + 4 = \sqrt{3z+10}$ 6. $2z - 3 = \sqrt{7z-3}$

7. $2x + 1 = \sqrt{10x+5}$ 8. $4x + 5 = \sqrt{3x+4}$ 9. $\sqrt{y+4} = y - 8$

10. $4\sqrt{x-4} = x$ 11. $\sqrt{2y-1} = \sqrt{3y-6}$ 12. $\sqrt{4y+1} = \sqrt{6y-3}$

13. $\sqrt{x} - 3\sqrt{x} = 2$ 14. $\sqrt{x}\sqrt{x-5} = 6$ 15. $\sqrt[3]{x} = -3$

16. $\sqrt[3]{x} = -4$ 17. $\sqrt[4]{x-1} = 2$ 18. $\sqrt[4]{x-1} = 3$

Example $\sqrt{y-5} - \sqrt{y} = 1$

Solution   Write an equivalent equation with $\sqrt{y-5}$ as the left-hand member.

$$\sqrt{y-5} = 1 + \sqrt{y}$$

Square each member.

$$(\sqrt{y-5})^2 = (1 + \sqrt{y})^2$$
$$y - 5 = 1 + 2\sqrt{y} + y$$

Write an equivalent equation with $\sqrt{y}$ as the right-hand member.

$$-3 = \sqrt{y}$$

[*Note:* At this point it is obvious that the equation has no solution, since $\sqrt{y}$ cannot be negative. We continue the solution process, however, for illustrative purposes.]

Square each member.

$$9 = y$$

Check Does $\sqrt{4} - \sqrt{9} = 1$? No.

9 is not a solution of the original equation. Hence, the solution set is $\varnothing$.

19. $\sqrt{y+4} = \sqrt{y+20} - 2$ 20. $4\sqrt{y} + \sqrt{1+16y} = 5$

21. $\sqrt{x} + \sqrt{2} = \sqrt{x+2}$ 22. $\sqrt{4x+17} = 4 - \sqrt{x+1}$

23. $(5+x)^{1/2} + x^{1/2} = 5$ 24. $(y+7)^{1/2} + (y+4)^{1/2} = 3$

25. $(y^2 - 3y + 5)^{1/2} - (y+2)^{1/2} = 0$ 26. $(z-3)^{1/2} + (z+5)^{1/2} = 4$

■ *Solve. Leave the results in the form of an equation. Assume that no variable takes a value for which any denominator is 0.*

Example $t = \sqrt{\dfrac{1+s^2}{g}}$, for s

Solution   Squaring each member yields

$$t^2 = \frac{1+s^2}{g},$$

from which

$$gt^2 = 1 + s^2,$$
$$gt^2 - 1 = s^2;$$

hence,

$$s = \pm\sqrt{gt^2 - 1}.$$

27. $r = \sqrt{\dfrac{A}{\pi}},$ for A

28. $t = \sqrt{\dfrac{2v}{g}},$ for g

29. $R\sqrt{RS} = 1,$ for S

30. $P = \pi\sqrt{\dfrac{l}{g}},$ for g

31. $r = \sqrt{t^2 - s^2},$ for t

32. $q - 1 = 2\sqrt{\dfrac{r^2 - 1}{3}},$ for r

33. $A = B + C\sqrt{D + E^2},$ for E

34. $A = B - C\sqrt{D - E^2},$ for E

B **35.** In Problem 27, specify the restrictions on each variable so that the solutions are real numbers.

36. In Problem 28, specify the restrictions on each variable so that the solutions are real numbers.

37. The base of an isosceles triangle is 6 centimeters. Find the altitude if the perimeter is 20 centimeters.

38. The longer leg of a right triangle is 1 centimeter shorter than the hypotenuse. Find the hypotenuse if the perimeter is 30 centimeters.

6.7

EQUATIONS THAT ARE QUADRATIC IN FORM

Some equations that are not quadratic equations are nevertheless quadratic in form—that is, they are of the form

$$au^2 + bu + c = 0, \tag{1}$$

where u represents some expression in terms of another variable. For example,

$$x^4 - 10x^2 + 9 = 0 \quad \text{is quadratic in } x^2, \tag{2}$$

$$(x^2 - 1)^2 + 3(x^2 - 1) + 2 = 0 \quad \text{is quadratic in } (x^2 - 1),$$

$$y + 2\sqrt{y} - 8 = 0 \quad \text{is quadratic in } \sqrt{y},$$

$$\left(\frac{z+1}{z}\right)^2 - 5\left(\frac{z+1}{z}\right) + 6 = 0 \quad \text{is quadratic in } \left(\frac{z+1}{z}\right).$$

We can solve an equation of the form (1) by first finding the equations

$$u = r_1 \quad \text{and} \quad u = r_2$$

and then solving these equations for the desired roots.

Example   Solve $x^4 - 10x^2 + 9 = 0$.   (2)

Solution   If we let $x^2 = u$, the equation reduces to

$$u^2 - 10u + 9 = 0.$$

Solving this equation, we obtain

$$(u - 9)(u - 1) = 0$$

$$u = 9 \qquad u = 1.$$

Since $u = x^2$, we have

$$x^2 = 9 \qquad x^2 = 1,$$

from which we obtain the solution set $\{3, -3, 1, -1\}$.

The method illustrated in the above example is commonly called *substitution of variables*.

As an alternative method of solving the equation in the above example, we could factor the left-hand member of (2) to obtain

$$(x^2 - 9)(x^2 - 1) = 0,$$

and then set each factor equal to 0. Thus,

$$x^2 - 9 = 0 \qquad \text{or} \qquad x^2 - 1 = 0$$

$$x = \pm 3 \qquad\qquad x = \pm 1,$$

and we obtain the solution set $\{3, -3, 1, -1\}$.

EXERCISE 6.7

A   ■ *Solve for x, y, or z.*

Example $y - 2\sqrt{y} - 8 = 0$

Solution   Let $\sqrt{y} = u$; therefore, $y = u^2$. Substitute for y and $\sqrt{y}$.

$$u^2 - 2u - 8 = 0$$

Solve for u.

$$(u - 4)(u + 2) = 0$$

$$u = 4 \quad \text{or} \quad u = -2$$

Replace u with $\sqrt{y}$ and solve for y. Since $\sqrt{y}$ cannot be negative, only 4 need be considered.

$$\sqrt{y} = 4$$

$$y = 16$$

Check Does $16 - 2\sqrt{16} - 8 = 0$? Yes.

Hence, $\{16\}$ is the solution set.

1. $x^4 - 5x^2 + 4 = 0$
2. $x^4 - 13x^2 + 36 = 0$
3. $2x^4 + 17x^2 - 9 = 0$
4. $z^4 - 2z^2 - 24 = 0$
5. $x - 2\sqrt{x} - 15 = 0$
6. $y + 3\sqrt{y} - 10 = 0$
7. $y^2 + 7 - \sqrt{y^2 + 7} - 12 = 0$
8. $y^2 - 5 - 5\sqrt{y^2 - 5} + 6 = 0$
9. $y^{2/3} - 2y^{1/3} - 8 = 0$
10. $z^{2/3} - 2z^{1/3} = 35$
11. $x^{2/3} - 3x^{1/3} = 4$
12. $2y^{2/3} + 5y^{1/3} = 3$
13. $x - 9x^{1/2} + 18 = 0$
14. $z + z^{1/2} = 72$
15. $2x - 9x^{1/2} = -4$
16. $8x^{1/2} + 7x^{1/4} = 1$

B
17. $y^{-2} - y^{-1} - 12 = 0$
18. $z^{-2} + 9z^{-1} - 10 = 0$
19. $(x - 1)^{1/2} - 2(x - 1)^{1/4} - 15 = 0$
20. $(x - 2)^{1/2} - 11(x - 2)^{1/4} + 18 = 0$

■ *Solve Problems 21–24 in two ways:*
a. By the method of Section 6.6.   *b. By the method of this section.*

21. $y - 7\sqrt{y} + 12 = 0$
22. $x + \sqrt{x} - 6 = 0$
23. $x + 18 = 11\sqrt{x}$
24. $y + 2\sqrt{y} = 15$

6.8

QUADRATIC INEQUALITIES

As in the case with first-degree inequalities, we can generate equivalent second-degree inequalities by applying Properties (1)–(3) and Equations (3)–(5) of Section 4.5. Additional procedures are necessary, however, to obtain the solution sets of such inequalities. For example, consider the inequality

$$x^2 + 4x < 5.$$

To determine values of x for which this condition holds, we might first rewrite the inequality equivalently as

$$x^2 + 4x - 5 < 0,$$

and then as

$$(x + 5)(x - 1) < 0.$$

In this case, the left-hand member will be negative (less than 0) only for those values of x for which the factors $x + 5$ and $x - 1$ are opposite in sign. These can be determined analytically by noting that $(x + 5)(x - 1) < 0$ implies either

$$x + 5 < 0 \quad \text{and} \quad x - 1 > 0 \tag{1}$$

or

$$x + 5 > 0 \quad \text{and} \quad x - 1 < 0. \tag{2}$$

Each of these two cases can be considered separately.

The inequalities for (1),

$$x + 5 < 0 \quad \text{and} \quad x - 1 > 0,$$

imply

$$x < -5 \quad \textit{and} \quad x > 1,$$

a condition which is not satisfied by any values of x. Therefore, the solution set of (1) is $\varnothing$.

The inequalities for (2),

$$x + 5 > 0 \quad \text{and} \quad x - 1 < 0,$$

imply

$$x > -5 \quad \textit{and} \quad x < 1,$$

which leads to the solution set

$$\{x \mid -5 < x < 1\}.$$

Sign arrays   The solution of the inequality $(x + 5)(x - 1) < 0$ in the above example can be interpreted through the use of an array of signs which are determined by the factors in the left-hand member. Note that in Figure 6.2 the product is negative (less than zero) wherever the factors have opposite signs.

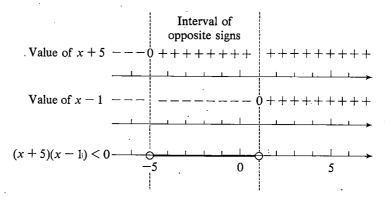

Figure 6.2

Critical
numbers

Sign arrays such as the one shown in Figure 6.2 can be used to solve quadratic inequalities. However, solutions of such inequalities can also be obtained by using the notion of a *critical number*.

Notice in the foregoing solution process that the numbers -5 and 1, which can be obtained by solving the equation $(x + 5)(x - 1) = 0$, separate the set of real numbers into the three intervals,

$$(-\infty, -5), \qquad (-5, 1), \quad \text{and} \quad (1, +\infty),$$

which are shown in Figure 6.3. Each of these intervals either is or is not a part of the solution set of the inequality $x^2 + 4x < 5$. To determine which, we need only

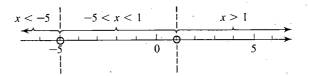

Figure 6.3

substitute an arbitrarily selected number from each interval and test it in the inequality. Let us use -6, 0, and 2.

$$(-6)^2 + 4(-6) \overset{?}{<} 5 \qquad\qquad 0^2 + 4(0) \overset{?}{<} 5 \qquad\qquad 2^2 + 4(2) \overset{?}{<} 5$$

$$36 - 24 \overset{?}{<} 5 \qquad\qquad\qquad 0 + 0 \overset{?}{<} 5 \qquad\qquad\qquad 4 + 8 \overset{?}{<} 5$$

$$12 \overset{?}{<} 5 \qquad\qquad\qquad\qquad 0 \overset{?}{<} 5 \qquad\qquad\qquad\qquad 12 \overset{?}{<} 5$$

$$\text{no} \qquad\qquad\qquad\qquad\qquad \text{yes} \qquad\qquad\qquad\qquad\qquad \text{no}$$

The only interval involved that is in the solution set is $(-5, 1)$, and hence this interval is the solution set. The graph is shown in Figure 6.4. If the inequality in this example were $x^2 + 4x - 5 \le 0$, the endpoints on the graph would be shown as closed dots.

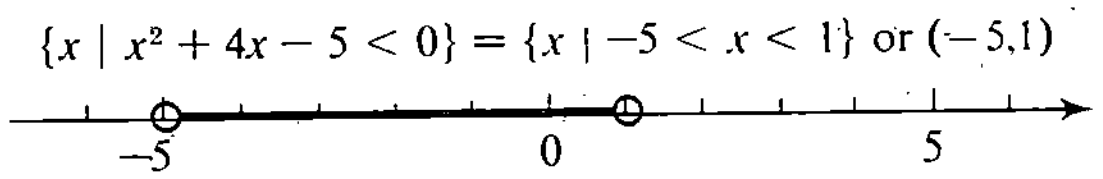

$$\{x \mid x^2 + 4x - 5 < 0\} = \{x \mid -5 < x < 1\} \text{ or } (-5, 1)$$

Figure 6.4

Real numbers such as -5 and 1 in the preceding example are called *critical numbers*. In general,

▶ *The values of x for which $P(x) = 0$ or $P(x)$ is undefined are called **critical numbers** of $P(x) < 0$ or $P(x) > 0$.*

Examples

a. Critical numbers for $2x^2 - x - 1 > 0$ are $-\frac{1}{2}$ and 1, because

$$2x^2 - x - 1 = (2x + 1)(x - 1) = 0$$

for these values.

b. Critical numbers for $\dfrac{1}{x^2 - 4} < 0$ are 2 and -2, because

$$\frac{1}{x^2 - 4} = \frac{1}{(x - 2)(x + 2)}$$

is not defined for either number.

Variable in a denominator

Inequalities involving fractions must be approached with care if any fraction contains a variable in the denominator. If each member of such an inequality is multiplied by an expression containing the variable, we must be careful either to distinguish between those values of the variable for which the expression denotes a positive and a negative number, or to make certain that the expression by which we multiply is always positive. However, we can use the notion of critical numbers to avoid these complications.

Example

Solve and graph the solution set of

$$\frac{x}{x - 2} \geq 5. \tag{3}$$

Solution

We first write (3) equivalently as

$$\frac{x}{x - 2} - 5 \geq 0,$$

from which

$$\frac{x - 5(x - 2)}{x - 2} \geq 0,$$

$$\frac{-4x + 10}{x - 2} \geq 0. \tag{3a}$$

In this case the critical numbers are $\frac{5}{2}$ and 2, because

$$\frac{-4x + 10}{x - 2}$$

equals 0 for $x = \frac{5}{2}$ and is undefined for $x = 2$. Thus, we want to check the intervals shown on the number line in Figure a. Substituting arbitrary values for the variable in (3) or (3a) in each of the three intervals, say, 0, ¾, and 3, we can identify the solution set

$$\left(2, \frac{5}{2}\right],$$

as we did in the example on page 207. The graph is shown in Figure b. Note that the left-hand endpoint is an open dot (2 is not a member of the solution set) because the left-hand member of (3) is undefined for $x = 2$.

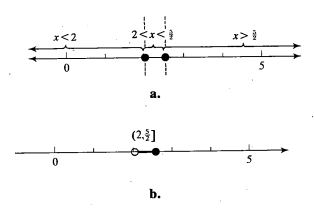

a.

b.

It is sometimes possible to determine the solution set of a quadratic inequality simply by inspection. For example, the solution set of

$$4x^2 + 6 > 0$$

is $\{x | x \text{ a real number}\}$ because the left-hand member is greater than zero for all real-number replacements of x. For the same reason, the solution set of

$$4x^2 + 6 < 0$$

is $\varnothing$.

EXERCISE 6.8

A ■ *Solve and represent each solution set on a line graph.*

Example $x^2 - 3x - 4 \geq 0$ (1)

Solution overleaf

Solution   Factoring the left-hand member yields

$$(x + 1)(x - 4) \geq 0,$$

and we observe that -1 and 4 are critical numbers because $(x + 1)(x - 4) = 0$ for these values. Hence, we check the intervals on the number line in Figure a. We substitute selected arbitrary values in each interval, say, $-2, 0$, and 5, for the variable in (1).

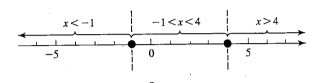

$$(-2)^2 - 3(-2) - 4 \overset{?}{\geq} 0 \qquad (0)^2 - 3(0) - 4 \overset{?}{\geq} 0 \qquad (5)^2 - 3(5) - 4 \overset{?}{\geq} 0$$
$$\text{yes} \qquad\qquad\qquad \text{no} \qquad\qquad\qquad \text{yes}$$

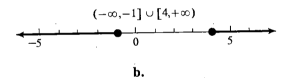

a.

Hence, the solution set is $(-\infty, -1] \cup [4, +\infty)$. The graph is shown in Figure b.

$$(-\infty, -1] \cup [4, +\infty)$$

b.

1. $(x + 1)(x - 2) > 0$ **2.** $(x + 2)(x + 5) < 0$ **3.** $x(x - 2) \leq 0$

4. $x(x + 3) \geq 0$ **5.** $x^2 - 3x - 4 > 0$ **6.** $x^2 - 5x - 6 \geq 0$

7. $x^2 < 5$ **8.** $4x^2 + 1 < 0$ **9.** $x^2 > -5$

10. $x^2 + 1 > 0$ **11.** $\dfrac{x + 3}{x - 2} < 0$ **12.** $\dfrac{x - 1}{x + 4} > 0$

13. $\dfrac{x}{x + 2} - 4 > 0$ **14.** $\dfrac{x + 2}{x - 2} - 6 \geq 0$ **15.** $\dfrac{x - 1}{x} \leq 3$

16. $\dfrac{x}{x - 1} \geq 5$ **17.** $\dfrac{1}{x - 1} \leq 1$ **18.** $\dfrac{x + 1}{x - 1} < 1$

B   **19.** $\dfrac{4}{x + 1} < \dfrac{3}{x}$ **20.** $\dfrac{3}{x - 1} \geq \dfrac{1}{x}$ **21.** $\dfrac{2}{x - 2} \geq \dfrac{4}{x}$

22. $\dfrac{3}{4x + 1} > \dfrac{2}{x - 5}$ **23.** $5 < x^2 + 1 < 10$ **24.** $-2 < y^2 - 3 < 13$

25. A projectile fired from ground level is at a height of $320t - 16t^2$ feet after t seconds. During what period of time is it higher than 1024 feet?

26. Using the information in Problem 25, find the total time that the projectile is in the air.

CHAPTER SUMMARY

[6.1] The **standard form** for a quadratic equation in one variable is

$$ax^2 + bx + c = 0,$$

where a, b, and c are constants with $a \neq 0$.

To solve an equation by **factoring**, we use the fact that the product of two factors is 0 if and only if one or both of the factors equals 0. That is,

$$ab = 0 \quad \text{if and only if} \quad a = 0 \quad \text{or} \quad b = 0.$$

[6.2] To solve an equation by the **extraction of roots** method, we use the fact that:

$$\text{If} \quad x^2 = b, \quad \text{then} \quad x = \sqrt{b} \quad \text{or} \quad x = -\sqrt{b}.$$

To solve an equation of the form $x^2 + bx = c$ by completing the square, we first add $(\frac{b}{2})^2$ to both members.

[6.3] A number of the form $a + bi$, where $a, b \in R$, is called a **complex number**. It is a real number if $b = 0$, it is an **imaginary number** if $b \neq 0$, and it is a **pure imaginary number** if $a = 0$ and $b \neq 0$.

The sum and the product of two complex numbers are defined so that they conform with the sum and the product of two binomials, respectively.

[6.4] The solutions of $ax^2 + bx + c = 0$ $(a \neq 0)$ are given by

$$x = \frac{-b + \sqrt{b^2 - 4ac}}{2a} \quad \text{and} \quad x = \frac{-b - \sqrt{b^2 - 4ac}}{2a}.$$

The number represented by $b^2 - 4ac$ is the **discriminant** of the quadratic equation $ax^2 + bx + c = 0$.

[6.5] Quadratic equations are sometimes useful as mathematical models in physical situations. In such cases, it may be that not all the solutions of an equation are meaningful.

[6.6] To solve equations containing radicals, we use the fact that if each member of an equation is raised to the same power, the solutions of the original equation are contained in the solution set of the resulting equation. That is:

For n a natural number, the solution set of $[P(x)]^n = [Q(x)]^n$ contains all the solutions of $P(x) = Q(x)$.

[6.7] An equation of the form

$$au^2 + bu + c = 0,$$

where u represents some expression in another variable, is said to be **quadratic in form**. Such equations may be solved by a method commonly called *substitution of variables*.

[6.8] Quadratic inequalities may be solved by using a sign array or by first identifying **critical numbers**.

The symbols introduced in this chapter are listed on the inside of the front cover.

REVIEW EXERCISES

A

[6.1] ■ *Solve for x by factoring.*

1. a. $x^2 - 2x = 0$ **b.** $x^2 - 5x + 6 = 0$

2. a. $(x - 6)(x + 4) = -9$ **b.** $x(x + 1) = 6$

3. a. $x - 1 = \frac{1}{4}x^2$ **b.** $x + \frac{3}{x} = 4$

■ *Given the solutions of a quadratic equation, write the equation in standard form with integral coefficients.*

4. a. -2 and 5 **b.** ⅓   and $-¼$

[6.2] ■ *Solve for x by the extraction of roots.*

5. a. $2x^2 = 50$ **b.** $3x^2 - 7 = 0$

6. a. $(x + 3)^2 = 25$ **b.** $(x - 4)^2 = 15$

■ *Solve by completing the square.*

7. a. $x^2 - 4x - 6 = 0$ **b.** $2x^2 + 3x - 3 = 0$

[6.3] ◾ *Write each expression in the form $a + bi$ or $a + ib$.*

8. a. $4 + 2\sqrt{-9}$ **b.** $5 - 3\sqrt{-12}$

9. a. $(2 - 3i) + (4 + 2i)$ **b.** $(6 + 2i) - (1 - i)$

10. a. $(5 + i)(2 + 3i)$ **b.** $i(3 - 4i)$

11. a. $\dfrac{4}{3i}$ **b.** $\dfrac{1 - 3i}{2 - i}$

[6.4] ◾ *Solve for x using the quadratic formula.*

12. a. $\dfrac{1}{2}x^2 + 1 = \dfrac{3}{2}x$ **b.** $x^2 - 3x + 7 = 0$

13. a. $x^2 - 3x + 1 = 0$ **b.** $2x^2 + x - 3 = 0$

[6.5] **14.** The base of a triangle is 1 inch longer than 2 times the length of its altitude, and its area is 18 square inches. Find the length of the base and the altitude.

15. The sum of a number and its reciprocal is $1\frac{3}{6}$. Find the number.

16. The distance d (in feet) that an object falls in a vacuum in t seconds is given by the formula

$$d = \frac{1}{2}gt^2,$$

where $g = 32$ feet per second per second, the acceleration due to gravity. Find the time it takes for an object to fall 96 feet.

17. The distance d (in meters) that an object travels in t seconds when given an initial velocity of 15 meters per second is given by

$$d = 15t + 2t^2.$$

Find the time necessary for an object to travel 63 meters.

[6.6] ◾ *Solve.*

18. a. $x - 3\sqrt{x} + 2 = 0$ **b.** $\sqrt{x + 1} + \sqrt{x + 8} = 7$

19. a. $p = \sqrt{\dfrac{1 - 2t^2}{s}}$ for t **b.** $R = \dfrac{1 + \sqrt{p^2 + 1}}{2}$ for p

[6.7] ◾ *Solve.*

20. a. $x^4 - 3x^2 - 4 = 0$ **b.** $x - x^{1/2} = 12$

[6.8]   ■ *Solve each inequality and graph the solution set on a line graph.*

21. **a.** $x^2 - 9x > 0$ **b.** $x^2 + 5x + 6 < 0$

22. **a.** $\dfrac{x+1}{x-3} \geq 0$ **b.** $\dfrac{x}{x+2} \leq 2$

B   23. Solve $(ax - b)^2 = 4$ for x in terms of a and b.

24. Express i^{-5} with a positive exponent and simplify.

25. Solve $2x^2 - kx + 1 = 0$ for x in terms of k.

26. Solve $2x^2 + xy + y^2 = 0$ for y in terms of x.

27. Determine k so that the solutions of $x^2 - 2x - k + 1 = 0$ are real numbers.

28. Solve $(x - 1)^{1/2} - (x - 1)^{1/4} - 6 = 0$.

29. Solve $\dfrac{1}{x-2} \leq \dfrac{2}{x}$.

30. A man drives 12 miles to work each day. His average rate coming home in congested traffic is 15 miles per hour less than his average rate going to work. What was his rate each way if he spent a total of 40 minutes driving each day?

7 FUNCTIONS, RELATIONS, AND THEIR GRAPHS: PART I

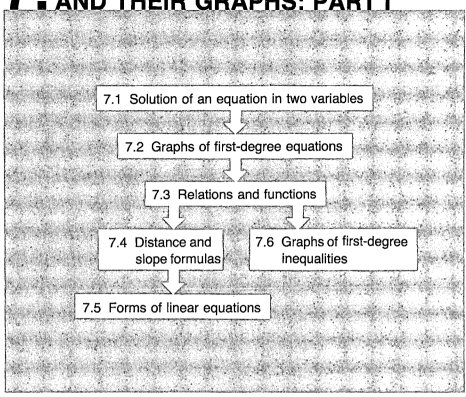

7.1 Solution of an equation in two variables

7.2 Graphs of first-degree equations

7.3 Relations and functions

7.4 Distance and slope formulas

7.6 Graphs of first-degree inequalities

7.5 Forms of linear equations

7.1

SOLUTION OF AN EQUATION IN TWO VARIABLES

Ordered pairs An equation in two variables, such as $y = 2x + 3$, is said to be *satisfied* if the variables are replaced with a pair of numbers—one from the replacement set of x and one from the replacement set of y—that make the resulting statement true. The pair of numbers, usually written in the form (x, y), is a **solution** of the equation or inequality.

The pair (x, y) is called an **ordered pair**, because it is understood that the numbers are considered in a particular order, x first and y second. These numbers are called the **first** and **second components** of the ordered pair, respectively. Although any letters may be used as variables, in this book we shall almost always use x and y. Note that parentheses have previously been used to denote an open interval. The intended meaning should be clear from the context in which it is used.

To find ordered pairs that are solutions of a given equation, we can assign any real-number value to one of the variables and then determine the related value, if any, of the other.

Example   For

$$y - x = 1,$$

we can obtain solutions (ordered pairs) by assigning to x any real number as a value and then determining the corresponding value of y. For example, substituting 2, 3, and 4 for x, we have

$$y - (2) = 1, \quad \text{from which} \quad y = 3;$$
$$y - (3) = 1, \quad \text{from which} \quad y = 4;$$
$$y - (4) = 1, \quad \text{from which} \quad y = 5.$$

Thus, $(2, 3)$, $(3, 4)$, and $(4, 5)$ are three solutions of $y - x = 1$. There is, of course, an infinite number of solutions to this equation.

Equivalent equations

In finding ordered pairs that satisfy a given equation, it is sometimes helpful if one variable is first expressed in terms of the other. The assertions in Section 4.1 that are used to generate equivalent equations in one variable apply also in transforming equations in two or more variables.

Example   Before assigning values to x in

$$y - 3x = 4, \tag{1}$$

we can transform the equation to the equivalent equation

$$y = 3x + 4. \tag{2}$$

Both equations in the example above have the same solution set. In (1) the variables x and y are said to be **implicitly** related, while in (2), y is said to be expressed **explicitly** in terms of x.

In our work we shall be interested primarily in the set of all ordered pairs that can be formed with real-number components; that is, the infinite set of ordered pairs

$$\{(x, y) \mid x \in R \ \text{and} \ y \in R\}.$$

This set is called the **Cartesian product** of R and R, and is denoted by $R \times R$ or R^2.

EXERCISE 7.1

A   ■ *Find each missing component so that each ordered pair is a solution of the equation.*

Examples $y - 2x = 4$
a. $(0, \underline{?})$ **b.** $(\underline{?}, 0)$ **c.** $(3, \underline{?})$

Solutions

a. $y - 2x = 4$
$y - 2(0) = 4$
$y = 4$
$(0, 4)$

b. $y - 2x = 4$
$(0) - 2x = 4$
$x = -2$
$(-2, 0)$

c. $y - 2x = 4$
$y - 2(3) = 4$
$y = 10$
$(3, 10)$

1. $y = x + 7$	**a.** $(0, \underline{?})$	**b.** $(2, \underline{?})$	**c.** $(-2, \underline{?})$
2. $y = 6 - 2x$	**a.** $(0, \underline{?})$	**b.** $(\underline{?}, 0)$	**c.** $(-1, \underline{?})$
3. $3x - 4y = 6$	**a.** $(0, \underline{?})$	**b.** $(\underline{?}, 0)$	**c.** $(-5, \underline{?})$
4. $x + 2y = 5$	**a.** $(0, \underline{?})$	**b.** $(5, \underline{?})$	**c.** $(-3, \underline{?})$

■ *List the ordered pairs that satisfy the given equation and have the given x-components.*

Example $y = 3x + 1;$ 1, 2, 3

Solution
For $x = 1,$ $y = 3(1) + 1 = 4.$
For $x = 2,$ $y = 3(2) + 1 = 7.$
For $x = 3,$ $y = 3(3) + 1 = 10.$

Thus, the desired solutions are $(1, 4)$, $(2, 7)$, and $(3, 10)$.

5. $y = x - 4;$ $-3, 0, 3$

6. $y = 2x + 6;$ $-2, 0, 2$

7. $y = \dfrac{3}{x + 2};$ 1, 2, 3

8. $y = \dfrac{4x}{x^2 - 1};$ 0, 2, 4

9. $y = \sqrt{x^2 - 1};$ 1, 2, 3

10. $y = \dfrac{1}{2}\sqrt{4 - x^2};$ 0, 1, 2

■ *Transform the given equation into one in which y is expressed explicitly in terms of x, and then find values for y for the given values of x.*

Examples

a. $3y - xy = 4;$ 2, 5

b. $x^2 + 4y^2 = 5;$ 0, 1

Solutions

a. $3y - xy = 4$
$y(3 - x) = 4$
$y = \dfrac{4}{3 - x}$

When $x = 2,$ $y = \dfrac{4}{3 - (2)} = 4;$

when $x = 5,$ $y = \dfrac{4}{3 - (5)} = -2.$

b. $x^2 + 4y^2 = 5$
$4y^2 = 5 - x^2$
$y^2 = \dfrac{1}{4}(5 - x^2)$
$y = \pm\dfrac{1}{2}\sqrt{5 - x^2}$

When $x = 0,$ $y = \pm\dfrac{1}{2}\sqrt{5};$

when $x = 1,$ $y = \pm 1.$

11. $2x + y = 6$; $2, 4$ **12.** $4x - y = 2$; $-2, -4$ **13.** $xy - x = 2$; $-2, 2$

14. $3x - xy = 6$; $1, 3$ **15.** $xy - y = 4$; $4, 8$ **16.** $x^2y - xy = -5$; $2, 4$

17. $x^2y - 4y = xy + 2$; $-1, 1$ **18.** $x^2y - xy + 3 = 5y$; $-2, 2$

19. $4 = \dfrac{x}{y^2 - 2}$; $-1, 3$ **20.** $3 = \dfrac{x}{y^2 + 1}$; $-2, 1$

21. $3x^2 - 4y^2 = 4$; $2, 3$ **22.** $3x^2 - 4y^2 = 4$; $2, 4$

B ■ *Find the solution of each equation for each specified component.*

23. $y = |x| - 3$,   for $x = -4$ **24.** $y = |x| + 5$,   for $x = -2$

25. $y = |x - 1| + |x|$,   for $x = -1$ **26.** $y = |2x + 1| - |x|$,   for $x = -3$

27. $y = 2^x$,   for $x = 3$ **28.** $y = 2^x$,   for $x = -3$

29. $y = 2^{-x}$,   for $x = 3$ **30.** $y = 2^{-x}$,   for $x = -3$

31. Find k if $(1, 2)$ is a solution of $x + ky = 8$.

32. Find k if $(-6, 1)$ is a solution of $kx - 2y = 6$.

7.2

GRAPHS OF
FIRST-DEGREE EQUATIONS

Cartesian coordinate system

The graph of $R \times R$, an infinite set of points, is the entire geometric plane. We shall assume that the ordered pairs (x, y) in $R \times R$ can be placed in a one-to-one correspondence with the geometric points in the plane by any arbitrary scaling of the axes of the familiar **Cartesian (or rectangular) coordinate system** shown in Figure 7.1. That is, for each point in the plane, there corresponds a unique ordered pair and vice versa.

This correspondence is established by using the first component, x, of the ordered pair to denote the directed (perpendicular) distance of the point from the vertical axis (the **abscissa** of the point); x is positive if it falls to the right of the vertical axis and negative if it falls to the left. The second component, y, of the ordered pair is used to denote the directed (perpendicular) distance of the point from the horizontal axis (the **ordinate** of the point); y is positive if it falls above the horizontal axis and negative if it falls below.

Each axis is labeled with the variable it represents. Most commonly, the horizontal axis is called the **x-axis** and the vertical axis is called the **y-axis**, and their point of intersection is called the **origin**. The components of the ordered pair that corresponds to a given point are called the **coordinates** of the point, and the point is called the **graph** of the ordered pair. Each of the four regions into which the axes divide the plane is called a **quadrant**, and they are referred to by Roman numerals, as illustrated in Figure 7.1.

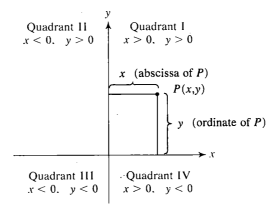

Figure 7.1

Graphs of linear equations

First-degree equations in two variables have infinitely many solutions that are ordered pairs of real numbers. Hence, these solutions can be displayed on a Cartesian coordinate system. For example, two solutions of

$$y = x + 2$$

are (1, 3) and (2, 4). The graphs of these two points are shown in Figure 7.2a. Figure 7.2b shows the location of the graphs of three additional solutions, namely, (−2, 0), (−1, 1), and (0, 2).

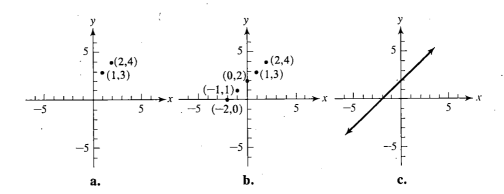

Figure 7.2

In general, we are interested in the graphs of equations in two variables over the set of all real numbers x for which y is a real number. The graph of $y = x + 2$ for all real numbers x (an infinite set of points) is shown in Figure 7.2c. It can be shown that the coordinates of each point on the line in part c satisfy $y = x + 2$ and, conversely, that every solution of the equation corresponds to a point on the line. The line is referred to as the **graph of the solution set of the equation** or as the **graph of the equation**.

Although we will not prove it here, any first-degree equation in two variables—that is, any equation that can be written equivalently in the form

$$ax + by + c = 0 \quad (a \text{ and } b \text{ not both } 0),$$

where a, b, and c are real numbers—has a graph that is a straight line. For this reason, such equations are often called **linear equations**. Since any two distinct points determine a straight line, it is evident that two solutions of such an equation determine its graph.

Intercept method of graphing

In practice, the two solutions that are easiest to find are those with second and first components, respectively, equal to 0—that is, the solutions $(x_1, 0)$ and $(0, y_1)$. Since these two points are the points where the graph intersects the x- and y-axes, they are easy to locate. The numbers x_1 and y_1 are, respectively, called the x- and y-**intercepts** of the graph. To find the x-intercept, substitute 0 for y in the equation and solve for x; to find the y-intercept, substitute 0 for x and solve for y.

Example   Graph $3x + 4y = 12$.

Solution   If $y = 0$, then

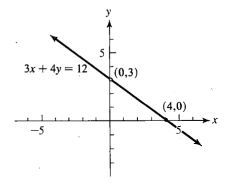

$$3x + 4(0) = 12,$$
$$3x = 12,$$
$$x = 4.$$

The x-intercept is 4 with coordinates $(4, 0)$. If $x = 0$, then

$$3(0) + 4y = 12,$$
$$4y = 12,$$
$$y = 3.$$

The y-intercept is 3 with coordinates $(0, 3)$. Thus, the graph of the equation appears as shown.

If the graph intersects the axes at or near the origin, the intercepts either do not represent two separate points, or the points are too close together to be of much use in drawing the graph. It is then necessary to plot *at least* one other point at a distance far enough removed from the origin to establish the line with accuracy.

Example   Graph $y = 3x$.

Solution   If $x = 0$, then $y = 0$, and both intercepts of the graph are at the point $(0, 0)$. Assigning any other replacement for x, say, 2, we obtain a second solution $(2, 6)$. We first graph the ordered pairs $(0, 0)$ and $(2, 6)$ and then complete the graph, as shown.

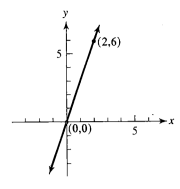

Special cases of linear equations   There are two special cases of linear equations worth noting. First, an equation such as

$$y = 4$$

can be considered an equation in two variables in $R \times R$,

$$0x + y = 4.$$

For each x, this equation assigns $y = 4$. That is, any ordered pair of the form $(x, 4)$ is a solution of the equation. For example,

$$(-1, 4), \quad (2, 4), \quad \text{and} \quad (4, 4)$$

are all solutions of the equation. If we draw a straight line through the graphs of these points, we obtain the graph shown in Figure 7.3.

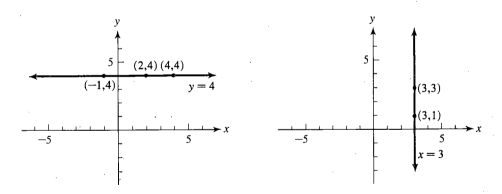

Figure 7.3   **Figure 7.4**

The other special case of a linear equation is of the type

$$x = 3,$$

which, in $R \times R$, may be looked upon as an equation in two variables,

$$x + 0y = 3.$$

Here, only one value is permissible for x, namely, 3, while any value may be assigned to y. That is, any ordered pair of the form $(3, y)$ is a solution of this equation. If we choose two solutions, say, $(3, 1)$ and $(3, 3)$, and draw a straight line through the graphs of these points, we have the graph of the equation shown in Figure 7.4 on page 221.

> *If k represents a constant (real number), then, in general, the graph of $y = k$ is a horizontal line and the graph of $x = k$ is a vertical line in $R \times R$.*

Examples   **a.** Graph $y = 2$.   **b.** Graph $x = 4$.

Solutions   **a.**   **b.**

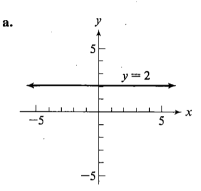

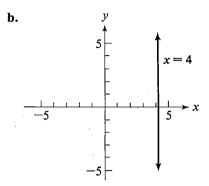

EXERCISE 7.2

A   ■ *Graph the ordered pairs that satisfy the given equation and have the given x-components.*

Example $y = 2x + 1$; $-2, 0, 2$

Solution   Substitute -2, 0, and 2 sequentially for x in $y = 2x + 1$.

$$y = 2(-2) + 1 = -3$$
$$y = 2(0) + 1 \quad = 1$$
$$y = 2(2) + 1 \quad = 5$$

Graph the ordered pairs $(-2, -3)$, $(0, 1)$, and $(2, 5)$.

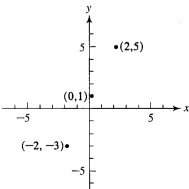

1. $y = x + 2$; $-4, -2, 0$
2. $y = x - 3$; $0, 2, 4$
3. $y = 2x - 1$; $-2, -1, 0, 1, 2$
4. $y = 3x - 2$; $-2, -1, 0, 1, 2$
5. $y = 4 - x$; $-2, -1, 0, 1, 2$
6. $y = 3 - 2x$; $-2, -1, 0, 1, 2$

■ *Graph each equation.*

Example $3x - 4y = 24$

Solution   Use the intercept method because it is possible to obtain two distinct intercepts.
If $x = 0$, then

$$3(0) - 4y = 24$$
$$y = -6.$$

If $y = 0$, then

$$3x - 4(0) = 24$$
$$x = 8.$$

Hence, -6 is the y-intercept and 8 is the x-intercept. Graph these intercepts and draw a line through the points.

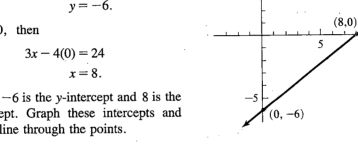

7. $y = x - 5$
8. $y = x + 3$
9. $y = 3x + 6$
10. $y = 4x - 8$
11. $x + 2y = 8$
12. $2x - y = 6$
13. $3x - 4y = 12$
14. $2x + 6y = 6$
15. $2x - y = 0$
16. $x + 3y = 0$
17. $y = -3x$
18. $x = 2y$
19. $4x - y = 0$
20. $4x + y = 0$

■ *Graph each equation in $R \times R$.*

Example $y = -2$

Solution   In $R \times R$, the equation is equivalent to $0x + y = -2$; the solutions are ordered pairs of the form $(x, -2)$. The y-value is -2 for all values of x.

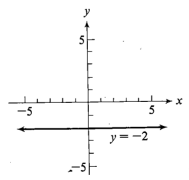

21. $y = -3$
22. $x = -2$
23. $2x = 8$
24. $3y = 15$
25. $x = 0$
26. $y = 0$

B ■ *Graph each equation.*

Example $y = x + 2$ $(x \geq 0)$

Solution   First sketch the graph of $y = x + 2$, as shown in Figure a. Then delete the part of the graph for which $x < 0$. The remaining graph of $y = x + 2$ for $x \geq 0$ is shown in Figure b.

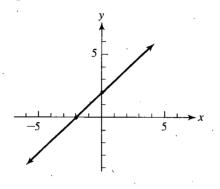

a.   b.

27. $y = x$ $(x \geq 0)$	**28.** $y = x$ $(x < 0)$
29. $y = x - 2$ $(x < 0)$	**30.** $y = x - 2$ $(x \geq 0)$
31. $y = x + 3$ $(x \geq -3)$	**32.** $y = x - 3$ $(x \geq -3)$
33. $y = -x - 3$ $(x < -3)$	**34.** $y = -x + 3$ $(x < -3)$

■ *Graph each equation.*

Example $y = |x + 3|$

Solution   By definition, $y = |x + 3|$ is equivalent to

$y = x + 3$ if $x + 3 \geq 0$ $(x \geq -3)$,

or

$y = -(x + 3)$ if $x + 3 < 0$ $(x < -3)$.

Graph each equation (see Problems 31 and 33) on the same set of axes. Note that the graph is entirely above or on the x-axis $(y \geq 0)$.

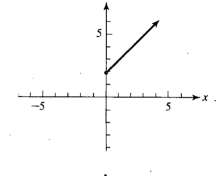

35. $y = |x|$	**36.** $y = 2|x|$	**37.** $y = |x + 2|$
38. $y = |x - 2|$	**39.** $y = |x| + 2$	**40.** $y = |x| - 2$

41. Graph $\{(x, y)|y = 2x + 4\} \cap \{(x, y)|y = -x\}$.

42. Graph $\{(x, y)|y = 2x\} \cap \{(x, y)|x = 3\}$.

7.3

RELATIONS AND FUNCTIONS

Relations

In mathematics and applications of mathematics we are often concerned with relationships that involve pairings of numbers. As an example, let us consider the two sets

$$A = \{1, 2, 3\}$$
$$\updownarrow \; \updownarrow \; \updownarrow$$
$$B = \{3, 4, 5\}$$

and pair their individual elements as shown. Because such associations display a relationship between the elements of the two sets involved, they are called **relations**.

Ordered pairs, which we introduced in Section 7.1, offer a clear way of specifying a relation. By using each element from set A as a first component for an ordered pair and the associated element in set B as a second component, we can exactly specify any relation between the elements of sets A and B. For example, the relation depicted above can be displayed as the set of ordered pairs

$$\{(1, 3), (2, 4), (3, 5)\}.$$

In fact, any set of ordered pairs is a relation.

The set of all first components of the ordered pairs in a relation is the **domain** of the relation, and the set of all second components of the ordered pairs is the **range** of the relation. For example, the relation

$$\{(1, 3), (2, 4), (3, 5)\}$$

— elements in the domain

— elements in the range

has domain $\{1, 2, 3\}$ and range $\{3, 4, 5\}$. Each element in the range of a relation is called a **value** of the relation.

Functions

In a relation, each element in the domain is paired with *one or more* elements in the range. Figure 7.5a illustrates a relation in which an element in the domain is paired with two elements in the range; two pairings, (3, 10) and (3, 15), have the same first component. Figure 7.5b illustrates a relation in which each element in the domain is associated with *only one* element in the range. In this relation no two pairings have the same first component. Such a relation is called a **function**.

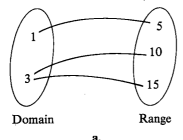

Domain   Range

a.

Not a function

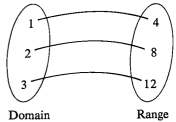

Domain   Range

b.

Function

Figure 7.5

▶ A *function* is a relation in which no two ordered pairs have the same first component.

Examples   **a.** {(2, 3), (3, 5), (4, 5)} is a relation; it is also a function because no two ordered pairs have the same first component.

b. {(3, 4), (3, 5), (4, 7)} is a relation but not a function, because two ordered pairs, (3, 4) and (3, 5), have the same first component.

Relations
defined by
equations

Since the solution set of an equation in two variables is a set of ordered pairs, such an equation clearly specifies a relation. If no value of x is paired with more than one value of y by an equation, then the relation specified is a function.

Examples   **a.** $y = 2x + 1$ defines a relation that is a function because only one value of y is paired with each real number replacement for x.

b. $y^2 - x^2 = 1$, or, equivalently, $y = \pm\sqrt{x^2 + 1}$, defines a relation; it does not define a function because two values of y are paired with each real number replacement for x.

Geometric
test for a
function

The graph of a relation gives us a means of mentally checking to see whether a particular graph does or does not represent a function. If we imagine a line parallel to the y-axis passing across the graph, we can determine whether or not it intersects the graph at more than one point at each position on the x-axis; that is, whether or not there are two or more different y-coordinates for any given x-coordinate. If there are, the graph is not the graph of a function. See Figure 7.6.

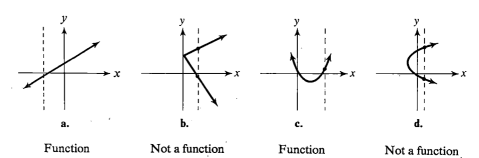

a.	b.	c.	d.
Function	Not a function	Function	Not a function

Figure 7.6

Note that Figure 7.6a is the graph of a first-degree equation. Because any first-degree equation (except for equations of the form $x = k$) has a graph that is a nonvertical straight line, such equations define functions.

Domain of a
function

Let us make the following agreement concerning functions defined by equations:

If the domain of a function (or relation) is not specified, we shall assume that the function (or relation) has as domain the set of real numbers for which real numbers exist in the range.

Thus, values of the variable which are associated with the first components of ordered pairs and which make a denominator zero or result in an imaginary value for y are excluded from the domain.

Examples

Specify the domain (the set of replacements for x) for each function.

a. $y = x + 5$ **b.** $\dfrac{3}{x-4}$ ′ **c.** $y = \sqrt{x-2}$

Solutions

a. The entire set of real numbers because y is a real number for all real numbers x.

b. The set of real numbers except $x = 4$, because $3/(x-4)$ is undefined for this value of x.

c. The set of real numbers $x \geq 2$ because $\sqrt{x-2}$ is a real number for these values only. (If $x < 2$, then $x - 2$ is negative and $\sqrt{x-2}$ is imaginary.)

Function notation

Functions are usually designated by means of a single symbol, P, R, f, g, or some other letter. The symbol for the function can be used in conjunction with the variable representing an element in the domain to represent an expression in that variable. Thus, we shall use $f(x)$ (read "f of x" or "the value of f at x") in precisely the same way that we used $P(x)$ in Section 2.1 when we discussed expressions and, in particular, polynomial expressions. This notation is commonly called **function notation**.

Examples

a. If f is the function defined by the equation

$$y = x - 3,$$

then we can just as well write

$$f(x) = x - 3,$$

where $f(x)$ plays the same role as y.

b. If $f(x) = x - 3$, then

$$f(6) = (6) - 3 = 3;$$
$$f(-2) = (-2) - 3 = -5;$$
$$f(-5) = (-5) - 3 = -8;$$
$$f(t) = (t) - 3 = t - 3.$$

For simplicity, when the intent is clear from the context, we shall often use the terminology "the function $y = x - 3$ or $f(x) = x - 3$," rather than the longer phrase "the function defined by . . . ." In Section 7.2 a line was referred to as the "graph of the solution set of the equation" or simply the "graph of the equation." The line is also referred to as the "graph of the function defined by the equation" or simply as the "graph of the function."

Function notation can be used to denote the ordinate associated with a given abscissa of a graph. For example, the graph of

$$f(x) = x - 3$$

is shown in Figure 7.7. Specific ordinates are shown as line segments labeled with their "lengths" in function notation. The "length" in this case is called a **directed length** because any function value is simply the ordinate of a point and can be positive, negative, or zero.

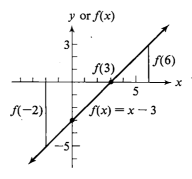

Figure 7.7

If f is a function and x is some element of the domain of f, then the corresponding element y of the range depends on the element x. Thus, we call x the **independent variable** and y the **dependent variable**. In such cases we say "y is a function of x."

EXERCISE 7.3

A   ■ *a. Specify the domain and the range of each relation.*
   b. State whether or not the relation is a function.

Example $\{(2, 3), (3, 5), (3, 6), (4, 6), (5, 6)\}$

Solution   **a.** Domain is set of first components: $\{2, 3, 4, 5\}$.
             Range is set of second components: $\{3, 5, 6\}$.

             b. Relation is not a function: Two ordered pairs, (3, 5) and (3, 6), have the same first
             component.

1. $\{(-2, 3), (-1, 4), (0, 5), (1, 6)\}$
2. $\{(-1, -1), (0, 0), (1, 1), (2, 2)\}$
3. $\{(5, 1), (6, 1), (7, 1), (8, 1)\}$
4. $\{(4, 1), (4, 2), (4, 3), (4, 4)\}$
5. $\{(2, 3), (2, 4), (3, 3), (3, 4)\}$
6. $\{(-5, 2), (2, -5), (3, 6), (6, 3)\}$
7. $\{(0, 0), (2, 1), (2, 2), (3, 3)\}$
8. $\{(2, -1), (3, -1), (3, 1), (4, 2)\}$

■ **a.** *Solve each equation explicitly for y in terms of x.*
   b. *Specify the domain of the relation defined by the equation.*
   c. *State whether the relation is or is not a function.*

Example $4x^2 + y^2 = 16$

Solution   **a.** $y^2 = 16 - 4x^2$

$$y = \pm\sqrt{16 - 4x^2} = \pm\sqrt{4(4 - x^2)}$$
$$= \pm 2\sqrt{4 - x^2}$$

   b. The values of x that make $4 - x^2 \geq 0$ give real values for y. The solution set of the inequality is the domain: $\{x|-2 \leq x \leq 2\}$.

   c. Relation is not a function because there are two values for y for each replacement of x in the domain between -2 and 2.

9. $2x + y = 6$
10. $2x + 3y = 12$
11. $x - 2y = 8$
12. $3x - 2y = 8$
13. $(x - 2)y = 4$
14. $(x + 3)y = 6$
15. $xy - 4y = 6$
16. $xy + 2y = 8$
17. $2x^2 + y^2 = 8$
18. $x^2 + 2y^2 = 8$
19. $4x^2 - y^2 = 16$
20. $y^2 - 4x^2 = 16$

■ *Find the value of each expression.*

Example $f(3) - f(1)$, if $f(x) = 5x + 2$

Solution $f(3) = 5(3) + 2$ $f(1) = 5(1) + 2$
          $= 15 + 2 = 17;$ $= 5 + 2 = 7$

   Hence,

$$f(3) - f(1) = 17 - 7 = 10.$$

21. $f(4)$, if $f(x) = x^2 - 2x + 1$
22. $g(3)$, if $g(x) = 2x^2 + 3x - 1$
23. $g(5) - g(2)$, if $g(x) = x + 3$
24. $f(2) - f(0)$, if $f(x) = 3x - 1$
25. $f(3) - f(-3)$, if $f(x) = x^2 - x + 1$
26. $f(0) - f(-2)$, if $f(x) = x^2 + 3x - 2$

B   ▪ *In Problems* 27–32, *find:*

  a. $f(x + h)$;   **b.** $f(x + h) - f(x)$;   **c.** $\dfrac{f(x + h) - f(x)}{h}$.

Example $f(x) = x^2 + 1$

Solution   **a.** $f(x + h) = (x + h)^2 + 1 = x^2 + 2xh + h^2 + 1$

     b. $f(x + h) - f(x) = (x^2 + 2xh + h^2 + 1) - (x^2 + 1) = 2xh + h^2$

     c. $\dfrac{f(x + h) - f(x)}{h} = \dfrac{2xh + h^2}{h} = 2x + h$

   27. $f(x) = 3x - 4$ **28.** $f(x) = 2x + 5$ **29.** $f(x) = x^2 - 3x + 5$

   30. $f(x) = x^2 + 2x$ **31.** $f(x) = x^3 + 2x - 1$ **32.** $f(x) = x^3 + 3x^2 - 1$

Example   Graph $f(x) = x - 1$. Represent $f(5)$ and $f(3)$ by drawing line segments from $(5, 0)$ to $(5, f(5))$ and from $(3, 0)$ to $(3, f(3))$.

Solution $f(5)$ is the ordinate at $x = 5$; $f(3)$ is the ordinate at $x = 3$.

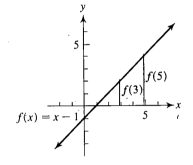

   33. a. Graph $f(x) = 2x + 1$.
     b. Represent $f(3)$ and $f(-2)$ by drawing line segments from $(3, 0)$ to $(3, f(3))$ and from $(-2, 0)$ to $(-2, f(-2))$.

   34. a. Graph $f(x) = 2x - 5$.
     b. Represent $f(4)$ and $f(0)$ by drawing line segments from $(4, 0)$ to $(4, f(4))$ and from $(0, 0)$ to $(0, f(0))$.

   35. Using function notation, represent the length of line segment CB in the figure in terms of $f(x)$ and $f(x + h)$.

   36. Using the results of Problem 35, represent the ratio of the length of line segment BC to the length of line segment AC in terms of $f(x)$ and $f(x + h)$.

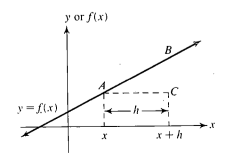

37. If $f(x) = 3x + 2$, show that $f(a + b)$ is not equivalent to $f(a) + f(b)$.

38. If $f(x) = x^2 + 3$, show that $f(a \cdot b)$ is not equivalent to $f(a) \cdot f(b)$.

39. Express the volume (V) of a cone as a function of its height (h) if the height of the cone is equal to three times the radius (r). [*Hint:* $V = \frac{1}{3}\pi r^2 h$.]

40. Express the area (A) of a circle as a function of the circumference (C) of the circle. [*Hint:* $A = \pi r^2$ and $C = 2\pi r$, where r is the radius of the circle.]

41. A circle with radius (r) is inscribed in a square. Express the area (A) of the region *outside* the circle and *inside* the square as a function of the radius.

42. The hypotenuse (c) of a right triangle is 12 centimeters long. Express the area (A) of the triangle as a function of the shortest side (b). [*Hint:* The square of the length of the hypotenuse of a right triangle equals the sum of the squares of the lengths of the other two sides.]

7.4

DISTANCE AND SLOPE FORMULAS

Distance formula

Any two distinct points in a plane can be looked upon as the endpoints of a line segment. We shall discuss two fundamental properties of a line segment—its length and its slope with respect to the x-axis. We first observe that any two distinct points P_1 and P_2 either lie on the same vertical line or one is to the right of the other; the point on the right has a greater x-coordinate than the point on the left. If we construct a line parallel to the y-axis through P_2 and a line parallel to the x-axis through P_1, the lines will meet at a point P_3, as shown in either part a or part b of Figure 7.8.

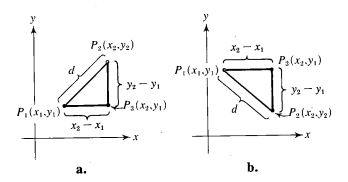

Figure 7.8

The x-coordinate of P_3 is evidently the same as the x-coordinate of P_2, while the y-coordinate of P_3 is the same as that of P_1; hence, the coordinates of P_3 are (x_2, y_1). By inspection, we observe that the distance between P_2 and P_3 is $(y_2 - y_1)$ and the distance between P_1 and P_3 is $(x_2 - x_1)$.

In general, since $(y_2 - y_1)$ is positive or negative as $y_2 > y_1$ or $y_2 < y_1$, respectively, and $(x_2 - x_1)$ is positive or negative as $x_2 > x_1$ or $x_2 < x_1$, respectively, we may designate the *directed* distances represented by $(x_2 - x_1)$ and $(y_2 - y_1)$ as positive or negative.

The **Pythagorean theorem** may be used to find the length of the line segment joining P_1 and P_2. This theorem asserts that the square of the length of the hypotenuse of any right triangle is equal to the sum of the squares of the lengths of the legs. Thus,

$$d^2 = (x_2 - x_1)^2 + (y_2 - y_1)^2.$$

If we consider only the positive square root of the right-hand member, we have the **distance formula**,

▶
$$d = \sqrt{(x_2 - x_1)^2 + (y_2 - y_1)^2}. \tag{1}$$

Example Find the distance between $(2, -1)$ and $(4, 3)$.

Solution Substituting $(2, -1)$ for $P_1(x_1, y_1)$ and $(4, 3)$ for $P_2(x_2, y_2)$ in the distance formula, we obtain

$$\begin{aligned} d &= \sqrt{(x_2 - x_1)^2 + (y_2 - y_1)^2} \\ &= \sqrt{[4 - 2]^2 + [3 - (-1)]^2} \\ &= \sqrt{4 + 16} \\ &= \sqrt{20} = 2\sqrt{5}. \end{aligned}$$

Notice that in the above example, we would obtain the same answer if we used $(4, 3)$ for P_1 and $(2, -1)$ for P_2:

$$\begin{aligned} d &= \sqrt{[2 - 4]^2 + [(-1) - 3]^2} \\ &= \sqrt{4 + 16} = 2\sqrt{5}. \end{aligned}$$

If the points P_1 and P_2 lie on the same horizontal line $(y_2 = y_1)$ and if we are concerned only with distance and not direction, then

$$d = \sqrt{(x_2 - x_1)^2 + 0^2} = |x_2 - x_1|.$$

If they lie on the same vertical line $(x_2 = x_1)$, then

$$d = \sqrt{0^2 + (y_2 - y_1)^2} = |y_2 - y_1|.$$

Examples **a.** The points $(6, 2)$ and $(-3, 2)$ lie on the same horizontal line, namely, $y = 2$. The distance between the points is

$$|x_2 - x_1| = |6 - (-3)| = 9.$$

b. The points $(4, -5)$ and $(4, -2)$ lie on the same vertical line, namely, $x = 4$. The distance between the points is

$$|y_2 - y_1| = |(-5) - (-2)| = 3.$$

Slope formula   A second useful property of a line segment that joins two points is its orientation in the plane. This property can be measured by comparing the **rise** (difference of y-coordinates) to a given **run** (difference of x-coordinates), as shown in Figure 7.9.

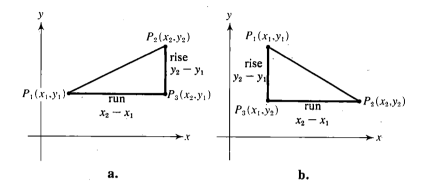

Figure 7.9

The ratio of rise to run is the **slope** of the segment and is designated by the letter m. Thus, since the rise is $(y_2 - y_1)$ and the run is $(x_2 - x_1)$, the slope of the segment joining P_1 and P_2 is given by

$$\text{slope:}\quad m = \frac{\text{rise}}{\text{run}} = \frac{y_2 - y_1}{x_2 - x_1}\qquad (x_2 \neq x_1). \tag{2}$$

Taking P_2 to the right of P_1, the run $(x_2 - x_1)$ is positive and the slope is positive or negative as the rise $(y_2 - y_1)$ is positive or negative—that is, according to whether the line slopes upward or downward from P_1 to P_2. A positive slope indicates that a line is rising to the right; a negative slope indicates that it is falling to the right. Since

$$\frac{y_2 - y_1}{x_2 - x_1} = \frac{-(y_1 - y_2)}{-(x_1 - x_2)} = \frac{y_1 - y_2}{x_1 - x_2},$$

the restriction that P_2 be to the right of P_1 is not necessary, and the order in which the points are considered is immaterial.

Example   Find the slope of the line segment joining the points $(2, -1)$ and $(4, 3)$.

Solution   Let $(2, -1)$ be (x_1, y_1) and $(4, 3)$ be (x_2, y_2). Then

$$m = \frac{y_2 - y_1}{x_2 - x_1} = \frac{3 - (-1)}{4 - 2}$$

$$= \frac{4}{2} = 2.$$

Several lines with different slopes are shown in Figure 7.10.

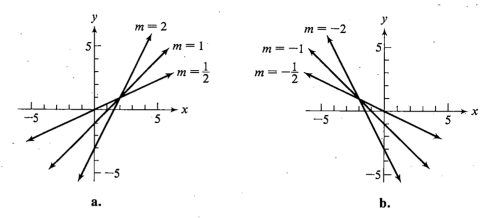

Figure 7.10

If a segment is parallel to the x-axis, as shown in Figure 7.11a, then $y_2 - y_1 = 0$ and it will have a slope of 0. If a segment is parallel to the y-axis, as shown in Figure 7.11b, $x_2 - x_1 = 0$ and its slope is not defined.

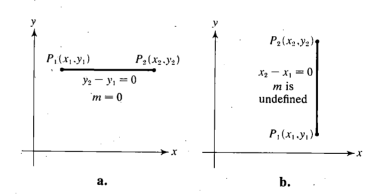

Figure 7.11

Parallel and perpendicular lines   The following two properties of line segments are important in mathematics and applications of mathematics.

▶ *Two line segments with slopes m_1 and m_2 are:*

Parallel if $m_1 = m_2$;

Perpendicular if $m_1 m_2 = -1$ *$(m_1 m_2 \neq 0)$.*

Examples

a. In Figure a, line segment AB with slope

$$m_1 = \frac{5-3}{5-2} = \frac{2}{3}$$

is parallel to line segment CD with slope

$$m_2 = \frac{1-(-1)}{4-1} = \frac{2}{3}.$$

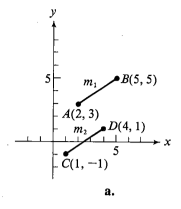

a.

b. In Figure b, line segment DE with slope

$$m_2 = \frac{1-7}{4-0} = \frac{-6}{4} = \frac{-3}{2}$$

is perpendicular to line segment AB with slope

$$m_1 = \frac{5-3}{5-2} = \frac{2}{3},$$

because

$$m_1 m_2 = \frac{2}{3}\left(\frac{-3}{2}\right) = -1.$$

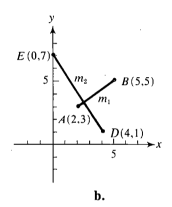

b.

EXERCISE 7.4

A ■ *Find the distance between each of the given pairs of points, and find the slope of the line segment joining them. Sketch each line segment in the coordinate plane.*

Example $(3, -5)$, $(2, 4)$

Solution   Consider $(3, -5)$ as P_1 and $(2, 4)$ as P_2.

$$d = \sqrt{(x_2 - x_1)^2 + (y_2 - y_1)^2}$$
$$= \sqrt{[2-3]^2 + [4-(-5)]^2}$$
$$= \sqrt{1+81} = \sqrt{82}$$

$$m = \frac{y_2 - y_1}{x_2 - x_1} = \frac{4-(-5)}{2-3}$$

$$= \frac{9}{-1} = -9$$

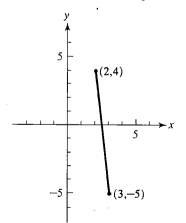

1. $(1, 1)$, $(4, 5)$ 2. $(-1, 1)$, $(5, 9)$ 3. $(-3, 2)$, $(2, 14)$

4. $(-4, -3)$, $(1, 9)$ 5. $(2, 1)$, $(4, 0)$ 6. $(-3, 2)$, $(0, 0)$

7. $(5, -4)$, $(-1, 1)$ 8. $(2, -3)$, $(-2, -1)$ 9. $(3, 5)$, $(-2, 5)$

10. $(2, 0)$, $(-2, 0)$ 11. $(0, 5)$, $(0, -5)$ 12. $(-2, -5)$, $(-2, 3)$

■ *Use the distance formula to find the perimeter of the triangle whose vertices are given. Sketch each triangle in the coordinate plane.*

13. $(0, 6)$, $(9, -6)$, $(-3, 0)$ 14. $(10, 1)$, $(3, 1)$, $(5, 9)$

15. $(5, 6)$, $(11, -2)$, $(-10, -2)$ 16. $(-1, 5)$, $(8, -7)$, $(4, 1)$

17. Show that the two line segments whose endpoints are $(5, 4)$, $(3, 0)$ and $(-1, 8)$, $(-4, 2)$ are parallel.

18. Show that the two line segments whose endpoints are $(-4, 2)$, $(2, -2)$ and $(3, 0)$, $(-3, 4)$ are parallel.

19. Show that the two line segments whose endpoints are $(0, -7)$, $(8, -5)$ and $(5, 7)$, $(8, -5)$ are perpendicular.

20. Show that the two line segments whose endpoints are $(8, 0)$, $(6, 6)$ and $(-3, 3)$, $(6, 6)$ are perpendicular.

■ *Sketch the line that goes through the given point and has the given slope.*

Examples   a. $(-3, -2)$; $m = -\dfrac{2}{3}$ b. $(3, 2)$; $m = 0$

Solutions   a.   b.

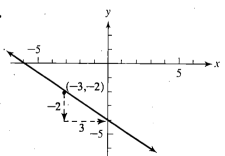

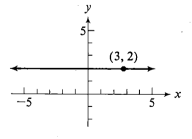

21. $(2, 1)$; $m = 4$ 22. $(-2, 3)$; $m = 5$ 23. $(5, 5)$; $m = -1$

24. $(-3, -2)$; $m = \dfrac{1}{2}$ 25. $(0, 0)$; $m = 3$ 26. $(-1, 0)$; $m = 1$

27. $(0, -1)$; $m = -\dfrac{1}{2}$

28. $(2, -1)$; $m = \dfrac{3}{4}$

29. $(-2, -2)$; $m = -\dfrac{3}{4}$

30. $(2, -3)$; $m = 0$

31. $(-4, 2)$; $m = 0$

32. $(-1, -2)$;
parallel to x-axis

■ *Specify the slope of each line in the figure. Assume that l_4 is parallel to the x-axis and l_8 is parallel to the y-axis.*

33. l_1

34. l_2

35. l_3

36. l_4

37. l_5

38. l_6

39. l_7

40. l_8

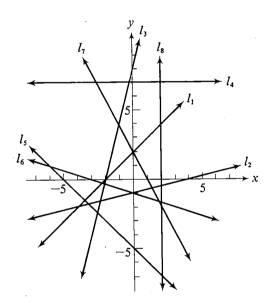

B **41.** Show that the triangle described in Problem 13 is a right triangle. [*Hint:* Use the converse of the Pythagorean theorem—that is, if $c^2 = a^2 + b^2$, the triangle is a right triangle. Alternatively, show that two sides are perpendicular.]

42. Show that the triangle with vertices at (0, 0), (6, 0), and (3, 3) is a right isosceles triangle—that is, a right triangle with two sides that have the same length.

43. Show that the points (2, 4), (3, 8), (5, 1), and (4, −3) are the vertices of a parallelogram. [*Hint:* A four-sided figure is a parallelogram if the opposite sides are parallel.]

44. Show that the points (−5, 4), (7, −11), (12, 25), and (0, 40) are the vertices of a parallelogram.

45. Given the points $P_1(4, -1)$, $P_2(2, 7)$, and $P_3(-3, 4)$, find the value of k in the ordered pair $P_4(5, k)$ that makes P_1P_2 parallel to P_3P_4.

46. Using the points in Problem 45, find the k that makes P_1P_2 perpendicular to P_3P_4.

7.5

FORMS OF LINEAR EQUATIONS

Let us designate

$$ax + by + c = 0 \quad \text{or} \quad ax + by = c \quad (b \neq 0) \tag{1}$$

as **standard form** for a linear equation. We shall now consider two alternative forms that display useful aspects.

Point-slope form

Assuming that the slope of the line segment joining any two points on a line does not depend upon the points, consider a line on the plane with given slope m and passing through a given point (x_1, y_1), as shown in Figure 7.12. If we choose any other point on the line and assign to it the coordinates (x, y), the slope of the line is given by

$$\frac{y - y_1}{x - x_1} = m \quad (x \neq x_1), \tag{2}$$

from which

$$y - y_1 = m(x - x_1) \quad (x \neq x_1). \tag{3}$$

The ordered pair (x_1, y_1) corresponds to a point on the line, and the coordinates of (x_1, y_1) also satisfy (3). Hence, the graph of (3) contains all points in the graph of (2) and, in addition, the point (x_1, y_1).

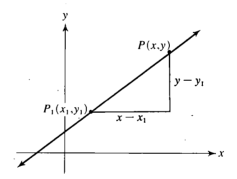

Figure 7.12

The equation

▶
$$y - y_1 = m(x - x_1) \tag{4}$$

is called the **point-slope form** for a linear equation.

Example   Find an equation of the line that goes through the point $(1, -4)$ and has slope $-\tfrac{3}{4}$.

Solution   Substituting -4 for y_1, 1 for x_1, and $-\tfrac{3}{4}$ for m in the point-slope form for a linear equation, we obtain

$$y - (-4) = -\frac{3}{4}(x - 1).$$

To change the equation to standard form we multiply each member by 4 to obtain

$$4y + 16 = -3(x - 1),$$
$$4y + 16 = -3x + 3,$$
$$3x + 4y + 13 = 0.$$

Slope-intercept form   Now consider the equation of the line passing through a given point on the y-axis with coordinates $(0, b)$ and slope m, as shown in Figure 7.13. Substituting $(0, b)$ in the point-slope form of a linear equation,

$$y - y_1 = m(x - x_1),$$

we obtain

$$y - b = m(x - 0),$$

from which

▶ $$y = mx + b. \tag{5}$$

Equation (5) is called the **slope-intercept form** for a linear equation. Note that b is the y-intercept of the graph of the equation.

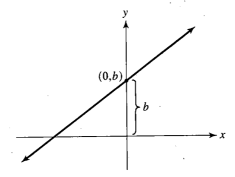

Figure 7.13

Example   Write $3x + 4y = 6$ in slope-intercept form and specify the slope of the line and the y-intercept.

Solution   We first solve the equation explicitly for y:

$$4y = -3x + 6$$
$$y = \frac{-3}{4}x + \frac{3}{2}.$$

Hence, the slope is $-\frac{3}{4}$, the coefficient of x, and the y-intercept is $\frac{3}{2}$.

EXERCISE 7.5

A   ■ *Find an equation of the line that goes through each of the given points and has the given slope. Write the equation in standard form.*

Example   (3, 2); $m = -\dfrac{2}{3}$

Solution   Substitute 3 for x, 2 for y, and $-\frac{2}{3}$ for m in the given values in the point-slope form of the linear equation.

$$y - y_1 = m(x - x_1)$$
$$y - 2 = -\frac{2}{3}(x - 3)$$
$$3(y - 2) = -2(x - 3)$$
$$3y - 6 = -2x + 6$$
$$2x + 3y - 12 = 0$$

1. $(-1, 3)$; $m = 2$ **2.** $(2, -5)$; $m = -3$ **3.** $(-2, 6)$; $m = -1$

4. $(-6, -1)$; $m = 4$ **5.** $(0, 3)$; $m = \dfrac{1}{2}$ **6.** $(2, 0)$; $m = -\dfrac{1}{3}$

7. $(-1, 2)$; $m = -\dfrac{3}{2}$ **8.** $(2, -1)$; $m = \dfrac{5}{3}$

9. $(-3, -5)$; $m = 0$ **10.** $(0, -6)$; $m = 0$

11. $(-3, 2)$;   parallel to y-axis   **12.** $(0, 0)$; $m = 1$

■ *a.* *Write each equation in slope-intercept form.*
  b. *Specify the slope of the line and the y-intercept.*

Example $2x - 5y = 5$

Solution   **a.** Solve explicitly for y.

$$-5y = 5 - 2x$$
$$5y = 2x - 5$$
$$y = \frac{2}{5}x + (-1)$$

b. Compare with the general slope-intercept form $y = mx + b$.

$$\text{slope: } \frac{2}{5}; \qquad y\text{-intercept: } -1$$

13. $x + y = 3$ **14.** $2x + y = -1$ **15.** $3x + 2y = 1$ **16.** $3x - y = 7$

17. $x - 3y = 2$ **18.** $2x - 3y = 0$ **19.** $8x - 3y = 0$ **20.** $-x = 2y - 5$

21. $y + 2 = 0$ **22.** $y - 3 = 0$

■ *Find an equation of the line whose graph includes the two given points. Write the equation in the form* $ax + by = c$.

Example $(2, 2), (-4, 1)$

Solution   First find the slope, selecting either point as (x_1, y_1) and the other point as (x_2, y_2).

$$m = \frac{y_2 - y_1}{x_2 - x_1} = \frac{1 - 2}{-4 - 2}$$

$$= \frac{-1}{-6} = \frac{1}{6}.$$

Then use the point-slope formula with either ordered pair. Using $(2, 2)$ for (x_1, y_1) in the formula $y - y_1 = m(x - x_1)$ yields

$$y - 2 = \frac{1}{6}(x - 2),$$

from which

$$6y - 12 = x - 2,$$
$$x - 6y = -10.$$

23. $(-4, 2), (3, 3)$ **24.** $(5, -1), (2, -3)$ **25.** $(-1, -3), (2, 0)$ **26.** $(0, 5), (3, -4)$

B

Example   Write an equation of the line that is perpendicular to the graph of $2x + 3y = 6$ and passes through $(3, 3)$.

Solution   First find the slope of the graph of $2x + 3y = 6$:

$$3y = -2x + 6$$

$$y = \frac{-2}{3}x + 2.$$

Hence, the slope is $-\tfrac{2}{3}$. The slope m_2 of the line perpendicular to the slope m_1 of the given line is

$$m_2 = \frac{-1}{m_1} = -\frac{1}{\dfrac{-2}{3}} = \frac{3}{2}.$$

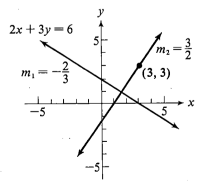

Solution continued overleaf

Hence, using the point-slope formula, the equation of the line with slope ³⁄₂ passing through the point (3, 3) is

$$y - 3 = \frac{3}{2}(x - 3),$$

$$2y - 6 = 3x - 9,$$

from which

$$3x - 2y = 3.$$

27. Write an equation of the line that is parallel to the graph of $x - 2y = 5$ and passes through the origin. Draw the graphs of both equations.

28. Write an equation of the line that passes through (0, 5) parallel to $2y - 3x = 5$. Draw the graphs of both equations.

29. Write an equation of the line that is perpendicular to the graph of $x - 2y = 5$ and passes through the origin. Draw the graphs of both equations.

30. Write an equation of the line that is perpendicular to the graph of $2y - 3x = 5$ and passes through (0, 5). Draw the graphs of both equations.

31. Water freezes at 0°C (32°F) and boils at 100°C (212°F). Find an equation in the form $F = aC + b$ that relates Celsius and Fahrenheit temperatures.

7.6

GRAPHS OF FIRST-DEGREE INEQUALITIES

The solutions of inequalities of the form

$$ax + by + c > 0 \quad \text{or} \quad ax + by + c < 0,$$

where a, b, and c are real numbers, are ordered pairs of real numbers. The solution sets can be graphed on the plane, but the graph will be a region of the plane rather than a straight line. As an example, consider the inequality

$$2x + y - 3 < 0. \tag{1}$$

Rewritten in the form

$$y < -2x + 3, \tag{2}$$

we have that y is less than $-2x + 3$ for each x. The graph of the equation

$$y = -2x + 3 \tag{3}$$

is simply a straight line, as illustrated in Figure 7.14. Therefore, to graph (2), we need only observe that any point below this line has x- and y-coordinates that satisfy (2). For example, if $x = 2$, then all ordered pairs of the form $(2, y)$ such that

$$y < -2(2) + 3$$

or

$$y < -1$$

are in the solution set and their graphs are below the line, as shown in Figure 7.14.

The solution set of (2) corresponds to the entire region below the line. The region is indicated on the graph (Figure 7.15) by shading. The broken line in

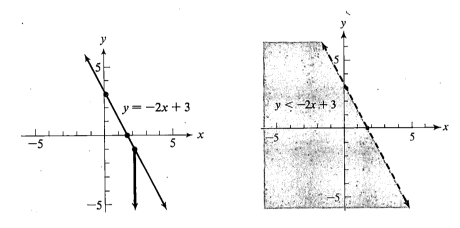

Figure 7.14   **Figure 7.15**

Figure 7.15 indicates that the points on the line do not correspond to elements in the solution set of the inequality. If the original inequality were

$$2x + y - 3 \leq 0,$$

the line would be a part of the graph of the solution set and would be shown as a solid line.

In general, if $b \neq 0$,

$$ax + by + c < 0 \qquad \text{or} \qquad ax + by + c > 0$$

will have as a solution set all ordered pairs associated with the points in a **half-plane** either above or below the line with equation

$$ax + by + c = 0,$$

depending upon the inequality symbols involved.

To determine which of the half-planes should be shaded, substitute the coordinates of any point not on the line into the original inequality and note whether or not they satisfy the inequality. If they do, then the half-plane containing the point is shaded; if they do not, then the other half-plane is shaded. A good point to use in this process is the origin, with coordinates $(0, 0)$, if the origin does not lie in the line.

Example   Graph $3x - 2y < 6$.

Solution   We first graph the equality
                   $3x - 2y = 6$ using the intercept
                   method.

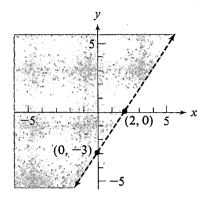

   Then, substituting 0 for x and 0
for y in the inequality, we obtain

$$3(0) - 2(0) < 6.$$

Since this is a true statement, we
shade the half-plane that contains
the origin.

   In this case, the edge of the half-
plane is a dashed line because the
original inequality does not contain
the "equal to" symbol.

   Inequalities do not define functions according to our definition in Section 7.3,
because each element in the domain is not associated with a unique element in the
range. For example, in the inequality

$$y < -2x + 3, \tag{4}$$

if $x = -3$,

$$y < -2(-3) + 3$$
$$y < 9,$$

which is certainly not a unique value for y. The infinite number of values for $y < 9$ is
indicated by the heavy red vertical line in Figure 7.16.

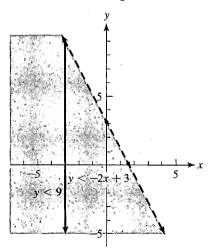

Figure 7.16

EXERCISE 7.6

A ■ *Graph each inequality.*

Example $2x + y \geq 4$

Solution   Graph the equality $2x + y = 4$. Substitute 0 for x and y in the inequality.

$$2(0) + 0 \geq 4$$

Since this is not a true statement, shade the half-plane not containing the origin.

The line is included in the graph, because the original inequality *does* contain the "equal to" symbol.

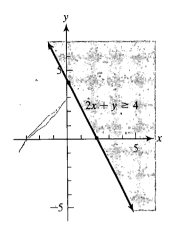

1. $y > x + 3$ 2. $y < x + 4$ 3. $y \leq x + 2$ 4. $y \geq x - 2$

5. $x + y < 5$ 6. $x - y < 3$ 7. $2x + y < 2$ 8. $x - 2y < 5$

9. $x \leq 2y + 4$ 10. $2x \leq y + 1$ 11. $0 \geq x - y$ 12. $0 \geq x + 3y$

■ *Graph each set of ordered pairs.*

Example $\{(x, y) | x > 2\}$

Solution   Graph the equality $x = 2$. The solution set consists of all ordered pairs where x is greater than 2. Hence, shade the region to the right of the line representing $x = 2$. The line is excluded from the graph.

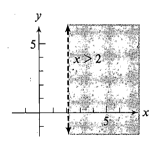

13. $\{(x, y) | x > 0\}$ 14. $\{(x, y) | y < 0\}$ 15. $\{(x, y) | y \geq 3\}$

16. $\{(x, y) | x < -2\}$ 17. $\{(x, y) | -1 < x < 5\}$ 18. $\{(x, y) | 0 \leq y \leq 1\}$

19. $\{(x, y) | x \geq 0$ and $y \geq 0\}$ 20. $\{(x, y) | x \leq 0$ and $y \geq 0\}$

21. $\{(x, y) | x \geq 0$ and $y \leq 0\}$ 22. $\{(x, y) | x \leq 0$ and $y \leq 0\}$

B   ■ *Graph each of the set intersections using double shading.*

Example $\{(x, y)|y > x\} \cap \{(x, y)|y > 2\}$

Solution   Graph each inequality. The region common to both solution sets is the graph of the intersection.

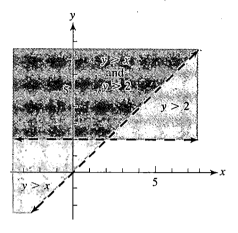

23. $\{(x, y)|x \geq 4\} \cap \{(x, y)|y \geq 2\}$ **24.** $\{(x, y)|x \leq 2\} \cap \{(x, y)|y \leq 2\}$

25. $\{(x, y)|x + y \leq 6\} \cap \{(x, y)|x + y \geq 4\}$ **26.** $\{(x, y)|y \geq x\} \cap \{(x, y)|y \leq -x\}$

■ *Graph each inequality.*

27. $y \geq |x|$ **28.** $y < |x|$ **29.** $y < |x| - 2$ **30.** $y \geq |x| + 5$

CHAPTER SUMMARY

[7.1]   The solution of an equation or inequality in two variables is an **ordered pair** of numbers. The replacement set for possible solutions is the infinite set of ordered pairs $\{(x, y)|x$ and y are real numbers$\}$, denoted by $R \times R$ or R^2.

[7.2]   We can use a **Cartesian** (or **rectangular**) **coordinate system** to graph ordered pairs of numbers on a plane. The components of a given ordered pair are called the **coordinates** of its graph. ·

The graph of a first-degree (linear) equation in two variables is a straight line. The *x*-coordinate of a point of intersection of a graph with the *x*-axis is an *x*-**intercept** of the graph, and the *y*-coordinate of a point of intersection of a graph with the *y*-axis is a *y*-**intercept** of the graph. These intercepts are obtained by setting $y = 0$ and $x = 0$, respectively, in the equation.

[7.3] A **relation** is a set of ordered pairs. A **function** is a relation in which no two ordered pairs have the same first component.

The **domain** of a relation (or function) is the set of all first components in the ordered pairs in the relation (or function). The **range** of a relation (or function) is the set of all second components in the ordered pairs in the relation (or function). The domain of a relation, defined by an equation $y = f(x)$, is the set of values of x which yields real values for y.

Symbols such as $f(x)$, $P(x)$, and $R(x)$ are called **function notations**.

[7.4] For any two points in a geometric plane corresponding to (x_1, y_1) and (x_2, y_2), the distance d between the points is given by the **distance formula**,

$$d = \sqrt{(x_2 - x_1)^2 + (y_2 - y_1)^2}.$$

The **slope** m of the line containing the points (x_1, y_1) and (x_2, y_2) is given by

$$m = \frac{y_2 - y_1}{x_2 - x_1} \qquad (x_2 \neq x_1).$$

The slopes of parallel lines are equal ($m_1 = m_2$), and the nonzero slopes of perpendicular lines are the negative reciprocals of each other ($m_1 = -1/m_2$).

[7.5] The **standard form** of an equation of a line is

$$ax + by + c = 0 \quad \text{or} \quad ax + by = c.$$

The **point-slope form** of an equation of a line is

$$y - y_1 = m(x - x_1),$$

where the line has slope m and contains the point (x_1, y_1).

The **slope-intercept form** of an equation of a line is

$$y = mx + b,$$

where the line has slope m and y-intercept b.

[7.6] The graph of a first-degree (linear) inequality in two variables,

$$ax + by < c \quad \text{or} \quad ax + by > c,$$

is a **half-plane**. The edge of the half-plane is not included; it is shown as a dashed line. The graph of

$$ax + by \leq c \quad \text{or} \quad ax + by \geq c$$

is also a half-plane; it includes the edge, which is shown as a solid line.

The symbols introduced in this chapter are listed on the inside of the front cover.

REVIEW EXERCISES

A

[7.1]   1. Find the missing component in each solution of $2x - 6y = 12$.

     a. $(0, ?)$ **b.** $(?, 0)$ **c.** $(3, ?)$

     2. List the solutions of $y = 2x - 3$, where x is 2, 4, and 6.

     3. In the equation $xy = 2x^2y + 3$, express y explicitly in terms of x.

[7.2]   ■ *Graph.*

     4. a. $y = 3x + 1$ **b.** $y = 2x - 5$

     5. a. $3x - y = 6$ **b.** $3x - y = 0$

[7.3]   6. Specify the domain and range of each relation. Is the relation a function?

     a. $\{(3, 5), (4, 5), (4, 6)\}$ **b.** $\{(2, 3), (3, 4), (4, 5), (5, 5)\}$

     7. Specify the domain of each relation and state whether the relation is a function.

     a. $xy - y = 6$ **b.** $4x^2 + y^2 = 16$

     8. Given that $f(x) = x - 4$, find each of the following:

     a. $f(6)$ **b.** $f(x + h) - f(x)$

     9. Given that $f(x) = 2x^2 - 3x + 1$, find each of the following:

     a. $f(3)$ **b.** $\dfrac{f(x + h) - f(x)}{h}$

[7.4]   10. **a.** Find the distance between the points $(3, -5)$ and $(6, 8)$.

        b. Find the slope of the line segment joining the points in part a.

     11. **a.** Find the distance between the points $(4, 2)$ and $(7, -4)$.

        b. Find the slope of the line segment joining the points in part a.

     12. **a.** Given points $P_1(2, 2)$, $P_2(-2, -2)$, $P_3(0, 4)$, and $P_4(-3, 1)$, show that the line segments P_1P_2 and P_3P_4 are parallel.

        b. Show that the line segments P_1P_2 and P_1P_3 are perpendicular.

[7.5]   13. Find the equation in standard form of the line through $(3, 5)$ with slope 2.

     14. Find the equation in standard form of the line through $(-4, 2)$ with slope $-\frac{1}{2}$.

     15. **a.** Write $3x - y = 4$ in slope-intercept form.

        b. Specify the slope and the y-intercept of its graph.

     16. **a.** Write $2x + 3y = 6$ in slope-intercept form.

        b. Specify the slope and the y-intercept of its graph.

[7.6]   ■ *Graph.*

17. **a.** $y > x + 2$ **b.** $y \leq 4x + 4$

18. **a.** $x - 2y < 6$ **b.** $2x - y \geq 6$

19. **a.** $\{(x, y)|y > -3\}$ **b.** $\{(x, y)|-2 < x \leq 3\}$

B   20. Graph
       a. $\{(x, y)|y \leq 4\} \cap \{(x, y)|x \geq 1\}$ **b.** $\{(x, y)|x + y \leq 4\} \cap \{(x, y)|y \geq 0\}$

21. Graph $y = |x| - 3$.

22. Graph $f(x) = 2x + 4$ and use line segments to represent the ordinates $f(-4)$ and $f(1)$.

23. Given the points $P_1(2, -3)$, $P_2(3, 4)$, and $P_3(4, -1)$, find the value of k in the ordered pair $P_4(6, k)$ that makes P_3P_4 perpendicular to P_1P_2.

24. Write an equation in standard form of the line that passes through $(3, -1)$ and $(4, 5)$.

25. Write an equation in standard form of the line that is perpendicular to the graph of $2y + x = 6$ and passes through $(2, -3)$.

26. Graph the inequality $y \geq |x + 2|$.

27. Express the area (A) of an equilateral triangle (three equal sides) as a function of the length of a side (s).

28. The sides of a rectangle are along the positive x- and y-axes of a rectangular coordinate system. Express the area (A) of the rectangle as a function of x if the corner of the rectangle that is not on an axis is on the graph of $2x + 4y = 7$ in the first quadrant.

8. FUNCTIONS, RELATIONS, AND THEIR GRAPHS: PART II

8.1 Graphs of quadratic relations

8.2 Sketching parabolas

8.3 Circles, ellipses, and hyperbolas

8.6 The inverse of a function

8.4 Sketching graphs of conic sections other than parabolas

8.5 Variation as a functional relationship

8.1

GRAPHS OF QUADRATIC RELATIONS

Consider the quadratic equation in two variables

$$y = x^2 - 4. \tag{1}$$

As with linear equations in two variables, solutions of this equation are ordered pairs. We need replacements for both x and y to obtain a statement we can judge to be true or false. Such ordered pairs can be found by arbitrarily assigning values to x and computing related values for y. For instance, assigning the value -3 to x in (1), we have

$$y = (-3)^2 - 4$$
$$= 5,$$

and so $(-3, 5)$ is a solution of Equation (1). Similarly, we find that

$$(-2, 0), \qquad (-1, -3), \qquad (0, -4), \qquad (1, -3), \qquad (2, 0), \quad \text{and} \quad (3, 5)$$

are also solutions of (1).

Plotting the corresponding points on the plane, we have the graph shown in Figure 8.1a. Clearly, these points do not lie on a straight line, and we might reasonably inquire whether the graph of

$$y = x^2 - 4$$

forms any kind of meaningful pattern on the plane. By plotting additional solutions with x-components between those already found, we may be able to obtain a clearer picture. Accordingly, we find the solutions

$$\left(\frac{-5}{2}, \frac{9}{4}\right), \left(\frac{-3}{2}, \frac{-7}{4}\right), \left(\frac{-1}{2}, \frac{-15}{4}\right), \left(\frac{1}{2}, \frac{-15}{4}\right), \left(\frac{3}{2}, \frac{-7}{4}\right), \left(\frac{5}{2}, \frac{9}{4}\right),$$

and, by plotting these points in addition to those found earlier, we have the graph shown in Figure 8.1b. It now appears reasonable to connect these points in sequence by a smooth curve, as in Figure 8.1c, and to assume that the curve is a good approximation to the graph of (1). This curve is an example of a **parabola**.

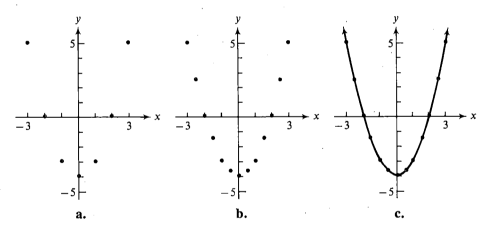

a.   b.   c.

Figure 8.1

More generally, the graph of any quadratic equation of the form

$$y = ax^2 + bx + c \quad (a \neq 0), \tag{2}$$

where a, b, and c are real numbers, is a parabola. Since with each x an equation of the form (2) associates only one y, such an equation defines a function whose domain is the set of real numbers and whose range is some subset of the real numbers. For example, by inspecting the graph in Figure 8.1c, we observe that the range of the function defined by (1) is

$$\{y \mid y \geq -4\}.$$

Observe that the coefficient, 1, of the second-degree term x^2 in $y = x^2 - 4$ is positive and that its graph, the parabola in Figure 8.1c, opens upward. In general, the graphs of all equations of the form $y = ax^2 + bx + c$ open upward if $a > 0$ and downward if $a < 0$.

Examples

a. The graph of $y = 2x^2 - x + 3$ will open upward because the coefficient of x^2 is positive.

b. The graph of $y = -2x^2 + x - 3$ will open downward because the coefficient of x^2 is negative.

The graph of a quadratic equation of the form

$$x = ay^2 + by + c \qquad (a \neq 0) \tag{3}$$

is also a parabola. In this case, in order to graph the equation, it is easier to assign arbitrary values to y to obtain associated values of x. For example, if y is assigned a value 2 in

$$x = y^2 - 4y + 3, \tag{4}$$

we have

$$x = (2)^2 - 4(2) + 3 = -1,$$

and $(-1, 2)$ is in the solution set of (4).

The graphs of $(-1, 2)$ and other ordered pairs in the solution set obtained in a similar way are shown in the graph of (4) in Figure 8.2. Note that for all but one value of x, (4) associates two values of y with each value of x in the domain. Hence, this equation and, more generally, equations of the form $x = ay^2 + by + c$ do not define functions.

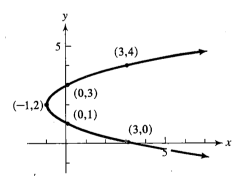

Figure 8.2

Observe that the graph of Equation (4), the parabola in Figure 8.2, opens to the right. As you graph the equations in Exercise 8.1, you will be able to observe that the graphs of parabolas are related to their equations as shown in the following chart.

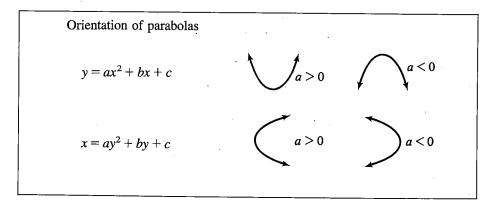

As in the case of the graph of a linear equation in Section 7.2, the absolute value of the ordinate of a point on the graph of a quadratic equation is the length of the line segment drawn perpendicular to the x-axis from the point on the curve to the x-axis. Graphing the equation

$$f(x) = -x^2 + 16$$

by first obtaining the ordered pairs $(-5, -9)$, $(-4, 0)$, $(-3, 7)$, $(-2, 12)$, $(-1, 15)$, $(0, 16)$, $(1, 15)$, $(2, 12)$, $(3, 7)$, $(4, 0)$, and $(5, -9)$, we have the parabola shown in Figure 8.3. Five representative ordinates are shown.

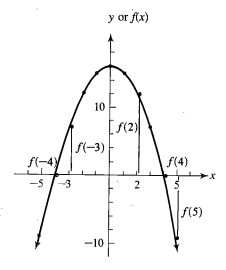

Figure 8.3

EXERCISE 8.1

A ■ **a.** *Find the solutions of the equation using integral values for x where* $-4 \leq x \leq 4$.
  b. *Use the solutions to graph the equation.*

Example $y = -x^2 + 3x$

Solution   Compute values of y for integral values of x in the interval.

a. $(-4, -28)$ **b.** y or $f(x)$
    $(-3, -18)$
    $(-2, -10)$
    $(-1, -4)$
    $(0, 0)$
    $(1, 2)$
    $(2, 2)$
    $(3, 0)$
    $(4, -4)$

1. $y = x^2 + 1$ **2.** $y = x^2 + 4$ **3.** $f(x) = x^2 - 3$

4. $f(x) = x^2 - 5$ **5.** $y = -x^2 + 4$ **6.** $y = -x^2 + 5$

7. $y = 3x^2 + x$ **8.** $y = 5x^2 - x$ **9.** $y = x^2 + 2x + 1$

10. $y = x^2 - 2x + 1$ **11.** $f(x) = -2x^2 + x - 3$ **12.** $f(x) = -x^2 + 2x - 1$

13. $f(x) = -x^2 + x + 1$ **14.** $f(x) = -3x^2 + x - 2$

■ **a.** *Find the solutions of the equation using integral values for y where* $-4 \leq y \leq 4$.
  b. *Use the solutions to graph the equation.*

Example $x = y^2 - 9$

Solution   Compute values of x for integral values of y in the interval.

a. $(7, -4)$ **b.**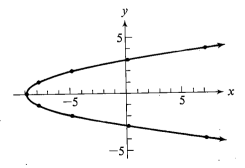
    $(0, -3)$
    $(-5, -2)$
    $(-8, -1)$
    $(-9, 0)$
    $(-8, 1)$
    $(-5, 2)$
    $(0, 3)$
    $(7, 4)$

15. $x = y^2$ **16.** $x = y^2 - 4$ **17.** $x = -4y^2 + 4$

18. $x = -2y^2 - 6$ **19.** $x = y^2 + 3y + 2$ **20.** $x = -y^2 + y + 2$

B   **21.** Graph the equation $f(x) = x^2 + 1$. Represent $f(0)$ and $f(4)$ by drawing line segments from $(0, 0)$ to $(0, f(0))$ and from $(4, 0)$ to $(4, f(4))$.

   22. Graph the equation $f(x) = x^2 - 1$. Represent $f(-3)$ and $f(2)$ by drawing line segments from $(-3, 0)$ to $(-3, f(-3))$ and from $(2, 0)$ to $(2, f(2))$.

8.2

SKETCHING PARABOLAS

When graphing a quadratic equation in two variables, it is desirable to select first components for the ordered pairs that ensure that the most significant parts of the graph are displayed. For a parabola, these parts include the intercepts, if they exist, and the **maximum** or **minimum** (highest or lowest) point on the curve.

Intercepts   In determining the x- and y-intercepts of the function

$$y = ax^2 + bx + c,$$

we let $x = 0$ to find the y-intercept and let $y = 0$ to find the x-intercepts.

Example   The y-intercept of

$$y = -2x^2 - 5x + 3 \tag{1}$$

can be found by assigning the value of 0 to x:

$$y = -2(0)^2 - 5(0) + 3 = 3.$$

Letting $y = 0$ in (1) yields

$$0 = -2x^2 - 5x + 3.$$

Thus, the x-intercepts of (1) are the solutions of

$$-2x^2 - 5x + 3 = 0.$$

These can be found by writing the equation equivalently as

$$-1(x + 3)(2x - 1) = 0.$$

By inspection, the solutions—and therefore the x-intercepts of the graph of (1)—are -3 and $\frac{1}{2}$.

For any function f, values of x for which $f(x) = 0$ are called **zeros** of the function. Thus, we have three different names for a single idea:

1. The elements of the solution set of the equation

$$ax^2 + bx + c = 0.$$

2. The zeros of the function defined by

$$y = ax^2 + bx + c.$$

3. The x-intercepts of the graph of

$$y = ax^2 + bx + c.$$

Recall from Section 6.4 that a quadratic equation in one variable may have no real solution, one real solution, or two real solutions. If the equation has no real solution, we find that the graph of the related quadratic equation in two variables does not touch the x-axis; if there is one solution, the graph is tangent to the x-axis at a point; if there are two real solutions, the graph crosses the x-axis at two distinct points. This is shown in Figure 8.4.

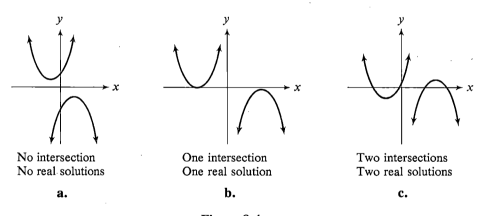

No intersection   One intersection   Two intersections
No real solutions   One real solution   Two real solutions
        a.   **b.**   **c.**

Figure 8.4

Low or high point   We can obtain the coordinates of the lowest (or the highest) point on a parabola by the procedure of completing the square that we considered in Section 6.2. For example, Equation (1) can be rewritten as

$$y = -2\left(x^2 + \frac{5}{2}x \qquad \right) + 3 \qquad\qquad (2)$$

and then as

$$y = -2\left(x^2 + \frac{5}{2}x + \frac{25}{16}\right) + 3 + 2\left(\frac{25}{16}\right),$$

where $2(^{25}\!/_{16})$ was both subtracted from and added to the right-hand member. Since

$$x^2 + \frac{5}{2}x + \frac{25}{16} = \left(x + \frac{5}{4}\right)^2$$

and

$$3 + 2\left(\frac{25}{16}\right) = \frac{49}{8},$$

Equation (2) is equivalent to

$$y = -2\left(x + \frac{5}{4}\right)^2 + \frac{49}{8}.$$

For $x = -^5\!/_4$, the expression $(x + ^5\!/_4)$ equals 0; hence y has a maximum value $^{49}\!/_8$. Therefore, the high point corresponds to $(-^5\!/_4, \,^{49}\!/_8)$. The graphs of the intercepts and high point are shown in Figure 8.5a. These points are sufficient to complete the graph in Figure 8.5b.

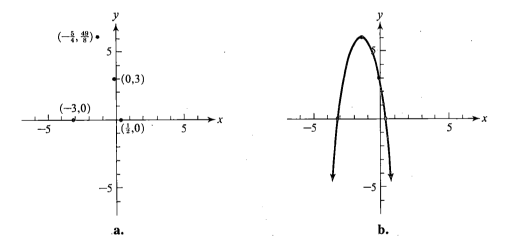

a.   b.

Figure 8.5

In general, if we complete the square on

$$y = ax^2 + bx + c,$$

we obtain

$$y = a\left(x + \frac{b}{2a}\right)^2 + \frac{4ac - b^2}{4a}.$$

Note that if $x = -\dfrac{b}{2a}$, the first term $a\left(x + \dfrac{b}{2a}\right)^2$ equals zero. For this value of x, y will have its greatest value or least value. Hence:

▶ *The graph of the equation $y = ax^2 + bx + c$ has a maximum or minimum point at*

$$x = -\frac{b}{2a}.$$ (3)

The maximum or minimum point is called the **vertex** of the parabola.

We can find the value of y at the vertex of any parabola by substituting the value of x obtained by Equation (3) in the equation of the parabola. The coordinates of the vertex, along with other information that we can obtain from an equation, enables us to sketch parabolas more efficiently than by plotting a large number of points.

Symmetry of a parabola Notice in Figure 8.6 that a line through the vertex of a parabola and parallel to an axis divides the parabola into two parts, each the mirror image of the other. This line is called the **axis** of the parabola, and we say that the curve is **symmetric** about this axis.

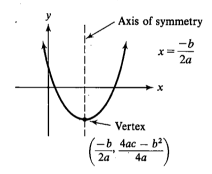

Figure 8.6

Example ﹒ Sketch the graph of $y = x^2 - 3x - 4$.

Solution We proceed as follows:

1. Since the coefficient of x^2 is positive, the curve opens upward.

2. The x-coordinate of the vertex is

$$-\frac{b}{2a} = -\frac{-3}{2} = \frac{3}{2}.$$

3. Substitute ³⁄₂ for x in the equation to obtain the y-coordinate of the vertex.

$$y = \left(\frac{3}{2}\right)^2 - 3\left(\frac{3}{2}\right) - 4 = -\frac{25}{4}$$

4. If $x = 0$,

$$y = 0^2 - 3(0) - 4 = -4,$$

and the y-intercept is -4. If $y = 0$, then

$$x^2 - 3x - 4 = 0$$
$$(x - 4)(x + 1) = 0$$
$$x = 4 \quad \text{or} \quad x = -1;$$

so the x-intercepts are 4 and -1.

5. Sketch the curve as shown.

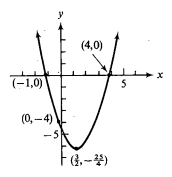

To sketch parabolas with equation $y = ax^2 + bx + c$:

1. Determine whether the curve opens upward (if $a > 0$) or downward (if $a < 0$).

2. Find the x-coordinate of the vertex, $-b/2a$.

3. Find the y-coordinate of the vertex by substituting the value of $-b/2a$ for x in the equation $y = ax^2 + bx + c$ and solving for y.

4. Identify and graph intercepts, if any.
   a. Let $x = 0$ to obtain y-intercept.
   b. Let $y = 0$ to obtain x-intercepts, if any.

5. **a.** If the x-intercepts are real numbers, you will have enough points to sketch the graph.
   b. If the x-values are imaginary when $y = 0$, locate one or two points on the curve and use symmetry to complete the sketch.

Quadratic inequalities

Quadratic inequalities of the form

$$y < ax^2 + bx + c$$

or

$$y > ax^2 + bx + c,$$

which define quadratic relations, can be graphed in the same manner in which we graphed linear inequalities in two variables in Section 7.6. We first graph the equation that has the same members and then shade an appropriate region as required.

Example   Graph $y < x^2 + 2$.   (4)

Solution overleaf

Solution

We first graph

$$y = x^2 + 2. \qquad (5)$$

Substituting the coordinates of the origin $(0, 0)$ in Inequality (4), we obtain

$$0 < 0 + 2,$$

which is true, and so the part of the plane including the origin is shaded. Since the graph of (5) is not part of the graph of (4), a broken curve is used.

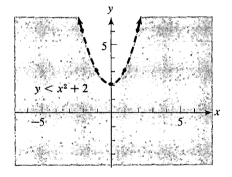

EXERCISE 8.2

A   ▪ *Use the steps on page 259 to sketch each graph.*

Example $y = -2x^2 + x - 1$

Solution

1. Since the coefficient of x^2 is -2 and $-2 < 0$, the curve opens downward.

2. The x-coordinate of the vertex is $-\dfrac{1}{2(-2)} = \dfrac{1}{4}$.

3. Substitute $\frac{1}{4}$ for x in the equation to obtain the y-value of the vertex.

$$y = -2\left(\frac{1}{4}\right)^2 + \frac{1}{4} - 1 = -\frac{7}{8}$$

4. If $x = 0$, then $y = -1$; so the y-intercept is -1.
If $y = 0$, then $-2x^2 + x - 1 = 0$.
The solutions are imaginary; so there are no x-intercepts.

5. To find two points, let $x = -1$ and $x = 1$.
If $x = -1$,

$$y = -2(-1)^2 + (-1) - 1 = -4;$$

so $(-1, -4)$ is on the curve. If $x = 1$,

$$y = -2(1)^2 + 1 - 1 = -2;$$

so $(1, -2)$ is on the curve.

Use symmetry to sketch the graph as shown.

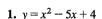

1. $y = x^2 - 5x + 4$ **2.** $y = x^2 + x - 6$ **3.** $f(x) = x^2 - 4x$

4. $f(x) = x^2 + 6x$ **5.** $g(x) = x^2 + 4x - 5$ **6.** $g(x) = x^2 + 6x + 8$

7. $y = 2x^2 + 3x - 9$ 8. $y = 3x^2 - 5x - 2$ 9. $f(x) = -x^2 + 7x - 6$

10. $f(x) = -x^2 - 4x + 5$ 11. $y = -2x^2 - 7x - 3$ 12. $y = -3x^2 + 7x - 2$

■ *Graph.*

Example $\{(x, y) \,|\, y = x^2 - 1\} \cap \{(x, y) \,|\, y = 1 - x^2\}$

Solution   Graph each equation.

The graph of the intersection of the given sets consists of the two points shown in red.

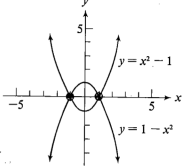

13. $\{(x, y) \,|\, y = -x^2\} \cap \{(x, y) \,|\, y = x^2 - 4\}$

14. $\{(x, y) \,|\, y = x^2\} \cap \{(x, y) \,|\, y = -x^2 + 4\}$

15. $\{(x, y) \,|\, y = x^2 - 2x + 1\} \cap \{(x, y) \,|\, y = -x^2 + 2x - 1\}$

16. $\{(x, y) \,|\, y = x^2 + 1\} \cap \{(x, y) \,|\, y = -x^2 - 1\}$

■ *Graph.*

Example $y \geq x^2 + 2x$

Solution   First, graph the equation $y = x^2 + 2x$. By inspecting the inequality $y \geq x^2 + 2x$, we can determine that the region to be shaded lies above the graph of $y = x^2 + 2x$.

Alternatively, we can find the region by selecting any ordered pair whose graph is not on the curve and determining whether or not the ordered pair is a solution of the inequality. Arbitrarily selecting $(0, 6)$ and substituting in $y \geq x^2 + 2x$, we obtain

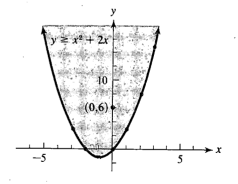

$$(6) \geq (0)^2 + 2(0),$$

which is a true statement. Hence, we shade the region above the graph of $y = x^2 + 2x$ because it includes the graph of $(0, 6)$.

17. $y \geq x^2 - 6x + 8$ 18. $y \leq x^2 - 6x + 8$ 19. $y < x^2 + x - 6$

20. $y > x^2 + x - 6$ 21. $y \leq x^2 + 3x + 2$ 22. $y \geq x^2 + 3x + 2$

B 23. Sketch the family of four curves $y = kx^2$ ($k = 1, 2, 3, 4$) on a single set of axes.

24. Sketch the family of four curves $y = kx^2$ ($k = -1, -2, -3, -4$) on a single set of axes.

25. Sketch the family of four curves $y = x^2 + k$ $(k = -2, 0, 2, 4)$ on a single set of axes.

26. Sketch the family of four curves $y = x^2 + kx$ $(k = -2, 0, 2, 4)$ on a single set of axes.

■ *Use graphical methods in Problems 27–30.*

Example   Find two numbers whose sum is 18 and whose product is as large as possible.

Solution

Let x represent the first number and $18 - x$ represent the second number. Then their product, P, is

$$P = x(18 - x)$$
$$= -x^2 + 18x.$$

The x-coordinate of the vertex is

$$-\frac{b}{2a} = -\frac{18}{2(-1)} = 9,$$

and hence $x = 9$ yields the maximum value for P. Since $18 - x = 9$, the numbers are 9 and 9.

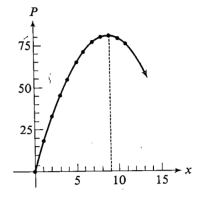

27. Find two numbers whose sum is 12 and whose product is a maximum.

28. Find the maximum area of a rectangle whose perimeter is 100 inches. [*Hint:* Let x represent the length; then $50 - x$ represents the width and $A = x(50 - x)$.]

29. The equation $d = 32t - 8t^2$ relates the distance d (feet) above the ground reached in time t (seconds) by an object thrown vertically upward. Sketch the graph of the equation for $0 \le t \le 4$ and estimate the time it will take the object to reach its greatest height.

30. In Problem 29, how many seconds is the object in the air?

■ *Sketch the graph of each function.*

Example

$$f(x) = \begin{cases} x^2 & \text{if } x \le 1 \\ -x^2 + 2 & \text{if } x > 1 \end{cases}$$

Solution   Graph the function separately over each part of the domain. First,

$$f(x) = x^2 \quad \text{for } x \le 1$$

is graphed, as shown in Figure a. Then,

$$f(x) = -x^2 + 2 \quad \text{for } x > 1$$

is graphed to obtain the complete graph shown in Figure b.

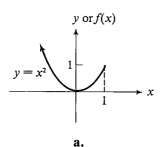

a.

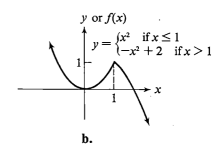

b.

31.
$$f(x) = \begin{cases} x^2 & \text{if } x \le 2 \\ x+2 & \text{if } x > 2 \end{cases}$$

32.
$$f(x) = \begin{cases} x^2 - 4 & \text{if } x \le 2 \\ 2 - x & \text{if } x > 2 \end{cases}$$

33.
$$f(x) = \begin{cases} x^2 & \text{if } x \le 0 \\ -x^2 & \text{if } x > 0 \end{cases}$$

34.
$$f(x) = \begin{cases} -x^2 + 1 & \text{if } x \le 0 \\ x^2 - 1 & \text{if } x > 0 \end{cases}$$

35.
$$f(x) = \begin{cases} x^2 & \text{if } x \le 0 \\ 3x & \text{if } 0 < x < 1 \\ -x^2 + 4 & \text{if } x \ge 1 \end{cases}$$

36.
$$f(x) = \begin{cases} x^2 - 4 & \text{if } x \le -2 \\ 0 & \text{if } -2 < x < 2 \\ -x^2 + 4 & \text{if } x \ge 2 \end{cases}$$

■ *If, for each x, $f(x) = f(-x)$, then the graph of f is symmetric with respect to the y-axis. In Problems 37 and 38, which functions have graphs that are symmetric with respect to the y-axis?*

37. a. $f(x) = x$ **b.** $f(x) = x^2$

38. a. $f(x) = \sqrt{x^2 + 1}$ **b.** $f(x) = |x|$

39. Graph $\{(x, y) | y > x^2\} \cap \{(x, y) | y \le x + 2\}$.

40. Graph $\{(x, y) | y < -x^2 + 4\} \cap \{(x, y) | y \ge x^2 - 1\}$.

8.3

CIRCLES, ELLIPSES, AND HYPERBOLAS

In addition to $y = ax^2 + bx + c$ and $x = ay^2 + by + c$, there are three other types of second-degree equations in two variables whose graphs are of particular interest. We shall discuss each of them separately.

Circles   First, consider the graph of the equation

$$x^2 + y^2 = 25. \tag{1}$$

Solving this equation explicitly for y, we have

$$y = \pm\sqrt{25 - x^2}. \tag{2}$$

Note that the domain of the relation defined by (2) is $\{x \mid -5 \leq x \leq 5\}$ because $25 - x^2$ is nonnegative for these values of x. Hence, it is not necessary to assign any values to x such that $x < -5$ or $x > 5$ because y is imaginary for these values.

Assigning values to x, we find the following ordered pairs in the solution set:

$$(-5, 0), \quad (-4, 3), \quad (-3, 4), \quad (0, 5), \quad (3, 4), \quad (4, 3),$$
$$(5, 0), \quad (-4, -3), \quad (-3, -4), \quad (0, -5), \quad (3, -4), \quad (4, -3).$$

Plotting these points, we have the graph shown in Figure 8.7a. Connecting these points with a smooth curve, we have the graph shown in Figure 8.7b. This graph is a circle with radius 5 and center at the origin. Since, except for -5 and 5, each permissible value for x is associated with two values for y—one positive and one negative— Equations (1) and (2) do not define functions.

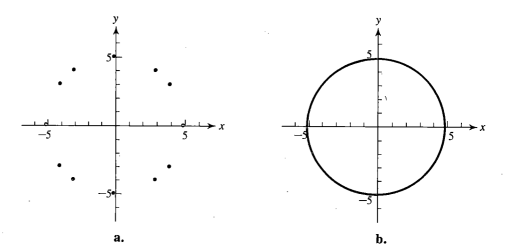

a.   b.

Figure 8.7

The number 25 in the right-hand member of Equation (1) determines the length of the radius of the circle. Hence, we can generalize and observe that any equation of the form

$$x^2 + y^2 = r^2$$

graphs into a circle with radius r and center at the origin. (See Problem 18, Exercise 8.3 for a more general approach.)

Ellipses

The second equation in two variables that is of special interest is typified by

$$4x^2 + 9y^2 = 36. \qquad (3)$$

We obtain solutions of (3) by first solving explicitly for y to obtain

$$y = \pm\frac{2}{3}\sqrt{9 - x^2}. \qquad (4)$$

We then note that the domain of the relation defined by (4) is $\{x \mid -3 \leq x \leq 3\}$ because $9 - x^2$ is nonnegative for these values of x. Then we assign to x some values in the domain to obtain part of the solution set:

$$(-3, 0), \quad \left(-2, \frac{2}{3}\sqrt{5}\right), \quad \left(-1, \frac{4}{3}\sqrt{2}\right), \quad (0, 2), \quad \left(1, \frac{4}{3}\sqrt{2}\right), \quad \left(2, \frac{2}{3}\sqrt{5}\right),$$

$$(3, 0), \quad \left(-2, -\frac{2}{3}\sqrt{5}\right), \quad \left(-1, -\frac{4}{3}\sqrt{2}\right), \quad (0, -2), \quad \left(1, -\frac{4}{3}\sqrt{2}\right), \quad \left(2, -\frac{2}{3}\sqrt{5}\right).$$

Locating the corresponding points on the plane (decimal approximations for irrational numbers can be obtained from a calculator or the table on page 431) and connecting them with a smooth curve, we have the graph shown in Figure 8.8. This curve is called an **ellipse**.

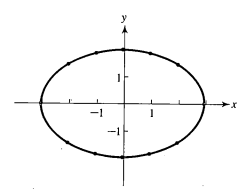

Figure 8.8

Hyperbolas

The third type of second-degree equation in two variables is typified by

$$x^2 - y^2 = 9.$$

Solving this equation for y, we obtain

$$y = \pm\sqrt{x^2 - 9}.$$

Solving $x^2 - 9 \geq 0$, we first observe that the domain of the relation is $\{x \mid x \geq 3 \text{ or } x \leq -3\}$ because members of the interval $-3 < x < 3$ yield imaginary values for y. We then obtain, as a part of the solution set, some ordered pairs:

$$(-5, 4), \quad (-4, \sqrt{7}), \quad (-3, 0), \quad (4, \sqrt{7}), \quad (5, 4),$$
$$(-5, -4), \quad (-4, -\sqrt{7}), \quad (3, 0), \quad (4, -\sqrt{7}), \quad (5, -4).$$

If we plot the corresponding points and connect them with a smooth curve, we obtain Figure 8.9. This curve is called a **hyperbola**.

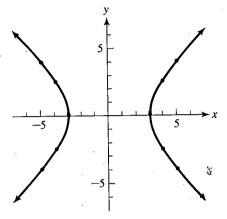

Figure 8.9

Graphs of $ax^2 - by^2 = 0$ A special type of quadratic equation in two variables that is of second degree is typified by

$$4x^2 - y^2 = 0.$$

Solving this equation for y, we obtain

$$y^2 = 4x^2.$$

By extraction of roots,

$$y = \sqrt{4x^2} = 2x \quad \text{or} \quad y = -\sqrt{4x^2} = -2x.$$

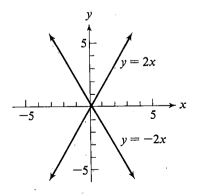

Figure 8.10

The graphs of these two first-degree equations are the straight lines shown in Figure 8.10. In general,

The graph of any equation of the form $ax^2 - by^2 = 0$ *is two straight lines that pass through the origin.*

The graphs of the equations dealt with in this section, together with the parabola of Sections 8.1 and 8.2, are called **conic sections**, or **conics**, because such curves are intersections of a plane and a cone, as shown in Figure 8.11.

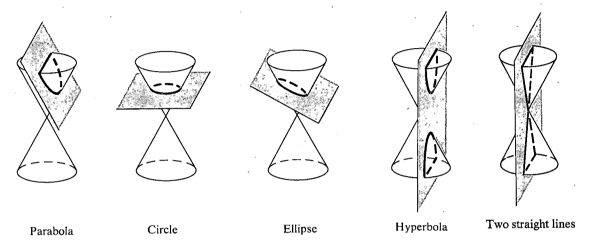

Parabola   Circle   Ellipse   Hyperbola   Two straight lines

Figure 8.11

EXERCISE 8.3

A ■ *a. Rewrite each equation with y as the left-hand member.*
     b. State the domain of the relation defined by the equation.
     c. Graph the equation.

Example $4x^2 + y^2 = 36$

Solution

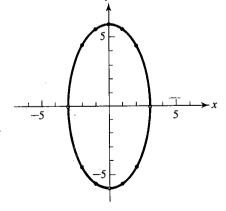

a. $y^2 = 36 - 4x^2$
    $y^2 = 4(9 - x^2)$
    $y = \pm 2\sqrt{9 - x^2}$

b. The domain is $\{x \mid -3 \le x \le 3\}$ because $9 - x^2$ is nonnegative for these
values of x.

c. Some solutions are

$(3, 0)$, $(2, \pm 2\sqrt{5})$, $(1, \pm 4\sqrt{2})$, $(0, \pm 6)$,
$(-3, 0)$, $(-2, \pm 2\sqrt{5})$, $(-1, \pm 4\sqrt{2})$.

1. $x^2 + y^2 = 4$

2. $x^2 + y^2 = 9$

3. $9x^2 + y^2 = 36$

4. $4x^2 + y^2 = 4$

5. $x^2 + 4y^2 = 16$

6. $x^2 + 9y^2 = 4$

7. $x^2 - y^2 = 1$

8. $4x^2 - y^2 = 1$

9. $y^2 - x^2 = 9$

10. $4y^2 - 9x^2 = 36$

11. $2x^2 + 3y^2 = 24$

12. $4x^2 + 3y^2 = 12$

13. $9x^2 - y^2 = 0$

14. $x^2 - 9y^2 = 0$

15. $4x^2 - 9y^2 = 0$

16. $9x^2 - 4y^2 = 0$

B 17. Graph $4x^2 + y^2 = 0$. Generalize from the result and discuss the graph of any equation of the form $ax^2 + by^2 = c$, where $a, b > 0$ and $c = 0$.

18. Use the distance formula to show that the graph of $\{(x, y) \mid x^2 + y^2 = r^2\}$ is the set of all points located a distance r from the origin.

19. Graph $x^2 - y^2 = 4$, $x^2 - y^2 = 1$, and $x^2 - y^2 = 0$ on the same set of axes.

20. Graph $4x^2 - y^2 = 16$, $4x^2 - y^2 = 4$, and $4x^2 - y^2 = 0$ on the same set of axes.

8.4

SKETCHING GRAPHS OF CONIC SECTIONS OTHER THAN PARABOLAS

In Section 8.2, we stressed the use of certain properties to help sketch parabolas. These properties include the x- and y-intercepts, the coordinates of the highest or lowest points on the graph, and symmetry. We can also use intercepts, together with the form of the equation, to help sketch other conic sections. To do this, we first observe that the ideas developed in Section 8.3 can be summarized as follows:

A quadratic equation of the form

$$ax^2 + by^2 = c \qquad (a \text{ and } b \text{ not both equal to } 0)$$

has a graph in the plane with center at the origin that is one of the following:

1. A circle if $a = b$ and a, b, and c have like signs. For example,

$$x^2 + y^2 = 9$$

or

$$4x^2 + 4y^2 = 12.$$

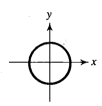

2. An ellipse if $a \neq b$ and a, b, and c have like signs. For example,

$$4x^2 + 9y^2 = 36 \qquad \text{(a)}$$

or

$$x^2 + 8y^2 = 12. \qquad \text{(b)}$$

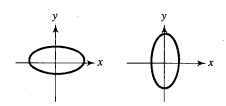

3. A hyperbola if a and b are opposite in sign and $c \neq 0$. For example,

$$2x^2 - 6y^2 = 9 \qquad \text{(c)}$$

or

$$4y^2 - x^2 = 2. \qquad \text{(d)}$$

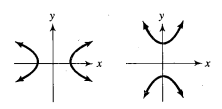

4. Two distinct lines through the origin if a and b are opposite in sign and $c = 0$. For example,

$$4x^2 - y^2 = 0$$

or

$$2y^2 - 7x^2 = 0.$$

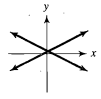

When you have recognized the general form of the curve, the graph of a few points should suffice to sketch the graph. The intercepts, for instance, are always easy to determine.

Example   Graph $x^2 + 4y^2 = 8$.

Solution   By comparing $x^2 + 4y^2 = 8$ to case 2 above, we note immediately that its graph is an ellipse.

If $y = 0$, then $x = \pm\sqrt{8}$, and if $x = 0$, then $y = \pm\sqrt{2}$. Hence, we have the solutions $(\sqrt{8},\ 0)$, $(-\sqrt{8},\ 0)$, $(0,\ \sqrt{2})$, and $(0,\ -\sqrt{2})$. We can then sketch the graph of the given equation as shown.

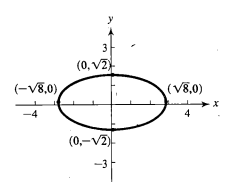

Example   Graph $x^2 - y^2 = 3$.

Solution   By comparing $x^2 - y^2 = 3$ to case 3c above, we see that its graph is a hyperbola.

Solution continued overleaf

If $y = 0$, then $x = \pm\sqrt{3}$, and if $x = 0$, then y is imaginary (the graph will not cross the y-axis).

By assigning a few other arbitrary values to one of the variables—say, $(4, \ )$ and $(-4, \ )$ to x—we can find the additional ordered pairs

$$(4, \sqrt{13}), \quad (4, -\sqrt{13}),$$
$$(-4, \sqrt{13}), \quad (-4, -\sqrt{13}),$$

which satisfy the given equation. The graph can then be sketched as shown.

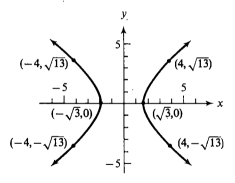

Asymptotes of hyperbolas

We can often simplify the process of graphing a hyperbola by first graphing two straight lines that are approached by the branches of the hyperbola. These lines are called **asymptotes** of the graph.

► *The graph of $ax^2 - by^2 = c$ will approach asymptotes that are the graph of $ax^2 - by^2 = 0$, and the graph of $ay^2 - bx^2 = c$ will approach asymptotes that are the graph of $ay^2 - bx^2 = 0$ for each $a, b, c > 0$.*

Example   Graph $4y^2 - 9x^2 = 36$.

Solution   We first compare this equation with case 3d above and note that its graph is a hyperbola.

If $x = 0$, $4y^2 = 36$, from which $y = \pm 3$. If $y = 0$, then x is imaginary.

We then note from the above property that the equation of the asymptotes to the graph of $4y^2 - 9x^2 = 36$ is

$$4y^2 - 9x^2 = 0,$$

which can be written as

$$y^2 = \frac{9}{4}x^2,$$

from which

$$y = \pm\frac{3}{2}x.$$

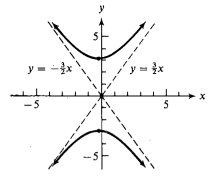

We first graph the asymptotes (dashed lines in the figure) and the y-intercepts of the hyperbola. Then we complete sketching the graph using the asymptotes as guides.

In Section 8.2 we noted that a parabola was symmetric about a line through its vertex. Perhaps you have noted that the graphs of quadratic equations of the form

$ax^2 + by^2 = c$ that we have considered in this section are symmetric about the y-axis, the x-axis, and about the origin. For this reason, such conic sections are called **central conics**.

EXERCISE 8.4

A ■ *a. Name the graph of each equation.* **b.** *Sketch the graph.* **c.** *Give the equations of the asymptotes for each hyperbola.*

Examples

a. $4x^2 = 36 - 9y^2$ **b.** $y^2 - 9x^2 = 4$

Solutions

a. Rewrite in the form $ax^2 + by^2 = c$.

$$4x^2 + 9y^2 = 36$$

By inspection, the graph is an ellipse.

The x-intercepts are 3 and -3; the y-intercepts are 2 and -2.

The graph is as shown.

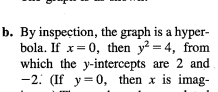

b. By inspection, the graph is a hyperbola. If $x = 0$, then $y^2 = 4$, from which the y-intercepts are 2 and -2. (If $y = 0$, then x is imaginary.) The graph can be completed by using the asymptotes as guides.

c. The equation of the asymptotes is

$$y^2 - 9x^2 = 0,$$

from which

$$y = 3x \quad \text{or} \quad y = -3x.$$

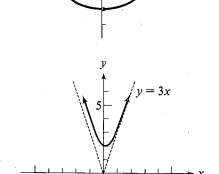

1. $x^2 + y^2 = 49$
2. $x^2 + y^2 = 64$
3. $4x^2 + 25y^2 = 100$
4. $x^2 + 2y^2 = 8$
5. $4x^2 - 4y^2 = 0$
6. $x^2 - 9y^2 = 0$
7. $x^2 - y^2 = 9$
8. $x^2 - 2y^2 = 8$
9. $x^2 + 4y^2 = 12$
10. $3x^2 + 2y^2 = 18$
11. $4x^2 = 1 - 4y^2$
12. $9y^2 = 2 - 9x^2$
13. $3x^2 = 12 - 4y^2$
14. $4x^2 = 12 - 3y^2$
15. $y^2 = 16 + 4x^2$
16. $x^2 = 25 + 5y^2$
17. $y^2 + 4 = -2x^2$
18. $4x^2 + 6 = -y^2$

19. Graph $\{(x, y)\,|\,y = x + 2\} \cap \{(x, y)\,|\,4x^2 + y^2 = 36\}$.

20. Graph $\{(x, y)\,|\,y = x - 2\} \cap \{(x, y)\,|\,4x^2 + 4y^2 = 36\}$.

B   ■ *Graph each inequality.*

21. $x^2 + 4y^2 \le 9$ 　　　　　　　　　　**22.** $x^2 - y^2 \ge 1$

23. $4x^2 + 9y^2 \ge 36$ 　　　　　　　　　**24.** $9x^2 - 4y^2 \le 36$

25. Show that the ordinates to the graph of $ax^2 - by^2 = c$ $(x \ne 0)$ for any value of x are given by

$$y = \pm \sqrt{\frac{a}{b}}x\left(\sqrt{1 - \frac{c}{ax^2}}\right).$$

26. Explain why, for large values of x, the value of the expression

$$\sqrt{1 - \frac{c}{ax^2}}$$

approaches 1 and hence, using the results of Exercise 25, why

$$y = \pm \sqrt{\frac{a}{b}}x$$

are equations for the asymptotes to the graph of $ax^2 - by^2 = c$.

8.5

VARIATION AS A
FUNCTIONAL RELATIONSHIP

There are two types of widely used functional relationships to which custom has assigned special names.

Direct variation

First, any function defined by the equation

$$y = kx \qquad (k \text{ a positive constant}) \tag{1}$$

is an example of **direct variation**. The variable y is said to **vary directly** as the variable x. Another example of direct variation is

$$y = kx^2 \qquad (k \text{ a positive constant}), \tag{1a}$$

where we say that y varies directly as the square of x. In general,

$$y = kx^n \qquad (k \text{ a positive constant and } n > 0) \tag{1b}$$

asserts that y varies directly as the nth power of x.

Examples

a. The circumference (C) of a circle is given by the equation

$$C = 2\pi r, \tag{2}$$

where r is the radius of the circle. Thus, the circumference of a circle *varies directly* as the radius.

b. The area (A) of a circle of radius r is given by the equation

$$A = \pi r^2. \tag{3}$$

Thus, the area of a circle *varies directly* as the square of the radius.

Since for each r, Equations (2) and (3) associate only one value of C or A, both of these equations define functions—Equation (2) a linear function and Equation (3) a quadratic function.

Inverse variation

The second important type of function is defined by the equation

$$xy = k \qquad (k \text{ a positive constant}), \tag{4}$$

where x (or y) is said to **vary inversely** as y (or x). When (4) is written in the form

▶

$$y = \frac{k}{x}, \tag{5}$$

y is said to vary inversely as x. Similarly, if

$$y = \frac{k}{x^2}, \tag{5a}$$

y is said to vary inversely as the square of x, and so on.

Example

Each rectangle in a set of rectangles has an area of 24 square units. Since the area of a rectangle is given by $lw = A$, we have

$$lw = 24,$$

and the length and width of the rectangle can be seen to vary inversely.

Equations (5) and (5a) associate only one y with x $(x \neq 0)$. Hence, an inverse variation defines a function with domain $\{x \mid x \neq 0\}$.

The names *direct* and *inverse* as applied to variation arise from the facts that in direct variation an assignment of increasing absolute values to x results in increasing absolute values of y, whereas in inverse variation an assignment of increasing absolute values to x results in decreasing absolute values of y.

Constant of variation

The constant involved in equations describing direct or inverse variation is called the **constant of variation**. If we know that one variable varies directly or inversely as another, and if we have one set of associated values for the variables, we can find the constant of variation involved.

Example

Given that y varies directly as x^2, and that $y = 4$ when $x = 7$, find the constant of variation.

Solution

The fact that y varies directly as x^2 tells us that

$$y = kx^2. \tag{6}$$

The fact that $y = 4$ when $x = 7$ tells us that $(7, 4)$ is a solution of Equation (6). Hence, by substitution,

$$4 = k(7)^2 = k(49),$$

from which

$$k = \frac{4}{49}.$$

Substituting $\frac{4}{49}$ for k in (6), we obtain

$$y = \frac{4}{49}x^2,$$

the equation that specifically expresses the direct variation.

Joint variation

In the event that one variable varies as the product of two or more other variables, we refer to the relationship as **joint variation**. Thus, if y varies jointly as u, v, and w, we have

▶
$$y = kuvw. \tag{7}$$

Two or more types of variation may take place concurrently. For example, y may vary directly as x and inversely as z, giving rise to the equation

$$y = k\frac{x}{z}.$$

It should be pointed out that the way in which the word *variation* is used in this section is technical, and when the ideas of direct, inverse, or joint variation are encountered, we should always think of equations of the form (1), (5), or (7). For instance, the equations

$$y = 2x + 1, \quad y = \frac{1}{x} - 2, \quad \text{and} \quad y = xz + 2$$

do not describe examples of variation within our meaning of the word.

Proportions

An alternative expression—**proportional to**—is used to describe the relationships discussed in this section. To say that "y is directly proportional to x" or "y is inversely proportional to x" is another way of describing direct and inverse variation. The use of the word *proportion* arises from the fact that any two solutions (a, b) and (c, d) of an equation expressing a direct variation satisfy an equation of the form

$$\frac{a}{b} = \frac{c}{d},$$

which is commonly called a **proportion**.

As an example, consider the situation in which the volume of a gas varies directly with the absolute temperature and inversely as the pressure; this can be represented by the relationship

$$V = \frac{kT}{P}. \tag{8}$$

For any set of values T_1, P_1, and V_1, we can solve for k to obtain

$$k = \frac{V_1 P_1}{T_1}, \tag{8a}$$

and for any other set of values T_2, P_2, and V_2,

$$k = \frac{V_2 P_2}{T_2}. \tag{8b}$$

Equating the right-hand members of (8a) and (8b), we obtain

$$\frac{V_1 P_1}{T_1} = \frac{V_2 P_2}{T_2},$$

from which the value of any variable can be determined if the values of the other variables are known.

EXERCISE 8.5

A ■ *Write an equation expressing the relationship between the variables, using k as the constant of variation.*

Example At a constant temperature, the volume (V) of a gas varies inversely as the pressure (P).

Solution
$$V = \frac{k}{P}$$

1. The distance (d) traveled by a car moving at a constant rate varies directly as the time (t).
2. The tension (T) on a spring varies directly as the distance (s) it is stretched.
3. The current (I) in an electrical circuit with constant voltage varies inversely as the resistance (R) of the circuit.

4. The time (t) required by a car to travel a fixed distance varies inversely as the rate (r) at which it travels.

5. The volume (V) of a rectangular box of fixed depth varies jointly as its length (l) and width (w).

6. The power (P) in an electric circuit varies jointly as the resistance (R) and the square of the current (I).

■ *Find the constant of variation for each of the stated conditions.*

Example V varies inversely as P, and $V = 100$ when $P = 30$.

Solution   Write an equation expressing the relationship between the variables.

$$V = \frac{k}{P}$$

Substitute the known values for the variables and solve for k.

$$100 = \frac{k}{30}$$
$$k = 3000$$

7. y varies directly as x, and $y = 6$ when $x = 2$.

8. y varies directly as x, and $y = 2$ when $x = 5$.

9. u varies inversely as the square of v, and $u = 2$ when $v = 10$.

10. r varies inversely as the cube of t, and $r = 8$ when $t = 10$.

11. z varies jointly as x and y, and $z = 8$ when $x = 2$ and $y = 2$.

12. p varies jointly as q and r, and $p = 5$ when $q = 2$ and $r = 7$.

13. z varies directly as the square of x and inversely as the cube of y, and $z = 4$ when $x = 3$ and $y = 2$.

14. z varies directly as the cube of x and inversely as the square of y, and $z = 2$ when $x = 2$ and $y = 4$.

15. z varies inversely as the sum of x^2 and y, and $z = 12$ when $x = 4$ and $y = 6$.

16. z varies directly as the sum of x and y and inversely as their product, and $z = 8$ when $x = 3$ and $y = 4$.

■ *Solve by first obtaining the constant of variation.*

Example   If V varies directly as T and inversely as P, and $V = 40$ when $T = 300$ and $P = 30$, find V when $T = 324$ and $P = 24$.

Solution

Write an equation expressing the relationship between the variables.

$$V = \frac{kT}{P} \tag{1}$$

Substitute the initially known values for V, T, and P. Solve for k.

$$40 = \frac{k\,300}{30}$$

$$4 = k$$

Rewrite Equation (1) with k replaced by 4.

$$V = \frac{4T}{P}$$

Substitute the second set of values for T and P and solve for V.

$$V = \frac{4(324)}{24} = 54$$

17. If y varies directly as x^2 and $y = 9$ when $x = 3$, find y when $x = 4$.

18. If r varies directly as s and inversely as t and $r = 12$ when $s = 8$ and $t = 2$, find r when $s = 3$ and $t = 6$.

19. The distance a particle falls in a certain medium is directly proportional to the square of the length of time it falls. If the particle falls 16 feet in 2 seconds, how far will it fall in 10 seconds?

20. In Problem 19, how far will the body fall in 20 seconds?

21. The pressure exerted by a liquid at a given point varies directly as the depth of the point beneath the surface of the liquid. If a certain liquid exerts a pressure of 40 pounds per square foot at a depth of 10 feet, what would be the pressure at 40 feet?

22. The volume (V) of a gas varies directly as its temperature (T) and inversely as its pressure (P). A gas occupies 20 cubic feet at a temperature of 300°K (Kelvin) and a pressure of 30 pounds per square inch. What will the volume be if the temperature is raised to 360°K and the pressure is decreased to 20 pounds per square inch?

23. The maximum safe uniformly distributed load (L) for a horizontal beam varies jointly as its breadth (b) and square of the depth (d) and inversely as the length (l). An 8-foot beam with $b = 2$ and $d = 4$ will safely support a uniformly distributed load of up to 750 pounds. How many uniformly distributed pounds will an 8-foot beam support if $b = 2$ and $d = 6$?

24. The resistance (R) of a wire varies directly as the length (l) and inversely as the square of its diameter (d); 50 feet of wire of diameter 0.012 inch has a resistance of 10 ohms. What is the resistance of 50 feet of the same type of wire if the diameter is increased to 0.015 inch?

25. Simple interest (I) varies jointly with the principal (P) and the time (t) in years for which P is invested. If $2400 interest is earned on a 4-year investment of $6000, how much money would have to be invested for 2 years in order to earn $800 interest?

26. Using the variation relationship in Problem 25, and given that $240 interest is earned on a 3-year investment of $1000, how long would it take to earn $240 on a $2000 investment?

■ *Represent the relationship of the problem cited as a proportion by eliminating the constant of variation. Then solve for the required variable.*

Example   If V varies directly as T and inversely as P, and $V = 40$ when $T = 300$ and $P = 30$, find V when $T = 324$ and $P = 24$.

Solution   Write an equation expressing the relationship between the variables.

$$V = \frac{kT}{P}$$

Solve for k.

$$k = \frac{VP}{T}$$

Write a proportion relating the variables for two different sets of conditions.

$$\frac{V_1 P_1}{T_1} = \frac{V_2 P_2}{T_2}$$

Substitute the known values of the variables and solve for V_2.

$$\frac{(40)(30)}{300} = \frac{V_2(24)}{324}$$

$$V_2 = \frac{(324)(40)(30)}{300(24)} = 54$$

27. Problem 17	**28.** Problem 18	**29.** Problem 19	**30.** Problem 20
31. Problem 21	**32.** Problem 22	**33.** Problem 23	**34.** Problem 24
35. Problem 25	**36.** Problem 26		

B   **37.** From the formula for the circumference of a circle, $C = \pi D$, show that the ratio of the circumference of two circles equals the ratio of their respective diameters.

38. From the formula for the area of a circle, $A = \pi r^2$, show that the ratio of the areas of two circles equals the ratio of the squares of their respective radii.

39. The intensity of light on a surface varies inversely as the square of its distance from a light source. What is the effect on the intensity of light on an object if the distance between the light source and the object is doubled?

40. The frequency of vibration of a guitar string varies directly as the square root of the tension and inversely as the length of the string. What is the effect on the frequency if the tension is increased fourfold and the length of the string is doubled?

41. Graph on the same set of axes the linear functions defined by $y = kx$, $x \geq 0$, when the constant $k = 1, 2, 3$. What effect does a variation in k have on the graph of $y = kx$?

42. Graph on the same set of axes the equations $y = kx$, $y = kx^2$, and $y = kx^3$, where $k = 2$ and $x \geq 0$. What effect does increasing the degree of the equation $y = kx^n$ have on the graph of the equation?

8.6

THE INVERSE
OF A FUNCTION

One-to-one
functions

A function f has been viewed as a relation in which each element in the domain is associated with just one element in its range. If each element in its range is also associated with just one element in its domain, the function f is called a **one-to-one function**.

Whether or not a function defined by an equation in two variables is a one-to-one function can be readily determined from its graph. Recall (page 226) that any vertical line will intersect the graph of a function in at most one point. If the function f is one-to-one, any horizontal line will also intersect its graph in at most one point, as shown in Figure 8.12a. If it is not one-to-one, a horizontal line will intersect the graph at more than one point, as shown in Figure 8.12b.

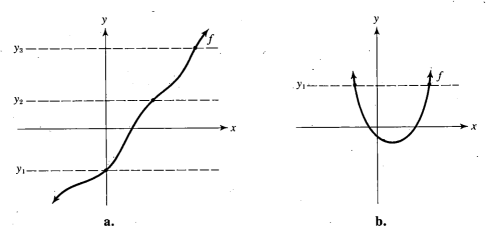

a.   b.

Figure 8.12

Inverse
functions

If the components of each ordered pair in a one-to-one function f are interchanged, the result will be a function, denoted by f^{-1} (read "f inverse" or "the inverse of f") and is called the **inverse function** of f.

Example

a. If $f = \{(1, 2), (3, 4), (5, 6)\}$, then $f^{-1} = \{(2, 1), (4, 3), (6, 5)\}$.

b. If $f = \{(1, 2), (2, 5), (3, 5)\}$, then its inverse $\{(2, 1), (5, 2), (5, 3)\}$ is not a function because f is not a one-to-one function.

It is evident from the definition of an inverse function that the domain and range of f^{-1} are the range and domain, respectively, of f. Hence, if a one-to-one function f is defined by an equation in two variables, the inverse f^{-1} can be obtained by interchanging the variables.

Example   The inverse of the function defined by

$$y = 4x - 3 \tag{1}$$

is defined by

$$x = 4y - 3, \tag{2}$$

where the variables in (1) have been interchanged. When y is expressed in terms of x we obtain

$$y = \frac{1}{4}(x + 3). \tag{2a}$$

Equations (2) and (2a) are equivalent.

Graphs of   The graphs of a function and its inverse are related in an interesting way. To see this,
inverse   we first observe in Figure 8.13 that the graphs of the ordered pairs (a, b) and (b, a) are
functions   always located symmetrically with respect to the graph of $y = x$. Therefore, because
                   for every ordered pair (a, b) in f, the ordered pair (b, a) is in f^{-1}, the graphs of
                   $y = f^{-1}(x)$ and $y = f(x)$ are reflections of each other in the graph of the equation
                   $y = x$. Figure 8.14 shows the graphs of

$$y = 4x - 3$$

and its inverse,

$$x = 4y - 3 \quad \text{or} \quad y = \frac{1}{4}(x + 3),$$

together with the graph of $y = x$.

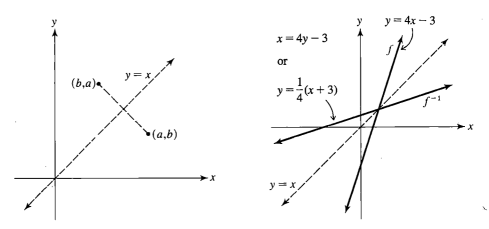

Figure 8.13   Figure 8.14

Of course, if a function defined by an equation in two variables is not a one-to-one function, its inverse is not a function.

Example

The figure shows the graph of the function

$$y = x^2,$$

together with the graph of its inverse,

$$x = y^2 \quad \text{or} \quad y = \pm\sqrt{x}.$$

Since, for all but one value in its domain ($x = 0$), the inverse associates two different values of y with each x, the inverse is not a function.

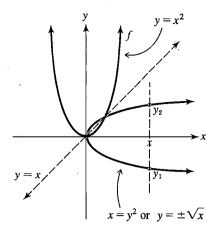

$x = y^2$ or $y = \pm\sqrt{x}$

$f^{-1}[f(x)] = x;$
$f[f^{-1}(x)] = x$

If f and f^{-1} are both functions, f associates the number a with the *unique* number b, and f^{-1} associates the number b with the *unique* number a. Therefore, it must be true that, for every x in the domain of f,

$$f^{-1}[f(x)] = x$$

(read "f inverse of f of x is equal to x"). And, for every x in the domain of f^{-1},

$$f[f^{-1}(x)] = x$$

(read "f of f inverse of x is equal to x"). For example, if $f = \{(2, 5)\}$, then $f(2) = 5$, $f^{-1} = \{(5, 2)\}$, and $f^{-1}(5) = 2$. Note that

$$f^{-1}[f(2)] = f^{-1}(5) = 2$$

and

$$f[f^{-1}(5)] = f(2) = 5.$$

As another example, consider Equation (1) and note that if f is the linear function defined by

$$f(x) = 4x - 3,$$

then f is a one-to-one function. Now f^{-1} is defined by

$$f^{-1}(x) = \frac{1}{4}(x + 3),$$

and we see that

$$f^{-1}[f(x)] = \frac{1}{4}[(4x - 3) + 3] = x$$

and

$$f[f^{-1}(x)] = 4\left[\frac{1}{4}(x + 3)\right] - 3 = x.$$

EXERCISE 8.6

A ■ *Find the inverse of each function and state whether the inverse is also a function.*

Examples   **a.** $f = \{(2, 1), (4, 6)\}$ **b.** $q = \{(-2, 1), (3, 4), (5, 4)\}$

Solutions   **a.** Interchanging the components in each   **b.** Interchanging the components in each
ordered pair, we obtain   ordered pair, we obtain

$$f^{-1} = \{(1, 2), (6, 4)\},$$ $$\{(1, -2), (4, 3), (4, 5)\}$$

which is a function because the or-   which is not a function because two
dered pairs do not contain the same   ordered pairs contain the same first
first component.   component, 4.

1. $f = \{(-2, -2), (2, 2)\}$ **2.** $f = \{(-5, 1), (5, 2)\}$
3. $q = \{(1, 3), (2, 3), (3, 4)\}$ **4.** $q = \{(2, 2), (3, 3), (4, 3)\}$
5. $g = \{(1, 1), (2, 2), (3, 3)\}$ **6.** $g = \{(-2, 0), (0, 0), (4, -2)\}$

■ *In Problems 7–16, each equation defines a function, f.*
a. *Write the equation defining its inverse.*
b. *Sketch the graphs of f and its inverse on the same set of axes.*
c. *State whether the inverse is a function.*

Example $y = x^2 - 9$

Solution   **a.** Interchange variables x and y.   **b.**

$$x = y^2 - 9.$$

c. The inverse is not a function
because for each $x > -9$,
there are two values of y.

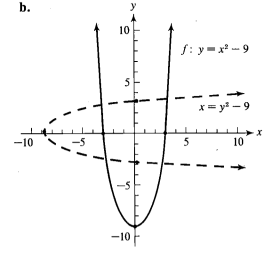

    7. $2x + 4y = 7$ **8.** $3x - 2y = 5$ **9.** $x - 3y = 6$

    10. $4x + y = 4$ **11.** $y = x^2 - 4x$ **12.** $y = x^2 - 4$

    13. $y = \sqrt{4 + x^2}$ **14.** $y = -\sqrt{x^2 - 4}$ **15.** $y = |x|$

    16. $y = |x| + 1$

B   ■ *In Problems 17–22, each equation defines a one-to-one function f. Find the equation defining f^{-1} and show that $f[f^{-1}(x)] = f^{-1}[f(x)] = x$. [Hint: Solve explicitly for y and let $y = f(x)$.]*

    17. $y = x$ **18.** $y = -x$ **19.** $2x + y = 4$

    20. $x - 2y = 4$ **21.** $3x - 4y = 12$ **22.** $3x + 4y = 12$

CHAPTER SUMMARY

[8.1]   The graph of $y = ax^2 + bx + c$, where a, b, and c are constants with $a \neq 0$, is a **parabola**. The parabola opens upward if $a > 0$ and downward if $a < 0$. The graph of $x = ay^2 + by + c$ is a parabola that opens to the right if $a > 0$ and to the left if $a < 0$.

[8.2]   A quick sketch of a parabola can be made by first identifying its form by inspecting the coefficient of the second-degree term, finding the y-intercepts and the x-intercepts if they exist, and then finding the coordinates of the **vertex** at $x = -b/2a$ and using the property of **symmetry.**

[8.3]   The graph of $ax^2 + by^2 = c$ can be an **ellipse**, a **circle**, a **hyperbola**, or a **pair of lines**. These graphs, together with parabolas, are called **conic sections**.

[8.4]   A quick sketch of the graph of an equation of the form $ax^2 + by^2 = c$ can be made by first identifying its form by inspecting the coefficients (see pages 268 and 269), and then using such aids as **intercepts** and **asymptotes** to draw the curve.

[8.5]   The equation $y = kx$ (k a positive constant) defines a function called a **direct variation**. The equation $xy = k$ (k a positive constant) defines a function called an **inverse variation**. The equation $y = kuv$ defines a relationship called a **joint variation**. In each case, the constant k is called the **constant of variation.**

[8.6]   The **inverse** of the function f can be obtained by interchanging the components of each ordered pair in f. If f is a one-to-one function, then the inverse f^{-1} is also a function. The graphs of f^{-1} and f are reflections of each other in the graph of $y = x$.

The symbols introduced in this chapter are listed on the inside of the front cover.

REVIEW EXERCISES

A

[8.1–8.2]   **1. a.** Use algebraic methods to find the x- and y-intercepts of the graph of $y = x^2 - 6x + 5$.
   b. Find the coordinates of the minimum point of the graph.

   2. Graph the equation in Problem 1.

   3. Graph $y = -x^2 + 6x - 8$.

   4. Graph $y \geq x^2 + 1$.

[8.3–8.4]   ■ *State the domain of each relation.*

   5. $9x^2 - y^2 = 9$ **6.** $x^2 + 4y^2 = 4$

   ■ *Name the graph of each equation.*

   7. a. $x^2 - 3y^2 = 8$ **b.** $x^2 = 4 - y^2$
   8. a. $x^2 - y + 4 = 0$ **b.** $2y^2 = 4 - x^2$

   ■ *Graph each equation.*

   9. $x^2 + y^2 = 36$ **10.** $4x^2 + y^2 = 36$
   11. $9x^2 - y^2 = 0$ **12.** $4x^2 - y^2 = 16$

[8.5]   **13.** If y varies inversely as t^2, and $y = 16$ when $t = 3$, find y when $t = 4$.

   14. If y varies directly as s^2 and inversely as t, and $y = 4$ when $s = 2$ and $t = 5$, find y when $s = 6$ and $t = 8$.

   15. The distance a particle falls in a certain medium is directly proportional to the square of the length of time it falls. If the particle falls 28 centimeters in 4 seconds, how far will it fall in 8 seconds?

   16. The weight of a body above the surface of the earth varies inversely as the square of its distance from the center of the earth. If we assume the radius of the earth to be 4000 miles, how much would a man weigh 500 miles above the earth's surface if he weighed 200 pounds on the surface?

[8.6]   **17.** Find the inverse of each function and state whether the inverse is also a function.
   a. $f = \{(3, 7), (4, 8), (8, 4)\}$ **b.** $g = \{(2, 6), (3, 8), (5, 8)\}$

   18. Find the inverse of the function defined by $y = x^2 + 9x$ and state whether the inverse is a function.

   19. Graph the function defined by $y = 2x + 6$ and its inverse on the same set of axes.

   20. Graph the function defined by $y = x^2 + 1$ and its inverse on the same set of axes.

B   21. Graph $\{(x, y)|y > x^2 + 1\} \cap \{(x, y)|y < x + 3\}$.

22. Graph $\{(x, y)|y \geq x^2 - 4\} \cap \{(x, y)|y < -x^2\}$.

23. Graph

$$f(x) = \begin{cases} x^2 & \text{if } x < -1, \\ -x^2 + 2 & \text{if } x \geq -1. \end{cases}$$

24. Graph the inequality $2x^2 + y^2 \leq 8$.

25. Graph $y = kx$, $y = kx^2$, and $y = kx^3$ on the same set of axes for $k = 2$ and $x \geq 0$.

26. A function f is defined by $y - 2x = 4$. Find f^{-1} and show that $f[f^{-1}(x)] = x$ and $f^{-1}[f(x)] = x$.

9. EXPONENTIAL AND LOGARITHMIC FUNCTIONS

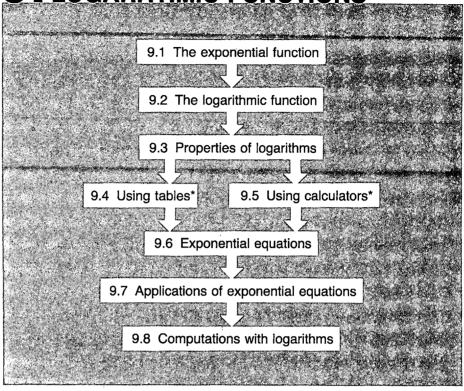

9.1 The exponential function

9.2 The logarithmic function

9.3 Properties of logarithms

9.4 Using tables*

9.5 Using calculators*

9.6 Exponential equations

9.7 Applications of exponential equations

9.8 Computations with logarithms

9.1

THE EXPONENTIAL FUNCTION

In Chapter 2, powers a^x were defined for any real number a and natural number x. In Chapter 5, the definition was extended to allow x to be a negative integer, or zero; also a rational number if $a > 0$. We now inquire whether we can interpret powers with irrational exponents, such as

$$b^\pi, \qquad b^{\sqrt{2}}, \quad \text{or} \quad b^{-\sqrt{3}},$$

to be real numbers.

In Section 5.5, we observed that irrational numbers may be approximated by rational numbers to as great a degree of accuracy as desired. That is, $\sqrt{2} \approx 1.4$ or

---

*Either Section 9.4 or Section 9.5 provides the prerequisites for the following sections.

$\sqrt{2} \approx 1.414$, and so on. Also, although we do not prove it here, if x and y are rational numbers such that $x > y$,

$$\text{if } b > 1, \quad \text{then } b^x > b^y,$$

and

$$\text{if } 0 < b < 1, \quad \text{then } b^x < b^y.$$

Now, because 2^x is defined for any rational number x, the following sequence of inequalities should seem plausible even though 2^x has not been defined for an irrational number x.

$$2^1 < 2^{\sqrt{2}} < 2^2$$
$$2^{1.4} < 2^{\sqrt{2}} < 2^{1.5}$$
$$2^{1.41} < 2^{\sqrt{2}} < 2^{1.42}$$
$$2^{1.414} < 2^{\sqrt{2}} < 2^{1.415},$$

and so on, where $2^{\sqrt{2}}$ is a number lying between the number on the left and that on the right.

It is clear that this process can be continued indefinitely and that the difference between the number on the left and that on the right can be made as small as we please. This being the case, we assume that there is just one number, $2^{\sqrt{2}}$, that will satisfy each inequality if this process is carried on indefinitely. Since we can produce the same type of argument for any irrational exponent x, we shall assume that b^x $(b > 0)$ is defined for all real values of x.

Since for each real x there is one and only one number b^x, the equation

$$f(x) = b^x \qquad (b > 0) \tag{1}$$

defines a function. Because $1^x = 1$ for all real values of x, Equation (1) defines a constant function if $b = 1$. If $b \neq 1$, we say that (1) defines an **exponential function**.

Note that b is restricted to be greater than 0. One reason is that b^x $(b < 0)$ is *imaginary* for $x = 1/n$, n even. For example, $(-4)^{1/2}$, $(-9)^{1/2}$, and $(-16)^{1/2}$ are imaginary numbers.

Graphs of exponential functions

Exponential functions can be visualized more clearly by considering their graphs. We illustrate two typical examples in which $0 < b < 1$ and $b > 1$, respectively on page 288. Assigning values to x in the equations

$$f(x) = \left(\frac{1}{2}\right)^x \quad \text{and} \quad f(x) = 2^x,$$

we find some ordered pairs in each function and sketch the graphs in Figure 9.1.

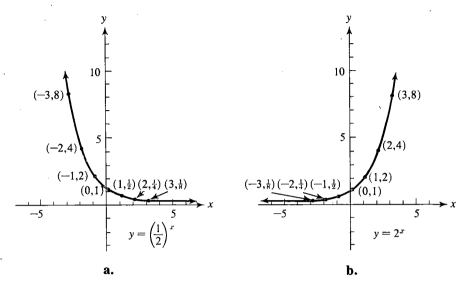

Figure 9.1

If $f(x) = \left(\dfrac{1}{2}\right)^x$, then

$$f(-3) = \left(\dfrac{1}{2}\right)^{-3} = 8;$$

$$\vdots \qquad \vdots \qquad \vdots$$

$$f(0) \ = \left(\dfrac{1}{2}\right)^{0} \ = 1;$$

$$\vdots \qquad \vdots \qquad \vdots$$

$$f(3) \ = \left(\dfrac{1}{2}\right)^{3} \ = \dfrac{1}{8}.$$

If $f(x) = 2^x$, then

$$f(-3) = 2^{-3} = \dfrac{1}{8};$$

$$\vdots \qquad \vdots \qquad \vdots$$

$$f(0) \ = 2^0 \ = 1;$$

$$\vdots \qquad \vdots \qquad \vdots$$

$$f(3) \ = 2^3 \ = 8.$$

Notice that the graph of the function determined by $f(x) = (\frac{1}{2})^x$ goes *down* to the right, and the graph of the function determined by $f(x) = 2^x$ goes *up* to the right. For this reason, we say that the former function is a **decreasing function** and the latter is an **increasing function.** In each case, the domain is the set of real numbers and the range is the set of positive real numbers. Notice that both functions are one-to-one functions.

The above properties are summarized as follows:

For the exponential function $f(x) = b^x$ $(b > 0,\ b \neq 1)$:

1. Domain: x a real number.

2. Range: $b^x > 0$.

3. If $b > 1$, the function is increasing;
if $0 < b < 1$, the function is decreasing.

EXERCISE 9.1

A   ■ *Find the second component of each of the ordered pairs that makes the pair a solution of the equation.*

Example $y = 2^x$; $(-4, \underline{?}), (0, \underline{?}), (4, \underline{?})$

Solution   For $x = -4$, $y = 2^{(-4)} = \dfrac{1}{2^4} = \dfrac{1}{16}$.

For $x = 0$, $y = 2^0 = 1$.
For $x = 4$, $y = 2^4 = 16$.

The ordered pairs are $(-4, \frac{1}{16})$, $(0, 1)$, and $(4, 16)$.

1. $y = 3^x$; $(0, \underline{?}), (1, \underline{?}), (2, \underline{?})$ **2.** $y = 4^x$; $(-\frac{1}{2}, \underline{?}), (0, \underline{?}), (\frac{1}{2}, \underline{?})$
3. $y = 2^x$; $(-4, \underline{?}), (0, \underline{?}), (4, \underline{?})$ **4.** $y = 5^x$; $(-2, \underline{?}), (0, \underline{?}), (2, \underline{?})$
5. $y = (\frac{1}{2})^x$; $(-4, \underline{?}), (0, \underline{?}), (4, \underline{?})$ **6.** $y = (\frac{1}{3})^x$; $(-3, \underline{?}), (0, \underline{?}), (3, \underline{?})$
7. $y = 10^x$; $(-2, \underline{?}), (-1, \underline{?}), (0, \underline{?})$ **8.** $y = 10^x$; $(0, \underline{?}), (1, \underline{?}), (2, \underline{?})$

■ *Graph each exponential equation. Use selected integral values.*

Example $y = 3^x$

Solution   For convenience, arbitrarily select integral values of x, say,

$(-2, \ ), (-1, \ ), (0, \ ), (1, \ ), (2, \ ), (3, \ )$.

Determine the y-components of each ordered pair.

$\left(-2, \dfrac{1}{9}\right), \left(-1, \dfrac{1}{3}\right), (0, 1), (1, 3), (2, 9), (3, 27)$.

Plot the points and connect them with a smooth curve.

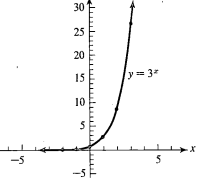

9. $y = 4^x$ **10.** $y = 6^x$ **11.** $y = 10^x$ **12.** $y = 2^{-x}$
13. $y = 3^{-x}$ **14.** $y = -3^x$ **15.** $y = -2^x$ **16.** $y = (\frac{1}{3})^x$
17. $y = (\frac{1}{4})^x$ **18.** $y = (\frac{1}{10})^x$ **19.** $y = (\frac{1}{2})^{-x}$ **20.** $y = (\frac{1}{3})^{-x}$

B   ■ *In Problems 21 and 22, use graphical methods to estimate the solutions of the exponential equations.*

21. $2^x = 5$ [*Hint:* Graph $y_1 = 2^x$ and $y_2 = 5$ and approximate the value of x at their point of intersection.]

22. $3^x = 4$ [See hint for Problem 21.]

23. For what set of positive real numbers a will $y = a^x$ define an increasing function? A decreasing function?

24. For what set of positive real numbers a will $y = a^{-x}$ define an increasing function? A decreasing function?

9.2

THE LOGARITHMIC FUNCTION

In Section 9.1, we were concerned with finding values for powers b^x ($b > 0$, $b \neq 1$) for given values of the exponent x. In this section, we shall first consider the inverse of the exponential function $y = b^x$. Then we shall see how the notion of this inverse will enable us to find values for the exponent x for given values of the power b^x.

Inverse of the exponential function

Recall from Section 8.6 that the inverse of a function can be obtained by interchanging the components in each ordered pair of the function and that this may be accomplished by interchanging the variables in the defining equation of the function. Recall also that the graph of the inverse of a function is symmetric to the graph of the function about the line with equation $y = x$.

Because the exponential function

$$f: \quad y = b^x \qquad (b > 0, \ b \neq 1) \tag{1}$$

with domain $\{x \mid x \in R\}$ and range $\{y \mid y > 0\}$ is a one-to-one function, there is only one y associated with each x as well as only one x associated with each y. Therefore, the inverse of function (1) is also a function, namely,

$$f^{-1}: \quad x = b^y \qquad (b > 0, \ b \neq 1). \tag{2}$$

Since the domain of (1) is the same as the range of (2) and the range of (1) is the same as the domain of (2), we have $\{x \mid x > 0\}$ for the domain of (2), while the range of (2) is the set of real numbers.

The graphs of functions of the form (2) can be illustrated by the example

$$x = 10^y \qquad (x > 0).$$

We assign arbitrary values to x, say, 0.01, 0.1, 1, 10, and 100, and obtain the ordered pairs which can be plotted and connected with a smooth curve as in Figure 9.2a.

If $x = 0.01$, then $0.01 = 10^{-2} = 10^y$; so $y = -2$.

If $x = 0.1$, then $0.1 = 10^{-1} = 10^y$; so $y = -1$.

If $x = 1$, then $1 = 10^0 = 10^y$; so $y = 0$.

If $x = 10$, then $10 = 10^1 = 10^y$; so $y = 1$.

If $x = 100$, then $100 = 10^2 = 10^y$; so $y = 2$.

Alternatively, we can reflect the graph of $y = 10^x$ about the graph of $y = x$ and obtain the same result as shown in Figure 9.2b.

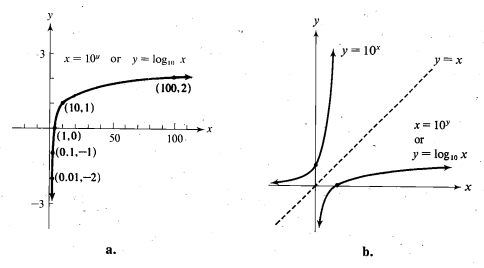

Figure 9.2

Logarithmic notation

It is always useful to be able to write an equation in the two variables x and y so that the variable y is expressed explicitly in terms of the variable x. To do this in equations such as (2), we use the notation

$$y = \log_b x \qquad (x > 0, \ b > 0, \ b \neq 1), \tag{3}$$

where $\log_b x$ is read "logarithm to the base b of x" or "logarithm of x to the base b." The functions defined by such equations are called **logarithmic functions**.

It should be recognized that

$$x = b^y \quad \text{and} \quad y = \log_b x \tag{4}$$

are two different forms of an equation defining the same function—in the same way that $x - y = 4$ and $y = x - 4$ define the same function—and that we may use whichever equation suits our purpose. The two equations in (4), which define $\log_b x$, enable us to write exponential equations in logarithmic form.

Examples   **a.** $5^2 = 25$ is equivalent to $\log_5 25 = 2$.

b. $8^{1/3} = 2$ is equivalent to $\log_8 2 = \frac{1}{3}$.

c. $3^{-2} = \frac{1}{9}$ is equivalent to $\log_3 \frac{1}{9} = -2$.

Also, logarithmic statements may be written in exponential form.

Examples   **a.** $\log_{10} 100 = 2$ is equivalent to $10^2 = 100$.

b. $\log_3 81 = 4$ is equivalent to $3^4 = 81$.

c. $\log_2 \frac{1}{2} = -1$ is equivalent to $2^{-1} = \frac{1}{2}$.

Note that because $x = b^y$ and $y = \log_b x$ are equivalent equations, $\log_b x$ (equal to y) is an *exponent*. In particular, $\log_b x$ is the exponent on b such that the power equals x. Thus,

$$b^{\log_b x} = x.$$

Examples   Express each logarithm as an integer.

a. $\log_{10} 100$ **b.** $\log_2 16$

Solutions   **a.** $\log_{10} 100$ is the *exponent* on 10 such that the power equals 100. Hence,

$$\log_{10} 100 = 2.$$

b. $\log_2 16$ is the *exponent* on 2 such that the power equals 16. Hence,

$$\log_2 16 = 4.$$

Note that because $b^0 = 1$,

▶ $\log_b 1 = 0$ $(b > 0,\ b \neq 1)$.

Furthermore, because $b^1 = b$,

▶ $\log_b b = 1$ $(b > 0,\ b \neq 1)$.

EXERCISE 9.2

A   ■ *Express each equation in logarithmic notation.*

| Examples | **a.** $3^2 = 9$ | **b.** $16^{1/4} = 2$ | **c.** $64^{-1/3} = \dfrac{1}{4}$ |

| Solutions | **a.** $\log_3 9 = 2$ | **b.** $\log_{16} 2 = \dfrac{1}{4}$ | **c.** $\log_{64} \dfrac{1}{4} = -\dfrac{1}{3}$ |

1. $4^2 = 16$ **2.** $5^3 = 125$ **3.** $3^3 = 27$ **4.** $8^2 = 64$

5. $\left(\dfrac{1}{2}\right)^2 = \dfrac{1}{4}$ **6.** $\left(\dfrac{1}{3}\right)^2 = \dfrac{1}{9}$ **7.** $8^{-1/3} = \dfrac{1}{2}$ **8.** $64^{-1/6} = \dfrac{1}{2}$

9. $10^2 = 100$ **10.** $10^0 = 1$ **11.** $10^{-1} = 0.1$ **12.** $10^{-2} = 0.01$

■ *Express each equation in exponential notation.*

| Examples | **a.** $\log_6 36 = 2$ | **b.** $\log_{1/5} 125 = -3$ | **c.** $\log_{10} 10{,}000 = 4$ |

| Solutions | **a.** $6^2 = 36$ | **b.** $\left(\dfrac{1}{5}\right)^{-3} = 125$ | **c.** $10^4 = 10{,}000$ |

13. $\log_2 64 = 6$ **14.** $\log_5 25 = 2$

15. $\log_3 9 = 2$ **16.** $\log_{16} 256 = 2$

17. $\log_{1/3} 9 = -2$ **18.** $\log_{1/2} 8 = -3$

19. $\log_{10} 1000 = 3$ **20.** $\log_{10} 1 = 0$

21. $\log_{10} 0.01 = -2$ **22.** $\log_{10} 0.0001 = -4$

■ *Find the value of each logarithm, using the fact that $\log_b x$ is the exponent on b such that the power is equal to x.*

| Examples | **a.** $\log_4 16$ | **b.** $\log_3 81$ |

| Solutions | **a.** 4 raised to what power equals 16? *Ans.* 2 | **b.** 3 raised to what power equals 81? *Ans.* 4 |

23. $\log_7 49$ **24.** $\log_2 32$ **25.** $\log_4 64$

26. $\log_3 27$ **27.** $\log_3 \sqrt{3}$ **28.** $\log_5 \sqrt{5}$

29. $\log_5 \dfrac{1}{5}$ **30.** $\log_3 \dfrac{1}{3}$ **31.** $\log_2 2$

32. $\log_{10} 10$ **33.** $\log_{10} 100$ **34.** $\log_{10} 1$

35. $\log_{10} 0.1$ **36.** $\log_{10} 0.01$

■ *Solve for the unknown value.*

Examples   **a.** $\log_2 x = 3$ **b.** $\log_b 2 = \dfrac{1}{2}$

Solutions   **a.** Write in exponential form.   **b.** Write in exponential form.

$$2^3 = x$$ $$b^{1/2} = 2$$

Solve for the variable.   Solve for the variable.

$$x = 8$$ $$(b^{1/2})^2 = 2^2$$
$$b = 4$$

37. $\log_3 9 = y$ **38.** $\log_5 125 = y$ **39.** $\log_b 8 = 3$ **40.** $\log_b 625 = 4$

41. $\log_4 x = 3$ **42.** $\log_{1/2} x = -5$ **43.** $\log_2 \dfrac{1}{8} = y$ **44.** $\log_5 \dfrac{1}{5} = y$

45. $\log_b 10 = \dfrac{1}{2}$ **46.** $\log_b 0.1 = -1$ **47.** $\log_2 x = 2$ **48.** $\log_{10} x = -3$

B ■ *Simplify each expression.*

Example   **a.** $\log_2(\log_3 3)$ **b.** $\log_5(\log_2 32)$

Solution   **a.** Since $\log_3 3 = 1$,   **b.** Since $\log_2 32 = 5$,

$$\log_2(\log_3 3) = \log_2 1$$ $$\log_5(\log_2 32) = \log_5 5$$
$$= 0.$$ $$= 1.$$

49. $\log_2(\log_4 16)$ **50.** $\log_5(\log_5 5)$ **51.** $\log_{10}[\log_3(\log_5 125)]$

52. $\log_{10}[\log_2(\log_3 9)]$ **53.** $\log_2[\log_2(\log_2 16)]$ **54.** $\log_4[\log_2(\log_3 81)]$

55. $\log_b(\log_b b)$ **56.** $\log_b(\log_a a^b)$

57. For what values of x is $\log_b(x - 9)$ defined?

58. For what values of x is $\log_b(x^2 - 4)$ defined?

9.3

PROPERTIES OF LOGARITHMS

Because a logarithm is an exponent by definition, the three laws given below are valid for positive real numbers b $(b \neq 1)$, x_1, x_2, and all real numbers m.

$$\log_b(x_1 x_2) = \log_b x_1 + \log_b x_2 \tag{1}$$

Example

$$\log_2 32 = \log_2(4 \cdot 8) = \log_2 4 + \log_2 8$$
$$5 \qquad\qquad\quad = \quad 2 \;+\; 3$$

$$\log_b \frac{x_2}{x_1} = \log_b x_2 - \log_b x_1 \tag{2}$$

Example

$$\log_2 8 = \log_2 \frac{16}{2} = \log_2 16 - \log_2 2$$
$$3 \qquad\qquad\quad = \quad 4 \;-\; 1$$

$$\log_b(x_1)^m = m \log_b x_1 \tag{3}$$

Example

$$\log_2 64 = \log_2 (4)^3 = 3 \log_2 4$$
$$6 \qquad\qquad\quad = 3 \cdot 2$$

We shall refer to the above equations as the first, second, and third laws of logarithms, respectively. These laws can be applied to rewrite expressions involving products, quotients, powers, and roots in forms that are sometimes more useful.

Example   Simplify $\log_b\left(\dfrac{x^3 y^{1/2}}{z}\right)^{1/5}$.

Solution   We can first write

$$\frac{1}{5} \log_b\left(\frac{x^3 y^{1/2}}{z}\right)$$

by using the third law of logarithms. Then, using the first and second laws, we have that

$$\frac{1}{5} \log_b\left(\frac{x^3 y^{1/2}}{z}\right) = \frac{1}{5}\left[\log_b x^3 + \log_b y^{1/2} - \log_b z\right],$$

from which, by using the third law again, we have that

$$\log_b\left(\frac{x^3 y^{1/2}}{z}\right)^{1/5} = \frac{1}{5}\left[3 \log_b x + \frac{1}{2}\log_b y - \log_b z\right].$$

We sometimes may want to use the laws of logarithms to write the sum or difference of two logarithms as a single term, as illustrated in the solution of an equation in the following example.

Example   Solve $\log_{10}(x + 1) + \log_{10}(x - 2) = 1$.

Solution   Using the first law of exponents to rewrite the left-hand member, we obtain

$$\log_{10}(x + 1)(x - 2) = 1.$$

Then, writing this equation in exponential form, we have

$$(x + 1)(x - 2) = 10^1,$$

from which

$$x^2 - x - 2 = 10$$

or

$$(x - 4)(x + 3) = 0.$$

Thus,

$$x = 4 \quad \text{or} \quad x = -3.$$

The number -3 is not a solution of the original equation: The terms $\log_{10}(x + 1)$ and $\log_{10}(x - 2)$ are not defined for $x = -3$. The solution set is $\{4\}$.

So far in this chapter, we have generalized the discussion of the properties of the exponential and logarithmic functions for any base $b > 0$, $b \neq 1$. In Sections 9.4 and 9.5, we shall find values for the powers and logarithms with base 10 and base e that occur most often in applied problems. In Section 9.4, we shall use tables to find such values, and, in Section 9.5, we shall see how a scientific calculator can be used for the same purpose.

EXERCISE 9.3

A   ■ *Use Properties* (1), (2), *and* (3) *on page* 295 *to write each expression in terms of simpler logarithmic quantities. Assume that all variables denote positive real numbers.*

Example $\log_b \sqrt{\dfrac{xy}{z}}$

Solution   First express $\sqrt{\dfrac{xy}{z}}$ using a fractional exponent.

$$\log_b \sqrt{\frac{xy}{z}} = \log_b \left(\frac{xy}{z}\right)^{1/2}$$

By the third law of logarithms,

$$\log_b \left(\frac{xy}{z}\right)^{1/2} = \frac{1}{2} \log_b \left(\frac{xy}{z}\right).$$

Now, by the first and second laws of logarithms,

$$\frac{1}{2}\log_b\left(\frac{xy}{z}\right) = \frac{1}{2}(\log_b x + \log_b y - \log_b z).$$

Therefore,

$$\log_b \sqrt{\frac{xy}{z}} = \frac{1}{2}(\log_b x + \log_b y - \log_b z).$$

1. $\log_b(2x)$ **2.** $\log_b(xy)$ **3.** $\log_b(3xy)$ **4.** $\log_b(4yz)$

5. $\log_b\left(\dfrac{x}{y}\right)$ **6.** $\log_b\left(\dfrac{y}{x}\right)$ **7.** $\log_b\left(\dfrac{xy}{z}\right)$ **8.** $\log_b\left(\dfrac{x}{yz}\right)$

9. $\log_b x^3$ **10.** $\log_b x^{1/3}$ **11.** $\log_b \sqrt{x}$ **12.** $\log_b \sqrt[5]{y}$

13. $\log_b \sqrt[3]{x^2}$ **14.** $\log_b \sqrt{x^3}$ **15.** $\log_b(x^2y^3)$ **16.** $\log_b(x^{1/3}z^2)$

17. $\log_b\left(\dfrac{x^{1/2}y}{z^2}\right)$ **18.** $\log_b\left(\dfrac{xy^3}{z^{1/2}}\right)$ **19.** $\log_{10} \sqrt[3]{\dfrac{xy^2}{z}}$ **20.** $\log_{10} \sqrt[5]{\dfrac{x^2y}{z^3}}$

21. $\log_{10}\left(x\sqrt{\dfrac{x}{y}}\right)$ **22.** $\log_{10}\left(2y\sqrt[3]{\dfrac{x}{y}}\right)$ **23.** $\log_{10}\left(2\pi\sqrt{\dfrac{l}{g}}\right)$ **24.** $\log_{10} \sqrt{\dfrac{2L}{R^2}}$

25. $\log_{10} \sqrt{(s-a)(s-b)}$ **26.** $\log_{10} \sqrt{s^2(s-a)^3}$

■ *Express as a single logarithm with a coefficient of* 1.

Example $\dfrac{1}{2}(\log_b x - \log_b y)$

Solution   By the second law of logarithms,

$$\frac{1}{2}(\log_b x - \log_b y) = \frac{1}{2}\log_b\left(\frac{x}{y}\right).$$

By the third law of logarithms,

$$\frac{1}{2}\log_b\left(\frac{x}{y}\right) = \log_b\left(\frac{x}{y}\right)^{1/2}.$$

Therefore,

$$\frac{1}{2}(\log_b x - \log_b y) = \log_b\left(\frac{x}{y}\right)^{1/2}.$$

27. $\log_b x + \log_b y$ **28.** $\log_b x - \log_b y$

29. $2\log_b x - 3\log_b y$ **30.** $\dfrac{1}{4}\log_b x + \dfrac{3}{4}\log_b y$

31. $3\log_b x + \log_b y - 2\log_b z$ **32.** $\dfrac{1}{3}(\log_b x + \log_b y - 2\log_b z)$

33. $\frac{1}{2}(\log_{10} y + \log_{10} x - 2 \log_{10} z)$ **34.** $\frac{1}{2}(\log_{10} x - 3 \log_{10} y - \log_{10} z)$

35. $-2 \log_b x$ **36.** $-\log_b x$

■ *Solve each logarithmic equation.*

Example $\log_{10}(x + 9) + \log_{10} x = 1$

Solution   Use the first law of logarithms.

$$\log_{10}[(x + 9)(x)] = 1$$

Write in exponential form.

$$x^2 + 9x = 10^1$$

Solve for x.

$$x^2 + 9x - 10 = 0$$
$$(x + 10)(x - 1) = 0$$
$$x = -10 \quad \text{or} \quad x = 1$$

Note that the left-hand member of the original equation is meaningful only if $x + 9 > 0$ and $x > 0$; -10 does not satisfy these conditions. Hence, the solution set is $\{1\}$.

37. $\log_{10} x + \log_{10} 2 = 3$ **38.** $\log_{10}(x - 1) - \log_{10} 4 = 2$

39. $\log_{10} x + \log_{10}(x + 21) = 2$ **40.** $\log_{10}(x + 3) + \log_{10} x = 1$

41. $\log_{10}(x + 2) + \log_{10}(x - 1) = 1$ **42.** $\log_{10}(x + 3) - \log_{10}(x - 1) = 1$

43. Show by an example that $\log_{10}(x + y)$ is not equivalent to $\log_{10} x + \log_{10} y$.

44. Show by an example that $\log_{10} \dfrac{x}{y}$ is not equivalent to $\dfrac{\log_{10} x}{\log_{10} y}$.

■ *Given that $\log_b 2 = 0.3010$, $\log_b 3 = 0.4771$, and $\log_b 5 = 0.6990$, find the value of each expression.*

45. $\log_b 6$ **46.** $\log_b 10$ **47.** $\log_b \dfrac{2}{5}$

48. $\log_b \dfrac{3}{2}$ **49.** $\log_b 9$ **50.** $\log_b 25$

51. $\log_b \dfrac{15}{2}$ **52.** $\log_b \dfrac{6}{5}$ **53.** $\log_b(0.002)^3$

54. $\log_b \sqrt{50}$ **55.** $\log_b 75$ **56.** $\log_b \dfrac{0.08}{15}$

B ■ *Verify that each statement is true.*

57. $\log_b 4 + \log_b 8 = \log_b 64 - \log_b 2$ **58.** $\log_b 24 - \log_b 2 = \log_b 3 + \log_b 4$

59. $2 \log_b 6 - \log_b 9 = 2 \log_b 2$ **60.** $4 \log_b 3 - 2 \log_b 3 = \log_b 9$

61. $\frac{1}{2} \log_b 12 - \frac{1}{2} \log_b 3 = \frac{1}{3} \log_b 8$ **62.** $\frac{1}{4} \log_b 8 + \frac{1}{4} \log_b 2 = \log_b 2$

9.4

USING TABLES

For $b > 0$ and $b \neq 1$, values for b^x ($x \in R$) and $\log_b x$ ($x > 0$) can be obtained by using prepared tables in conjunction with the laws of logarithms. Because we are familiar with the number 10 as the base of our numeration system, we shall first give our attention to logarithms and powers to the base 10. This was the base first used in the invention of logarithms to perform computations in astronomy and navigation.

Common logarithms

Recall from Equation (4) on page 291 that $\log_{10} x$ is the exponent that must be placed on 10 so that the resulting power is x. Values for $\log_{10} x$ are sometimes called **common logarithms.**

Values for $\log_{10} 10^k$, $k \in J$

Some values of $\log_{10} x$ can be obtained simply by considering the definition of a logarithm, while other values require tables. Let us first consider values of $\log_{10} x$ for all values of x that are integral powers of 10. These can be obtained by inspection.

$$\begin{array}{lll} \text{Since} & 10^3 = 1000, & \log_{10} 1000 = 3; \\ \text{since} & 10^2 = 100, & \log_{10} 100 = 2; \\ \text{since} & 10^1 = 10, & \log_{10} 10 = 1; \\ \text{since} & 10^0 = 1, & \log_{10} 1 = 0; \\ \text{since} & 10^{-1} = 0.1, & \log_{10} 0.1 = -1; \\ \text{since} & 10^{-2} = 0.01, & \log_{10} 0.01 = -2; \\ \text{since} & 10^{-3} = 0.001, & \log_{10} 0.001 = -3. \end{array}$$

Notice that the logarithm of a power of 10 is simply the exponent on the base 10. For example,

$$\log_{10} 100 = \log_{10} 10^2 = 2,$$

$$\log_{10} 0.01 = \log_{10} 10^{-2} = -2,$$

and so on.

Values for
$\log_{10} x$,
$1 < x < 10$

Table II in Appendix D gives values for $\log_{10} x$ for $1 < x < 10$. Consider the following excerpt from the table. Each number in the column headed x represents the first two digits of the numeral for x, while each of the other column-head numbers represents the third significant digit of the numeral for x. The number located at the intersection of a row and a column forms the logarithm of x. For example, to find $\log_{10} 4.25$, we look at the intersection of the row containing 4.2 under x and the column containing 5. Thus,

$$\log_{10} 4.25 = 0.6284.$$

Similarly,

$$\log_{10} 4.02 = 0.6042,$$

$$\log_{10} 4.49 = 0.6522,$$

and so on.

x	0	1	2	3	4	$\boxed{5}$	6	7	8	9
3.8	.5798	.5809	.5821	.5832	.5843	.5855	.5866	.5877	.5888	.5899
3.9	.5911	.5922	.5933	.5944	.5955	.5966	.5977	.5988	.5999	.6010
4.0	.6021	.6031	.6042	.6053	.6064	.6075	.6085	.6096	.6107	.6117
4.1	.6128	.6138	.6149	.6160	.6170	.6180	.6191	.6201	.6212	.6222
$\boxed{4.2}$	.6232	.6243	.6253	.6263	.6274	.6284	.6294	.6304	.6314	.6325
4.3	.6335	.6345	.6355	.6365	.6375	.6385	.6395	.6405	.6415	.6425
4.4	.6435	.6444	.6454	.6464	.6474	.6484	.6493	.6503	.6513	.6522
4.5	.6532	.6542	.6551	.6561	.6571	.6580	.6590	.6599	.6609	.6618
4.6	.6628	.6637	.6646	.6656	.6665	.6675	.6684	.6693	.6702	.6712

The values in the tables are rational-number approximations of irrational numbers. We shall follow customary usage and write $=$ instead of $\approx$.

Values for
$\log_{10} x$,
$x > 10$

Now suppose we wish to find $\log_{10} x$ for values of x outside the range of the table—that is, for $x > 10$ or $0 < x < 1$ (see Figure 9.3). This can be done quite readily by first representing the number in scientific notation and then applying the first law of logarithms.

Examples

a. $\log_{10} 42.5 = \log_{10}(4.25 \times 10^1)$
$\qquad = \log_{10} 4.25 + \log_{10} 10^1$
$\qquad = 0.6284 + 1$
$\qquad = 1.6284$

b. $\log_{10} 425 = \log_{10}(4.25 \times 10^2)$
$\qquad = \log_{10} 4.25 + \log_{10} 10^2$
$\qquad = 0.6284 + 2$
$\qquad = 2.6284$

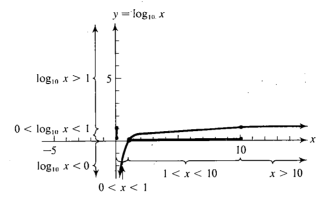

Figure 9.3

c. $\log_{10} 4250 = \log_{10}(4.25 \times 10^3)$

$= \log_{10} 4.25 + \log_{10} 10^3$

$= 0.6284 + 3$

$= 3.6284$

d. $\log_{10} 42{,}500 = \log_{10}(4.25 \times 10^4)$

$= \log_{10} 4.25 + \log_{10} 10^4$

$= 0.6284 + 4$

$= 4.6284$

Observe that the decimal portion of the logarithms in Examples a–d is always 0.6284 and *the integral portion is the exponent on* 10 *when the number is written in scientific notation.*

This process can be reduced to a mechanical one by considering $\log_{10} x$ to consist of two parts, an *integral part* (called the **characteristic**) and a *nonnegative decimal fraction part* (called the **mantissa**). Thus, the table of values for $\log_{10} x$ for $1 < x < 10$ can be looked upon as a table of mantissas for $\log_{10} x$ for all $x > 0$.

In Examples a, b, c, and d above, where $x > 10$, the mantissa in each case is 0.6284 and the characteristics are 1, 2, 3, and 4, respectively.

Values for
log_{10} x,
$0 < x < 1$

Now consider an example of the form $\log_{10} x$ for $0 < x < 1$. To find $\log_{10} 0.00425$, we write

$$\log_{10} 0.00425 = \log_{10}(4.25 \times 10^{-3})$$
$$= \log_{10} 4.25 + \log_{10} 10^{-3}.$$

We find from the table that $\log_{10} 4.25 = 0.6284$. Upon adding 0.6284 to the characteristic -3, we obtain

$$\log_{10} 0.00425 = 0.6284 + (-3)$$
$$= -2.3716,$$

where the decimal part of the logarithm is no longer 0.6284. The decimal part is -0.3716, a negative number.

If we want to use the table, which contains only positive entries, it is customary to write the logarithm in a form in which the decimal part is positive. In the above example, we write

$$\log_{10} 0.00425 = 0.6284 - 3,$$

where the decimal part is positive. Because -3 can be written as $1 - 4, 2 - 5, 3 - 6,$ $7 - 10,$ and so on, the forms $1.6284 - 4, 2.6284 - 5, 3.6284 - 6, 7.6284 - 10,$ and so on, are equally valid representations of the desired logarithm. It will sometimes be convenient to use these alternative forms when we use the tables.

Examples

a. $\log_{10} 0.294 = \log_{10}(2.94 \times 10^{-1})$ **b.** $\log_{10} 0.00294 = \log_{10}(2.94 \times 10^{-3})$
$\qquad\qquad = \log_{10} 2.94 + \log_{10} 10^{-1} \qquad\qquad\qquad = \log_{10} 2.94 + \log_{10} 10^{-3}$
$\qquad\qquad = 0.4683 - 1 \qquad\qquad\qquad\qquad\qquad\qquad = 0.4683 - 3$

Antilog₁₀ N

Given a value for an exponent, $\log_{10} x$, we can use Table II to find the power x by reversing the process described to find the logarithm of a number. In this case, the power x is sometimes called the **antilogarithm** of $\log_{10} x$. For example, if

$$\log_{10} x = 0.4409,$$

then

$$x = \text{antilog}_{10} 0.4409,$$

which can be obtained by locating 0.4409 in the body of Table II and observing that

$$\text{antilog}_{10} 0.4409 = 2.76.$$

If the $\log_{10} x$ is greater than one, it can first be written as the sum of a positive decimal (the mantissa) and a positive integer (the characteristic). Antilog₁₀ x can then be written as the product of a number between one and ten and a power of ten.

Example

If $\log_{10} x = 2.4409,$ then

$$x = \text{antilog}_{10} 2.4409$$
$$= \text{antilog}_{10} (0.4409 + 2)$$
$$= 2.76 \times 10^2$$
$$= 276.$$

If the decimal part of $\log_{10} x$ is negative and we wish to use the table to obtain x, we cannot use the table directly. However, we can first write $\log_{10} x$ equivalently with a positive decimal part. For example, to find

$$\text{antilog}_{10}(-0.4522) \quad \text{or} \quad \text{antilog}_{10}(-2.4522),$$

we can first add $(+1 - 1)$ to write -0.4522 as

$$-0.4522 + 1 - 1 = 0.5478 - 1$$

and add $(+3 - 3)$ to write -2.4522 as

$$-2.4522 + 3 - 3 = 0.5478 - 3,$$

and then use the tables. Thus,

$$\text{antilog}_{10}(-0.4522) = \text{antilog}_{10}(\underbrace{0.5478}_{} - 1)$$

$$= 3.53 \times 10^{-1} = 0.353$$

and

$$\text{antilog}_{10}(-2.4522) = \text{antilog}_{10}(\underbrace{0.5478}_{} - 3)$$

$$= 3.53 \times 10^{-3} = 0.00353.$$

Examples   Use Table II to find the value of x.

a. $\log_{10} x = -0.7292$ **b.** $\log_{10} x = -1.4634$

Solutions   **a.** $x = \text{antilog}_{10}(-0.7292)$

$= \text{antilog}_{10}(\underbrace{-0.7292 + 1}_{} - 1)$

$= \text{antilog}_{10}(0.2708 - 1)$

$= 1.87 \times 10^{-1} = 0.187$

b. $x = \text{antilog}_{10}(-1.4634)$

$= \text{antilog}_{10}(\underbrace{-1.4634 + 2}_{} - 2)$

$= \text{antilog}_{10}(0.5366 - 2)$

$= 3.44 \times 10^{-2} = 0.0344$

In the above examples, the mantissas, 0.2718 and 0.5366, were listed in Table II. If we seek the common logarithm of a number that is not an entry in the table (for example, $\log_{10} 23.42$) or if we seek x when $\log_{10} x$ is not an entry in the table, it is possible to use a procedure called **linear interpolation**. A detailed explanation of this process is given in Appendix C. In the examples and exercises of this chapter we shall simply use the entry in the table which is closest to the value that we seek.

***Powers to the
base* 10**

By the definition of a logarithm,

$$P = 10^E$$

can be written in logarithmic form as

$$\log_{10} P = E,$$

from which we see that the power P is the antilogarithm of the exponent E,

$$P = \text{antilog}_{10} E.$$

Since $P = 10^E$,

$$10^E = \text{antilog}_{10} E$$

and we can obtain a power 10^E simply by finding the antilogarithm of the exponent E.

Examples   Compute each power.

    a. $10^{0.2148}$ **b.** $10^{-1.6345}$

Solutions   **a.** $10^{0.2148}$ **b.** $10^{-1.6345}$

$$= \text{antilog}_{10}\ 0.2148$$

$$= 1.64$$

$$= \text{antilog}_{10}(-1.6345)$$

$$= \text{antilog}_{10}(\underbrace{-1.6345 + 2} - 2)$$

$$= \text{antilog}_{10}(0.3655 - 2)$$

$$= 2.32 \times 10^{-2}$$

Powers to the base e

The number $e \approx 2.7182818*$ is an irrational number that has applications in business, biological and physical sciences, and in engineering. Because of its importance, special tables have been prepared for both e^x and $\log_e x$.

    Table III in Appendix D gives approximations for powers e^x and e^{-x} for $0 \le x \le 1.00$ in 0.01 intervals and for $1.00 < x \le 10.00$ in 0.1 intervals.

Examples   Using Table III:

    a. $e^{2.4} = 11.023$ **b.** $e^{-4.7} = 0.0091$

    Although we can obtain values for e^x and e^{-x} outside the interval 0 to 10 by using the table along with the first law of exponents, at this time the function values in the table over this interval will be adequate for our work.

Natural logarithms, ln x

The most-used values for $\log_e x$ $(x > 0)$ are printed in Table IV of Appendix D. These values, like those for e^x, e^{-x}, and $\log_{10} x$, are *approximations* accurate to the number of decimals shown. The symbol $\log_e x$ is often written as **ln x** and read as "**natural logarithm of x**." Unlike Table II for common logarithms, which only provides the decimal part of $\log_{10} x$, the table for natural logarithms gives the entire value for $\ln x$, both the integral and decimal portions.

Examples   Using Table IV:

    a. $\ln 6.6 = 1.8871$ **b.** $\ln 0.7 = -0.3567$

    If we seek a value for e^x or $\ln x$ for a value of x between two entries in the tables, it is possible to use linear interpolation (see Appendix C).

Powers to the base e, anti-logarithms

Given an exponent $\ln x$, we can obtain the power x by using the definition of a logarithm and Table III.

Examples   **a.** $\ln x = 1.3$ **b.** $\ln x = -0.47$

---

*The number e is discussed in more detail in Section 11.4.

Solutions   **a.** $\ln x = 1.3$ is equivalent to   **b.** $\ln x = -0.47$ is equivalent to

$$x = e^{1.3}$$
$$= 3.6693$$

$$x = e^{-0.47}$$
$$= 0.6250$$

As noted above, a power to the base b is called the antilog$_b$ of the exponent. Thus in Example a,

$$x = e^{1.3} = \text{antilog}_e \ 1.3$$
$$= 3.6693.$$

EXERCISE 9.4

A   ■ *Find each logarithm by inspection.*

Example $\log_{10} 10^3$

Solution   By definition, $\log_{10} 10^3$ is the exponent on 10 so that the power equals 10^3. Hence,

$$\log_{10} 10^3 = 3.$$

1. $\log_{10} 10^2$ **2.** $\log_{10} 10^4$ **3.** $\log_{10} 10^{-4}$

4. $\log_{10} 10^{-6}$ **5.** $\log_{10} 10^0$ **6.** $\log_{10} 10^n$

■ *Find an approximation for each logarithm using Table II.*

Examples   **a.** $\log_{10} 16.8$ **b.** $\log_{10} 0.043$

Solutions   **a.** Represent the number in scientific no-   **b.** Represent the number in scientific notation.   tation.

$$\log_{10}(1.68 \times 10^1)$$ $$\log_{10}(4.3 \times 10^{-2})$$

Use 1.68 to determine the mantissa   Use 4.3 to determine the mantissa
from the table.   from the table.

0.2253   0.6335

Add the characteristic 1 as determined   Add the characteristic -2 as deter-
by the exponent on the base 10.   mined by the exponent on the base 10.

$$0.2253 + 1 = 1.2253$$ $$0.6335 - 2$$

7. $\log_{10} 6.73$ **8.** $\log_{10} 891$ **9.** $\log_{10} 83.7$

10. $\log_{10} 21.4$ **11.** $\log_{10} 317$ **12.** $\log_{10} 219$

13. $\log_{10} 0.813$	**14.** $\log_{10} 0.00214$	**15.** $\log_{10} 0.08$
16. $\log_{10} 0.000413$	**17.** $\log_{10}(2.48 \times 10^2)$	**18.** $\log_{10}(5.39 \times 10^{-3})$

■ *Solve for x using Table II.*

Example $\log_{10} x = 2.7364$

Solution $x = \text{antilog}_{10} 2.7364 = \text{antilog}_{10}(0.7364 + 2)$

Locate the mantissa 0.7364 in the body of Table II and determine the associated antilog_{10} (a number between 1 and 10). Write the characteristic 2 as an exponent on the factor with base 10.

$$x = \text{antilog}_{10}(0.7364 + 2)$$
$$= 5.45 \times 10^2$$
$$= 545$$

19. $\log_{10} x = 0.6128$	**20.** $\log_{10} x = 0.2504$	**21.** $\log_{10} x = 1.5647$
22. $\log_{10} x = 3.9258$	**23.** $\log_{10} x = 0.8075 - 2$	**24.** $\log_{10} x = 0.9722 - 3$
25. $\log_{10} x = 7.8562 - 10$	**26.** $\log_{10} x = 1.8155 - 4$	

Example $\log_{10} x = -1.3420$

Solution   In this case, both the characteristic (-1) and decimal part (-0.3420) are negative. Hence, first add $(2 - 2)$ to write -1.3420 as

$$-1.3420 + 2 - 2 = 0.6580 - 2.$$

Thus,

$$x = \text{antilog}_{10}(-1.3420) = \text{antilog}_{10}(0.6580 - 2)$$

Using Table II,

$$x = 4.55 \times 10^{-2} = 0.0455.$$

27. $\log_{10} x = -0.5272$	**28.** $\log_{10} x = -0.4123$	**29.** $\log_{10} x = -1.2984$
30. $\log_{10} x = -1.0545$	**31.** $\log_{10} x = -2.6882$	**32.** $\log_{10} x = -2.0670$

■ *Compute each power.*

Examples   **a.** $10^{1.4728}$ **b.** $10^{-0.3215}$

Solutions   **a.** $10^{1.4728}$

$$= \text{antilog}_{10} \, 1.4728$$
$$= 2.97 \times 10^1$$
$$= 29.7$$

b. $10^{-0.3215}$

$$= \text{antilog}_{10}(-0.3215)$$
$$= \text{antilog}_{10}(-0.3215 + 1 - 1$$
$$= \text{antilog}_{10}(0.6785 - 1)$$
$$= 4.77 \times 10^{-1}$$
$$\doteq 0.477$$

33. $10^{0.8762}$ **34.** $10^{1.6405}$ **35.** $10^{2.8943}$

36. $10^{4.3766}$ **37.** $10^{-1.4473}$ **38.** $10^{-2.0958}$

■ *Find each power.*

Examples   **a.** $e^{0.21}$ **b.** $e^{-1.5}$

Solutions   Using Table III:

a. $e^{0.21} = 1.2337$ **b.** $e^{-1.5} = 0.2231$

39. $e^{0.43}$ **40.** $e^{0.62}$ **41.** $e^{-0.57}$

42. $e^{-0.08}$ **43.** $e^{1.5}$ **44.** $e^{2.6}$

45. $e^{-2.4}$ **46.** $e^{-1.2}$

■ *Find each logarithm.*

Examples   **a.** $\ln 2.3$ **b.** $\ln 0.3$

Solutions   Using Table IV:

a. $\ln 2.3 = 0.8329$ **b.** $\ln 0.3 = 0.7960 - 2$
$$= -1.2040$$

47. $\ln 3.9$ **48.** $\ln 6.3$ **49.** $\ln 16$

50. $\ln 55$ **51.** $\ln 0.4$ **52.** $\ln 0.7$

■ *Find each value of x.*

Examples   **a.** $\ln x = 0.21$ **b.** $\ln x = 3.8$

Solutions overleaf

Solutions   From the definition of a logarithm:

a. $\ln x = 0.21$ is equivalent to

$x = e^{0.21}$.

Using Table III,

$x = 1.2337$.

b. $\ln x = 3.8$ is equivalent to

$x = e^{3.8}$.

Using Table III,

$x = 44.701$.

53. $\ln x = 0.16$ **54.** $\ln x = 0.25$ **55.** $\ln x = 1.8$

56. $\ln x = 2.4$ **57.** $\ln x = 4.5$ **58.** $\ln x = 6.0$

9.5

USING CALCULATORS

Scientific calculators are useful in a variety of computations. Furthermore, a calculator enables us to obtain function values for exponential and logarithmic functions more efficiently than we could obtain by using tables.

A great variety of calculators exist. We shall need one that has at least one of the keys $\boxed{\text{LOG}}$, $\boxed{\text{LN } x}$, or $\boxed{e^x}$ and the inverse operation connected with that key. Many scientific calculators contain all three keys and may contain $\boxed{10^x}$ and $\boxed{y^x}$.*

Base **10**

log_{10} **10**^{k},
$k \in J$

Values of $\log_{10} x$, called **logarithms to the base** 10 or **common logarithms** can readily be obtained for all $x > 0$ by using a calculator. However, values for $\log_{10} x$, where x *is an integral power of* 10, can be obtained directly from the definition of a logarithm. You should try to obtain such values *without using your calculator*.

$$\begin{array}{lll} \text{Since} & 10^3 = 1000, & \log_{10} 1000 = 3; \\ \text{since} & 10^2 = 100, & \log_{10} 100 = 2; \\ \text{since} & 10^1 = 10, & \log_{10} 10 = 1; \\ \text{since} & 10^0 = 1, & \log_{10} 1 = 0; \\ \text{since} & 10^{-1} = 0.1, & \log_{10} 0.1 = -1; \\ \text{since} & 10^{-2} = 0.01, & \log_{10} 0.01 = -2; \\ \text{since} & 10^{-3} = 0.001, & \log_{10} 0.001 = -3. \end{array}$$

Notice that the logarithm of a power of 10 is simply the exponent on the base 10. For example,

$$\log_{10} 100 = \log_{10} 10^2 = 2,$$
$$\log_{10} 0.01 = \log_{10} 10^{-2} = -2,$$

and so on.

*The calculator instructions shown in this section apply to calculators with algebraic logic. See the instruction manual for your calculator if it is programmed with a different logic.

Your ability to determine the logarithms of powers of 10 by inspection will help you estimate function values $\log_{10} x$ when x is not a power of 10. For example,

$$\text{if } 1 < x < 10, \quad \text{then } 0 < \log_{10} x < 1,$$

$$\text{if } 100 < x < 1000, \quad \text{then } 2 < \log_{10} x < 3,$$

and so on.

$log_{10}\, x$,
$x > 0$

If a calculator is used to find values for $\log_{10} x$ ($x > 0$) when x is not an integral power of 10, the function key $\boxed{\text{LOG}}$ is ordinarily used to obtain these values. The values are generally produced to five or more significant digits. We shall round off readings of $\log_{10} x$ to five significant digits.

Some calculators do not have a $\boxed{\text{LOG}}$ key but do have a $\boxed{\text{LN}}$ key. In such cases, function values $\log_{10} N$ can be obtained by finding the quotient $\ln N/\ln 10$. The reason why this is valid is discussed in Section 9.6.

Examples

Find each logarithm.

a. $\log_{10} 23.4$ 　　　　　　　　　　　　　**b.** $\log_{10} 0.00402$

Solutions

a. Using the $\boxed{\text{LOG}}$ key,

$$23.4 \,\boxed{\text{LOG}} = 1.3692.$$

Or, using the fact that $\log_{10} 23.4 = \ln 23.4/\ln 10$,

$$23.4 \,\boxed{\text{LN}}\,\boxed{\div}\, 10 \,\boxed{\text{LN}}\,\boxed{=}\, 1.3692.$$

b. Using the $\boxed{\text{LOG}}$ key,

$$0.00402 \,\boxed{\text{LOG}} = -2.3958.$$

Or, using the fact that $\log_{10} 0.00402 = \ln 0.00402/\ln 10$,

$$0.00402 \,\boxed{\text{LN}}\,\boxed{\div}\, 10 \,\boxed{\text{LN}}\,\boxed{=}\, -2.3958.$$

Note that in Example b the integer part of the logarithm (-2) is negative and *the decimal part (-0.3958) is also negative.* Contrary to our procedure when using tables, when we use a calculator it is not necessary to maintain a positive decimal part of a logarithm.

Furthermore, it should be understood that the values obtained by using a calculator are, in general, approximations for irrational numbers even though we shall, as is customary, continue to use the " = " sign.

10^x, $x \in R$

Some calculators have a $\boxed{10^x}$ key. In this case, the power is readily computed. If a calculator does not have this key, the power can be computed by using the $\boxed{\text{LOG}}$ key in conjunction with the $\boxed{\text{INV}}$ key (or $\boxed{\text{2ND}}$ function key). This is possible because the logarithmic function is the inverse of the exponential function. Alternatively, the $\boxed{y^x}$ key can be used. In this case, the $\boxed{=}$ key must be pressed as shown in the following example. We shall round off readings of powers of 10^x to five significant digits.

Example   Compute $10^{2.34}$.

Solution   Using the $\boxed{10^x}$ key,

$$2.34 \;\boxed{10^x} = 218.78.$$

Or, using the $\boxed{\text{INV}}$ (or second function) key,

$$2.34 \;\boxed{\text{INV}}\; \boxed{\text{LOG}} = 218.78.$$

Or, using the $\boxed{y^x}$ key,

$$10 \;\boxed{y^x}\; 2.34 \;\boxed{=}\; 218.78.$$

Given a value for $\log_{10} x$, we can find x by first expressing the logarithm in exponential form.

Example   If $\log_{10} x = 2.34$, then, by the definition of a logarithm,

$$x = 10^{2.34} = 218.78.$$

A power to the base b is sometimes called the **antilog$_b$** of the exponent. Hence, in the above example,

$$x = 10^{2.34} = \text{antilog}_{10}\, 2.34.$$

Examples   Find each value of x.

a. $\log_{10} x = 2.4211$ **b.** $\log_{10} x = -1.2147$

Solutions   **a.** $\log_{10} x = 2.4211$ is equivalent to   **b.** $\log_{10} x = -1.2147$ is equivalent to

$x = 10^{2.4211}$ $x = 10^{-1.2147}$

(antilog$_{10}$ 2.4211)   (antilog$_{10}$ -1.2147)

$= 263.69.$ $= 0.06100.$

Base e   The second base in general use with exponential and logarithmic functions is the irrational number $e \approx 2.7182818.$* This number has applications in business, biological and physical sciences, and engineering.

ln x,
x > 0   Approximations for $\log_e x$, commonly written as **ln x**, are called **natural logarithms** and can be obtained on most calculators by using the $\boxed{\text{LN}}$ key. We shall round off the readings on a calculator to five significant digits.

---

*This number is considered further in Section 11.4.

Examples   Find each logarithm.

a. ln 6.6   b. ln 0.7

Solutions   a. 6.6 $\boxed{\text{LN}}$ = 1.8871   b. 0.7 $\boxed{\text{LN}}$ = −0.35667

$e^x, x \in R$ Some calculators have a special key for e^x. If your calculator does not, this power can be obtained by using the $\boxed{\text{LN}}$ key in conjunction with the $\boxed{\text{INV}}$ key (or $\boxed{\text{2ND}}$ function key).

Examples   Find each power.

a. $e^{2.4}$ b. $e^{-4.7}$

Solutions   a. 2.4 $\boxed{e^x}$ = 11.023   b. −4.7 $\boxed{e^x}$ = 0.0090952

    or   or

    2.4 $\boxed{\text{INV}}$ $\boxed{\text{LN}}$ = 11.023   −4.7 $\boxed{\text{INV}}$ $\boxed{\text{LN}}$ = 0.0090952

   Given a value for an exponent in the form of a natural logarithm, we can obtain the power by first expressing the logarithm in exponential form. This is the same procedure we used on page 310 for exponents that were written as common logarithms.

Examples   Find each value of x.

a. ln $x = 2.4$ b. ln $x = -4.7$

Solutions   a. ln $x = 2.4$ is equivalent to   b. ln $x = -4.7$ is equivalent to

$$x = e^{2.4} = 11.023.$$ $$x = e^{-4.7} = 0.0090952.$$

EXERCISE 9.5

A ▪ *Find each logarithm by inspection.*

Example $\log_{10} 10^3$

Solution   By definition, $\log_{10} 10^3$ is the exponent on 10 so that the power equals 10^3. Hence,

$$\log_{10} 10^3 = 3.$$

1. $\log_{10} 10^2$ 2. $\log_{10} 10^4$ 3. $\log_{10} 10^{-4}$ 4. $\log_{10} 10^{-6}$
5. $\log_{10} 10$ 6. $\log_{10} 1$ 7. $\log_{10} 10,000$ 8. $\log_{10} 0.001$

■ *Use a calculator in all of the following exercises. Round off each reading to five significant digits.*

■ *Find each logarithm.*

9. $\log_{10} 54.3$ **10.** $\log_{10} 27.9$ **11.** $\log_{10} 2344$

12. $\log_{10} 1476$ **13.** $\log_{10} 0.073$ **14.** $\log_{10} 0.00614$

15. $\log_{10} 0.6942$ **16.** $\log_{10} 0.0104$

■ *Find each power.*

17. $10^{1.62}$ **18.** $10^{0.43}$ **19.** $10^{-0.87}$

20. $10^{-1.31}$ **21.** $10^{2.113}$ **22.** $10^{3.141}$

23. $10^{-0.2354}$ **24.** $10^{-2.0413}$

■ *Find each x.*

Examples

    a. $\log_{10} x = 1.69$ **b.** $\log_{10} x = -0.43$

Solutions

    a. $\log_{10} x = 1.69$ **b.** $\log_{10} x = -0.43$

       is equivalent to   is equivalent to

$$x = 10^{1.69}$$
$$= 48.978.$$

$$x = 10^{-0.43}$$
$$= 0.37153.$$

25. $\log_{10} x = 1.41$ **26.** $\log_{10} x = 2.3$ **27.** $\log_{10} x = 0.52$

28. $\log_{10} x = 0.8$ **29.** $\log_{10} x = -1.3$ **30.** $\log_{10} x = -1.69$

■ *Find each logarithm.*

31. $\ln 3.9$ **32.** $\ln 6.3$ **33.** $\ln 16$

34. $\ln 55$ **35.** $\ln 6.4$ **36.** $\ln 0.7$

■ *Find each power.*

37. $e^{0.4}$ **38.** $e^{0.73}$ **39.** $e^{2.34}$

40. $e^{3.16}$ **41.** $e^{-1.2}$ **42.** $e^{-2.3}$

43. $e^{-0.4}$ **44.** $e^{-0.62}$

■ *Find each x.*

Examples

a. $\ln x = 1.73$ **b.** $\ln x = -2.4$

Solutions

a. $\ln x = 1.73$ is equivalent to

$$x = e^{1.73} = 5.6407.$$

b. $\ln x = -2.4$ is equivalent to

$$x = e^{-2.4} = 0.090718.$$

45. $\ln x = 1.42$ **46.** $\ln x = 2.03$ **47.** $\ln x = 0.63$

48. $\ln x = 0.59$ **49.** $\ln x = -2.6$ **50.** $\ln x = -3.4$

9.6

EXPONENTIAL EQUATIONS

In Section 9.1, we obtained by inspection the solutions to exponential equations of the form $y = b^x$ for integral values of x. For example, the solution of $8 = 2^x$ is 3 and the solution of $100 = 10^x$ is 2. Our work with logarithmic functions in the previous sections has now provided us with the mathematics to solve such an equation when the solution is not evident by inspection.

We can use the definition of a logarithm and Table II or Table IV or a calculator to obtain solutions to some simple equations in which the variable is part of the exponent. Such equations are called **exponential equations**. We shall express the solutions to such equations in the examples in this section to three significant digits.

Examples

Solve each equation.

a. $10^x = 2.73$ **b.** $e^x = 0.24$

Solutions

a. $10^x = 2.73$ is equivalent to

$$x = \log_{10} 2.73 = 0.436.$$

b. $e^x = 0.24$ is equivalent to

$$x = \ln 0.24 = -1.43.$$

In more complicated equations, it is sometimes necessary to first rewrite the equation in an equivalent form so that the power is the only term in one member and has a coefficient of one. The equation can then be written in logarithmic form.

Example

Solve $4.31 = 1.73 + 2 \cdot 10^{1.2x}$.

Solution

Adding -1.73 to each member, we have

$$2.58 = 2 \cdot 10^{1.2x},$$

from which by dividing each member by 2,

$$1.29 = 10^{1.2x}.$$

Solution continued overleaf

The equivalent logarithmic equation is

$$1.2\, x = \log_{10} 1.29.$$

Hence, by dividing each member by 1.2 and using Table II or a calculator,

$$x = \frac{\log_{10} 1.29}{1.2} = 0.0922.$$

This result may be obtained on some calculators as follows:

$$1.29\ \boxed{\text{LOG}}\ \boxed{\div}\ 1.2\ \boxed{=}\ 0.092158 \approx 0.0922.$$

Example

Solve $140 = 20e^{0.4x}$.

Solution

We first divide each member by 20 to obtain

$$7 = e^{0.4x},$$

from which the equivalent logarithmic equation is

$$0.4x = \ln 7.$$

Hence, by dividing each member by 0.4 and using Table IV or a calculator, we have

$$x = \frac{\ln 7}{0.4} = 4.86.$$

This result may be obtained on some calculators as follows:

$$7\ \boxed{\text{LN}}\ \boxed{\div}\ 0.4\ \boxed{=}\ 4.8648 \approx 4.86.$$

Changing bases

A procedure similar to the one used in the previous two examples would be applicable to any exponential equation involving the base e or the base 10 because values for $\ln x$ and $\log_{10} x$ can be obtained from tables or a scientific calculator. However, the solution of an exponential equation that involves a base other than e or 10 and for which a table of logarithmic function values is not available requires other methods.

One method involves writing a logarithm to one base in terms of logarithms to a different base. To express $\log_b N$ in terms of another base, say a, we first let

$$y = \log_b N. \tag{1}$$

Then, from the definition of a logarithm, we have that

$$N = b^y.$$

Equating the logarithm of each member using the base a, we have

$$\log_a N = \log_a b^y.$$

From the third law of logarithms,

$$\log_a N = y \log_a b,$$

from which

$$y = \frac{\log_a N}{\log_a b}.$$

Hence, substituting $\log_b N$ for y from (1) above, we obtain the following equation:*

$$\log_b N = \frac{\log_a N}{\log_a b}. \tag{2}$$

Example

Compute a value for $\log_5 16$.

Solution

From Equation (2) above, with $a = 10$,

$$\log_5 16 = \frac{\log_{10} 16}{\log_{10} 5}$$

$$= \frac{1.2041}{0.69897} = 1.72.$$

Calculator sequence:

$$16 \boxed{\text{LOG}} \boxed{\div} 5 \boxed{\text{LOG}} \boxed{=} 1.7227 \approx 1.72.$$

Alternatively, with $a = e$,

$$\log_5 16 = \frac{\ln 16}{\ln 5}$$

$$= \frac{2.7726}{1.6094} = 1.72.$$

Calculator sequence:

$$16 \boxed{\text{LN}} \boxed{\div} 5 \boxed{\text{LN}} \boxed{=} 1.7227 \approx 1.72.$$

Equation (2) above can be used to solve exponential equations in which the base is a number other than 10 or e.

Example

Solve $5^x = 7$.

Solution

We can first write this equation in logarithmic form as

$$x = \log_5 7.$$

Now, from Equation (2) and using the base 10, we have

$$x = \log_5 7$$

$$= \frac{\log_{10} 7}{\log_{10} 5} = \frac{0.84510}{0.69897} = 1.21.$$

The use of natural logarithms will yield the same result.

---

*A special case of Equation (2) with $b = 10$ is the relationship that we used on page 309 to find values of $\log_{10} N$ where we used a calculator that did not have a $\boxed{\text{LOG}}$ key.

Note that the quotient $\log_{10} 7/\log_{10} 5$ in the above example could have been obtained by first equating the logarithm of each member of $5^x = 7$ to the base 10 to obtain

$$\log_{10} 5^x = \log_{10} 7.$$

Then, from the third law of logarithms,

$$x \log_{10} 5 = \log_{10} 7,$$

from which

$$x = \frac{\log_{10} 7}{\log_{10} 5}.$$

Common error: Note that when we seek a numerical approximation to the solution by the procedure in the above example, the logarithms are *divided*, not subtracted:

$$\frac{\log_{10} 7}{\log_{10} 5} \neq \log_{10} 7 - \log_{10} 5.$$

Equations with several variables

We can use the definition of a logarithm in exponential equations and logarithmic equations of more than one variable to solve for one of the variables in terms of the others.

Examples

a. Solve $P = Cb^{kt}$ for t.

b. Solve $N = N_0 \log_b(ks)$ for s.

Solutions

a. First express the power b^{kt} in terms of the other variables.

$$b^{kt} = \frac{P}{C} \qquad (C \neq 0)$$

Write the exponential equation in logarithmic form.

$$kt = \log_b \frac{P}{C}$$

Multiply each member by $1/k$.

$$t = \frac{1}{k} \log_b \frac{P}{C} \qquad (k \neq 0)$$

b. First express $\log_b(ks)$ in terms of the other variables.

$$\log_b(ks) = \frac{N}{N_0} \qquad (N_0 \neq 0)$$

Write the logarithmic equation in exponential form.

$$ks = b^{N/N_0}$$

Multiply each member by $1/k$.

$$s = \frac{1}{k} b^{N/N_0} \qquad (k \neq 0)$$

EXERCISE 9.6

A ■ *Compute solutions of Problems 1–30 to three significant digits. A calculator was used to obtain the answers for this section. Answers obtained by using tables may differ slightly from those given.*

■ *Solve each exponential equation.*

Examples   **a.** $10^x = 48.7$ **b.** $10^x = 5870$

Solutions   From the definition of a logarithm:

a. $x = \log_{10} 48.7$ **b.** $x = \log_{10} 5870$

Use Table II or a calculator and round off answers to three significant digits.

$x = 1.69$ $x = 3.77$

1. $10^x = 4.93$ **2.** $10^x = 8.07$
3. $10^x = 23.4$ **4.** $10^x = 182.4$
5. $10^x = 6832.3$ **6.** $10^x = 9480.2$

Examples   **a.** $e^x = 4.7$ **b.** $e^x = 0.6$

Solutions   From the definition of a logarithm:

a. $x = \ln 4.7$.   **b.** $x = \ln 0.6$.

Use Table IV or a calculator.

$x = 1.55$ $x = -0.511$

7. $e^x = 1.9$ **8.** $e^x = 2.1$
9. $e^x = 45$ **10.** $e^x = 60$
11. $e^x = 0.3$ **12.** $e^x = 0.9$

Examples   **a.** $110 = 1.8(10^{0.3x})$ **b.** $16.4 = 4.9 + 10^{1.2x}$

Solutions   **a.** $\dfrac{110}{1.8} = 10^{0.3x}$ **b.** $16.4 - 4.9 = 10^{1.2x}$

$0.3x = \log_{10} \dfrac{110}{1.8}$ $1.2x = \log_{10}(16.4 - 4.9)$

$x = \dfrac{1}{0.3} \log_{10} \dfrac{110}{1.8}$ $x = \dfrac{1}{1.2} \log_{10}(16.4 - 4.9)$

$= 5.95$ $= 0.884$

13. $26.1 = 1.4(10^{1.3x})$ **14.** $140 = 63.1(10^{0.2x})$
15. $14.8 = 1.72 + 10^{-0.3x}$ **16.** $180 = 64 + 10^{-1.3x}$
17. $12.2 = 2(10^{1.4x}) - 11.6$ **18.** $163 = 3(10^{0.7x}) - 49.3$
19. $3(10^{-1.5x}) - 14.7 = 17.1$ **20.** $4(10^{-0.6x}) + 16.1 = 28.2$

Examples   **a.** $12 = 4e^{2x}$ **b.** $16.4 = 12.1 + 2e^{-1.2x}$

Solutions   **a.** $3 = e^{2x}$ **b.** $4.3 = 2e^{-1.2x}$

$2x = \ln 3$ $2.15 = e^{-1.2x}$

$x = \dfrac{1}{2} \ln 3$ $-1.2x = \ln 2.15$

$= 0.549$ $x = -\dfrac{1}{1.2} \ln 2.15 = -0.638$

21. $6.21 = 2.3\, e^{1.2x}$ **22.** $22.26 = 5.3\, e^{0.4x}$

23. $7.74 = 1.72\, e^{0.2x}$ **24.** $14.105 = 4.03\, e^{1.4x}$

25. $6.4 = 20\, e^{0.3x} - 1.8$ **26.** $4.5 = 4\, e^{2.1x} + 3.3$

27. $46.52 = 3.1\, e^{1.2x} + 24.2$ **28.** $1.23 = 1.3\, e^{2.1x} - 17.1$

29. $16.24 = 0.7\, e^{-1.3x} - 21.7$ **30.** $55.68 = 0.6\, e^{-0.7x} + 23.1$

■ *Use Equation* (2) *on page* 315 *to find the value of each of the following logarithms to the nearest hundredth. Use base* 10 *or base e.*

31. $\log_3 18$ **32.** $\log_3 24$ **33.** $\log_2 7.43$

34. $\log_2 14.3$ **35.** $\log_4 17.3$ **36.** $\log_4 28.1$

■ *Solve using logarithms to the base* 10.

Example $3^{x-2} = 16$

Solution   Write the equation in logarithmic form.

$$x - 2 = \log_3 16$$

Use Equation (2) to change $\log_3 16$ to a logarithmic expression to the base 10.

$$x - 2 = \log_3 16 = \frac{\log_{10} 16}{\log_{10} 3}$$

$$x = \frac{\log_{10} 16}{\log_{10} 3} + 2 = 4.52 \tag{1}$$

Alternative Solution   Equate the logarithm of each member of $3^{x-2} = 16$ to the base 10.

$$\log_{10} 3^{x-2} = \log_{10} 16$$

Use the third law of logarithms.

$$(x - 2)\log_{10} 3 = \log_{10} 16$$

$$x - 2 = \frac{\log_{10} 16}{\log_{10} 3}$$

The solution 4.52 can now be obtained as in (1) above.

37. $2^x = 7$ **38.** $3^x = 4$ **39.** $3^{x+1} = 8$ **40.** $2^{x-1} = 9$

41. $4^{x^2} = 15$ **42.** $3^{x^2} = 21$ **43.** $3^{-x} = 10$ **44.** $2.13^{-x} = 8.1$

B   ■ *Solve each exponential or logarithmic equation for the specified variable. Leave the results in the form of an equation equivalent to the given equation.*

45. $y = e^{kt}$,   for t using the base e **46.** $y = k(1 - e^{-t})$,   for t using the base e

47. $\dfrac{T}{R} = e^{t/2}$,   for t using the base e

48. $B - 2 = (A + 3)e^{-t/3}$,   for t using the base e

49. $T = T_0 \ln(k + 10)$,   for k

50. $P = P_0 + \ln 10k$,   for k

51. Show that $\ln N \approx 2.303 \log_{10} N$.

52. Show that $\ln 10 = \dfrac{1}{\log_{10} e}$.

53. Show that $y^x = 10^{x \log_{10} y}$.

54. Use the results of Problem 53 to show that $3^{1/2} = 10^{(1/2)\log_{10} 3}$.

9.7

APPLICATIONS OF EXPONENTIAL EQUATIONS

Exponential equations involving powers with base 10 or base e can be used as models for a variety of real-world phenomena. In equations such as $A = Be^{kt} + C$ or $A = B \cdot 10^{kt} + C$, we want to find values for a particular variable in an expression that may or may not be "part" of the exponent when we know the values of the other variables.

The first case, in which we want to find a value for a variable that is *not* part of the exponent, is illustrated by the following example.

Example   A scientist starts an experiment with 25 grams of a radioactive element. The number of grams remaining at any time t is given by $y = 25e^{-0.5t}$, where t is in seconds. How much of the element (to the nearest hundredth of a gram) is remaining after three seconds?

Solution   Substituting 3 for t, we have

$$y = 25e^{-0.5(3)}$$
$$= 25e^{-1.5} = 5.57825.$$

There are approximately 5.58 grams of material remaining after three seconds.

The second case, in which we want to find a value for a variable that is in the exponent, is illustrated by the following example.

Example

The atmospheric pressure P, in inches of mercury, is given approximately by

$$P = 30(10)^{-0.09a}, \tag{1}$$

where a is the altitude in miles above sea level. How high above the earth (to the nearest hundredth of a mile) is the atmospheric pressure 26.4 inches of mercury?

Solution

Substituting 26.4 for P, we obtain

$$26.4 = 30(10)^{-0.09a},$$

which is equivalent to

$$\frac{26.4}{30} = 10^{-0.09a}.$$

Writing the equation in logarithmic form, we have

$$-0.09a = \log_{10} \frac{26.4}{30},$$

from which

$$a = -\frac{1}{0.09} \log_{10} \frac{26.4}{30}$$

$$= 0.61686.$$

Hence, the pressure is 26.4 inches of mercury at approximately 0.62 miles above the earth.

EXERCISE 9.7

A   ■ *A calculator was used to obtain the answers for this section: Answers obtained by using tables may differ slightly from those given.*

■ *Solve Problems 1–6 using the relationship $P = 30(10)^{-0.09a}$ between altitude (a) in miles and atmospheric pressure (P) in inches of mercury. Round off results to the nearest hundredth.*

1. The elevation of Mt. Everest, the highest mountain in the world, is 29,028 feet. What is the atmospheric pressure at the top? [*Hint:* One mile equals 5280 feet.]

2. What is the atmospheric pressure at sea level? 50,000 feet? 100,000 feet?

3. How high above sea level is the atmospheric pressure 20.2 inches of mercury?

4. How high above sea level is the atmospheric pressure 16.1 inches of mercury?

5. Find the height above sea level at which the pressure is equal to one-half of the pressure at sea level. (Pressure at sea level is approximately 30 inches of mercury.)

6. Find the height above sea level at which the pressure is equal to one-fourth of the pressure at sea level.

■ *Solve Problems 7–12 using the fact that population growth is given approximately by $P = P_0 e^{rt}$, where an initial population P_0 increases at an annual rate r (expressed as a decimal) to a population P after t years.*

7. The population of the state of California increased from 1960 to 1970 at a rate of approximately 2.39% per year. The population in 1960 was 15,717,000.
   a. Approximately what was the population in 1970?
   b. Assuming the same rate of growth, estimate the population in the years 1980, 1990, and 2000.

8. The population of the state of New York increased from 1960 to 1970 at a rate of approximately 0.83% per year. The population in 1960 was 16,782,000.
   a. Approximately what was the population in 1970?
   b. Assuming the same rate of growth, estimate the population in the years 1980, 1990, and 2000.

9. The population of the state of Texas in 1960 was 9,579,700. In 1970 the population was 11,196,700. What was the annual rate of growth to the nearest hundredth of a percent?

10. The population of the state of Florida in 1960 was 4,951,600. In 1970 the population was 6,789,400. What was the annual rate of growth to the nearest hundredth of a percent?

11. If the annual rate of growth of a country is 3.7%, how long will it take for the population to double?

12. The population of a country doubled in 20 years. What was the annual rate of growth (to the nearest hundredth of a percent)?

13. The amount of a radioactive element present at any time t is given by $y = y_0 e^{-0.4t}$, where t is measured in seconds and y_0 is the amount present initially. How much of the element (to the nearest hundredth of a gram) would remain after 3 seconds if 40 grams were present initially?

14. The number N of bacteria present in a culture is given by $N = N_0 e^{0.04t}$, where N_0 is the number of bacteria present at time $t = 0$, and t is time in hours. If 6000 bacteria were present at $t = 0$, how many were present 10 hours later?

15. The voltage V across a capacitor in a certain circuit is given by $V = 100(1 - e^{-0.5t})$, where t is the time in seconds. What is the voltage (to the nearest tenth of a volt) after 10 seconds?

16. The intensity I (in lumens) of a light beam after passing through a thickness t (in centimeters) of a medium having an absorption coefficient of 0.1 is given by $I = 1000e^{-0.1t}$. What is the intensity (to the nearest tenth) of a light beam passing through 0.6 centimeters of the medium?

17. The voltage V across a capacitor in a certain circuit is given by $V = 100(1 - e^{-0.5t})$, where t is the time in seconds. How much time must elapse (to the nearest hundredth of a second) for the voltage to reach 75 volts?

18. The amount of a radioactive element present at any time t is given by $y = y_0 e^{-0.4t}$, where t is measured in seconds and y_0 is the amount present initially. How much time must elapse (to the nearest hundredth of a second) for 40 grams to be reduced to 12 grams?

19. The number N of bacteria present in a culture is given by $N = N_0 e^{0.04t}$, where N_0 is the number of bacteria present at time $t = 0$, and t is time in hours. How much time must elapse (to the nearest tenth of an hour) for 2500 bacteria to increase to 10,000?

20. The intensity I (in lumens) of a light beam after passing through a thickness t (in centimeters) of a medium having an absorption coefficient of 0.1 is given by $I = 1000e^{-0.1t}$. How many centimeters (to the nearest tenth) of the material would reduce the illumination to 800 lumens?

B   ■ *Solve Problems 21–28 using the following information: P dollars invested at an annual interest rate r (expressed as a decimal) compounded yearly yields an amount A after n years given by $A = P(1 + r)^n$. If the interest is compounded t times yearly, the amount is given by*

$$A = P\left(1 + \frac{r}{t}\right)^{tn}.$$

Example   One dollar compounded annually for 12 years yields $1.127. What is the rate of interest to the nearest ½%?

Solution   Using the given equation, we obtain $(1 + r)^{12} = 1.127$. Equate $\log_{10}$ of each member and apply the third law of logarithms.

$$12 \log_{10}(1 + r) = \log_{10} 1.127 = 0.0519$$

Divide each member by 12.

$$\log_{10}(1 + r) = \frac{1}{12}(0.0519) = 0.0043$$

Write $\log_{10}(1 + r) = 0.0043$ in exponential form.

$$1 + r = 10^{0.0043} = 1.01$$

Hence,

$$r = 0.01 \quad \text{or} \quad r = 1\%.$$

21. One dollar compounded annually for 10 years yields $5.12. What is the rate of interest to the nearest ½%?

22. How many years (to the nearest year) would it take for $1000 to "grow" to $2000 if compounded annually at 12%?

23. What rate of interest (to the nearest ½%) is required so that $100 would yield $190 after 5 years if the money were compounded semiannually?

24. What rate of interest (to the nearest ½%) is required so that $40 would yield $60 after 3 years if the money were compounded quarterly?

25. Find the compounded amount of $5000 invested at 12% for 10 years when compounded annually. When compounded semiannually.

26. How many years (to the nearest year) would it take for a sum of money to double if invested at 10% and compounded quarterly?

27. How many years (to the nearest year) would it take for a sum of money to increase fivefold if invested at 10% and compounded quarterly?

28. Two investors, A and B, each invested $10,000 at 8% for 20 years with a bank that computed interest quarterly. Investor A withdrew interest at the end of each 3 month period, but B allowed the investment to be compounded. How much more than A did B earn over the period of 20 years?

9.8

COMPUTATIONS WITH LOGARITHMS

The advent of electronic computing devices has removed the need to perform routine numerical computations with pencil and paper using logarithms. Nevertheless, we introduce the techniques involved in making such computations because they shed light on the properties of the logarithmic function. They also provide an insight into the algorithms that calculators use to perform such computations.

Laws of logarithms

We first reproduce the laws of logarithms here, using the base 10. These laws will be used in conjunction with tables or calculator.

If x_1 and x_2 are positive real numbers, then

$$\log_{10}(x_1 x_2) = \log_{10} x_1 + \log_{10} x_2, \tag{1}$$

$$\log_{10}\frac{x_2}{x_1} = \log_{10} x_2 - \log_{10} x_1, \tag{2}$$

$$\log_{10}(x_1)^m = m \log_{10} x_1. \tag{3}$$

We also name two assumptions that we have been using which are helpful in stating the instructions for computations using logarithms.

> **L-1** If $x_1 = x_2$ $(x_1, x_2 > 0)$, then $\log_{10} x_1 = \log_{10} x_2$.
>
> **L-2** If $\log_{10} x_1 = \log_{10} x_2$, then $x_1 = x_2$.

Using Tables

In addition to working with the properties of logarithms, some of the computations provide an opportunity to see the difficulties that can arise when tables are employed.

The examples below have been solved by using the tables of logarithms and *reading to the nearest entry when a number is not an entry*.

Example   Multiply: $(3.82)(0.00729)$.

Solution   Set

$$N = (3.82)(0.00729).$$

By assumption L-1,

$$\log_{10} N = \log_{10}[(3.82)(0.00729)].$$

By the first law of logarithms,

$$\log_{10} N = \log_{10} 3.82 + \log_{10} 0.00729.$$

From Table II,

$$\log_{10} 3.82 = 0.5821 \;\text{———}\Big\rceil$$
$$+$$
$$\log_{10} 0.00729 = \underline{0.8627 - 3} \;\text{——}\Big\rfloor$$
$$\log_{10} N = 1.4448 - 3.$$

Hence, $N = \text{antilog}_{10}(1.4448 - 3)$ or, equivalently, $\text{antilog}_{10}(0.4448 - 2)$.

From the table, 0.4448 is halfway between 0.4456 and 0.4440. We choose the greater number. Hence,

$$N = \text{antilog}_{10}(0.4448 - 2)$$
$$\approx \text{antilog}_{10}(0.4456 - 2)$$
$$= 2.79 \times 10^{-2} = 0.0279.$$

Now consider a more complicated example.

Example   Compute $\dfrac{(8.21)^{1/2}(2.17)^{2/3}}{(3.14)^{1/3}}$.

Solution   Set

$$N = \frac{(8.21)^{1/2}(2.17)^{2/3}}{(3.14)^{1/3}}.$$

We have

$$\log_{10} N = \log_{10}\left[\frac{(8.21)^{1/2}(2.17)^{2/3}}{(3.14)^{1/3}}\right]$$
$$= \log_{10}(8.21)^{1/2} + \log_{10}(2.17)^{2/3} - \log_{10}(3.14)^{1/3}$$
$$= \frac{1}{2}\log_{10} 8.21 + \frac{2}{3}\log_{10} 2.17 - \frac{1}{3}\log_{10} 3.14.$$

From Table II,

$$\frac{1}{2}\log_{10} 8.21 = \frac{1}{2}(0.9143) = 0.4572$$

$$\frac{2}{3}\log_{10} 2.17 = \frac{2}{3}(0.3365) = \underline{0.2243}$$

$$0.6815$$

$$\frac{1}{3}\log_{10} 3.14 = \frac{1}{3}(0.4969) = \underline{0.1656}$$

$$\log_{10} N = 0.1559$$

Therefore,

$$N = \text{antilog}_{10}\ 0.5159 = 3.28.$$

If we wish to use the tables in computations with logarithms, it is sometimes necessary to use alternative forms for characteristics in order to keep the decimal part (mantissa) of the logarithm positive.

Example   Compute $\dfrac{2.43}{7.83}$.

Solution   Set

$$Q = \frac{2.43}{7.83}.$$

We have

$$\log_{10} Q = \log_{10} \frac{2.43}{7.83}$$

$$= \log_{10} 2.43 - \log_{10} 7.83$$

$$= 0.3856 - 0.8938$$

$$= -0.5082.$$

Because -0.5082 is negative, it is not an entry in the table. Hence, we first write

$$\log_{10} Q = (-0.5082 + 1) - 1 = 0.4918 - 1.$$

From Table II,

$$Q = \text{antilog}_{10}(0.4918 - 1)$$

$$\approx \text{antilog}_{10}(0.4914 - 1)$$

$$= 3.10 \times 10^{-1}$$

$$= 0.310.$$

When using the tables, it is also necessary that the characteristic of the logarithm be an integer.

Example   Compute $\sqrt[3]{0.043}$.

Solution   Set

$$N = \sqrt[3]{0.043} = (0.043)^{1/3}.$$

We have

$$\log_{10} N = \log_{10}(0.043)^{1/3} = \frac{1}{3}\log_{10} 0.043.$$

From Table II,

$$\log_{10} N = \frac{1}{3}(0.6335 - 2).$$

Now, if we use $1.6335 - 3$, $4.6335 - 6$, or $7.6335 - 9$ in place of $0.6335 - 2$, the negative portion of the characteristic is exactly divisible by 3. Thus, we have

$$\log_{10} N = \frac{1}{3}(1.6335 - 3) = 0.5445 - 1$$

instead of

$$\log_{10} N = \frac{1}{3}(0.6335 - 2) = 0.2112 - \frac{2}{3},$$

where the latter has a nonintegral characteristic. Hence,

$$N = \text{antilog}_{10}(0.5445 - 1)$$

Using Calculators
$$= 3.50 \times 10^{-1}$$
$$= 0.350.$$

In order to practice using the properties of logarithms and to see how logarithms were used to aid computation before the advent of calculators, we shall perform multiplication without using the multiplication key of the calculator, division without using the division key, and take roots without using a root or power key. However, we shall use the calculator to obtain function values. The following example is the same product that we computed on page 324 using the tables. The computation is also shown here in a parallel solution using laws of exponents to show how the laws of logarithms are related to the laws of exponents.

Example   Multiply: $(3.82)(0.00729)$.

Solution

Laws of logarithms

Set

$$N = (3.82)(0.00729).$$

By assumption L-1,

$$\log_{10} N = \log_{10}[(3.82)(0.00729)].$$

By the first law of logarithms,

$$\log_{10} N = \log_{10} 3.82 + \log_{10} 0.00729.$$

$$\log_{10} 3.82 = 0.5821 \;\rule[0.5ex]{0.8cm}{0.4pt}$$

$$+$$

$$\log_{10} 0.00729 = -2.1373 \;\rule[0.5ex]{0.8cm}{0.4pt}$$

$$\log_{10} N = -1.5552$$

Hence,

$$N = \text{antilog}_{10}(-1.5552)$$
$$= 10^{-1.5552} = 0.02785.$$

Laws of exponents

We first use a calculator to obtain $\log_{10} 3.82$ and $\log_{10} 0.00729$. Then we write each factor as a power of 10 and use the first law of exponents.

$$(3.82)(0.00729) = 10^{0.5821} \cdot 10^{-2.1373}$$
$$= 10^{-1.5552}$$
$$= 0.02784$$

Note that the use of a calculator in the above example avoids the requirement—necessary when using tables—that the decimal part of the logarithm be positive.

The following examples also show the use of a calculator instead of Table II to find function values. (See similar problems on pages 325 and 326.)

Example   Compute $\dfrac{2.43}{7.83}$.

Solution

Laws of logarithms

Set

$$Q = \frac{2.43}{7.83}.$$

We have

$$\log_{10} Q = \log_{10} \frac{2.43}{7.83}$$
$$= \log_{10} 2.43 - \log_{10} 7.83$$
$$= 0.3856 - 0.8938$$
$$= -0.5082;$$
$$Q = \text{antilog}_{10}(-0.5082)$$
$$= 10^{-0.5082} = 0.310.$$

Laws of exponents

We first use a calculator to obtain $\log_{10} 2.43$ and $\log_{10} 7.83$. Then we write each factor as a power of 10 and use the second law of exponents.

$$\frac{2.43}{7.83} = \frac{10^{0.3856}}{10^{0.8938}}$$
$$= 10^{(0.3856-0.8938)}$$
$$= 10^{-0.5082}$$
$$= 0.310$$

Example   Compute $\sqrt[3]{0.043}$.

Solution

Laws of logarithms	*Laws of exponents*
Set	We first use a calculator to obtain $\log_{10} 0.043$. Then we write 0.043 as a power of 10 and use the third law of exponents.
$$N = \sqrt[3]{0.043} = (0.043)^{1/3}.$$	
We then have	
$$\log_{10} N = \log_{10}(0.043)^{1/3}$$	$$\sqrt[3]{0.043} = (0.043)^{1/3}$$
$$= \frac{1}{3}\log_{10}(0.043)$$	$$= (10^{-1.3665})^{1/3}$$
$$= \frac{1}{3}(-1.3665) = -0.4555;$$	$$= 10^{-0.4555}$$
$$N = \text{antilog}_{10}(-0.4555)$$	$$= 0.350$$
$$= 10^{-0.4555} = 0.350.$$	

EXERCISE 9.8

A   ■ *Compute by first using the properties of logarithms and then using Table II (read values to the nearest table entry) or a calculator. Round off results to three significant digits.*

1. $(2.32)(1.73)$

2. $(82.3)(6.12)$

3. $\dfrac{3.15}{1.37}$

4. $\dfrac{1.38}{2.52}$

5. $\dfrac{0.0149}{32.3}$

6. $\dfrac{0.00214}{3.17}$

7. $(2.3)^5$

8. $(4.62)^3$

9. $\sqrt[3]{8.12}$

10. $\sqrt[5]{75}$

11. $(0.0128)^5$

12. $(0.0021)^6$

13. $\sqrt{0.0021}$

14. $\sqrt[5]{0.0471}$

15. $\sqrt[3]{0.0214}$

16. $\sqrt[4]{0.0018}$

17. $\dfrac{(8.12)(8.74)}{7.19}$

18. $\dfrac{(0.421)^2(84.3)}{\sqrt{21.7}}$

19. $\dfrac{(6.49)^2\sqrt[3]{8.21}}{17.9}$

20. $\dfrac{(2.61)^2(4.32)}{\sqrt{7.83}}$

21. $\dfrac{(0.349)(27.1)}{6.81}$

22. $\dfrac{(4.81)^2(20.1)}{3.61}$

23. $\sqrt{\dfrac{(4.71)(0.00481)}{(0.0432)^2}}$

24. $\sqrt{\dfrac{(2.85)^3(0.97)}{0.035}}$

25. $\sqrt{25.1(25.1 - 18.7)(25.1 - 4.3)}$

26. $\sqrt{\dfrac{(4.17)^3(68.1 - 4.7)}{(68.1 - 52.9)}}$

27. $\dfrac{\sqrt{(23.4)^3(0.0064)}}{\sqrt[3]{69.1}}$

28. $\dfrac{\sqrt{38.7}\sqrt[3]{491}}{\sqrt[4]{9.21}}$

■ *Solve.*

29. A formula for computing kilowatt-hours (kwh) of electricity sold to a consumer is given by

$$\text{kwh} = \frac{E \cdot I \cdot t}{1000},$$

where E is the voltage, I is the amperage, and t is the time in hours. Find to the nearest tenth the number of kilowatt-hours used by a circuit drawing 3.5 amps from a 110-volt line for 12 hours.

30. Using the formula in Problem 29, find the number of kilowatt-hours (to the nearest tenth) used by a circuit drawing 4.1 amps from a 220-volt line for 90 minutes.

31. The horsepower (hp) of a gasoline engine is sometimes given by the formula

$$\text{hp} = \frac{D^2 \cdot N}{2.5},$$

where D is the bore of each cylinder in inches and N is the number of cylinders. Find to the nearest tenth the horsepower of a 4-cylinder engine if the bore of each cylinder is 3.42 inches.

32. The force F necessary to raise a weight w by a chain hoist consisting of three pulleys is given by

$$F = \frac{w}{2}\left(\frac{D_1 - D_2}{D_1}\right),$$

where the two pulleys on top have diameters D_1 and D_2 and the pulley on the bottom has diameter D_1. Find the force necessary to lift 2500 pounds when the diameters of the pulleys D_1 and D_2 are 22.4 inches and 18.1 inches, respectively.

33. The period T of a simple pendulum is given by the formula

$$T = 2\pi\sqrt{\frac{L}{g}},$$

where T is measured in seconds, L is the length of the pendulum in feet, $\pi \approx 3.14$, and $g \approx 32$ feet per second per second. Compute the period of a pendulum 1 foot long.

34. The area A of a triangle in terms of the sides is given by the formula

$$A = \sqrt{s(s - a)(s - b)(s - c)},$$

where a, b, and c are the lengths of the sides of the triangle and s equals one-half the perimeter. Use logarithms to compute the area of a triangle in which the lengths of the three sides are 2.31 inches, 4.21 inches, and 5.62 inches.

CHAPTER SUMMARY

[9.1] A function defined by an equation of the form

$$f(x) = b^x \qquad (b > 0,\ b \neq 1),$$

is called an **exponential function**. If $b > 1$, it is an **increasing function**; if $0 < b < 1$, it is a **decreasing function**. The domain of the function is the set of real numbers; the range is the set of positive real numbers.

[9.2] The inverse of an exponential function is called a **logarithmic function** and is defined by an equation of the form

$$x = b^y \quad \text{or} \quad y = \log_b x \quad (b > 0, \ b \neq 1).$$

$\text{Log}_b x$ is the **exponent** on b such that the power equals x. The domain of a logarithmic function is the set of positive real numbers; the range is the set of real numbers.

[9.3] The properties of logarithms given below follow from the definition of a logarithm and the properties of exponents developed in Chapter 5.

$$\log_b(x_1 x_2) = \log_b x_1 + \log_b x_2 \tag{1}$$

$$\log_b \frac{x_2}{x_1} = \log_b x_2 - \log_b x_1 \tag{2}$$

$$\log_b(x_1)^m = m \log_b x_1 \tag{3}$$

[9.4] Values of $\log_{10} x$, $x > 0$ are known as **common logarithms**.

Values of $\log_{10} x$, where x is a power of 10 with an integer exponent, can be obtained by inspection directly from the definition of a logarithm.

Values of $\log_{10} x$ $(1 < x < 10)$ are between 0 and 1 and can be obtained directly from Table II.

Values of $\log_{10} x$ $(x > 10$ and $0 < x < 1)$ can be obtained from Table II in conjunction with the first law of logarithms. The *integral part* of a logarithm to the base 10 is the **characteristic** of the logarithm, and the *positive decimal part* is the **mantissa**.

Values for e^x can be obtained by using Table III; values for $\log_e x$ or $\ln x$, called **natural logarithms**, can be obtained by using Table IV.

A power b^x is called the **antilogarithm** of x. Thus,

$$10^x = \text{antilog}_{10} x \quad \text{and} \quad e^x = \text{antilog}_e x.$$

[9.5] A scientific calculator can be used to find values for the powers 10^x and e^x $(x \in R)$ and the exponents $\log_{10} x$ and $\ln x$ $(x > 0)$. (See tables on page 331.)

[9.6] Exponential equations can be solved by using the properties of logarithms plus a table of logarithms or a calculator.

[9.7] Exponential equations can be used as models for many real-world problems.

[9.8] Logarithms can be used in computations involving multiplication, division, powers, and roots.

The symbols introduced in this chapter are listed on the inside of the front cover.

The following tables summarize the different ways of finding the exponential and logarithmic function values that have been considered in this chapter.

To find the power for a given value of an exponent:

	10^x (antilog$_{10}$ x)	e^x (antilog$_e$ x)	b^x $(b>0, b \neq 1)$
Table	II*	III	
Calculator	$\boxed{10^x}$ or $\boxed{y^x}$ or $\frac{\boxed{\text{INV}}}{\boxed{\text{2ND}}}$ $\boxed{\text{LOG}}$	$\boxed{e^x}$ or $\boxed{y^x}$ or $\frac{\boxed{\text{INV}}}{\boxed{\text{2ND}}}$ $\boxed{\text{LN}}$	$\boxed{y^x}$

To find the exponent for a given value of a power:

	$10^x = N$ $x = \log_{10} N$	$e^x = N$ $x = \ln N$	$b^x = N$ $(b>0, b \neq 1)$
Table	II*	IV*	II* or IV*
Calculator	$\boxed{\text{LOG}}$ or $x = \log_{10} N = \frac{\ln N}{\ln 10}$ $N \boxed{\text{LN}} \boxed{\div} 10 \boxed{\text{LN}} \boxed{=} x$	$\boxed{\text{LN}}$	$x = \log_b N = \frac{\ln N}{\ln b}$ $N \boxed{\text{LN}} \boxed{\div} b \boxed{\text{LN}} \boxed{=} x$

Note: When using Table II or IV, the decimal part (mantissa) of a logarithm must be positive; the characteristic must be an integer.

REVIEW EXERCISES

A

[9.1]   **1.** Sketch the graph of each equation.

      a. $y = 5^x$ **b.** $y = 5^{-x}$

[9.2]   2. Write each statement in logarithmic notation.

a. $9^{3/2} = 27$ b. $\left(\dfrac{4}{9}\right)^{1/2} = \dfrac{2}{3}$

3. Write each statement in exponential notation.

a. $\log_5 625 = 4$ b. $\log_{10} 0.0001 = -4$

4. Find a value for each variable.

a. $\log_4 16 = x$ b. $\log_2 x = 3$

[9.3]   5. Express each as the sum or difference of simpler logarithmic quantities.

a. $\log_b 3x^2 y$ b. $\log_b \dfrac{y\sqrt{x}}{z^2}$

6. Write each expression as a single logarithm with a coefficient of 1.

a. $5 \log_b x - 2 \log_b y$ b. $\dfrac{1}{3}(\log_b x - 4 \log_b y + 2 \log_b z)$

7. a. Solve $\log_{10}(x+9) - \log_{10} x = 1$.   b. Solve $\log_{10} x + \log_{10}(x-3) = 1$.

8. Given that $\log_b 2 = 0.3010$ and $\log_b 3 = 0.4771$, find the value for each logarithm.

a. $\log_b 12$ b. $\log_b \sqrt{18}$

■ *In Problems 9–24, use the tables or a calculator (round off readings to three significant digits).*

[9.4–9.5]   9. Find a value for each logarithm.

a. $\log_{10} 0.713$ b. $\log_{10} 1810$

10. Solve for x.

a. $\log_{10} x = 2.6345$ b. $\log_{10} x = -1.4214$

11. Compute each power.

a. $10^{1.2347}$ b. $10^{-0.5453}$

12. Find the value of each power.

a. $e^{0.83}$ b. $e^{-1.3}$

13. Find the value of each logarithm.

a. $\ln 7$ b. $\ln 0.4$

14. Solve for x.

a. $\ln x = 0.73$ b. $\ln x = 2.7$

[9.6]   ■ *Solve each exponential equation.*

15. $10^x = 1.3$ 16. $e^x = 62$

17. $7.35 = 2.1(10)^{1.2x}$ 18. $12.4 = 2e^{0.3x} - 4.2$

19. Given that $N = N_0 10^{0.4t}$, find t if $N = 280$ and $N_0 = 4$.

20. Given that $y = y_0 e^{-0.2t}$, find t if $y = 4$ and $y_0 = 20$.

■ *Use Equation* (2) *on page* 315 *to find each logarithm.*

21. $\log_2 23.1$ **22.** $\log_3 7.04$

■ *Solve for x using logarithms to the base* 10 *or the base e.*

23. $3^x = 15$ **24.** $2^{x-4} = 10$

[9.7]   **25.** The concentration C of a certain drug in the bloodstream at any time t is given by $C(t) = 10 - 10e^{-0.5t}$, where C is in milligrams and t is in minutes. Determine the concentration to the nearest tenth of a milligram at $t = 0$ and $t = 1$.

         26. The intensity I of a light beam after passing through a thickness t of a certain medium is given by $I = 500e^{-0.2t}$, where I is in lumens and t is in centimeters. What is the intensity (to the nearest tenth of a lumen) of a light beam passing through 0.4 centimeters of the medium?

         27. Using the formula for population growth on page 321, how long (to the nearest tenth of a year) will it take for a city with an annual rate of growth of 3% to grow from a population of 200,000 to 300,000?

         28. Using the formula for radioactive decay in Problem 13 on page 321, how much time (to the nearest tenth of a second) must elapse for 100 grams to be reduced to 50 grams?

[9.8]   ■ *Compute (to three significant digits) by using properties of logarithms.*

29. $(3.17)(8.23)$ **30.** $\dfrac{\sqrt{18.72}}{3.12}$ **31.** $\dfrac{(2.12)^2(3.42)^3}{147}$

32. $\dfrac{\sqrt{(10.6)(2.31)^3}}{(7.016)}$ **33.** $\dfrac{2.045}{7.312}$ **34.** $\dfrac{(31.24)(0.02)^3}{1.003}$

B   **35.** Simplify $\log_{10} \log_2 (\log_5 25)$.

        36. Verify that $2 \log_b 8 - \log_b 4 = 4 \log_b 2$ is a true statement.

        37. Solve $N = N_0 e^{-kt}$ for t, using natural logarithms.

        38. Solve $N = N_0 \ln(t/k) + c$ for t.

■ *Chemists define the* pH *(hydrogen potential) of a solution by* $\mathrm{pH} = \log_{10} \dfrac{1}{[H^+]}$, *where* $[H^+]$ *represents a numerical value for the concentration of hydrogen ions in aqueous solution in moles per liter.*

39. Calculate to the nearest tenth the pH of a solution with hydrogen ion concentration 6.3×10^{-7}.

40. Calculate the hydrogen ion concentration of a solution with pH 5.6.

10. SYSTEMS OF EQUATIONS

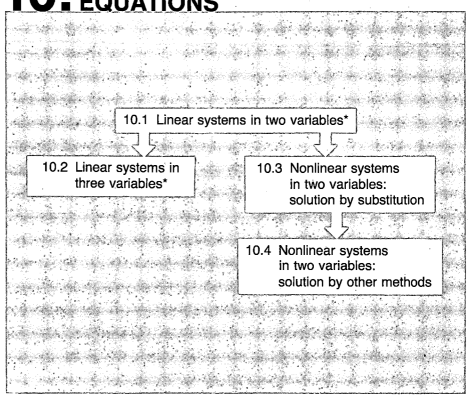

10.1 Linear systems in two variables*

10.2 Linear systems in three variables*

10.3 Nonlinear systems in two variables: solution by substitution

10.4 Nonlinear systems in two variables: solution by other methods

10.1

LINEAR SYSTEMS IN TWO VARIABLES

In Chapters 7 and 8, we observed that the solution set of an equation in two variables, such as

$$ax + by = c,$$
$$y = ax^2 + bx + c,$$

or

$$ax^2 + by^2 = c,$$

might contain infinitely many ordered pairs of numbers. It is often necessary to consider *pairs* of such equations and to inquire whether or not the solution sets of the two equations contain ordered pairs in common. More specifically, we are interested in determining the members of the *intersection* of their solution sets.

*Other methods of solving linear systems are considered in Appendix A.

Solution set of a system

We shall begin by considering the system (in standard form),

$$a_1 x + b_1 y = c_1 \qquad (a_1,\ b_1 \text{ not both } 0)$$
$$a_2 x + b_2 y = c_2 \qquad (a_2,\ b_2 \text{ not both } 0).$$

In a geometric sense, because the graphs of both of these equations are straight lines, we are confronted with three possibilities, as illustrated in Figure 10.1:

a. The graphs are the same line.

b. The graphs are parallel but distinct lines.

c. The graphs intersect at one and only one point.

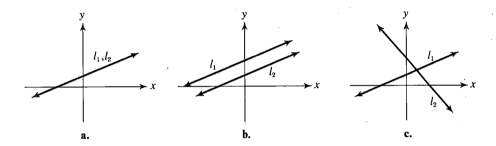

Figure 10.1

These possibilities lead, correspondingly, to the conclusion that one and only one of the following is true for any given system of two such linear equations in x and y:

a. The solution sets of the equations are equal, and their intersection contains all (an infinite number of) ordered pairs found in either one of the solution sets.

b. The intersection of the two solution sets is the null set.

c. The intersection of the two solution sets contains one and only one ordered pair.

In case a, the linear equations in x and y are said to be **dependent**, and, in case b, the equations are said to be **inconsistent**. In case c, the equations are **consistent** and **independent** and the system has one and only one solution.

Consider the system

$$x + y = 5$$
$$x - y = 1.$$

From the graph of the system in Figure 10.2, it is evident that the ordered pair $(3, 2)$ is the only ordered pair common to the solution sets of both equations. We can verify that $(3, 2)$ is indeed the ordered pair in question by substituting $(3, 2)$ into each equation and observing that a true statement results in each case.

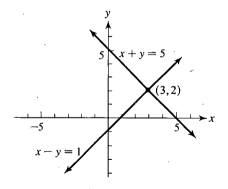

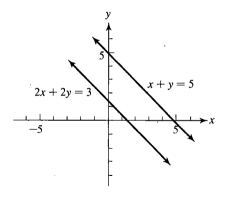

Figure 10.2   Figure 10.3

As another example, consider the system

$$x + y = 5$$
$$2x + 2y = 3,$$

and the graphs of these equations in Figure 10.3, where the lines appear to be parallel. We conclude from this that the solution set of this system is $\varnothing$.

Linear combinations   Because graphing equations is a time-consuming process, and, more important, because graphic results are not always precise, solutions to systems of linear equations are usually sought by algebraic methods. One such method depends on the following property:

▶ *Any ordered pair* (x, y) *that satisfies the equations*

$$a_1x + b_1y = c_1 \tag{1}$$

$$a_2x + b_2y = c_2 \tag{2}$$

will also satisfy the equation

$$A(a_1x + b_1y) + B(a_2x + b_2y) = Ac_1 + Bc_2 \tag{3}$$

for all real numbers A and B.

We can see the validity of this property if we first rewrite the system of Equations (1) and (2) as

$$a_1x + b_1y - c_1 = 0 \tag{1a}$$

$$a_2x + b_2y - c_2 = 0, \tag{2a}$$

and Equation (3) as

$$A(a_1x + b_1y - c_1) + B(a_2x + b_2y - c_2) = 0. \tag{3a}$$

Note that the replacement of the variables in (3a) with the components of any ordered pair (x, y) that satisfies both (1a) and (2a) results in

$$A(0) + B(0) = 0,$$

which is clearly true for any values of A and B. The left-hand member of (3a) is called a **linear combination** of the left-hand members of (1a) and (2a).

The property stated above asserts that any ordered pair satisfying both (1) and (2) must also satisfy the sum of any real-number multiples of (1) and (2). This fact can be used to identify any such ordered pairs.

The concept of a linear combination can be used to solve a system by choosing multipliers A and B so that the coefficients of *one* of the variables, x or y, are additive inverses of each other, resulting in an equation free of *one* of the variables.

Example   Solve the system

$$2x + 3y = 8 \tag{4}$$

$$3x - 4y = -5. \tag{5}$$

Solution   We first want to rewrite one or both of the equations in equivalent forms so that the coefficients of the same variable (either x or y) will be additive inverses of each other. In this example, we shall obtain the coefficients of x as additive inverses by multiplying each member of (4) by 3 and each member of (5) by -2 to obtain

$$6x + 9y = 24 \tag{4a}$$

$$-6x + 8y = 10. \tag{5a}$$

Adding the corresponding members of (4a) and (5a), we obtain

$$17y = 34 \tag{6}$$

$$y = 2, \tag{6a}$$

which must contain in its solution set any solution common to the solution sets of the two original equations. But any solution of (6) or (6a) is of the form $(x, 2)$—that is, it has y-component 2 for any value of x. Now, substituting 2 for y in either (4) or (5), we can determine the x-component for the ordered pair $(x, 2)$ that satisfies both (4) and (5). If (4) is used, we have

$$2x + 3(2) = 8$$
$$x = 1;$$

and if (5) is used, we have

$$3x - 4(2) = -5$$
$$x = 1.$$

Since the ordered pair $(1, 2)$ satisfies both (4) and (5), the required solution set is $\{(1, 2)\}$.

Test for
a unique
solution

We can readily determine whether a system has a unique solution or whether the equations are inconsistent or dependent by first writing each equation in slope-intercept form. In general, the system

$$a_1x + b_1y = c_1$$

$$a_2x + b_2y = c_2$$

can be written equivalently as

$$y = \frac{-a_1}{b_1}x + \frac{c_1}{b_1}$$

$$y = \frac{-a_2}{b_2}x + \frac{c_2}{b_2}.$$

Now, if the slopes $-a_1/b_1$ and $-a_2/b_2$ are unequal, the graphs will intersect at one point and there will be one and only one solution. If $a_1/b_1 = a_2/b_2$, then the graphs are either the same line (if $c_1/b_1 = c_2/b_2$) or parallel lines (if $c_1/b_1 \neq c_2/b_2$). These conclusions are summarized in the following property:

► *Any system of the form*

$$a_1x + b_1y = c_1$$

$$a_2x + b_2y = c_2$$

has one and only one solution if

$$\frac{a_1}{a_2} \neq \frac{b_1}{b_2}, \tag{7}$$

has no solution if

$$\frac{a_1}{a_2} = \frac{b_1}{b_2} \neq \frac{c_1}{c_2}, \tag{8}$$

and has an infinite number of solutions if

$$\frac{a_1}{a_2} = \frac{b_1}{b_2} = \frac{c_1}{c_2}. \tag{9}$$

Examples

a. For the system

$$2x + 3y = 2$$

$$4x + 6y = 7,$$

we note that

$$\frac{2}{4} = \frac{3}{6} \neq \frac{2}{7}.$$

Hence, from Property (8) above, the system does not have a solution.

b. For the system

$$2x + 3y = 5$$
$$4x + 7y = 8,$$

we note that

$$\frac{2}{4} \neq \frac{3}{7}.$$

Hence, from Property (7) above, the system has one and only one solution.

Systems of equations are quite useful in expressing relationships in practical applications. By the assignment of separate variables to represent separate physical quantities, the difficulty encountered in symbolically representing these relationships can usually be decreased. In writing systems of equations, we must be careful that the conditions giving rise to one equation are independent of the conditions giving rise to any other equation.

EXERCISE 10.1

A ■ *Solve each system by linear combinations. In Problems 1–16 sketch the graphs of the equations. If the system has no solution or an infinite number of solutions, so state.*

Example

$$\frac{2}{3}x - y = 2 \qquad (1)$$

$$x + \frac{1}{2}y = 7 \qquad (2)$$

Solution

Multiply each member of Equation (1) by 3 and each member of Equation (2) by 2.

$$2x - 3y = 6 \qquad (1a)$$
$$2x + \ y = 14 \qquad (2a)$$

Add −1 times Equation (1a) to 1 times Equation (2a) and solve for *y*.

$$4y = 8$$
$$y = 2$$

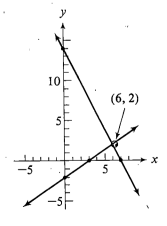

(6, 2)

Solution continued overleaf

Substitute 2 for y in (1), (2), (1a), or (2a) and solve for x. In this example, Equation (2) is used.

$$x + \frac{1}{2}(2) = 7$$

$$x = 6$$

The solution set is $\{(6, 2)\}$.

1. $x - y = 1$
$\quad\; x + y = 5$

2. $2x - 3y = 6$
$\quad\;\; x + 3y = 3$

3. $3x + \; y = 7$
$\quad\; 2x - 5y = -1$

4. $2x - \; y = 7$
$\quad\; 3x + 2y = 14$

5. $5x - \; y = -29$
$\quad\; 2x + 3y = 2$

6. $\quad x + 4y = -14$
$\quad\; 3x + 2y = -2$

7. $3x + 2y = 7$
$\quad\; x + \; y = 3$

8. $2x - 3y = 8$
$\quad\;\; x + \; y = -1$

9. $- \; x + 3y = -1$
$\quad\; -6x + \; y = -6$

10. $\quad 3x - 3y = -3$
$\quad\; -6x + 2y = 14$

11. $5x + 3y = 19$
$\quad\; 2x - \; y = 12$

12. $3x - 5y = -1$
$\quad\;\; x + 2y = 18$

13. $5x + 2y = 3$
$\quad\quad\quad x = 0$

14. $2x - y = 0$
$\quad\quad\quad x = -3$

15. $3x - 2y = 4$
$\quad\quad\quad\; y = -1$

16. $x + 2y = 6$
$\quad\quad\quad x = 2$

17. $\frac{1}{4}x - \frac{1}{3}y = -\frac{5}{12}$

$\quad\; \frac{1}{10}x + \frac{1}{5}y = \frac{1}{2}$

18. $\frac{2}{3}x - y = 4$

$\quad\; x - \frac{3}{4}y = 6$

19. $\frac{1}{7}x - \frac{3}{7}y = 1$

$\quad\; 2x - y = -4$

20. $\frac{1}{3}x - \frac{2}{3}y = 2$

$\quad\; x - 2y = 6$

21. $\quad x + 3y = 6$
$\quad\; 2x + 6y = 12$

22. $3x - 2y = 6$
$\quad\; 6x - 4y = 8$

23. $2x - \; y = 1$
$\quad\; 8x - 4y = 3$

24. $\quad 6x + 2y = 1$
$\quad\; 12x + 4y = 2$

■ *Solve each problem using a system of equations.*

Example

The sum of two numbers is 17 and one of the numbers is 4 less than 2 times the other. Find the numbers.

Solution

Represent each number by a separate variable.

One number: x
Other number: y

Represent the two independent conditions stated in the problem by two equations.

$$x + y = 17$$
$$x = 2y - 4$$

Rewrite the equations in the form

$$x + y = 17$$
$$x - 2y = -4.$$

Solve the resulting system. The numbers are 10 and 7.

25. The sum of two numbers is 24 and one of the numbers is 6 less than the other. Find the numbers.

26. The difference of two numbers is 14 and one of the numbers is 1 more than 2 times the other. Find the numbers.

27. If ⅓ of an integer is added to ½ the next consecutive integer, the sum is 33. Find the integers.

28. If ½ of an integer is added to ⅕ the next consecutive integer, the sum is 17. Find the integers.

29. The admission at a baseball game was $1.50 for adults and $0.85 for children. The receipts were $93.10 for 82 paid admissions. How many adults and how many children attended the game?

30. In an election, 7179 votes were cast for two candidates. If 6 votes had switched from the winner to the loser, the loser would have won by 1 vote. How many votes were cast for each candidate?

31. A sum of $2000 is invested, part at 10% and the remainder at 8%. Find the amount invested at each rate if the yearly income from the two investments is $184.

32. A woman has $1200 invested in two stocks, one of which returns 8% per year and the other 12% per year. How much has she invested in each stock if the income from the 8% stock is $3 more than the income from the 12% stock?

33. In 1983 a vintner has a white wine that is 4 years older than a certain red wine. In 1973, the white wine was 2 times as old as the red wine. In what years were the wines produced?

34. In 1980, Mr. Evans died leaving a will saying that when his son was 2 times as old as his daughter, both would receive the funds from a trust. If the boy was 3 times as old as his sister in 1980 and if they receive their funds in 1985, how old was each when their father died?

35. A record company determines that each production run to manufacture a record involves an initial set-up cost of $20 and $0.40 for each record produced. The records sell for $1.20 each.

    a. Express the cost C of production in terms of the number x of records produced.

    b. Express the revenue R in terms of the number of records sold.

    c. How many records must be sold for the record company to break even (no profit and no loss) on a particular production?

36. How many records must the record company in Problem 35 sell in order to break even if the price of each record is lowered to $0.60?

B ■ *In Problems 37–42, set* $u = \dfrac{1}{x}$ *and* $v = \dfrac{1}{y}$, *solve for u and v; then solve for x and y.*

Example

$$\frac{4}{x} + \frac{3}{y} = 1$$

$$\frac{2}{x} - \frac{3}{y} = 2$$

Solution

Substituting u for $\dfrac{1}{x}$ and v for $\dfrac{1}{y}$ in each equation, we obtain

$$4u + 3v = 1$$
$$2u - 3v = 2.$$

Solving for u and v, we have

$$6u = 3 \quad \text{or} \quad u = \frac{1}{2},$$

and then we obtain

$$v = -\frac{1}{3}.$$

Since $u = \dfrac{1}{x}$ and $v = \dfrac{1}{y}$, we get $x = 2$ and $y = -3$. The solution set is $\{(2, -3)\}$.

37. $\dfrac{1}{x} + \dfrac{1}{y} = 7$

$\dfrac{2}{x} + \dfrac{3}{y} = 16$

38. $\dfrac{1}{x} + \dfrac{2}{y} = -\dfrac{11}{12}$

$\dfrac{1}{x} + \dfrac{1}{y} = -\dfrac{7}{12}$

39. $\dfrac{5}{x} - \dfrac{6}{y} = -3$

$\dfrac{10}{x} + \dfrac{9}{y} = 1$

40. $\dfrac{1}{x} + \dfrac{2}{y} = 11$

$\dfrac{1}{x} + \dfrac{2}{y} = -1$

41. $\dfrac{1}{x} - \dfrac{1}{y} = 4$

$\dfrac{2}{x} - \dfrac{1}{2y} = 11$

42. $\dfrac{2}{3x} + \dfrac{3}{4y} = \dfrac{7}{12}$

$\dfrac{4}{x} - \dfrac{3}{4y} = \dfrac{7}{4}$

43. Find a and b so that the graph of $ax + by + 3 = 0$ passes through the points $(-1, 2)$ and $(-3, 0)$. [*Hint:* If the graph of the equation passes through the points, the components of each ordered pair must be valid replacements for x and y. Substitute -1 for x and 2 for y and -3 for x and 0 for y to obtain a system in a and b.]

44. Find a and b so that the solution set of the system below is $\{(1, 2)\}$.

$$ax + by = 4$$
$$bx - ay = -3$$

45. Solve the system below for x and y in terms of the coefficients and the constants c_1 and c_2.

$$a_1 x + b_1 y = c_1$$
$$a_2 x + b_2 y = c_2$$

46. Use the results of Problem 45 to find the solution set of the system of Problem 3.

10.2

LINEAR SYSTEMS
IN THREE VARIABLES

A solution of an equation in three variables, such as

$$x + 2y - 3z = -4,$$

is an ordered triple of numbers (x, y, z), because all three of the variables must be replaced by numerals before we can decide whether the result is an equality. Thus, $(0, -2, 0)$ and $(-1, 0, 1)$ are solutions of this equation, while $(1, 1, 1)$ is not. There are, of course, infinitely many members in the solution set.

The solution set of a system of three linear equations in three variables, such as

$$x + 2y - 3z = -4 \tag{1}$$

$$2x - y + z = 3 \tag{2}$$

$$3x + 2y + z = 10, \tag{3}$$

is the intersection of the solution sets of all three equations in the system. We seek solution sets of systems such as these by methods similar to those used in solving linear systems in two variables.

Linear equations in three variables can be represented by planes, and their common point(s) of intersection, if any, would represent the solution set. Since a graphical treatment would therefore be three-dimensional and difficult to represent, we shall consider algebraic solutions only.

In the system presented here, we might begin by multiplying Equation (1) by -2 and adding the result to 1 times Equation (2), to produce

$$-5y + 7z = 11, \tag{4}$$

which is satisfied by any ordered triple (x, y, z) that satisfies (1) and (2). Similarly, we can add -3 times Equation (1) to 1 times Equation (3) to obtain

$$-4y + 10z = 22, \tag{5}$$

which is satisfied by any ordered triple (x, y, z) that satisfies both (1) and (3).

We can now argue that any ordered triple satisfying the system (1), (2), and (3) will also satisfy the system

$$-5y + 7z = 11 \tag{4}$$

$$-4y + 10z = 22. \tag{5}$$

Since the system (4) and (5) does not depend on x, the problem has been reduced to one of finding only the y- and z-components of the solution. The system (4) and (5) can be solved by the method of Section 10.1, which leads to the values $y = 2$ and $z = 3$. Now, since any solution of (1), (2), and (3) must be of the form $(x, 2, 3)$, we can substitute 2 for y and 3 for z in (1) to obtain $x = 1$. The desired solution set is

$$\{(1, 2, 3)\}.$$

If at any step in this procedure the resulting linear combination vanishes or yields a contradiction, then the system contains dependent equations or else two or three inconsistent equations, and it has either an infinite number of members or no members in its solution set.

The process of solving a system of equations can be reduced to a series of mechanical procedures, as illustrated by the first example in Exercise 10.2.

EXERCISE 10.2

A   ■ *Solve. If the system does not have a unique (one and only one) solution, so state.*

Example

$$x + 2y - z = -3 \tag{1}$$

$$x - 3y + z = 6 \tag{2}$$

$$2x + y + 2z = 5 \tag{3}$$

Solution   First obtain a system of two equations in two variables. Multiply Equation (1) by -1 and add the result to 1 times Equation (2) to get (4); multiply Equation (1) by -2 and add the result to 1 times Equation (3) to get (5).

$$-5y + 2z = 9 \tag{4}$$

$$-3y + 4z = 11 \tag{5}$$

Multiply Equation (4) by -2 and add the result to 1 times Equation (5).

$$7y = -7$$

$$y = -1$$

Substitute -1 for y in either (4) or (5)—we shall use (4)—and solve for z.

$$-5(-1) + 2z = 9$$

$$z = 2$$

Substitute -1 for y and 2 for z in (1), (2), or (3)—we shall use (1)—and solve for x.

$$x + 2(-1) - 2 = -3$$

$$x = 1$$

The solution set is $\{(1, -1, 2)\}$.

1. $x + y + z = 2$ **2.** $x + y + z = 1$ **3.** $x + y + 2z = 0$
     $2x - y + z = -1$ $2x - y + 3z = 2$ $2x - 2y + z = 8$
     $x - y - z = 0$ $2x - y - z = 2$ $3x + 2y + z = 2$

4. $x - 2y + 4z = -3$
$3x + y - 2z = 12$
$2x + y - 3z = 11$

5. $x - 2y + z = -1$
$2x + y - 3z = 3$
$3x + 3y - 2z = 10$

6. $x + 5y - z = 2$
$3x - 9y + 3z = 6$
$x - 3y + z = 4$

7. $x - 2y + 3z = 4$
$2x - y + z = 1$
$3x - 3y + 4z = 5$

8. $2x - 3y + z = 3$
$x - y - 2z = -1$
$-x + 2y - 3z = -4$

9. $2x + z = 7$
$y - z = -2$
$x + y = 2$

10. $5y - 8z = -19$
$5x - 8z = 6$
$3x - 2y = 12$

11. $x - \dfrac{1}{2}y - \dfrac{1}{2}z = 4$

$x - \dfrac{3}{2}y - 2z = 3$

$\dfrac{1}{4}x + \dfrac{1}{4}y - \dfrac{1}{4}z = 0$

12. $x + 2y + \dfrac{1}{2}z = 0$

$x + \dfrac{3}{5}y - \dfrac{2}{5}z = \dfrac{1}{5}$

$4x - 7y - 7z = 6$

■ *Solve each problem using a system of equations.*

Example

The sum of three numbers is 12. Twice the first number is equal to the second, and the third is equal to the sum of the other two. Find the numbers.

Solution

Represent each number by a separate variable.

First number: x

Second number: y

Third number: z

Write the three conditions stated in the problem as three equations.

$$x + y + z = 12$$
$$2x = y$$
$$x + y = z$$

Rewrite the equations in the form

$$x + y + z = 12 \qquad (1)$$
$$2x - y \quad\quad = 0 \qquad (2)$$
$$x + y - z = 0. \qquad (3)$$

Multiply Equation (1) by 1 and add the result to 1 times Equation (3) to obtain

$$2x + 2y = 12.$$

Multiply this Equation by ½ and add the result to 1 times Equation (2).

$$3x = 6$$

Solve this equation to get $x = 2$, and substitute the x-value in (2) to obtain $y = 4$. Substitute these values in (1) or (3) to obtain $z = 6$.

13. The sum of three numbers is 15. The second equals 2 times the first and the third equals the second. Find the numbers.

14. The sum of three numbers is 2. The first number is equal to the sum of the other two, and the third number is the result of subtracting the first from the second. Find the numbers.

15. A box contains $6.25 in nickels, dimes, and quarters. There are 85 coins in all with 3 times as many nickels as dimes. How many coins of each kind are there?

16. The perimeter of a triangle is 155 inches. Side x is 20 inches shorter than side y, and side y is 5 inches longer than size z. Find the lengths of the sides of the triangle.

B 17. The equation for a circle can be written $x^2 + y^2 + ax + by + c = 0$. Find the equation of the circle whose graph contains the points $(2, 3)$, $(3, 2)$, and $(-4, -5)$.

18. Find values for a, b, and c such that the graph of $x^2 + y^2 + ax + by + c = 0$ will contain the points $(-2, 3)$, $(1, 6)$, and $(2, 4)$.

19. Find values for a, b, and c such that the graph of $y = ax^2 + bx + c$ will contain the points $(-1, 2)$, $(1, 6)$, and $(2, 11)$.

20. Find values for a, b, and c such that the graph of $y = ax^2 + bx + c$ will contain the points $(1, 0)$, $(3, -2)$, and $(5, 4)$.

10.3

NONLINEAR SYSTEMS IN TWO VARIABLES: SOLUTION BY SUBSTITUTION

*Real
solutions
obtained
from
graphs*

Approximate solutions of systems of equations in two variables, where one or both of the equations are quadratic, may often be found by graphing both equations and estimating the coordinates of any points they have in common. For example, to find the solution set of the system

$$x^2 + y^2 = 26 \qquad (1)$$

$$x + y = 6, \qquad (2)$$

we graph the equations on the same set of axes, as shown in Figure 10.4, and observe that the graphs appear to intersect at $(1, 5)$ and $(5, 1)$. The solution set of the system (1) and (2) is, in fact, $\{(1, 5), (5, 1)\}$.

However, solving second-degree systems graphically on the real plane may produce only approximations to real solutions, and we cannot expect to locate solutions in

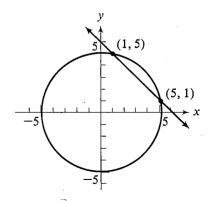

Figure 10.4

which one or both of the components are imaginary numbers. It is therefore more practical to concentrate on algebraic methods of solution since the results are exact and we can obtain imaginary solutions. It is suggested that, whenever feasible, you sketch the graphs of the equations as a rough check on an algebraic solution.

Solving a system by substitution

One of the most useful techniques available for finding solution sets for systems of equations is **substitution**. This technique is particularly helpful with systems containing one first-degree and one higher-degree equation.

Example

Solve

$$x^2 + y^2 = 26 \tag{1}$$

$$x + y = 6 \tag{2}$$

using algebraic methods.

Solution

Equation (2) can be written in the form

$$y = 6 - x, \tag{3}$$

and we can argue that for any ordered pair (x, y) in the solution set of both (1) and (2), x and y in (1) represent the same numbers as x and y in (3). Hence, the substitution property may be used to replace y in (1) by its equal $(6 - x)$ from (3). This will produce

$$x^2 + (6 - x)^2 = 26, \tag{4}$$

which will have as a solution set those values of x for which the ordered pair (x, y) is a common solution of (1) and (2). Rewriting (4) equivalently, we have

$$x^2 + 36 - 12x + x^2 = 26$$
$$2x^2 - 12x + 10 = 0$$
$$x^2 - 6x + 5 = 0$$
$$(x - 5)(x - 1) = 0,$$

from which x is either 1 or 5. Now, by replacing x in (3) with each of these numbers, we have

$$y = 6 - (1) = 5 \quad \text{and} \quad y = 6 - (5) = 1;$$

so the solution set of the system (1) and (2) is $\{(1, 5), (5, 1)\}$.

Check this solution and notice that these ordered pairs are also solutions of (1).

Note that in the above example if we used (1) rather than (2) or (3) to obtain values for the y-component, we would have

$$(1)^2 + y^2 = 26 \qquad (5)^2 + y^2 = 26$$
$$y = \pm 5 \qquad\qquad y = \pm 1$$

and the solutions obtained would be $(1, 5)$, $(1, -5)$, $(5, 1)$, and $(5, -1)$. However, $(1, -5)$ and $(5, -1)$ are not solutions of (2). The solution set is again $\{(1, 5), (5, 1)\}$.

The foregoing example suggests: *if the degrees of equations differ, one component of a solution should be substituted in the equation of <u>lower</u> degree in order to find <u>only</u> those ordered pairs that are solutions of <u>both</u> equations.*

Imaginary solutions

In the foregoing example, the components of each solution are real numbers, and their graphs are the points of intersection of the graphs of each equation, as shown in Figure 10.4. If one or more of the components of the solutions of a system are imaginary numbers, we can find these solutions but, in such cases, the graphs in the real plane do not have points of intersection.

Example   Solve

$$x^2 + y^2 = 26 \qquad\qquad (5)$$

$$x + y = 8. \qquad\qquad (6)$$

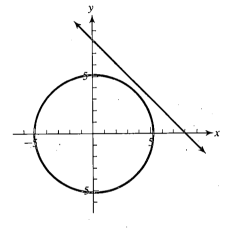

Solution   Solving (6) for y, we have

$$y = 8 - x. \qquad\qquad (6a)$$

Substituting $(8 - x)$ for y in (5) and simplifying yields

$$x^2 + (8 - x)^2 = 26$$
$$x^2 + 64 - 16x + x^2 = 26$$
$$2x^2 - 16x + 38 = 0$$
$$x^2 - 8x + 19 = 0.$$

Using the quadratic formula to solve for x, we obtain

$$x = \frac{8 \pm \sqrt{64 - 76}}{2(1)}$$

$$= \frac{8 \pm \sqrt{-12}}{2}$$

$$= \frac{8 \pm 2i\sqrt{3}}{2}$$

$$= \frac{2(4 \pm i\sqrt{3})}{2}$$

$$= 4 \pm i\sqrt{3}.$$

Then, substituting $(4 + i\sqrt{3})$ for x in (6a) gives $y = 4 - i\sqrt{3}$, and substituting $(4 + i\sqrt{3})$ for x in (6a) gives $y = 4 + i\sqrt{3}$. Hence, the solution set is

$$\{(4 + i\sqrt{3}, 4 - i\sqrt{3}), (4 - i\sqrt{3}, 4 + i\sqrt{3})\};$$

the graphs of the equations are shown in the figure.

EXERCISE 10.3

A   ■ *Solve by the method of substitution. In Problems 1–12, sketch the graphs of the equations.*

Example

$$y = x^2 + 2x + 1 \qquad (1)$$

$$y - x = 3 \qquad (2)$$

Solution   Solve (2) explicitly for y.

$$y = x + 3 \qquad (2a)$$

Substitute $(x + 3)$ for y in (1).

$$x + 3 = x^2 + 2x + 1$$

Solve for x.

$$x^2 + x - 2 = 0$$

$$(x + 2)(x - 1) = 0$$

$$x = -2 \quad \text{or} \quad x = 1$$

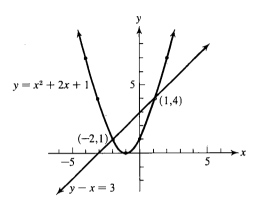

Substitute each of these values in (2a) to determine values for y. If $x = -2$, then $y = 1$, and if $x = 1$, then $y = 4$. The solution set is $\{(-2, 1), (1, 4)\}$.

1. $y = x^2 - 5$
   $y = 4x$

2. $\quad y = x^2 - 2x + 1$
   $y + x = 3$

3. $x^2 + y^2 = 13$
   $x + y = 5$

4. $x^2 + 2y^2 = 12$
   $2x - y = 2$

5. $x + y = 1$
   $xy = -12$

6. $2x - y = 9$
   $xy = -4$

7. $\quad xy = 4$
   $x^2 + y^2 = 8$

8. $x^2 - y^2 = 35$
   $xy = 6$

9. $x^2 + y^2 = 9$
   $y = 4$

10. $2x^2 - 4y^2 = 12$
    $x = 4$

11. $x^2 + y = 4$
    $x - y = -1$

12. $x^2 + 9y^2 = 36$
    $x - 2y = -8$

13. $x^2 - xy - 2y^2 = 4$
    $x - y = 2$

14. $x^2 - 2x + y^2 = 3$
    $2x + y = 4$

15. $2x^2 - 5xy + 2y^2 = 5$
    $2x - y = 1$

16. $2x^2 + xy + y^2 = 9$
    $-x + 3y = 9$

■ *Solve each problem using a system of equations.*

17. The sum of the squares of two positive numbers is 13. If 2 times the first number is added to the second, the sum is 7. Find the numbers.

18. The sum of two numbers is 6 and their product is $^{35}/_4$. Find the numbers.

19. The perimeter of a rectangle is 26 inches and the area is 12 square inches. Find the dimensions of the rectangle.

20. The area of a rectangle is 216 square feet. If the perimeter is 60 feet, find the dimensions of the rectangle.

21. Consider the system

$$x^2 + y^2 = 8 \qquad (1)$$

$$xy = 4. \qquad (2)$$

We can solve this system by substituting $4/x$ for y in (1) to obtain

$$x^2 + \frac{16}{x^2} = 8,$$

from which we have $x = 2$ or $x = -2$. Now, if we obtain the y-components of the solution from (2), we find that:

For $x = 2$, $y = 2$.   For $x = -2$, $y = -2$.

But if we seek y-components from (1), we have:

For $x = 2$, $y = \pm 2$.   For $x = -2$, $y = \pm 2$.

Discuss the fact that we seem to obtain more solutions from (1) than from (2). What is the solution set of the system?

22. Graph Equations (1) and (2) of Problem 21 and relate your discussion of Problem 21 to these graphs.

10.4

NONLINEAR SYSTEMS IN TWO VARIABLES: SOLUTION BY OTHER METHODS

If both the equations in a system are second-degree in both variables, the use of linear combinations of members of the equations often provides a simpler means of solution than does substitution.

Example   Solve

$$4x^2 + y^2 = 25 \qquad (1)$$

$$x^2 - y^2 = -5. \qquad (2)$$

Solution   By forming a linear combination using 1 times Equation (1) and 1 times Equation (2), we have

$$5x^2 = 20,$$

from which

$$x = 2 \quad \text{or} \quad x = -2.$$

We now have the x-components of the members of the solution set of the system (1) and (2). Substituting 2 for x in either (1) or (2)—we shall use (1)—we obtain

$$4(2)^2 + y^2 = 25$$
$$y^2 = 25 - 16$$
$$y^2 = 9,$$

from which

$$y = 3 \quad \text{or} \quad y = -3.$$

Thus, the ordered pairs (2, 3) and (2, -3) are in the solution set of the system. Substituting -2 for x in (1) or (2)—this time we shall use (2)—gives us

$$(-2)^2 - y^2 = -5$$
$$-y^2 = -5 - 4$$
$$y^2 = 9,$$

from which

$$y = 3 \quad \text{or} \quad y = -3.$$

Thus, the ordered pairs $(-2, 3)$ and $(-2, -3)$ are also solutions of the system, and the complete solution set is

$$\{(2, 3), (2, -3), (-2, 3), (-2, -3)\}.$$

The following example shows another means of solving a system of two second-degree equations.

Example   Solve

$$x^2 + y^2 = 5 \tag{3}$$
$$x^2 - 2xy + y^2 = 1. \tag{4}$$

Solution   By forming a linear combination using 1 times Equation (3) and -1 times Equation (4), we have

$$2xy = 4$$
$$xy = 2, \tag{5}$$

which has a solution set containing all the ordered pairs that satisfy both (3) and (4). Therefore, forming the new system

$$x^2 + y^2 = 5 \tag{3}$$
$$xy = 2, \tag{5}$$

we can be sure that the solution set of this system is the same as the solution set of the system (3) and (4).

This latter system can be solved by substitution. We have, from (5),

$$y = \frac{2}{x}.$$

Solution continued overleaf

Replacing y in (3) by $2/x$, we find

$$x^2 + \left(\frac{2}{x}\right)^2 = 5,$$

from which

$$x^2 + \frac{4}{x^2} = 5. \tag{6}$$

Multiplying each member by x^2, we have

$$x^4 + 4 = 5x^2 \tag{7}$$

$$x^4 - 5x^2 + 4 = 0, \tag{7a}$$

which is quadratic in x^2. Factoring the left-hand member of (7a), we obtain

$$(x^2 - 1)(x^2 - 4) = 0,$$

from which

$$x^2 - 1 = 0 \quad \text{or} \quad x^2 - 4 = 0.$$

Solving these equations (by factorization), we obtain

$$x = 1, \quad x = -1 \quad \text{and} \quad x = 2, \quad x = -2.$$

Since we multiplied Equation (6) by a variable, we are careful to note that these values of x ($x \neq 0$) all satisfy (6).

Now, substituting $1, -1, 2$, and -2 for x in the equation $xy = 2$ or $y = 2/x$, we have:

For $x = 1, \quad y = 2.$　　　For $x = -1, \quad y = -2.$

For $x = 2, \quad y = 1.$　　　For $x = -2, \quad y = -1.$

The solution set of either system (3) and (5) or system (3) and (4) is

$$\{(1, 2), (-1, -2), (2, 1), (-2, -1)\}.$$

There are other techniques involving substitution in conjunction with linear combinations that are useful in handling systems of higher-degree equations, but they are all similar to those illustrated. Each system should be scrutinized for some means of finding an equivalent system that will lend itself to solution by linear combination or substitution.

Approximating solutions by graphical methods

When the degree of an equation in a set of equations is greater than two, or when one of the members of an equation is not a polynomial, it may be difficult or impossible to find common solutions algebraically. In this case, we can obtain approximations to any *real* solutions by graphical methods. Several such systems are included in the exercises.

EXERCISE 10.4

A   ■ *Solve each system.*

Example

$$3x^2 + y^2 = 15 \qquad\qquad (1)$$
$$11x^2 - 2y^2 = 4 \qquad\qquad (2)$$

Solution

Obtain a linear combination using 2 times Equation (1) and 1 times Equation (2) and solve for x.

$$6x^2 + 2y^2 = 30$$
$$11x^2 - 2y^2 = 4$$
$$17x^2 = 34$$
$$x^2 = 2$$

$$x = \sqrt{2} \quad \text{or} \quad x = -\sqrt{2}$$

Substitute values for x in (1) or (2)—we shall use (1)—to obtain associated values for y.

$$3(\sqrt{2})^2 + y^2 = 15 \qquad\qquad 3(-\sqrt{2})^2 + y^2 = 15$$
$$6 + y^2 = 15 \qquad\qquad 6 + y^2 = 15$$
$$y^2 = 9 \qquad\qquad\qquad y^2 = 9$$
$$y = \pm 3 \qquad\qquad\qquad y = \pm 3$$

The solution set is $\{(\sqrt{2}, 3), (\sqrt{2}, -3), (-\sqrt{2}, 3), (-\sqrt{2}, -3)\}$.

1. $x^2 + y^2 = 10$
$\quad 9x^2 + y^2 = 18$

2. $x^2 + 4y^2 = 52$
$\quad x^2 + y^2 = 25$

3. $x^2 + 4y^2 = 17$
$\quad 3x^2 - y^2 = -1$

4. $9x^2 + 16y^2 = 100$
$\quad x^2 + y^2 = 8$

5. $x^2 - y^2 = 7$
$\quad 2x^2 + 3y^2 = 24$

6. $x^2 + 4y^2 = 25$
$\quad 4x^2 + y^2 = 25$

7. $3x^2 + 4y^2 = 16$
$\quad x^2 - y^2 = 3$

8. $4x^2 + 3y^2 = 12$
$\quad x^2 + 3y^2 = 12$

9. $4x^2 - 9y^2 + 132 = 0$
$\quad x^2 + 4y^2 - 67 = 0$

10. $16y^2 + 5x^2 - 26 = 0$
$\quad 25y^2 - 4x^2 - 17 = 0$

11. $2x^2 + xy - 4y^2 = -12$
$\quad x^2 - 2y^2 = -4$

12. $x^2 + 2xy - y^2 = 14$
$\quad x^2 - y^2 = 8$

13. $x^2 + 3xy - y^2 = -3$
$\quad x^2 - xy - y^2 = 1$

14. $2x^2 + xy - 2y^2 = 16$
$\quad x^2 + 2xy - y^2 = 17$

B   ■ *Solve by graphing. Approximate components of solutions.*

15. $y = 10^x$ 16. $y = 2^x$
   $x + y = 6$ $y - x = 2$

17. $y = \log_{10} x$ 18. $y = 10^{-x}$
   $y = x^2 - 4$ $y = \log_{10} x$

19. How many *real* solutions are possible for systems of *independent* equations that consist of:

   a. Two linear equations in two variables?

   b. One linear equation and one quadratic equation in two variables?

   c. Two quadratic equations in two variables?

   Support your answers with sketches.

CHAPTER SUMMARY

[10.1]   The system of linear equations (in standard form)

$$a_1 x + b_1 y = c_1$$

$$a_2 x + b_2 y = c_2$$

has no solution if the equations are **inconsistent**, infinitely many solutions if the equations are **dependent**, and exactly one solution when the equations are **consistent** and **independent**.

The system has exactly one solution if $\dfrac{a_1}{a_2} \neq \dfrac{b_1}{b_2}$.

The equations are inconsistent if $\dfrac{a_1}{a_2} = \dfrac{b_1}{b_2} \neq \dfrac{c_1}{c_2}$.

The equations are dependent if $\dfrac{a_1}{a_2} = \dfrac{b_1}{b_2} = \dfrac{c_1}{c_2}$.

[10.2]   The solution of a system of three linear equations in three variables (if a solution exists) can be obtained by first using linear combinations to form a system of two equations in two variables. The solution to this latter system contains components that are the respective components of the solution of the original system in three variables. The third component can be obtained by substituting these two values into any one of the equations of the original system.

[10.3–10.4]   Systems of equations in two variables in which either or both equations are second-degree in one or both variables may have solutions with real components, imaginary components, or solutions of both kinds. Such systems can be solved by using substitution methods or by using linear combinations of the members of the equations in the system.

REVIEW EXERCISES

A

[10.1]   ■ *Solve each system by linear combinations.*

1. $x + 5y = 18$
$\quad x - y = -3$

2. $x + 5y = 11$
$\quad 2x + 3y = 8$

3. $\frac{2}{3}x - 3y = 8$

$\quad x + \frac{3}{4}y = 12$

4. $\frac{3}{x} - \frac{1}{y} = \frac{7}{2}$

$\quad \frac{2}{x} + \frac{3}{y} = 7$

■ *State whether the equations in each system have a unique solution, are dependent, or inconsistent.*

5. a. $2x - 3y = 4$
$\quad\quad x + 2y = 7$

b. $2x - 3y = 4$
$\quad\quad 6x - 9y = 12$

6. a. $2x - 3y = 4$
$\quad\quad 6x - 9y = 4$

b. $x - y = 6$
$\quad\quad x + y = 6$

[10.2]   ■ *Solve each system by linear combinations.*

7. $x + 3y - z = 3$
$\quad 2x - y + 3z = 1$
$\quad 3x + 2y + z = 5$

8. $x + y + z = 2$
$\quad 3x - y + z = 4$
$\quad 2x + y + 2z = 3$

9. $x + z = 5$
$\quad y - z = -8$
$\quad 2x + z = 7$

10. $x + 4y + 4z = -20$
$\quad 3x - 2y + z = -4$
$\quad 2x - 4y + z = -4$

11. $\frac{1}{2}x + y + z = 3$

$\quad x - 2y - \frac{1}{3}z = -5$

$\quad \frac{1}{2}x - 3y - \frac{2}{3}z = -6$

12. $\frac{3}{4}x - \frac{1}{2}y + 6z = 2$

$\quad \frac{1}{2}x + y - \frac{3}{4}z = 0$

$\quad \frac{1}{4}x + \frac{1}{2}y - \frac{1}{2}z = 0$

[10.3–10.4]   ■ *Solve each system by substitution or linear combination.*

13. $x + 3y^2 = 4$
$\quad\quad x = 3$

14. $x^2 + 2y^2 = -8$
$\quad\quad y = -2$

15. $x^2 + y = 3$
$5x + y = 7$

16. $x^2 + 3xy + x = -12$
$2x - y = 7$

17. $6x^2 - y^2 = 1$
$3x^2 + 2y^2 = 13$

18. $2x^2 + 5y^2 - 53 = 0$
$4x^2 + 3y^2 - 43 = 0$

19. $x^2 - 2xy + 3y^2 = 17$
$2x^2 + xy + 6y^2 = 24$

20. $x^2 - xy - y^2 = 1$
$x^2 + 3xy - y^2 = 9$

21. The perimeter of a rectangle is 34 centimeters long and the area is 70 square centimeters. Find the dimensions of the rectangle.

22. A rectangle has a perimeter of 18 feet. If the length is decreased by 5 feet and the width is increased by 12 feet, the area is doubled. Find the dimensions of the original rectangle.

B **23.** The equation for a parabola can be written $y = ax^2 + bx + c$. Find the equation of a parabola whose graph contains the points $(-1, -4)$, $(0, -6)$, and $(4, 6)$.

24. The equation for a circle can be written $x^2 + y^2 + ax + by + c = 0$. Find the equation of the circle whose graph contains the points $(-1, 2)$, $(1, 4)$, and $(3, -2)$.

25. What relationship must exist between the numbers a and b so that the solution set of the system

$$y = x^2 - 4$$

$$y = ax + b$$

will have exactly one solution?

26. Approximate the solution of the system

$$y = 2^x$$

$$x + y = 5$$

by graphical methods.

11. NATURAL-NUMBER FUNCTIONS

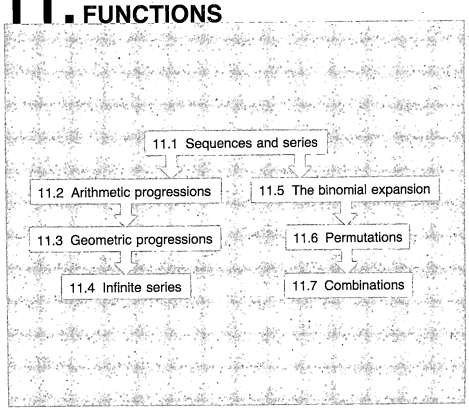

11.1 Sequences and series

11.2 Arithmetic progressions

11.5 The binomial expansion

11.3 Geometric progressions

11.6 Permutations

11.4 Infinite series

11.7 Combinations

11.1

SEQUENCES AND SERIES

Sequences

A function whose domain is a set of successive positive integers, for example, a function defined by an equation such as

$$s(n) = 2n - 1 \qquad (n \in \{3, 4, 5\}) \tag{1}$$

or

$$s(n) = n + 3 \qquad (n \in \{1, 2, 3, \ldots\}), \tag{2}$$

is called a **sequence function**. The function defined by (1) is called a **finite sequence**, and the function defined by (2) is called an **infinite sequence**. The elements in the range of such functions arranged in the order

$$s(3), \ s(4), \ s(5) \quad \text{or} \quad s(1), \ s(2), \ s(3), \ \ldots$$

are said to form a **sequence,** and the elements are referred to as the **terms** of the sequence. Thus, the sequence associated with (2) is found by successively substituting the numbers 1, 2, 3, ... for n:

$$s(1) = (1) + 3 = 4$$
$$s(2) = (2) + 3 = 5$$
$$s(3) = (3) + 3 = 6$$

$s(1)$ or 4 is called the first term, $s(2)$ or 5 is called the second term, $s(3)$ or 6 is called the third term, etc. The expression $n + 3$ is called the **general term** or **nth term.**

Example   The first three terms of the sequence with the general term $\dfrac{3}{2n - 1}$ are:

$$s(1) = \frac{3}{2(1) - 1} = \frac{3}{1}$$

$$s(2) = \frac{3}{2(2) - 1} = \frac{3}{3}$$

$$s(3) = \frac{3}{2(3) - 1} = \frac{3}{5}.$$

The twenty-fifth term is

$$s(25) = \frac{3}{2(25) - 1} = \frac{3}{49}.$$

The notation ordinarily used for the terms in a sequence is not function notation as such; rather, it is customary to denote a term in a sequence by means of a subscript. Thus, we will use s_n rather than $s(n)$, and the sequence $s(1), s(2), s(3), \ldots$ will appear as $s_1, s_2, s_3, \ldots$.

As we have seen in the examples above, for any given general term s_n, we can obtain the elements of an associated sequence by successively substituting the numbers 1, 2, 3, ... for n. However, it is usually difficult to obtain a general term for a given sequence. In Sections 11.2 and 11.3 we shall consider some formulas that will enable us to find general terms for several particular sequences. For now, we might attempt to find general terms by inspection (trial and error). Several such examples and problems are included in the exercises.

Series   Associated with any sequence is a **series;** the series is defined as the sum of the terms in the sequence and is denoted by S_n.

Examples   **a.** Associated with the finite sequence

$$4, 7, 10, \ldots, 3n + 1 \tag{3}$$

is the finite series

$$S_n = 4 + 7 + 10 + \cdots + (3n + 1). \tag{4}$$

b. Associated with the sequence

$$x, x^2, x^3, x^4, \ldots, x^n$$

is the series

$$S_n = x + x^2 + x^3 + x^4 + \cdots + x^n.$$

Since the terms in the series are the same as those in the corresponding sequence, we can refer to the first term or the second term or the general term of a series in the same manner as we do for a sequence.

Sigma notation

A series with a general term that is known can be represented in a very convenient, compact way by using the symbol Σ (the Greek letter **sigma**) in conjunction with the general term in sigma or summation notation; this denotes the sum of all the terms in the series. For example, series (4) can be written

$$S_n = \sum_{i=1}^{n} (3i + 1),$$

where we understand that S_n is the series with terms obtained by successively replacing i in the expression $3i + 1$ with the numbers, 1, 2, 3, ... , n. Thus,

$$S_6 = \sum_{i=1}^{6} (3i + 1)$$

appears in **expanded form** as

$$[3(1) + 1] + [3(2) + 1] + [3(3) + 1] + [3(4) + 1] + [3(5) + 1] + [3(6) + 1]$$
$$= 4 + 7 + 10 + 13 + 16 + 19.$$

The variable used in conjunction with summation notation (in the above case, i) is called the **index of summation**; the set of integers over which we sum in this case is $\{1, 2, 3, 4, 5, 6\}$. The use of the symbol i as an index of summation should not be confused with its use as an imaginary unit in the set of complex numbers. Alternatively, the summation index can be any letter such as j, k, l, and so on:

$$\sum_{i=1}^{6} (3i + 1) = \sum_{j=1}^{6} (3j + 1) = \sum_{k=1}^{6} (3k + 1) = \cdots .$$

The first member of the replacement set for the index of summation is not necessarily 1.

Example

$$\sum_{i=3}^{6} (3i + 1) = [3(3) + 1] + [3(4) + 1] + [3(5) + 1] + [3(6) + 1]$$

$$= 10 + 13 + 16 + 19,$$

where the first replacement for i is 3; the set of integers over which we sum is $\{3, 4, 5, 6\}$.

Note that the series in the above example contains four terms, which is one more than the difference between the last and the first replacement for i. In general, the series $\sum_{i=a}^{b} s_i$ contains $(b - a + 1)$ terms.

To show that a series has an infinite number of terms—that is, has no last term—we adopt a special notation. For example,

$$S_\infty = \sum_{i=4}^{\infty} (3i + 1)$$

denotes the series that would appear in expanded form as

$$S_\infty = 13 + 16 + 19 + 22 + \cdots, \tag{5}$$

where, in this case, i has been replaced by 4, 5, 6, 7, ...

A series can be represented in summation notation by various general terms and different ranges. For example, both

$$\sum_{i=5}^{\infty} (3i - 2) \quad \text{and} \quad \sum_{i=6}^{\infty} (3i - 5)$$

also represent (5). This can be verified by writing the first few terms in each series.

EXERCISE 11.1

A   ■ *Find the first four terms in a sequence with the general term as given.*

Examples   **a.** $s_n = \dfrac{n(n + 1)}{2}$ **b.** $s_n = (-1)^n 2^n$

Solutions   **a.** $s_1 = \dfrac{1(1 + 1)}{2} = 1$ **b.** $s_1 = (-1)^1 2^1 = -2$

$s_2 = \dfrac{2(2 + 1)}{2} = 3$ $s_2 = (-1)^2 2^2 = 4$

$s_3 = \dfrac{3(3 + 1)}{2} = 6$ $s_3 = (-1)^3 2^3 = -8$

$s_4 = \dfrac{4(4 + 1)}{2} = 10$ $s_4 = (-1)^4 2^4 = 16$

The first four terms are 1, 3, 6, 10.   The first four terms are -2, 4, -8, 16.

1. $s_n = n - 5$ **2.** $s_n = 2n - 3$ **3.** $s_n = \dfrac{n^2 - 2}{2}$

4. $s_n = \dfrac{3}{n^2 + 1}$ **5.** $s_n = 1 + \dfrac{1}{n}$ **6.** $s_n = \dfrac{n}{2n - 1}$

7. $s_n = \dfrac{n(n - 1)}{2}$ **8.** $s_n = \dfrac{5}{n(n + 1)}$ **9.** $s_n = (-1)^n$

10. $s_n = (-1)^{n+1}$ **11.** $s_n = \dfrac{(-1)^n(n - 2)}{n}$ **12.** $s_n = (-1)^{n-1}3^{n+1}$

■ *Write in expanded form.*

Examples   **a.** $\displaystyle\sum_{i=2}^{4} (i^2 + 1)$ **b.** $\displaystyle\sum_{k=1}^{\infty} (-1)^k 2^{k+1}$

Solutions   **a.** i takes values 2, 3, 4.

$i = 2,\quad (2)^2 + 1 = 5$
$i = 3,\quad (3)^2 + 1 = 10$
$i = 4,\quad (4)^2 + 1 = 17$

Expanded form: $5 + 10 + 17$.

b. k takes values 1, 2, 3, $\ldots$.

$k = 1,\quad (-1)^1 2^{1+1} = (-1)(4) = -4$
$k = 2,\quad (-1)^2 2^{2+1} = (1)(8) = 8$
$k = 3,\quad (-1)^3 2^{3+1} = (-1)(16) = -16$

Expanded form: $-4 + 8 - 16 + \cdots$.

13. $\displaystyle\sum_{i=1}^{4} i^2$ **14.** $\displaystyle\sum_{i=1}^{3} (3i - 2)$ **15.** $\displaystyle\sum_{j=5}^{7} (j - 2)$

16. $\displaystyle\sum_{j=2}^{6} (j^2 + 1)$ **17.** $\displaystyle\sum_{k=1}^{4} k(k + 1)$ **18.** $\displaystyle\sum_{i=2}^{6} \dfrac{i}{2}(i + 1)$

19. $\displaystyle\sum_{i=1}^{4} \dfrac{(-1)^i}{2^i}$ **20.** $\displaystyle\sum_{i=3}^{5} \dfrac{(-1)^{i+1}}{i - 2}$ **21.** $\displaystyle\sum_{i=1}^{\infty} (2i - 1)$

22. $\displaystyle\sum_{j=1}^{\infty} \dfrac{1}{j}$ **23.** $\displaystyle\sum_{k=0}^{\infty} \dfrac{1}{2^k}$ **24.** $\displaystyle\sum_{k=0}^{\infty} \dfrac{k}{1 + k}$

B   ■ *Write in summation notation. (There are no unique solutions.)*

Examples   **a.** $5 + 8 + 11 + 14$ **b.** $x^2 + x^4 + x^6 + \cdots + x^{2n}$

Solutions   By inspection (trial and error), we first find a general term for the sequence associated with each series.

a. $3i + 2$

Hence,

$$5 + 8 + 11 + 14 = \sum_{i=1}^{4} (3i + 2).$$

b. x^{2i}

Hence,

$$x^2 + x^4 + x^6 + \cdots + x^{2n} = \sum_{i=1}^{n} x^{2i}.$$

25. $1 + 2 + 3 + 4$

26. $2 + 4 + 6 + 8$

27. $x + x^3 + x^5 + x^7$

28. $x^3 + x^5 + x^7 + x^9 + x^{11}$

29. $1 + 4 + 9 + 16 + 25$

30. $1 + 8 + 27 + 64 + 125$

Examples

a. $3 + 6 + 9 + 12 + \cdots$

b. $\dfrac{3}{5} + \dfrac{5}{7} + \dfrac{7}{9} + \dfrac{9}{11} + \cdots$

Solutions

Find a general term.

a. $3i$

b. $\dfrac{2i + 1}{2i + 3}$

Hence,

Hence,

$$3 + 6 + 9 + 12 + \cdots = \sum_{i=1}^{\infty} 3i.$$

$$\dfrac{3}{5} + \dfrac{5}{7} + \dfrac{7}{9} + \dfrac{9}{11} + \cdots = \sum_{i=1}^{\infty} \dfrac{2i + 1}{2i + 3}.$$

31. $\dfrac{1}{2} + \dfrac{2}{3} + \dfrac{3}{4} + \dfrac{4}{5} + \cdots$

32. $\dfrac{2}{1} + \dfrac{3}{2} + \dfrac{4}{3} + \dfrac{5}{4} + \cdots$

33. $\dfrac{1}{1} + \dfrac{2}{3} + \dfrac{3}{5} + \dfrac{4}{7} + \cdots$

34. $\dfrac{3}{1} + \dfrac{5}{3} + \dfrac{7}{5} + \dfrac{9}{7} + \cdots$

35. $\dfrac{1}{1} + \dfrac{2}{2} + \dfrac{4}{3} + \dfrac{8}{4} + \cdots$

36. $\dfrac{1}{2} + \dfrac{3}{4} + \dfrac{9}{6} + \dfrac{27}{8} + \cdots$

11.2
ARITHMETIC PROGRESSIONS

Any sequence with a general term that is linear in n has the property that each term except the first can be obtained from the preceding term by added a common number called the **common difference**. A sequence with this property is called an **arithmetic progression**, or **arithmetic sequence**. We can state the definition for such a sequence symbolically:

$$s_1 = a,$$
$$s_{n+1} = s_n + d,$$

where d is the common difference. Definitions of this sort are called **recursive definitions**. It is customary to denote the first term in such a sequence by the letter a, the common difference between successive terms by d, the number of terms in the sequence (when finite) by n, and the nth term by s_n.

We can verify that a finite sequence is an arithmetic progression simply by subtracting each term from its successor and noting that the difference in each case is the same.

Example The sequence 7, 18, 29, 40 is an arithmetic progression, because

$$18 - 7 = 11, \qquad 29 - 18 = 11, \quad \text{and} \quad 40 - 29 = 11.$$

If at least two consecutive terms in an arithmetic progression are known, we can determine the common difference and generate as many terms as we wish.

Example The third and fourth terms of an arithmetic progression are 6 and 11, respectively. Write the next three terms of the sequence.

Solution Given $s_3 = 6$ and $s_4 = 11$, we can obtain

$$d = s_4 - s_3$$
$$= 11 - 6 = 5.$$

Thus, $s_5 = 11 + 5 = 16$, $s_6 = 16 + 5 = 21$, and $s_7 = 21 + 5 = 26$. Hence, the next three terms following 6 and 11 are 16, 21, and 26.

nth term of an arithmetic progression Now consider the general arithmetic progression with first term a and common difference d. The

first term is	a,
second term is	$a + d$,
third term is	$a + d + d = a + 2d$,
fourth term is	$a + d + d + d = a + 3d$,
$\vdots$	$\vdots$
nth term is	$a + d + d + \cdots + d = a + (n - 1)d$.

Thus,

$$s_n = a + (n - 1)d. \tag{1}$$

Here we have used an informal inductive process to obtain Equation (1). We shall assume its validity for all natural numbers n.

Equation (1) provides us with a formula to find the nth term of any arithmetic progression when the first term and common difference are known. For example, if the first term of an arithmetic progression is 7 and the common difference is 2, we have from Equation (1) that the nth term is

$$s_n = 7 + (n - 1)2$$
$$= 7 + 2n - 2$$
$$= 2n + 5.$$

Observe that Equation (1) defines a *linear function* for any values of a and d. As we have noted, the domain (replacement set of n) of this function is a set of natural numbers.

Example   The twenty-third term $(n = 23)$ of the sequence

$$4, 7, 10, \ldots,$$

which has first term 4 and common difference 3, is given by

$$s_{23} = 4 + (23 - 1)3 = 70.$$

Sum of n   The problem of finding an explicit representation for the sum of n terms of a sequence
terms   in terms of n is, in general, very difficult. However, we can obtain such a representation for the sum of n terms in an arithmetic progression. Consider the series of n
terms associated with the general arithmetic progression

$$a, (a + d), (a + 2d), \ldots, a + (n - 1)d;$$

that is,

$$S_n = a + (a + d) + (a + 2d) + \cdots + [a + (n - 1)d]. \tag{2}$$

Then consider the same series written as

$$S_n = s_n + (s_n - d) + (s_n - 2d) + \cdots + [s_n - (n - 1)d], \tag{3}$$

where the terms are in reverse order. Adding (2) and (3) term-by-term, we have

$$S_n + S_n = (a + s_n) + (a + s_n) + (a + s_n) + \cdots + (a + s_n),$$

where the term $(a + s_n)$ occurs n times. It follows that

$$2S_n = n(a + s_n).$$

Thus,

$$\blacktriangleright \qquad\qquad S_n = \frac{n}{2}(a + s_n). \tag{4}$$

Example   The sum of the eight terms of an arithmetic progression with first term -7 and last term
14 is given by

$$S_8 = \frac{8}{2}(-7 + 14) = 28.$$

If (4) is rewritten as

$$S_n = n\left(\frac{a + s_n}{2}\right),$$

we observe that the sum is given by the product of the number of terms in the series and
the average of the first and last terms.

An alternative form for (4) is obtained by substituting in (4) the value for s_n equal to
$a + (n - 1)d$ as given by (1) to obtain

$$S_n = \frac{n}{2}(a + [a + (n - 1)d]).$$

Thus,

$$S_n = \frac{n}{2}[2a + (n-1)d].\qquad(5)$$

In (5) the sum is now expressed in terms of a, n, and d.

Example

The sum of the first twelve terms of the arithmetic progression $-5, -2, 1, \ldots$, which has first term -5 and common difference 3, is given by

$$S_{12} = \frac{12}{2}[2(-5) + (12-1)3]$$
$$= 6(-10 + 33) = 138.$$

EXERCISE 11.2

A ■ *Write the next three terms in each arithmetic progression. Find an expression for the general term.*

Examples   **a.** $5, 9, \ldots$ **b.** $x, x-a, \ldots$

Solutions

a. Find the common difference and then continue the sequence.

$$d = 9 - 5 = 4$$
$$13, 17, 21$$

Use $s_n = a + (n-1)d$ to find an expression for the general term.

$$s_n = 5 + (n-1)4$$
$$s_n = 4n + 1$$

b. Find the common difference and then continue the sequence.

$$d = (x-a) - x = -a$$
$$x - 2a,\ x - 3a,\ x - 4a$$

Use $s_n = a + (n-1)d$ to find an expression for the general term.

$$s_n = x + (n-1)(-a)$$
$$s_n = x - a(n-1)$$

1. $3, 7, \ldots$ **2.** $-6, -1, \ldots$ **3.** $-1, -5, \ldots$ **4.** $-10, -20, \ldots$

5. $x, x+1, \ldots$ **6.** $a, a+5, \ldots$ **7.** $x+a, x+3a, \ldots$ **8.** $y-2b, y, \ldots$

9. $2x+1, 2x+4, \ldots$ **10.** $a+2b, a-2b, \ldots$ **11.** $x, 2x, \ldots$ **12.** $3a, 5a, \ldots$

Example   Find the fourteenth term of the arithmetic progression $-6, -1, 4, \ldots$.

Solution   Find the common difference.

$$d = -1 - (-6) = 5$$

Solution continued overleaf

Use $s_n = a + (n-1)d$.

$$s_{14} = -6 + (14-1)5 = 59$$

13. Find the seventh term in the arithmetic progression 7, 11, 15, ...
14. Find the tenth term in the arithmetic progression $-3, -12, -21, \ldots$.
15. Find the twelfth term in the arithmetic progression 2, $\frac{5}{2}$, 3, ...
16. Find the seventeenth term in the arithmetic progression $-5, -2, 1, \ldots$.
17. Find the twentieth term in the arithmetic progression 3, -2, -7, ...
18. Find the tenth term in the arithmetic progression $\frac{3}{4}$, 2, $\frac{13}{4}$, ...

Example

Find the first term in an arithmetic progression in which the third term is 7 and the eleventh term is 55.

Solution

A diagram of the situation is helpful here.

$$n: \ 1\ ,\ 2\ ,\ 3\ ,\ 4\ ,\ 5\ ,\ 6\ ,\ 7\ ,\ 8\ ,\ 9\ ,10,11$$

$$s_n: \ \frac{?}{} ,\!\!-\!\!,\frac{7}{7},\!\!-\!\!,\!-\!,\!-\!,\!-\!,\!-\!,\!-\!,\!-\!,\frac{55}{}$$

Find a common difference by considering an arithmetic progression with first term 7 and ninth term 55. Use $s_n = a + (n-1)d$.

$$s_9 = 7 + (9-1)d$$
$$55 = 7 + 8d$$
$$d = 6$$

Use this difference to find the first term in an arithmetic progression in which the third term is 7. Use $s_n = a + (n-1)d$.

$$s_3 = a + (3-1)6$$
$$7 = a + 12$$
$$a = -5$$

The first term is -5.

Alternative Solution

Use $s_n = a + (n-1)d$, with $s_3 = 7$.

$$7 = a + (3-1)d$$
$$7 = a + 2d \tag{1}$$

Use $s_n = a + (n-1)d$ with $s_{11} = 55$.

$$55 = a + (11-1)d$$
$$55 = a + 10d \tag{2}$$

Solve the system (1) and (2) to obtain $a = -5$.

19. If the third term in an arithmetic progression is 7 and the eighth term is 17, find the common difference. What is the first term? What is the twentieth term?

20. If the fifth term of an arithmetic progression is -16 and the twentieth term is -46, what is the twelfth term?

21. What term in the arithmetic progression 4, 1, -2, ... is -77?

22. What term in the arithmetic progression 7, 3, -1, ... is -81?

■ *Terms between given terms in an arithmetic progression are called* **arithmetic means** *of the given terms. Insert the given number of arithmetic means between the given two numbers.*

Example   Three between 4 and -8.

Solution   If there are three terms between 4 and -8, then the difference between 4 and -8 must be 4 times the common difference (d), as suggested by

$$\underset{d}{\overset{4}{\rule{0pt}{0pt}}} \quad \underset{d}{\overset{?}{\rule{0pt}{0pt}}} \quad \underset{d}{\overset{?}{\rule{0pt}{0pt}}} \quad \underset{d}{\overset{?}{\rule{0pt}{0pt}}} \quad \overset{-8}{\rule{0pt}{0pt}}.$$

Therefore,

$$4d = (-8) - (4) = -12$$
$$d = -3.$$

The three requested arithmetic means can then be obtained by successive additions of -3. We obtain 1, -2, and -5.

23. Two between -6 and 15.   **24.** Four between 10 and 65.

25. One between 12 and 20.   **26.** One between -11 and 7.

27. Three between 24 and 4.   **28.** Six between -12 and 23.

■ *Find the sum of each finite series.*

Example $\displaystyle\sum_{i=1}^{12} (4i + 1)$

Solution   Write the first two or three terms in expanded form.

$$5 + 9 + 13 + \cdots$$

By inspection, the first term is 5 and the common difference is 4.

Use $S_n = \dfrac{n}{2}[2a + (n-1)d]$ with $n = 12$.

$$S_{12} = \frac{12}{2}[2(5) + (12-1)4] = 324$$

29. $\displaystyle\sum_{i=1}^{7} (2i + 1)$ **30.** $\displaystyle\sum_{i=1}^{21} (3i - 2)$ **31.** $\displaystyle\sum_{j=3}^{15} (7j - 1)$

32. $\displaystyle\sum_{j=10}^{20} (2j - 3)$ **33.** $\displaystyle\sum_{k=1}^{8} \left(\frac{1}{2}k - 3\right)$ **34.** $\displaystyle\sum_{k=1}^{100} k$

B **35.** Find the sum of all even integers n, where $13 < n < 89$.

36. Find the sum of all integral multiples of 7 between 8 and 110.

37. How many bricks will there be in a stack one brick deep if there are 27 bricks in the first row, 25 in the second row, ..., and 1 in the top row?

38. If there is a total of 256 bricks in a stack arranged in the manner of those in Problem 37, how many bricks are there in the third row from the bottom?

39. Find three numbers that form an arithmetic sequence such that their sum is 21 and their product is 168.

40. Find three numbers that form an arithmetic sequence such that their sum is 21 and their product is 231.

41. Find k if $\displaystyle\sum_{j=1}^{5} kj = 14$.

42. Find p and q if $\displaystyle\sum_{i=1}^{4} (pi + q) = 28$ and $\displaystyle\sum_{i=2}^{5} (pi + q) = 44$.

43. Show that the sum of the first n odd natural numbers is n^2.

44. Show that the sum of the first n even natural numbers is $n^2 + n$.

11.3

GEOMETRIC PROGRESSIONS

Any sequence in which each term except the first is obtained by multiplying the preceding term by a common multiplier is called a **geometric progression**, or **geometric sequence**, and is defined by the recursive equations

$$s_1 = a,$$

$$s_{n+1} = rs_n,$$

where r is called the **common ratio**.

Example   The sequence

$$2, 6, 18, 54, \ldots$$

is a geometric progression in which each term except the first is obtained by multiplying the preceding term by 3.

nth term of a geometric progression   Now, if we designate the first term of a geometric progression by a, then the

second term is ar,
third term is $ar \cdot r = ar^2$,
fourth term is $ar^2 \cdot r = ar^3$,

and it appears that the nth term will take the following form:

$$\blacktriangleright \qquad s_n = ar^{n-1}. \tag{1}$$

We assume the validity of this expression for all natural numbers n. The general geometric progression will appear as

$$a, ar, ar^2, ar^3, ar^4, \ldots, ar^{n-1}, \ldots .$$

Example   Find the ratio r and general term s_n for the geometric progression

$$2, 6, 18, \ldots .$$

Solution   By writing the ratio of any term to its predecessor, say, $^{18}\!/\!_6$, we find that $r = 3$. A representation for s_n of this sequence can now be written in terms of n by substituting 2 for a and 3 for r in (1). Thus,

$$s_n = ar^{n-1} = 2(3)^{n-1}.$$

Observe that Equation (1) defines an *exponential function* for all values of a and r. As we noted, the domain (replacement set of n) of this function is a set of natural numbers.

Sum of n terms   To find an explicit representation for the sum of a given number of terms in a geometric progression in terms of a, r, and n, we employ a device somewhat similar to the one used in finding the sum of an arithmetic series. Consider the geometric series (2) containing n terms and the series (3) obtained by multiplying both members of (2) by r:

$$S_n = a + ar + ar^2 + ar^3 + \cdots + ar^{n-2} + ar^{n-1}, \tag{2}$$

$$rS_n = \quad ar + ar^2 + ar^3 + ar^4 + \cdots + ar^{n-1} + ar^n. \tag{3}$$

Subtracting (3) from (2), we find that all terms in the right-hand member vanish except the first term in (2) and the last term in (3), and therefore

$$S_n - rS_n = a - ar^n.$$

Factoring S_n from the left-hand member yields

$$(1 - r)S_n = a - ar^n.$$

Thus,

$$\blacktriangleright \qquad S_n = \frac{a - ar^n}{1 - r} \qquad (r \neq 1). \qquad (4)$$

This is a general formula for the sum of n terms of a geometric progression.

Example   The sum of four terms of a geometric progression with first term 5 and common ratio -3 is given by

$$S_4 = \frac{5 - 5(-3)^4}{1 - (-3)}$$

$$= \frac{5 - 5(81)}{4} = -100.$$

An alternative expression for (4) can be obtained by noting that it may be written

$$S_n = \frac{a - r(ar^{n-1})}{1 - r},$$

and, since $s_n = ar^{n-1}$, we have the following relationship, where the sum is now given in terms of a, s_n, and r:

$$\blacktriangleright \qquad S_n = \frac{a - rs_n}{1 - r} \qquad (r \neq 1). \qquad (5)$$

Example   The sum of the geometric progression

$$6 + 3 + \frac{3}{2} + \frac{3}{4} + \frac{3}{8}$$

with first term 6, common ratio ½, and fifth term ⅜, is given by

$$S_5 = \frac{6 - \left(\frac{1}{2}\right)\left(\frac{3}{8}\right)}{1 - \frac{1}{2}} = \frac{6 - \frac{3}{16}}{\frac{1}{2}}$$

$$= 2\left(6 - \frac{3}{16}\right) = 12 - \frac{3}{8} = \frac{93}{8}.$$

EXERCISE 11.3

A ■ *Write the next three terms in each geometric progression. Find the general term.*

Examples **a.** 3, 6, 12, . . .   **b.** $x, 2, \dfrac{4}{x}, \ldots$

Solutions **a.** Find the common ratio.   **b.** Find the common ratio.

$$r = \frac{6}{3} = 2$$

$$r = \frac{2}{x}$$

Multiply successively by r to determine the following terms:

Multiply successively by r to determine the following terms:

24, 48, 96.

$$\frac{8}{x^2}, \frac{16}{x^3}, \frac{32}{x^4}.$$

Use $s_n = ar^{n-1}$ to find the general term.

Use $s_n = ar^{n-1}$ to find the general term.

$$s_n = 3(2)^{n-1}$$

$$s_n = x\left(\frac{2}{x}\right)^{n-1} = \frac{2^{n-1}}{x^{n-2}}$$

1. 2, 8, 32, . . .   **2.** 4, 8, 16, . . .   **3.** $\dfrac{2}{3}, \dfrac{4}{3}, \dfrac{8}{3}, \ldots$ **4.** $6, 3, \dfrac{3}{2}, \ldots$

5. 4, −2, 1, . . .   **6.** $\dfrac{1}{2}, -\dfrac{3}{2}, \dfrac{9}{2}, \ldots$ **7.** $\dfrac{a}{x}, -1, \dfrac{x}{a}, \ldots$ **8.** $\dfrac{a}{b}, \dfrac{a}{bc}, \dfrac{a}{bc^2}, \ldots$

Example Find the ninth term of the geometric progression −24, 12, −6, ...

Solution Find the common ratio.

$$r = \frac{12}{-24} = -\frac{1}{2}$$

Use $s_n = ar^{n-1}$.

$$s_9 = -24\left(-\frac{1}{2}\right)^8 = -\frac{3}{32}$$

9. Find the sixth term in the geometric progression 48, 96, 192, ...

10. Find the eighth term in the geometric progression −3, ³⁄₂, −¾, ...

11. Find the seventh term in the geometric progression $-\frac{1}{3}a^2, a^5, -3a^8, \ldots$.

12. Find the ninth term in the geometric progression $-81a, -27a^2, -9a^3, \ldots$.

13. Find the first term of a geometric progression with fifth term 48 and ratio 2.

14. Find the first term of a geometric progression with fifth term 1 and ratio −½.

■ *Terms between two given terms in a geometric progression are called* **geometric means**. *Insert the given number of geometric means between the two given numbers.*

Example   Two between 3 and 24.

Solution   Since there are three multiplications by the common ratio between 3 and 24, the quotient when 24 is divided by 3 must be the third power of the common ratio, as suggested by the following:

$$\frac{3}{\searrow \times r} \quad \frac{?}{\nearrow \times r} \quad \frac{?}{\nearrow \times r} \quad \frac{24}{\nearrow}.$$

Hence, $r^3 = {}^{24}\!/_3 = 8$; so $r = \sqrt[3]{8} = 2$. Therefore, the missing terms can be determined by successive multiplications by 2, and we have $3 \times 2 = 6$ and $6 \times 2 = 12$. Thus, the geometric means are 6 and 12.

15. Two between 1 and 27.

16. Two between -4 and -32.

17. One between 36 and 9. (Two answers are possible.)

18. One between -12 and $-\dfrac{1}{12}$. (Two answers are possible.)

19. Three between 32 and 2. (Two answers are possible.)

20. Three between -25 and $-\dfrac{1}{25}$. (Two answers are possible.)

■ *Find each sum.*

Example $\displaystyle\sum_{i=2}^{7} \left(\frac{1}{3}\right)^i$

Solution   Write the first two or three terms in expanded form.

$$\left(\frac{1}{3}\right)^2 + \left(\frac{1}{3}\right)^3 + \cdots$$

By inspection, the first term is $\frac{1}{9}$, the ratio is $\frac{1}{3}$, and $n = 6$ [recall from page 360 that the series $\displaystyle\sum_{i=a}^{b}$ contains $(b - a + 1)$ terms]. Use $S_n = \dfrac{a - ar^n}{1 - r}$.

$$S_6 = \frac{\dfrac{1}{9} - \dfrac{1}{9}\left(\dfrac{1}{3}\right)^6}{1 - \dfrac{1}{3}} = \frac{\dfrac{1}{9}\left(1 - \dfrac{1}{729}\right)}{\dfrac{2}{3}} = \frac{1}{9} \cdot \frac{728}{729} \cdot \frac{3}{2} = \frac{364}{2187}$$

21. $\displaystyle\sum_{i=1}^{6} 3^i$

22. $\displaystyle\sum_{j=1}^{4} (-2)^j$

23. $\displaystyle\sum_{k=3}^{7} \left(\frac{1}{2}\right)^{k-2}$

24. $\displaystyle\sum_{i=3}^{12} (2)^{i-5}$

25. $\displaystyle\sum_{j=1}^{6} \left(\frac{1}{3}\right)^j$

26. $\displaystyle\sum_{k=1}^{5} \left(\frac{1}{4}\right)^k$

B **27.** Graph the geometric progression defined by $s_n = 2^n$ for $1 \le n \le 4$. Use the horizontal axis for n and the vertical axis for s_n.

28. Graph the geometric progression defined by $s_n = 2^{n-4}$ for $1 \le n \le 7$. Use the horizontal axis for n and the vertical axis for s_n.

29. A culture of bacteria doubles every hour. If there were 10 bacteria in the culture originally, how many are there after 1 hour? After 2 hours? After 3 hours? After 4 hours? After n hours?

30. A certain radioactive substance has a half-life of 2400 years (50% of the original material is present at the end of 2400 years). If 100 grams were produced today, how many grams would be present in 2400 years? In 4800 years? In 7200 years? In 9600 years? In $2400n$ years?

11.4

INFINITE SERIES

Infinite geometric series

Consider the infinite geometric series

$$\frac{1}{2} + \frac{1}{4} + \frac{1}{8} + \frac{1}{16} + \cdots$$

and the partial sums of terms of the series,

$$S_1 = \frac{1}{2}$$

$$S_2 = \frac{1}{2} + \frac{1}{4} = \frac{3}{4}$$

$$S_3 = \frac{1}{2} + \frac{1}{4} + \frac{1}{8} = \frac{7}{8}$$

$$S_4 = \frac{1}{2} + \frac{1}{4} + \frac{1}{8} + \frac{1}{16} = \frac{15}{16}$$

$$\vdots$$

Note that the nth term of the sequence of partial sums

$$\frac{1}{2}, \frac{3}{4}, \frac{7}{8}, \frac{15}{16}, \ldots, S_n, \ldots$$

appears to be "approaching" 1. That is, as the number n becomes very large, S_n is very close to 1. In fact, we can make the difference between S_n and 1 as small as we like by using a sufficiently large value for n.

Recall from Section 11.3 that the sum of n terms of a geometric progression is given by

$$S_n = \frac{a - ar^n}{1 - r} \qquad (r \neq 1). \tag{1}$$

If $|r| < 1$, that is, if $-1 < r < 1$, then r^n becomes smaller and smaller for increasingly large n. For example, if $r = \frac{1}{2}$, then

$$r^2 = \left(\frac{1}{2}\right)^2 = \frac{1}{4}, \qquad r^3 = \left(\frac{1}{2}\right)^3 = \frac{1}{8}, \qquad r^4 = \left(\frac{1}{2}\right)^4 = \frac{1}{16},$$

and so on, and we can make $(\frac{1}{2})^n$ as small as we please by taking n sufficiently large. Writing (1) in the form

$$S_n = \frac{a}{1 - r}(1 - r^n), \tag{2}$$

we see that the value of the factor $(1 - r^n)$ can be made as close as we please to 1, providing $|r| < 1$ and n is taken large enough. Since this asserts that the sum (2) can be made to approximate

$$\frac{a}{1 - r}$$

as close as we please, we define the sum of an infinite geometric series with $|r| < 1$ as follows:

$$\blacktriangleright \qquad S_\infty = \frac{a}{1 - r}. \tag{3}$$

Example   The sum of the infinite geometric series

$$-18 + 6 - 2 + \frac{2}{3} - \frac{2}{9} + \cdots,$$

with first term -18 and common ratio $-\frac{1}{3}$, is given by

$$S_\infty = \frac{-18}{1 - \left(-\frac{1}{3}\right)} = \frac{-18}{\frac{4}{3}}$$

$$= -18\left(\frac{3}{4}\right) = -\frac{27}{2}.$$

Limit of a sequence   A sequence that has an nth term (and all terms after the nth) that can be made to approximate a fixed number L as closely as desired by simply taking n large enough is said to *approach the limit L* as n increases without bound. We can indicate this in terms of symbols,

$$\lim_{n \to \infty} S_n = L,$$

where S_n is the nth term of the sequence of sums $S_1, S_2, S_3, \ldots, S_n, \ldots$. Thus, (3) might be written

$$S_\infty = \lim_{n \to \infty} S_n = \frac{a}{1-r}.$$

If $|r| \geq 1$, then r^n in (2) does not approach 0 and $\lim_{n \to \infty} S_n$ does not exist.

Repeating decimals An interesting application of this sum arises in connection with repeating decimals—that is, decimal numerals that, after a finite number of decimal places, have endlessly repeating groups of digits. For example,

$$0.2121\overline{21}, \qquad 0.333\overline{3} \quad \text{and} \quad 0.138512512\overline{512}$$

are repeating decimals, where in each case the bar indicates the repeating digits. Consider the problem of expressing such a decimal numeral as a fraction. We illustrate the process involved with the first repeating decimal above:

$$0.2121\overline{21}. \tag{4}$$

This decimal can be written either as

$$0.21 + 0.0021 + 0.000021 + \cdots \tag{5}$$

or

$$\frac{21}{100} + \frac{21}{10,000} + \frac{21}{1,000,000} + \cdots, \tag{6}$$

which are sums with terms that form a geometric progression with a ratio 0.01 (or $\frac{1}{100}$). Since the ratio is less than 1 in absolute value, we can use (3) to find the sum of an infinite number of terms of (5) or (6). Thus, using (6) we have

$$S_\infty = \frac{a}{1-r} = \frac{\dfrac{21}{100}}{1 - \dfrac{1}{100}} = \frac{\dfrac{21}{100}}{\dfrac{99}{100}} = \frac{21}{99} = \frac{7}{33},$$

and the given decimal numeral, $0.2121\overline{21}$, is equivalent to $\frac{7}{33}$.

The number e The number e that we introduced in Chapter 9 is the limit of the sequence

$$\left(1 + \frac{1}{1}\right)^1, \left(1 + \frac{1}{2}\right)^2, \left(1 + \frac{1}{3}\right)^3, \ldots,$$

and the general term is given by

$$\left(1+\frac{1}{t}\right)^t,$$

where t is a natural number. We can use the $\boxed{y^x}$ key on a calculator to obtain a decimal approximation for the expression for different values of t. The first several terms of this sequence, rounded to 4 decimal places, are

$$2.000, \ 2.2500, \ 2.3704, \ 2.4414, \ 2.4883, \ \cdots ,$$

and the 1000th term is

$$\left(1+\frac{1}{1000}\right)^{1000} = 2.7169.$$

The question arises whether we would approach some number as $t \to \infty$. It is shown in courses in advanced mathematics that

$$\lim_{t \to \infty} \left(1+\frac{1}{t}\right)^t = e,$$

where, as we have noted, e is an irrational number approximately equal to 2.7182818.

The expression $[1 + (1/t)]^t$ is a special case of the right-hand member of the formula

$$A = P\left(1+\frac{r}{t}\right)^{tn} \tag{7}$$

for compound interest that we used in Exercise 9.7. This formula is sometimes written in the form

$$A = P(1 + r)^n. \tag{8}$$

Here, r is the rate and n is the number of compounding periods. It can be shown that if $t \to \infty$ in (7), which implies continuous compounding, we obtain

$$A = Pe^{rn}, \tag{9}$$

the formula introduced on page 321.

Example   Five thousand dollars is invested for 1 year at 10% interest. What is the total value of the investment at the end of the year if the interest is compounded quarterly? If the interest is compounded continuously?

Solution   If the interest is compounded quarterly, we have, from Equation (7),

$$A = 5000\left(1+\frac{0.10}{4}\right)^{4(1)}$$
$$= 5000(1.025)^4$$
$$= 5519.06.$$

The total value is \$5519.06. If the interest is compounded continuously, we have, from Equation (9),

$$A = 5000e^{0.10(1)} = 5525.85.$$

The total value is \$5525.85.

The notion of compounding interest continuously does not provide a practical application of Equation (9). However, different forms of this equation do have wide application in a variety of areas. Some of these applications were included in Exercise 9.7.

EXERCISE 11.4

A ■ *Find the sum of each infinite geometric series. If the series has no sum, so state.*

Examples

a. $3 + 2 + \dfrac{4}{3} + \cdots$

b. $\dfrac{1}{81} - \dfrac{1}{54} + \dfrac{1}{36} + \cdots$

Solutions

a. $r = \dfrac{2}{3}$; the series has a sum, since $|r| < 1$.

$$S_\infty = \frac{a}{1-r} = \frac{3}{1 - \dfrac{2}{3}} = 9$$

b. $r = -\dfrac{1}{54} \div \dfrac{1}{81} = -\dfrac{3}{2}$; the series does not have a sum, since $|r| > 1$.

1. $12 + 6 + 3 + \cdots$

2. $2 + 1 + \dfrac{1}{2} + \cdots$

3. $\dfrac{1}{36} + \dfrac{1}{30} + \dfrac{1}{25} + \cdots$

4. $1 + \dfrac{2}{3} + \dfrac{4}{9} + \cdots$

5. $\dfrac{3}{4} - \dfrac{1}{2} + \dfrac{1}{3} - \cdots$

6. $\dfrac{1}{16} - \dfrac{1}{8} + \dfrac{1}{4} - \cdots$

7. $\dfrac{1}{49} + \dfrac{1}{56} + \dfrac{1}{64} + \cdots$

8. $2 - \dfrac{3}{2} + \dfrac{9}{8} - \cdots$

9. $\displaystyle\sum_{i=1}^{\infty} \left(\dfrac{2}{3}\right)^i$

10. $\displaystyle\sum_{i=1}^{\infty} \left(-\dfrac{1}{4}\right)^i$

■ *Find a fraction equivalent to each of the given decimal numerals.*

Example

$2.045045\overline{045}$

Solution

Rewrite as a series.

$$2 + \frac{45}{1000} + \frac{45}{1,000,000} + \cdots *$$

*Alternatively, the series can be written in decimal notation as $2 + 0.045 + 0.000045 + \cdots$ with $r = 0.001$ and $a = 0.045$.

Solution continued overleaf

For series beginning with $\dfrac{45}{1000}$, $r = \dfrac{1}{1000}$. Use $S_\infty = \dfrac{a}{1-r}$.

$$S_\infty = \frac{\dfrac{45}{1000}}{1 - \dfrac{1}{1000}} = \frac{\dfrac{45}{1000}}{\dfrac{999}{1000}} = \frac{45}{999} = \frac{5}{111}$$

Hence,

$$2.045045\overline{045} = 2\frac{5}{111} = \frac{227}{111}.$$

11. $0.3333\overline{3}$ **12.** $0.6666\overline{6}$ **13.** $0.3131\overline{31}$ **14.** $0.4545\overline{45}$

15. $2.410\overline{410}$ **16.** $3.027\overline{027}$ **17.** $0.12888\overline{8}$ **18.** $0.8333\overline{3}$

■ *Use the formulas for compound interest given in Exercise 9.7 on page 376.*

19. One hundred dollars is invested for 10 years at 8% interest. What is the total value of the investment if the interest is compounded quarterly? If the interest is compounded continuously?

20. Which investment would produce the greatest annual income: five thousand dollars at 12% compounded semiannually or five thousand dollars at 11% compounded continuously?

B   **21.** A force is applied to a particle moving in a straight line in such a fashion that each second it moves only one-half of the distance it moved the preceding second. If the particle moves 10 centimeters the first second, approximately how far will it move before coming to rest?

22. The arc length through which the bob on a pendulum moves is nine-tenths of its preceding arc length. Approximately how far will the bob move before coming to rest if the first arc length is 12 inches?

23. A ball returns two-thirds of its preceding height on each bounce. If the ball is dropped from a height of 6 feet, approximately what is the total distance the ball travels before coming to rest?

24. If a ball is dropped from a height of 10 feet and returns three-fifths of its preceding height on each bounce, approximately what is the total distance the ball travels before coming to rest?

25. If P dollars is invested at an interest rate r *compounded annually*, show that the amount A present after two years is given by $A = P(1 + r)^2$ and that the amount present after three years is $A = P(1 + r)^3$. Make a conjecture about the amount that is present after n years.

26. If P dollars is invested at an interest rate r, show that the amount A present in one year when the interest is compounded semiannually for n years is given by $A = P[1 + (r/2)]^{2n}$ and when compounded quarterly for n years is given by $A = P[1 + (r/4)]^{4n}$. Make a conjecture about the amount that is present if the interest is compounded t times yearly.

11.5

THE BINOMIAL EXPANSION

Factorial notation

Sometimes it is necessary to write the product of consecutive positive integers. To do this in some cases, we use a special symbol, $n!$ (read "n factorial" or "factorial n"), which is defined by:

▶

$$n! = n(n-1)(n-2) \cdot \cdots \cdot 1.$$

Examples   **a.** $5! = 5 \cdot 4 \cdot 3 \cdot 2 \cdot 1$ **b.** $8! = 8 \cdot 7 \cdot 6 \cdot 5 \cdot 4 \cdot 3 \cdot 2 \cdot 1$ **c.** $1! = 1$

Since

$$n! = n(n-1)(n-2)(n-3) \cdot \cdots \cdot 5 \cdot 4 \cdot 3 \cdot 2 \cdot 1$$

and

$$(n-1)! = (n-1)(n-2)(n-3) \cdot \cdots \cdot 5 \cdot 4 \cdot 3 \cdot 2 \cdot 1,$$

we can, for $n > 1$, write the recursive relationship:

▶

$$n! = n(n-1)!. \tag{1}$$

Examples   **a.** $7! = 7 \cdot 6!$ **b.** $27! = 27 \cdot 26!$ **c.** $(n+2)! = (n+2)(n+1)!$

If $n = 1$ in Equation (1) above, we have

$$1! = 1 \cdot (1-1)!$$
$$1! = 1 \cdot 0!.$$

Therefore, for consistency, we shall make the following definition:

▶

$$0! = 1.$$

Note that both 1! and 0! are equal to 1.

Binomial expansion

The series obtained by expanding a binomial of the form

$$(a+b)^n$$

is particularly useful in certain branches of mathematics. Starting with familiar examples where n takes the value 1, 2, 3, 4, and 5, we can show by direct multiplication that

$$(a + b)^1 = a + b,$$
$$(a + b)^2 = a^2 + 2ab + b^2,$$
$$(a + b)^3 = a^3 + 3a^2b + 3ab^2 + b^3,$$
$$(a + b)^4 = a^4 + 4a^3b + 6a^2b^2 + 4ab^3 + b^4,$$
$$(a + b)^5 = a^5 + 5a^4b + 10a^3b^2 + 10a^2b^3 + 5ab^4 + b^5.$$

We observe that in each case:

1. The first term may be considered to be $a^n b^0$. The exponent on a *decreases* by 1 and the exponent on b *increases* by 1 in each of the following terms. The last term may be considered to be $a^0 b^n$. For example,

$$
\underbrace{\frac{5}{1!}}_{\downarrow} \quad \underbrace{\frac{5 \cdot 4}{2!}}_{\downarrow} \quad \underbrace{\frac{5 \cdot 4 \cdot 3}{3!}}_{\downarrow} \quad \underbrace{\frac{5 \cdot 4 \cdot 3 \cdot 2}{4!}}_{\downarrow} \quad \underbrace{\frac{5 \cdot 4 \cdot 3 \cdot 2 \cdot 1}{5!}}_{\downarrow}
$$
$$(a + b)^5 = a^5 + 5a^4b + 10a^3b^2 + 10a^2b^3 + 5ab^4 + b^5.$$

Exponents of a decrease by 1
Exponents of b increase by 1

2. The variable factors of the second term are $a^{n-1}b^1$ and the coefficient is n, which can be written in the form

$$\frac{n}{1!}.$$

3. The variable factors of the third term are $a^{n-2}b^2$, and the coefficient can be written in the form

$$\frac{n(n-1)}{2!}.$$

4. The variable factors of the fourth term are $a^{n-3}b^3$, and the coefficient can be written in the form

$$\frac{n(n-1)(n-2)}{3!}.$$

The above results can be generalized to obtain the **binomial expansion**:

▶ $$(a + b)^n = a^n + \frac{n}{1!}a^{n-1}b + \frac{n(n-1)}{2!}a^{n-2}b^2 + \frac{n(n-1)(n-2)}{3!}a^{n-3}b^3 + \cdots$$

$$+ \frac{n(n-1)(n-2) \cdot \cdots \cdot (n-r+2)}{(r-1)!}a^{n-r+1}b^{r-1} + \cdots + b^n, \quad (2)$$

where r is the number of the term.

Example   Expand $(x - 2)^4$.

Solution   In this case, $a = x$ and $b = -2$ in the binomial expansion:

$$(x - 2)^4 = x^4 + \frac{4}{1!}x^3(-2)^1 + \frac{4 \cdot 3}{2!}x^2(-2)^2 + \frac{4 \cdot 3 \cdot 2}{3!}x(-2)^3 + \frac{4 \cdot 3 \cdot 2 \cdot 1}{4!}(-2)^4$$

$$= x^4 - 8x^3 + 24x^2 - 32x + 16$$

rth term of   Note in (2) that the rth term in a binomial expansion is given by
an expansion

▶

$$\frac{n(n - 1)(n - 2) \cdot \cdots \cdot (n - r + 2)}{(r - 1)!}a^{n-r+1}b^{r-1}. \tag{3}$$

Examples   **a.** Find the fifth term in the expansion of $(x - 2)^{10}$.

                  b. Find the seventh term in the expansion of $(x - 2)^{10}$.

Solutions   **a.** We can use (3) above, where $n = 10$, $r = 5$, $r - 1 = 4$, and $n - r + 2 = 7$.

$$\frac{10 \cdot 9 \cdot 8 \cdot 7}{4 \cdot 3 \cdot 2 \cdot 1}x^6(-2)^4 = 3360x^6$$

                  b. We can use (3) above, where $n = 10$, $r = 7$, $r - 1 = 6$, and $n - r + 2 = 5$.

$$\frac{10 \cdot 9 \cdot 8 \cdot 7 \cdot 6 \cdot 5}{6 \cdot 5 \cdot 4 \cdot 3 \cdot 2 \cdot 1}x^4(-2)^6 = 13{,}440x^4$$

When using Equation (3), it is helpful to note that:

1. The exponent on b, $(r - 1)$, is one less than r, the number of the term.

2. The sum of the exponents $(n - r + 1) + (r - 1)$ equals n, the exponent of the power.

3. The factor $(r - 1)$ in the denominator of the coefficient equals the exponent on b.

4. The number of factors in the numerator of the coefficient is the same as the number of factors in the denominator.

EXERCISE 11.5

A   ▪ *Write each expression in expanded form.*

Examples   **a.** $5n!$ for $n = 4$ **b.** $(2n - 1)!$ for $n = 4$

Solutions overleaf

Solutions   **a.** $5n! = 5 \cdot (4 \cdot 3 \cdot 2 \cdot 1)$ **b.** $(2n - 1)! = [2(4) - 1]!$
                          $= 120$ $= (8 - 1)!$
                                                                            $= 7! = 7 \cdot 6 \cdot 5 \cdot 4 \cdot 3 \cdot 2 \cdot 1$

1. $(2n)!$ for $n = 4$ **2.** $(3n)!$ for $n = 4$

3. $2n!$ for $n = 4$ **4.** $3n!$ for $n = 4$

5. $n(n - 1)!$ for $n = 6$ **6.** $2n(2n - 1)!$ for $n = 2$

■ *Write in expanded form and simplify.*

Examples   **a.** $\dfrac{7!}{4!}$ **b.** $\dfrac{4!6!}{8!}$

Solutions   **a.** $\dfrac{7!}{4!} = \dfrac{7 \cdot 6 \cdot 5 \cdot 4!}{4!} = 210$ **b.** $\dfrac{4!6!}{8!} = \dfrac{4 \cdot 3 \cdot 2 \cdot 1 \cdot 6!}{8 \cdot 7 \cdot 6!} = \dfrac{3}{7}$

7. $5!$ **8.** $7!$ **9.** $\dfrac{9!}{7!}$ **10.** $\dfrac{12!}{11!}$

11. $\dfrac{5!7!}{8!}$ **12.** $\dfrac{12!8!}{16!}$ **13.** $\dfrac{8!}{2!(8 - 2)!}$ **14.** $\dfrac{10!}{4!(10 - 4)!}$

■ *Write each product in factorial notation.*

Examples   **a.** $1 \cdot 2 \cdot 3 \cdot 4 \cdot 5 \cdot 6$ **b.** $11 \cdot 12 \cdot 13 \cdot 14$ **c.** 150

Solutions   **a.** $1 \cdot 2 \cdot 3 \cdot 4 \cdot 5 \cdot 6 = 6!$ **b.** $11 \cdot 12 \cdot 13 \cdot 14 = \dfrac{14!}{10!}$ **c.** $150 = \dfrac{150!}{149!}$

15. $1 \cdot 2 \cdot 3$ **16.** $1 \cdot 2 \cdot 3 \cdot 4 \cdot 5$ **17.** $3 \cdot 4 \cdot 5 \cdot 6$

18. 7 **19.** $8 \cdot 7 \cdot 6$ **20.** $28 \cdot 27 \cdot 26 \cdot 25 \cdot 24$

■ *Write each expression in factored form and show the first three factors and the last three factors.*

Example $(2n + 1)!$

Solution $(2n + 1)! = (2n + 1)(2n)(2n - 1) \cdot \cdots \cdot 3 \cdot 2 \cdot 1$

21. $n!$ **22.** $(n + 4)!$ **23.** $(3n)!$

24. $3n!$ **25.** $(n - 2)!$ **26.** $(3n - 2)!$

■ *Expand.*

Example $(a - 3b)^4$

Solution From the binomial expansion (2) on page 380,

$$(a - 3b)^4 = a^4 + \frac{4}{1!}a^3(-3b) + \frac{4 \cdot 3}{2!}a^2(-3b)^2 + \frac{4 \cdot 3 \cdot 2}{3!}a(-3b)^3 + \frac{4 \cdot 3 \cdot 2 \cdot 1}{4!}(-3b)^4$$

$$= a^4 - 12a^3b + 54a^2b^2 - 108ab^3 + 81b^4.$$

27. $(x + 3)^5$ **28.** $(2x + y)^4$ **29.** $(x - 3)^4$ **30.** $(2x - 1)^5$

31. $\left(2x - \dfrac{y}{2}\right)^3$ **32.** $\left(\dfrac{x}{3} + 3\right)^5$ **33.** $\left(\dfrac{x}{2} + 2\right)^6$ **34.** $\left(\dfrac{2}{3} - a^2\right)^4$

■ *Write the first four terms in each expansion. Do not simplify the terms.*

Example $(x + 2y)^{15}$

Solution $$x^{15} + \frac{15}{1!}x^{14}(2y) + \frac{15 \cdot 14}{2!}x^{13}(2y)^2 + \frac{15 \cdot 14 \cdot 13}{3!}x^{12}(2y)^3$$

35. $(x + y)^{20}$ **36.** $(x - y)^{15}$ **37.** $(a - 2b)^{12}$

38. $(2a - b)^{12}$ **39.** $(x - \sqrt{2})^{10}$ **40.** $\left(\dfrac{x}{2} + 2\right)^8$

■ *Find each specified term.*

Example $(x - 2y)^{12}$,   seventh term

Solution Use expression (3) on page 381. Set $n = 12$ and $r = 7$.

$$\frac{12 \cdot 11 \cdot 10 \cdot 9 \cdot 8 \cdot 7}{6 \cdot 5 \cdot 4 \cdot 3 \cdot 2 \cdot 1}x^6(-2y)^6 = 59{,}136x^6y^6$$

41. $(a - b)^{15}$,   sixth term

42. $(x + 2)^{12}$,   fifth term

43. $(x - 2y)^{10}$,   fifth term

44. $(a^3 - b)^9$,   seventh term

45. $(x^2 - y^2)^7$,   third term

46. $\left(x - \dfrac{1}{2}\right)^8$,   fourth term

47. $\left(\dfrac{a}{2} - 2b\right)^9$,   fifth term

48. $\left(\dfrac{x}{2} + 4\right)^{10}$,   eighth term

B **49.** Given that the binomial formula holds for $(1 + x)^n$, where n is a negative integer:

   a. Write the first four terms of $(1 + x)^{-1}$.

   b. Find the first four terms of the quotient $\dfrac{1}{1 + x}$ by dividing $(1 + x)$ into 1.

   c. Compare the results of parts a and b.

50. Given that the binomial formula holds as an infinite "sum" for $(1 + x)^n$, where n is a noninteger rational number and $|x| < 1$, find to two decimal places:

   a. $\sqrt{1.02}$ **b.** $\sqrt{0.99}$

11.6

PERMUTATIONS

An arrangement of the elements of a set in specified order (where no element is repeated) is called a **permutation** of the elements. For example, the possible permutations of the elements of $\{a, b, c\}$ are

$$abc, \quad bac, \quad cab, \quad acb, \quad bca, \quad \text{and} \quad cba.$$

To count the permutations of a given set containing n elements, think of placing them in order as follows:

1. Select one member to be first. There are n possible selections.

2. Having selected a first element, select a second. Since one element has been placed in the first position, there remain $(n - 1)$ elements to be placed in the second position. The total number of possible first and second selections is then given by the product $n(n - 1)$.

3. Select a third element. Since one has been used for the first position and one for the second, there are $(n - 2)$ possibilities for the third. The total number of possible first, second, and third selections is $n(n - 1)(n - 2)$.

4. Continue this process until the last element is put in the last position. There are then $n(n - 1)(n - 2) \cdot \cdots \cdot 3 \cdot 2 \cdot 1$ possible permutations of n elements.

Formally, we represent the number of permutations of n things by means of the symbol $P(n, n)$ (read "the number of permutations of n things taken n at a time"). From the foregoing discussion, it follows that:

$$P(n, n) = n! \tag{1}$$

For example, suppose we wish to determine how many different signals can be formed with 5 different signal flags on a pole by altering the location of the flags. A helpful way to picture the situation is to first draw 5 dashes to represent the positions of the flags:

$$— \quad — \quad — \quad — \quad —.$$

Next, since we can place any of the 5 flags in the first position, we write 5 on the first dash:

$$\underline{5} \quad — \quad — \quad — \quad —.$$

Having selected 1 flag for the first position, we have a choice from among 4 flags for the second position.

$$\underline{5} \quad \underline{4} \quad — \quad — \quad —.$$

Similarly, 3 flags remain for the third position, 2 for the fourth, and 1 for the fifth. Thus, we fill the dashes:

$$\underline{5} \quad \underline{4} \quad \underline{3} \quad \underline{2} \quad \underline{1}.$$

Then, the total number of different signals is

$$5 \cdot 4 \cdot 3 \cdot 2 \cdot 1 = 120.$$

Counting principle

Counting permutations of distinct things in this way makes use of a more general counting principle, namely:

▶ *If one thing can be done in m ways, another thing in n ways, another in p ways, and so on, then the number of ways all the things can be done is*

$$m \cdot n \cdot p \cdot \cdots .$$

Example

The number of different license plates containing four-digit numerals using the digits 1, 2, 3, 4, 5, 6, 7, 8, and 9, is given by

$$9 \cdot 9 \cdot 9 \cdot 9 = 6561,$$

since nothing prevents the use of the same digit in two, three, or four positions.

Frequently, it is important to be able to compute the number of possible permutations of a set contained in a given set. Thus, we might wish to know the number of permutations of n different things taken r $(r \leq n)$ at a time. The symbol representing this number is $P(n, r)$, and, by reasoning in exactly the same way as we did to compute $P(n, n)$, we find that:

▶ $$P(n, r) = n(n - 1)(n - 2) \cdot \cdots \cdot (n - r + 1). \tag{2}$$

Now, because

$$n! = n(n - 1)(n - 2) \cdot \cdots \cdot (n - r + 1)(n - r)!,$$

Equation (2) can be written simply as

▶
$$P(n, r) = \frac{n!}{(n - r)!}.$$ (3)

Example   Find the number of permutations of 5 things taken 3 at a time using Equation (3).

Solution
$$P(5, 3) = \frac{5!}{2!} = \frac{5 \cdot 4 \cdot 3 \cdot 2 \cdot 1}{2 \cdot 1}$$
$$= 5 \cdot 4 \cdot 3 = 60$$

Again, if we wish to use dashes to help visualize the situation, we could first draw 3 dashes,

— — —

and then fill each dash in order,

5   4   3.
— — —

The product of these numbers then gives us the total number of possibilities for permuting 5 things taken 3 at a time.

Distin-   Sometimes the elements with which we wish to form permutations are not all different.
guishable   Thus, to find the number of distinguishable permutations of the letters in the word
permutations *toast*, we must take into consideration the fact that we cannot distinguish between the 2
t's in any permutation. The number of permutations of the 5 letters in the word is clearly 5!, but since in any one of these the 2 *t*'s can be permuted in 2! ways without producing a different result, the number of *distinguishable* permutations P is given by

$$2!P = 5!.$$

Hence,

$$P = \frac{5!}{2!},$$

where the number of permutations of five letters, 5!, is divided by the number of permutations, 2!, of the repeat letter *t*. Thus,

$$P = \frac{5!}{2!} = \frac{5 \cdot 4 \cdot 3 \cdot 2 \cdot 1}{2 \cdot 1} = 60.$$

Example   Find the number of distinguishable permutations of the letters in the word *pineapple*.

Solution   Here there is a total of 9! permutations. However, there are 3 *p*'s and 2 *e*'s which can be permuted 3! and 2! ways, respectively, in each permutation of the 9 letters without altering the result. Accordingly, the number of distinguishable permutations P is given by

$$P = \frac{9!}{3!2!} = \frac{9 \cdot 8 \cdot 7 \cdot 6 \cdot 5 \cdot 4 \cdot 3 \cdot 2 \cdot 1}{3 \cdot 2 \cdot 1 \cdot 2 \cdot 1} = 30{,}240.$$

EXERCISE 11.6

A

Example

In how many ways can a jury of 12 persons be seated in a jury box containing 12 chairs?

Solution

We want to find $P(12, 12)$.

$$P(12, 12) = 12!$$
$$= 12 \cdot 11 \cdot 10 \cdot 9 \cdot 8 \cdot 7 \cdot 6 \cdot 5 \cdot 4 \cdot 3 \cdot 2 \cdot 1 = 479{,}001{,}600$$

1. In how many different ways can 4 books be arranged between bookends?
2. In how many ways can 7 students be seated at 7 desks?
3. In how many ways can 9 players be assigned to the 9 positions on a baseball team?
4. In how many ways can 10 floats be arranged for a parade?

Examples

How many different three-digit numerals for whole numbers can be formed from the digits 0, 1, 2, 3, 4, 5, 6, if:

a. No restrictions are placed on the repetition of digits?

b. No digit can be used more than once?

c. The last digit is 6 and no digit can be used more than once?

Solutions

a. Since the numeral must contain 3 digits, the first digit cannot be 0. Drawing 3 dashes, we see that the first place can be filled with any 1 of the digits from 1 to 6, a total of 6 digits:

$$\underline{6} \quad \underline{\quad} \quad \underline{\quad}.$$

The remaining places can be filled with any of the 7 digits; so we have

$$\underline{6} \quad \underline{7} \quad \underline{7}.$$

Then the answer to the original question is

$$6 \cdot 7 \cdot 7 = 294.$$

Solutions continued overleaf

b. If no digit can be used more than once, then we can select any one of the digits from 1 to 6 for the first digit, but, having fixed the first digit, we have 6 possibilities for the second, and 5 for the third. Our diagram appears as follows:

$$\underline{6} \quad \underline{6} \quad \underline{5}.$$

The answer to the original question then is

$$6 \cdot 6 \cdot 5 = 180.$$

c. If the last digit is specified, in this case 6, our diagram begins like this:

$$\underline{} \quad \underline{} \quad \underline{1}.$$

Next, having fixed one digit, and being unable to use 0 as the first digit, we have

$$\underline{5} \quad \underline{5} \quad \underline{1}.$$

The answer to the original question is then

$$5 \cdot 5 \cdot 1 = 25.$$

■ *How many different four-digit numerals for whole numbers can be formed from the digits* 0, 1, 2, 3, 4, 5, 6, *if:*

5. No restrictions are placed on the repetition of digits?

6. No restrictions are placed on the repetition of digits and the last digit is 3?

7. No digit can be used more than once in each number?

8. No digit may be used more than once in each number, and the last digit is 5?

■ *Find the number of three-digit numerals, using the digits* 0, 1, 2, 3, 4, 5, 6, 7, 8 *for the problem cited.*

9. Problem 5   **10.** Problem 6   **11.** Problem 7   **12.** Problem 8

13. How many three-letter permutations can be formed from the twenty-six letter English alphabet if no restrictions are placed on the repetition of letters?

14. How many five-letter permutations can be formed from the twenty-six letter English alphabet if no restrictions are placed on the repetition of letters?

15. How many license numbers can be formed if each license contains 2 letters of the alphabet followed by 3 digits?

16. How many license numbers can be formed if each license contains 1 digit followed by 1 letter of the alphabet followed by 2 digits?

■ *How many distinguishable permutations are there of the letters of the given word?*

Example   Seventeen

Solution   There are 9 letters in the word, but there are 2 n's and 4 e's. Hence, the number of distinguishable permutations is given by

$$P = \frac{9!}{2!4!} = \frac{9 \cdot 8 \cdot 7 \cdot 6 \cdot 5 \cdot 4 \cdot 3 \cdot 2 \cdot 1}{2 \cdot 4 \cdot 3 \cdot 2} = 7560.$$

17. Hurry   **18.** Sonnet   **19.** Stress   **20.** Between

21. Committee   **22.** Consists   **23.** Selected   **24.** Permutation

25. How many distinguishable arrangements can be made by aligning five identically shaped flags, two blue, two red, and one white?

26. How many distinguishable arrangements can be made by aligning seven identically shaped flags, two blue, two red, and three white?

11.7

COMBINATIONS

If the order in which the elements of a set are listed is unimportant, then the set is called a **combination**. Thus, while abc, acb, bac, bca, cab, and cba are six permutations of the elements of $\{a, b, c\}$, the set constitutes a *single* combination, abc. Selecting the elements two at a time would yield the following:

permutations   combinations

ab
 ab
ba

ac
 ac
ca

bc
 bc
cb

The symbol $\binom{n}{r}$, or sometimes $C(n, r)$, is used to designate the number of combinations of r elements that can be formed from a set of n elements. Thus, in the example above

$$P(3, 2) = 6 \quad \text{and} \quad \binom{3}{2} = 3.$$

We can use permutations to count combinations. Since each r-element combination can be permuted in $P(r, r)$, or $r!$, ways, the total number of permutations of n things taken r at a time can be computed by multiplying $P(r, r)$ by $\binom{n}{r}$. That is,

$$P(r, r)\binom{n}{r} = P(n, r).$$

Multiplying both members by $1/P(r, r)$, we have the following:

$$\binom{n}{r} = \frac{P(n, r)}{P(r, r)}. \tag{1}$$

Since $P(n, r) = n(n-1)(n-2) \cdot \cdots \cdot (n-r+1)$ and $P(r, r) = r!$, it follows that:

$$\binom{n}{r} = \frac{n(n-1)(n-2) \cdot \cdots \cdot (n-r+1)}{r!}. \tag{2}$$

For example, the number of different committees of 3 people that could be appointed in a club having 8 members is the number of combinations of 8 things taken 3 at a time, rather than the number of permutations of 8 things taken 3 at a time, because the order in which the members of a committee are considered is of no consequence. Thus, from Equation (2) we would have

$$\binom{8}{3} = \frac{8 \cdot 7 \cdot 6}{3 \cdot 2 \cdot 1} = 56,$$

and 56 such committees could be appointed in the club. Notice that in computing $\binom{n}{r}$, it is easier to first write $r!$ in the denominator, because we can then simply write the same number of factors in the numerator, rather than actually determining $(n-r+1)$ as the final factor of the numerator.

In Section 11.6 we observed that

$$P(n, r) = \frac{n!}{(n-r)!}$$

Substituting the right-hand member of this equation for $P(n, r)$ in Equation (1) above, we have

$$\binom{n}{r} = \frac{P(n, r)}{P(r, r)} = \frac{\dfrac{n!}{(n-r)!}}{r!},$$

from which we have the formula:

$$\binom{n}{r} = \frac{n!}{r!(n-r)!}. \tag{3}$$

Example

Using Equation (3), we have

$$\binom{8}{3} = \frac{8!}{3!(8-3)!} = \frac{8!}{3!5!} = 56.$$

As expected, the result in the above example is the same result we obtained using Equation (2).

EXERCISE 11.7

A **1.** How many different committees of 4 people can be appointed in a club containing 15 members?

2. How many different committees of 5 people can be appointed from a group containing 9 people?

Example How many different amounts of money can be formed from a penny, a nickel, a dime, and a quarter?

Solution Using 1 coin, we can form $\binom{4}{1} = 4$ different amounts; using 2 coins, we obtain $\binom{4}{2} = \dfrac{4 \cdot 3}{1 \cdot 2} = 6$; using 3 coins, we have $\binom{4}{3} = \dfrac{4 \cdot 3 \cdot 2}{1 \cdot 2 \cdot 3} = 4$; using all the coins, we obtain $\binom{4}{4} = \dfrac{4 \cdot 3 \cdot 2 \cdot 1}{1 \cdot 2 \cdot 3 \cdot 4} = 1$; so the total number of different amounts of money is

$$\binom{4}{1} + \binom{4}{2} + \binom{4}{3} + \binom{4}{4} = 4 + 6 + 4 + 1 = 15.$$

3. How many different amounts of money can be formed from a nickel, a dime, and a quarter?

4. How many different amounts of money can be formed from a nickel, a dime, a quarter, a penny, and a half-dollar?

5. In how many different ways can a set of 5 cards be selected from a deck of 52 cards?

6. In how many different ways can a set of 13 cards be selected from a deck of 52 cards?

7. How many straight lines are determined by 6 points, no 3 of which are collinear?

8. How many diagonals does a regular octagon (8 sides) have?

9. In how many ways can 3 marbles be drawn from an urn containing 9 marbles?

10. In how many ways can 2 face cards be drawn from a standard deck of 52 cards?

11. In how many ways can a bridge hand consisting of 6 hearts, 3 spades, and 4 diamonds be selected from a deck of 52 cards?

12. In how many ways can a hand consisting of 4 aces and 1 card that is not an ace be selected from a deck of 52 cards?

13. In how many ways can a committee of 2 men and 3 women be selected from a club with 10 men and 11 women members?

14. In how many ways can a committee containing 8 men and 8 women be selected from the club in Problem 13?

B **15.** Use Equation (2) or (3) on page 390 to show that $\binom{n}{r} = \binom{n}{n-r}$.

16. Use the results of Problem 15 to compute $\binom{100}{98}$.

17. Given $\dbinom{n}{3} = \dbinom{n}{4}$, find n.

18. Given $\dbinom{n}{7} = \dbinom{n}{5}$, find n.

19. Show that the coefficients of the terms in the binomial expansion of $(a+b)^4$ are $\dbinom{4}{0}, \dbinom{4}{1}, \dbinom{4}{2}, \dbinom{4}{3},$ and $\dbinom{4}{4}$.

20. Show that the coefficients of the first five terms in the binomial expansion of $(a+b)^n$ are $\dbinom{n}{0}, \dbinom{n}{1}, \dbinom{n}{2}, \dbinom{n}{3},$ and $\dbinom{n}{4}$.

CHAPTER SUMMARY

[11.1]   The elements in the range of a function whose domain is a set of successive positive integers form a **sequence**. The sequence is **finite** if it has a last member; otherwise it is **infinite**.

A **series** is the indicated sum of the terms in a sequence. We can use **sigma**, or **summation**, notation to represent a series.

[11.2]   A sequence in which each term after the first is obtained by adding a constant to the preceding term is an **arithmetic progression**. The constant is called the **common difference** of the terms.

The nth term of an arithmetic progression is given by

$$s_n = a + (n-1)d,$$

and the sum of n terms is given by

$$S_n = \frac{n}{2}(a + s_n) = \frac{n}{2}[2a + (n-1)d],$$

where a is the first term, n is the number of terms, and d is the common difference.

[11.3]   A sequence in which each term after the first is obtained by multiplying its predecessor by a constant is called a **geometric progression**. The constant is called the **common ratio**.

The nth term of a geometric progression is given by

$$s_n = ar^{n-1},$$

and the sum of n terms is given by

$$S_n = \frac{a - ar^n}{1 - r} = \frac{a - rs_n}{1 - r} \qquad (r \neq 1),$$

where a is the first term, n is the number of terms, and r is the common ratio.

[11.4] An infinite geometric series has a sum if the common ratio has an absolute value less than 1. This sum is given by

$$S_\infty = \lim_{n \to \infty} S_n = \frac{a}{1-r}.$$

The number e is defined by

$$\lim_{t \to \infty} \left(1 + \frac{1}{t}\right)^t.$$

[11.5] Factorial notation is convenient to represent special kinds of products. For n a natural number,

$$n! = n(n-1)(n-2) \cdot \cdots \cdot (3)(2)(1)$$
$$= n(n-1)!.$$

The binomial power $(a+b)^n$ can be expanded into a series containing $(n+1)$ terms for n a natural number:

$$(a+b)^n = a^n + \frac{n}{1!}a^{n-1}b + \frac{n(n-1)}{2!}a^{n-2}b^2 + \frac{n(n-1)(n-2)}{3!}a^{n-3}b^3 + \cdots$$
$$+ \frac{n(n-1)(n-2) \cdot \cdots \cdot (n-r+2)}{(r-1)!}a^{n-r+1}b^{r-1} + \cdots + b^n,$$

where r is the number of the term.

[11.6] An arrangement of the elements in a set (in which no element is repeated) is called a **permutation** of the elements. To count permutations, we can use the general counting principle:

> *If one thing can be done in m ways, another thing in n ways, another in p ways, and so on, then the number of ways all the things can be done is* $m \cdot n \cdot p \cdot \cdots$.

For n, r natural numbers $(r \leq n)$,

$$P(n, n) = n!$$

and

$$P(n, r) = n(n-1)(n-2) \cdot \cdots \cdot (n-r+1)$$
$$= \frac{n!}{(n-r)!}.$$

[11.7] Any r-element set contained in an n-element set is called a **combination**. For n, r natural numbers,

$$\binom{n}{r} = \frac{n(n-1)(n-2) \cdot \cdots \cdot (n-r+1)}{r!} = \frac{n!}{r!(n-r)!}.$$

The symbols introduced in this chapter are listed on the inside of the front cover.

REVIEW EXERCISES

A

[11.1]   1. Find the first four terms in a sequence with the general term $s_n = \dfrac{(-1)^{n-1}}{n}$.

2. Write $\displaystyle\sum_{k=2}^{5} k(k-1)$ in expanded form.

[11.2]   3. Given that 5 and 9 are the first two terms of an arithmetic progression, find an expression for the general term.

4. **a.** Find the twenty-third term of the arithmetic progression $-82, -74, -66, \ldots$.

   b. Find the sum of the first twenty-three terms.

5. The first term of an arithmetic progression is 8 and the twenty-eighth term is 89. Find the twenty-first term.

[11.3]   6. Given that 5 and 9 are the first two terms of a geometric progression, find an expression for the general term.

7. **a.** Find the eighth term of the geometric progression $\frac{16}{27}, -\frac{8}{9}, \frac{4}{3}, \ldots$.

   b. Find the sum of the first four terms.

8. Find $\displaystyle\sum_{j=1}^{5} \left(\frac{1}{3}\right)^j$.

[11.4]   9. Find $\displaystyle\sum_{i=1}^{\infty} \left(\frac{1}{3}\right)^i$.

10. Find a fraction equivalent to $0.44\overline{4}$.

[11.5]   11. Simplify $\dfrac{8!}{3!5!}$.

12. Write the first four terms of the binomial expansion of $(x - 2y)^{10}$.

13. Find the eighth term in the expansion of $(x - 2y)^{10}$.

[11.6]   14. How many different three-digit numerals for whole numbers can be formed from the digits 1, 2, 3, 4, 5 when:

   a. No restrictions are placed on the repetition of digits?

   b. No digit can be used more than once?

15. In how many ways can 6 men be assigned to a 4-man relay team?

16. How many distinguishable permutations are there in the letters of the word *fifteen*?

[11.7]   17. How many different doubles combinations could be formed from a tennis team containing 6 members?

18. How many different committees of 4 persons can be formed from a group of 9 persons?

19. In how many ways can 1 ace and 3 kings be selected from a deck of 52 cards?

20. In how many ways can 3 aces and 2 kings be selected from a deck of 52 cards?

B   **21.** Find the sum of all integer multiples of 3 between 17 and 97.

22. Find three numbers that form an arithmetic sequence such that their sum is 15 and their product is 80.

23. The bacteria in a culture doubles every 4 hours. If there were 2 bacteria in the culture originally, how many are there at the end of n hours?

24. A ball returns three-fourths of its preceding height on each bounce. If the ball is dropped from a height of 8 feet, approximately what is the total distance the ball travels before coming to rest?

■ *Simplify each expression.*

25. $\dfrac{2n!(n+3)}{(n+3)!}$

26. $\dfrac{(2n-1)!(n+1)!}{n!(2n-3)!}$

A. SYNTHETIC DIVISION; POLYNOMIAL AND RATIONAL FUNCTIONS

A.1 Synthetic division

A.2 Remainder theorem; polynomial functions

A.3 Rational functions

A.1

SYNTHETIC DIVISION

In Section 3.3, we rewrote quotients of polynomials of the form $P(x)/D(x)$ using a long division algorithm (process). If the divisor $D(x)$ is of the form $(x - a)$—a first-degree polynomial where the coefficient of x is 1 and $a \neq 0$—this algorithm may be simplified by a procedure known as **synthetic division**. Consider the quotient

$$\frac{x^4 + x^2 + 2x - 1}{x + 3}.$$

The division can be accomplished as follows:

$$
\begin{array}{r}
x^3 - 3x^2 + 10x - 28 \\
x + 3 \,\overline{\smash{\big)}\, x^4 + 0x^3 + x^2 + 2x - 1} \\
\underline{x^4 + 3x^3} \\
-3x^3 + x^2 \\
\underline{-3x^3 - 9x^2} \\
10x^2 + 2x \\
\underline{10x^2 + 30x} \\
-28x - 1 \\
\underline{-28x - 84} \\
83 \quad \text{(remainder)}.
\end{array}
$$

We see that, for $x \neq -3$,

$$\frac{x^4 + x^2 + 2x - 1}{x + 3} = x^3 - 3x^2 + 10x - 28 + \frac{83}{x + 3}$$

or

$$x^4 + x^2 + 2x - 1 = (x^3 - 3x^2 + 10x - 28)(x + 3) + 83.$$

If we omit the variables, writing only the coefficients of the terms, and use 0 for the coefficient of any missing power, we have

$$
\begin{array}{r}
1 - 3 + 10 \; -28 \\
\hline
1 + 3 \overline{\smash{\big)}\, 1 + 0 + \;\; 1 + \;\; 2 \; - \;\; 1} \\
\underline{1 + 3 } \\
-3 + (1) \\
\underline{-3 - \;\; 9 } \\
10 + (2) \\
\underline{10 + 30 } \\
-28 - (1) \\
\underline{-28 - 84} \\
83 \quad \text{(remainder).}
\end{array}
$$

Now, observe that the numbers shown in color are repetitions of the numbers written immediately above and are also repetitions of the coefficients of the associated variable in the quotient. The numbers shown in parentheses are repetitions of the coefficients of the dividend. Therefore, the whole process can be written in compact form as

$$
\begin{array}{c|rrrrr}
3 & 1 & 0 & 1 & 2 & -1 \\
  &   & 3 & -9 & 30 & -84 \\
\hline
  & 1 & -3 & 10 & -28 & 83
\end{array}
\quad \text{(remainder: 83),}
$$

     (1)
     (2)
     (3)

where the repetitions are omitted and where 1, the coefficient of x in the divisor, has also been omitted.

The numbers in line (3), which are the coefficients of the variables in the quotient and the remainder, have been obtained by *subtracting* the **detached coefficients** in line (2) from the detached coefficients of terms of the same degree in line (1). We could obtain the same result by replacing 3 with -3 in the divisor and *adding* instead of subtracting at each step. This is what is done in the *synthetic division* process. The final form then is

$$
\begin{array}{c|rrrrr}
-3 & 1 & 0 & 1 & 2 & -1 \\
   &   & -3 & 9 & -30 & 84 \\
\hline
   & 1 & -3 & 10 & -28 & 83
\end{array}
\quad \text{(remainder: 83),}
$$

     (1)
     (2)
     (3)

Note that the numbers in line (2) can be obtained by multiplying the preceding number to the left in line (3) by -3.

Comparing the results of using synthetic division with the same process using long division, we observe that the numbers in line (3) are the coefficients of the polynomial

$$x^3 - 3x^2 + 10x - 28,$$

and that there is a remainder of 83.

As another example, let us write the quotient

$$\frac{3x^3 - 4x - 1}{x - 2}$$

in the form $Q(x) + r/(x - a)$. Using synthetic division, we begin by writing

$$\underline{2 \rfloor \quad 3 \qquad 0 \qquad -4 \qquad -1}$$

where 0 has been inserted in the position corresponding to the coefficient of a second-degree term. The divisor is the negative of -2, or 2. Then, we have

$$
\begin{array}{rrrrl}
2 \rfloor \ 3 & 0 & -4 & -1 & \qquad\qquad\qquad (1) \\
& 6 & 12 & 16 & \qquad\qquad\qquad (2) \\
\hline
3 & 6 & 8 & 15 & \text{(remainder: 15).} \qquad (3)
\end{array}
$$

This process employs these steps:

1. 3 is "brought down" from line (1) to line (3).

2. 6, the product of 2 and 3, is written in the next position on line (2).

3. 6, the sum of 0 and 6, is written on line (3).

4. 12, the product of 2 and 6, is written in the next position on line (2).

5. 8, the sum of -4 and 12, is written on line (3).

6. 16, the product of 2 and 8, is written in the next position on line (2).

7. 15, the sum of -1 and 16, is written on line (3).

We can use the first three numbers on line (3) as coefficients to write a polynomial of degree one less than the degree of the dividend. This polynomial is the quotient lacking the remainder. The last number is the remainder. Thus, for $x - 2 \neq 0$, the quotient of $3x^3 - 4x - 1$ divided by $x - 2$ is

$$3x^2 + 6x + 8$$

with a remainder of 15; that is,

$$\frac{3x^3 - 4x - 1}{x - 2} = 3x^2 + 6x + 8 + \frac{15}{x - 2} \qquad (x \neq 2).$$

EXERCISE A.1

A ▪ *Use synthetic division to write each quotient P(x)/(x − a) in the form Q(x) or Q(x) + r/(x − a), where r is a constant.*

Examples **a.** $\dfrac{2x^4 + x^3 - 1}{x + 2}$ **b.** $\dfrac{x^4 - 2x^2 + 3}{x - 2}$

Solutions **a.** Begin by writing

$$-2 \rfloor \quad 2 \quad 1 \quad 0 \quad 0 \quad -1$$

where 0 is the coefficient of any missing power. The divisor is the negative of 2, i.e., −2. Use the following steps.

b. Begin by writing

$$2 \rfloor \quad 1 \quad 0 \quad -2 \quad 0 \quad 3$$

where 0 is the coefficient of any missing power. The divisor is the negative of −2, i.e., 2. Use the following steps.

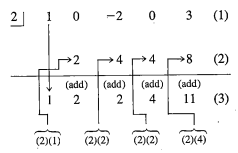

Use the first four numbers on line (3) as the coefficients of Q(x). The last number is the remainder:

$$2x^3 - 3x^2 + 6x - 12 + \frac{23}{x + 2} \quad (x \neq -2)$$

Use the first four numbers on line (3) as the coefficients of Q(x). The last number is the remainder:

$$x^3 + 2x^2 + 2x + 4 + \frac{11}{x - 2} \quad (x \neq 2)$$

1. $\dfrac{x^2 - 8x + 12}{x - 6}$ **2.** $\dfrac{a^2 + a - 6}{a + 3}$ **3.** $\dfrac{x^2 + 4x + 4}{x + 2}$

4. $\dfrac{x^2 + 6x + 9}{x + 3}$ **5.** $\dfrac{x^4 - 3x^3 + 2x^2 - 1}{x - 2}$ **6.** $\dfrac{x^4 + 2x^2 - 3x + 5}{x - 3}$

7. $\dfrac{2x^3 + x - 5}{x + 1}$ **8.** $\dfrac{3x^3 + x^2 - 7}{x + 2}$ **9.** $\dfrac{2x^4 - x + 6}{x - 5}$

10. $\dfrac{3x^4 - x^2 + 1}{x - 4}$ **11.** $\dfrac{x^3 + 4x^2 + x - 2}{x + 2}$ **12.** $\dfrac{x^3 - 7x^2 - x + 3}{x + 3}$

13. $\dfrac{x^6 + x^4 - x}{x - 1}$ **14.** $\dfrac{x^6 + 3x^3 - 2x - 1}{x - 2}$ **15.** $\dfrac{x^5 - 1}{x - 1}$

16. $\dfrac{x^5 + 1}{x + 1}$ **17.** $\dfrac{x^6 - 1}{x - 1}$ **18.** $\dfrac{x^6 + 1}{x + 1}$

A.2

REMAINDER THEOREM; POLYNOMIAL FUNCTIONS

Remainder theorem

In Section 3.3, we observed that the quotient

$$\frac{x^2 + 2x + 2}{x + 1}$$

can be expressed as

$$x + 1 + \frac{1}{x + 1} \qquad (x \ne -1).$$

In general, we have that

$$\frac{P(x)}{x - a} = Q(x) + \frac{r}{x - a}, \tag{1}$$

as indicated in Section A.1, or

$$P(x) = (x - a)Q(x) + r, \tag{2}$$

where $P(x)$ is a polynomial of degree $n \ge 1$ with real coefficients, $Q(x)$ is a polynomial of degree $(n - 1)$ with real coefficients, and r is a real number.

Now in (2), if $x = a$, we have

$$P(a) = (a - a)Q(a) + r$$
$$= 0 \cdot Q(a) + r$$
$$= r.$$

This result, called the **Remainder Theorem**, is stated as follows:

► *If $P(x)$ is divided by $(x - a)$, the remainder is $P(a)$.*

Since synthetic division offers a means of finding values of $r = P(a)$, we can sometimes find such values more quickly by synthetic division than by direct substitution. For example, if $P(x) = 2x^3 - 3x^2 + 2x + 1$, then we can find $P(2)$ by synthetically dividing $2x^3 - 3x^2 + 2x + 1$ by $x - 2$:

$$
\begin{array}{r|rrrr}
2 & 2 & -3 & 2 & 1 \\
  &   & 4 & 2 & 8 \\
\hline
  & 2 & 1 & 4 & 9
\end{array}
$$

By inspection, we note that $r = P(2) = 9$.

Graphs of polynomial functions

In Section 7.3, we graphed linear functions defined by

$$f(x) = a_0 x + a_1,$$

and in Sections 8.1 and 8.2, we graphed quadratic functions defined by

$$f(x) = a_0x^2 + a_1x + a_2.$$

We can graph any polynomial function defined by

$$f(x) = a_0x^n + a_1x^{n-1} + \cdots + a_n,$$

where $a_0, a_1, \ldots, a_n$, and x are real numbers, by obtaining a number of solutions (ordered pairs) sufficient to determine the behavior of its graph. We can obtain the ordered pairs $(x, f(x))$ by direct substitution of arbitrary values of x, as we did earlier in this book, or by synthetic division. In this appendix we shall use synthetic division.

For example, we can obtain solutions of

$$P(x) = x^3 - 2x^2 - 5x + 6$$

for selected values of x, say, -3, -2, -1, 0, 1, 2, 3, and 4, by dividing $x^3 - 2x^2 - 5x + 6$ by $x + 3$, $x + 2$, and so on. Dividing synthetically by $x + 3$, we have

$$
\begin{array}{r|rrrr}
-3 & 1 & -2 & -5 & 6 \\
   &   & -3 & 15 & -30 \\
\hline
   & 1 & -5 & 10 & -24
\end{array}
$$

and $(-3, -24)$ is a solution of (3). Dividing by $x + 2$, we have

$$
\begin{array}{r|rrrr}
-2 & 1 & -2 & -5 & 6 \\
   &   & -2 & 8 & -6 \\
\hline
   & 1 & -4 & 3 & 0
\end{array}
$$

and $(-2, 0)$ is a solution of (3). Dividing by $x + 1$, we have

$$
\begin{array}{r|rrrr}
-1 & 1 & -2 & -5 & 6 \\
   &   & -1 & 3 & 2 \\
\hline
   & 1 & -3 & -2 & 8
\end{array}
$$

and $(-1, 8)$ is a solution of (3). Similarly, we find that $(0, 6)$, $(1, 0)$, $(2, -4)$, $(3, 0)$, and $(4, 18)$ are solutions, and their graphs are on the graph of the function.

The points, shown in Figure A.1a on page 402, suggest the appearance of the graph and enable us to complete the graph as shown in Figure A.1b. Any additional values of x less than -3 and greater than 4 would not change the general appearance of the graph. Note that the graph of this third-degree polynomial function changes direction twice. In fact, it can be shown, although we will not do so, that *the maximum number of direction changes of the graph of a polynomial function is one less than the degree of the polynomial.*

Observe that the high and low points in the curve in Figure A.1 *appear* to correspond with the ordered pairs $(-1, 8)$ and $(2, -4)$, respectively. This may or may not be the case. To find the exact values of the components of the ordered pairs that are the high and low points of the graphs of polynomial functions of degree greater than 2 requires methods considered only in more advanced courses. However, we can obtain approximations for these points by plotting ordered pairs as demonstrated in the example above.

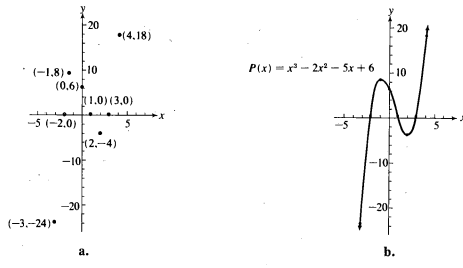

Figure A.1

Factor theorem

We observe from (2) on page 400 that if $r = P(a) = 0$, then

$$P(x) = (x - a) \cdot Q(x) \qquad (4)$$

and $(x - a)$ is a factor of $P(x)$. This result, called the **Factor Theorem**, is stated as follows:

▶ *If* $P(a) = 0$, *then* $(x - a)$ *is a factor of* $P(x)$.

In the example on page 401 we established that $(x + 2)$, $(x - 1)$, and $(x - 3)$ are factors of $x^3 - 2x^2 - 5x + 6$, because in each synthetic division by these factors the remainder was 0.

Number of solutions of $P(x) = 0$

From (4) we note that a is a solution of $P(x) = 0$ if and only if $(x - a)$ is a factor of $P(x)$. This suggests that *an equation* $P(x) = 0$, *where $P(x)$ is of nth degree, has n solutions*. This is indeed the case. For example, we note that the third-degree polynomial $x^3 - 2x^2 - 5x + 6$ is equivalent to $(x + 2)(x - 1)(x - 3)$, and that there are exactly three solutions of

$$x^3 - 2x^2 - 5x + 6 = (x + 2)(x - 1)(x - 3) = 0,$$

namely, -2, 1, and 3.

Of course, it may be that one or more factors of such an expression are the same. When this happens, we count the solution as many times as the factor involved occurs. Thus, if

$$P(x) = x^4 + 2x^3 - 2x - 1$$
$$= (x + 1)(x + 1)(x + 1)(x - 1),$$

we can see that -1 and 1 are the only solutions of $P(x) = 0$, but we say that -1 is a solution of multiplicity three.

EXERCISE A.2

A ■ *Find the designated value.*

Example If $P(x) = 4x^4 - 2x^3 + 3x - 2$, find $P(-1)$ and $P(2)$.

Solution

$$
\begin{array}{r|rrrrr}
-1 & 4 & -2 & 0 & 3 & -2 \\
   &   & -4 & 6 & -6 & 3 \\
\hline
   & 4 & -6 & 6 & -3 & 1
\end{array}
\qquad
\begin{array}{r|rrrrr}
2 & 4 & -2 & 0 & 3 & -2 \\
  &   & 8 & 12 & 24 & 54 \\
\hline
  & 4 & 6 & 12 & 27 & 52
\end{array}
$$

$$P(-1) = 1 \qquad\qquad\qquad P(2) = 52$$

1. If $P(x) = 3x^3 - 2x^2 + 5x - 4$, find $P(3)$ and $P(-2)$.
2. If $P(x) = 4x^4 - 2x^3 + 3x^2 - 5$, find $P(1)$ and $P(-1)$.
3. If $P(x) = 2x^5 - 3x^3 + x^2 - x + 2$, find $P(-1)$ and $P(2)$.
4. If $P(x) = x^4 - 10x^3 + 5x^2 - 3x + 6$, find $P(-2)$ and $P(3)$.

■ *Use synthetic division and the remainder theorem to find solutions to each equation, and then graph the equation.*

5. $y = x^3 + x^2 - 6x$ 6. $y = x^3 + 5x^2 + 4x$
7. $y = x^3 - 2x^2 + 1$ 8. $y = x^3 - 4x^2 + 3x$
9. $y = 2x^3 + 9x^2 + 7x - 6$ 10. $y = x^3 - 3x^2 - 6x + 8$
11. $y = x^4 - 4x^2$ 12. $y = x^4 - x^3 - 4x^2 + 4x$

■ *Use the remainder in synthetic division to determine whether or not the given binomial is a factor of the given polynomial.*

13. $x - 2;\ x^3 - 3x^2 + 2x + 2$ 14. $x - 1;\ 2x^3 - 5x^2 + 4x - 1$
15. $x + 3;\ 3x^3 + 11x^2 + x - 15$ 16. $x + 1;\ 2x^3 - 5x^2 + 3x + 3$

B 17. Verify that 1 is a solution of $x^3 + 2x^2 - x - 2 = 0$, and find the other solutions.
18. Verify that 3 is a solution of $x^3 - 6x^2 - x + 30 = 0$, and find the other solutions.
19. Verify that -3 is a solution of $x^4 - 3x^3 - 10x^2 + 24x = 0$, and find the other solutions.
20. Verify that -5 is a solution of $x^4 + 5x^3 - x^2 - 5x = 0$, and find the other solutions.

A.3

RATIONAL FUNCTIONS

A function defined by an equation of the form

$$y = \frac{P(x)}{Q(x)},$$

where $P(x)$ and $Q(x)$ are polynomials and $Q(x) \neq 0$, is called a **rational function**. We can graph such a function by obtaining a number of solutions (ordered pairs) sufficient to determine its behavior by direct substitution of arbitrary values of x. It is also helpful to first obtain any asymptotes to the curve that may exist.

Vertical asymptotes

Vertical asymptotes can be found by using the following property:

▶ *The graph of the function defined by $y = P(x)/Q(x)$ has a vertical asymptote $x = a$ for each value a at which $Q(x) = 0$ and $P(x) \neq 0$.*

Examples

Find the vertical asymptotes of the graphs of each function.

a. $y = \dfrac{2}{x-2}$ 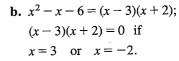 **b.** $y = \dfrac{4}{x^2 - x - 6}$

Solutions

a. $x - 2 = 0$ if

$\quad x = 2.$

b. $x^2 - x - 6 = (x-3)(x+2);$

$\quad (x-3)(x+2) = 0$ if

$\quad x = 3 \quad$ or $\quad x = -2.$

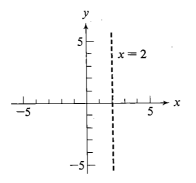

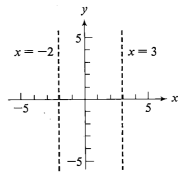

Horizontal asymptotes

The graphs of some rational functions have horizontal asymptotes which can be identified by using the following property:

▶ *If ax^n is the term of highest degree of a polynomial $P(x)$ and bx^m is the term of highest degree of a polynomial $Q(x)$, then the graph of the function $y = P(x)/Q(x)$ has a horizontal asymptote*

$$at \quad y = 0 \quad \quad if \quad n < m;$$
$$at \quad y = a/b \quad if \quad n = m;$$
$$nowhere \quad \quad if \quad n > m.$$

Examples   Determine any horizontal asymptotes of the graphs of each function.

a. $y = \dfrac{3x}{x^2 - 5x + 4}$ **b.** $y = \dfrac{4x^2}{2x^2 - x}$ **c.** $y = \dfrac{x^4 + 1}{x^2 + 2}$

Solutions   **a.** The degree of $3x$ is less than the degree of x^2. Hence, the graph has a horizontal asymptote at $y = 0$.

b. $4x^2$ and $2x^2$ are of the same degree. Hence, the graph has a horizontal asymptote at $y = 4/2 = 2$.

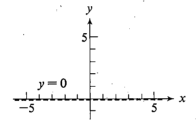
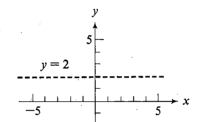

c. The degree of x^4 is greater than the degree of x^2. Hence, the graph does not have a horizontal asymptote.

After vertical and horizontal asymptotes have been found, a few additional ordered pairs associated with points on opposite sides of each asymptote will usually be sufficient to complete the graph.

Example   Graph $y = \dfrac{2}{x - 2}$.

Solution   Because $x - 2 = 0$ at $x = 2$, the vertical asymptote is $x = 2$. Because the degree of the numerator is less than the degree of the denominator, the horizontal asymptote is $y = 0$. We note that if $x = 0$, the y-intercept is -1.

Solution continued overleaf

Several additional ordered pairs,

$$(-2, -\tfrac{1}{2}), \quad (1, -2), \quad (3, 2), \quad \text{and} \quad (4, 1),$$

enable us to complete the graph as we use the asymptotes to direct the branches of the curve.

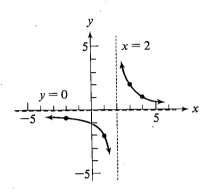

EXERCISE A.3

A ▪ *Determine the vertical asymptotes of the graph of each function.*

1. $y = \dfrac{2}{x+3}$ **2.** $y = \dfrac{1}{x-4}$ **3.** $y = \dfrac{3}{(x-2)(x+3)}$

4. $y = \dfrac{4}{(x+1)(x-4)}$ **5.** $y = \dfrac{2x}{x^2-x-6}$ **6.** $y = \dfrac{2x+1}{x^2-3x+2}$

▪ *Determine any vertical or horizontal asymptotes of the graphs of each function.*

7. $y = \dfrac{x}{x^2-9}$ **8.** $y = \dfrac{2x-4}{x^2+5x+4}$ **9.** $y = \dfrac{x-4}{2x-1}$

10. $y = \dfrac{2x+1}{x-3}$ **11.** $y = \dfrac{2x^2}{x^2-3x-4}$ **12.** $y = \dfrac{x^2}{x^2-x-12}$

▪ *Graph each function after first identifying all asymptotes and intercepts.*

Example $y = \dfrac{2x-4}{x^2-9}$

Solution   Because $x^2 - 9 = (x - 3)(x + 3)$, which is zero for $x = 3$ and $x = -3$, the vertical asymptotes are $x = 3$ and $x = -3$. Because the degree of $2x$ is less than the degree of x^2, the horizontal asymptote is $y = 0$. If $x = 0$, the y-intercept is $\frac{4}{9}$. If $y = 0$, $2x - 4 = 0$; hence, the x-intercept is 2. Several additional ordered pairs,

$$\left(-4, -\frac{12}{7}\right), \qquad \left(-2, \frac{8}{5}\right), \quad \text{and} \quad \left(4, \frac{4}{7}\right),$$

enable us to complete the graph as we use the asymptotes to direct the branches of the curve.

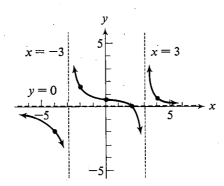

13. $y = \dfrac{1}{x + 3}$

14. $y = \dfrac{1}{x - 3}$

15. $y = \dfrac{2}{(x - 4)(x + 1)}$

16. $y = \dfrac{4}{(x + 2)(x - 1)}$

17. $y = \dfrac{2}{x^2 - 5x + 4}$

18. $y = \dfrac{4}{x^2 - x - 6}$

19. $y = \dfrac{x}{x + 3}$

20. $y = \dfrac{x}{x - 2}$

21. $y = \dfrac{x + 1}{x + 2}$

22. $y = \dfrac{x - 1}{x - 3}$

23. $y = \dfrac{2x}{x^2 - 4}$

24. $y = \dfrac{x}{x^2 - 9}$

25. $y = \dfrac{x - 2}{x^2 + 5x + 4}$

26. $y = \dfrac{x + 1}{x^2 - x - 6}$

27. Graph $xy = k$ for $k = 4$ and $k = 12$ on the same set of axes.

28. Graph $xy = k$ for $k = -4$ and $k = -12$ on the same set of axes.

APPENDIX A SUMMARY

[A.1]   **Synthetic division** is a condensation of the division algorithm using only coefficients.

[A.2]   A quotient of the form $P(x)/(x - a)$, where $P(x)$ is a polynomial of degree $n \geq 1$ with real coefficients, can be expressed in the form $Q(x) + r/(x - a)$, where $Q(x)$ is a polynomial of degree $(n - 1)$ with real coefficients and r is a real number. Furthermore, $r = P(a)$ (**remainder theorem**).

Synthetic division can be used to find values $P(a)$ of a polynomial $P(x)$ in order to determine the behavior of its graph.

If $P(x)/(x - a)$ yields a remainder $r = P(a) = 0$, then $(x - a)$ is a factor of $P(x)$ (**factor theorem**).

An equation of the form $P(x) = 0$, where $P(x)$ is of nth degree, has n solutions.

[A.3]   A function defined by $y = P(x)/Q(x)$, where $P(x)$ and $Q(x)$ are polynomials and $Q(x) \neq 0$, is called a **rational function**.

The process of graphing a rational function can be facilitated by first finding any vertical and/or horizontal asymptotes that exist.

REVIEW EXERCISES

[A.1]   ■ *Use synthetic division to write each quotient as a polynomial in simple form.*

1. $\dfrac{y^3 + 3y^2 - 2y - 4}{y + 1}$ **2.** $\dfrac{y^7 - 1}{y - 1}$

[A.2]   **3.** If $P(x) = 2x^3 - x^2 + 3x + 1$, find $P(2)$ and $P(-2)$.

4. If $P(x) = x^4 + 3x^2 - 2x + 2$, find $P(1)$ and $P(-1)$.

■ *Use synthetic division and the remainder theorem to find solutions to each equation, and then graph the equation.*

5. $y = x^3 + x^2 - 2x$ **6.** $y = x^4 + 2x^3 - 5x^2 - 6x$

7. Is $(x + 2)$ a factor of $3x^3 - 2x^2 - x + 4$?

8. Verify that $(x - 2)$ is a factor of $x^3 - 4x^2 + x + 6$, and find the other factors.

[A.3]   ■ *Graph each function.*

9. $y = \dfrac{1}{x - 4}$ **10.** $y = \dfrac{2}{x^2 - 3x - 10}$

11. $y = \dfrac{x - 2}{x + 3}$ **12.** $y = \dfrac{x - 1}{x^2 - 2x - 3}$

B. MATRICES AND DETERMINANTS

B.1

MATRICES

A **matrix** is a rectangular array of elements or **entries** (in this book, real numbers). These entries are ordinarily displayed using brackets or parentheses (we shall use brackets). Thus,

$$\begin{bmatrix} 1 & 2 & 3 \\ 4 & 5 & 6 \\ 7 & 8 & 9 \end{bmatrix}, \qquad \begin{bmatrix} 2 & -1 & 3 \\ 4 & 0 & 2 \end{bmatrix}, \qquad \begin{bmatrix} 4 \\ 5 \\ 6 \end{bmatrix}$$

are matrices with real-number elements. The **order**, or **dimension**, of a matrix is the ordered pair having as first component the number of (horizontal) rows and as second component the number of (vertical) columns in the matrix. Thus, the matrices above are 3×3 (read "three-by-three"), 2×3 (read "two-by-three"), and 3×1 (read "three-by-one"), respectively. Note that the first matrix—where the number of rows is equal to the number of columns—is an example of a **square** matrix.

Elementary transformations

We can transform one matrix into another in a variety of ways. However, here we are only concerned with the following kinds of transformations:

1. Multiplying the entries of any row by a nonzero real number.

2. Interchanging two rows.

3. Multiplying the entries of any row by a real number and adding the results to the corresponding elements of another row.

409

Such transformations are called **elementary transformations**, and, if a matrix A is transformed into a matrix B by a finite succession of such transformations, then we say that A and B are **row-equivalent**. We represent this by writing $A \sim B$ (read "A is row-equivalent to B"). For example,

$$A = \begin{bmatrix} 1 & 3 & -1 \\ 2 & 1 & 4 \\ 6 & 2 & -1 \end{bmatrix} \quad \text{and} \quad B = \begin{bmatrix} 1 & 3 & -1 \\ 6 & 3 & 12 \\ 6 & 2 & -1 \end{bmatrix}$$

are row-equivalent, because we can multiply each entry in row 2 of A by 3 to obtain B;

$$A = \begin{bmatrix} 3 & -1 & 2 \\ 2 & 1 & 4 \\ 3 & 1 & 9 \end{bmatrix} \quad \text{and} \quad B = \begin{bmatrix} 3 & 1 & 9 \\ 2 & 1 & 4 \\ 3 & -1 & 2 \end{bmatrix}$$

are row-equivalent, because we can interchange rows 1 and 3 of A to obtain B; and

$$A = \begin{bmatrix} 1 & 2 & 1 \\ 2 & 0 & -1 \\ 3 & 1 & 2 \end{bmatrix} \quad \text{and} \quad B = \begin{bmatrix} 1 & 2 & 1 \\ 0 & -4 & -3 \\ 3 & 1 & 2 \end{bmatrix}$$

are row-equivalent, because we can multiply each entry in row 1 of A by -2 and add the results to the corresponding entries in row 2 of A to obtain B.

It is often convenient to perform more than one elementary transformation on a given matrix. For example, if in the matrix

$$\begin{bmatrix} 1 & -2 & 1 \\ 2 & 1 & 3 \\ -3 & 0 & 0 \end{bmatrix}$$

we add $-2 \cdot (\text{row } 1)$ to row 2, and $3 \cdot (\text{row } 1)$ to row 3, we obtain the row-equivalent matrix

$$\begin{bmatrix} 1 & -2 & 1 \\ 0 & 5 & 1 \\ 0 & -6 & 3 \end{bmatrix}.$$

In a system of linear equations of the form

$$a_1 x + b_1 y + c_1 z = d_1$$
$$a_2 x + b_2 y + c_2 z = d_2$$
$$a_3 x + b_3 y + c_3 z = d_3,$$

the matrices

$$\begin{bmatrix} a_1 & b_1 & c_1 \\ a_2 & b_2 & c_2 \\ a_3 & b_3 & c_3 \end{bmatrix} \quad \text{and} \quad \begin{bmatrix} a_1 & b_1 & c_1 & \vdots & d_1 \\ a_2 & b_2 & c_2 & \vdots & d_2 \\ a_3 & b_3 & c_3 & \vdots & d_3 \end{bmatrix}$$

are called the **coefficient matrix** and the **augmented matrix**, respectively.

Solution of linear systems

By performing elementary transformations on the augmented matrix of a system of equations, we can obtain a matrix from which the solution set of the system is readily determined. The validity of the method, as illustrated in the examples below, stems from the fact that performing elementary row transformations on the augmented matrix of a system corresponds to forming equivalent systems of equations.

For example, the augmented matrix of

$$\begin{aligned} x + 2y &= 7 \\ 2x - y &= 4 \end{aligned} \quad \text{is} \quad \begin{bmatrix} 1 & 2 & | & 7 \\ 2 & -1 & | & 4 \end{bmatrix}.$$

Performing elementary transformations, we have:

$$-2(\text{row } 1) + \text{row } 2 \quad \begin{bmatrix} 1 & 2 & | & 7 \\ 0 & -5 & | & -10 \end{bmatrix}.$$

This matrix corresponds to the system

$$\begin{aligned} x + 2y &= 7 \\ -5y &= -10. \end{aligned}$$

From the last equation, $y = 2$. Substituting 2 for y in the first equation, we obtain $x = 3$. Hence, the solution set is $\{(3, 2)\}$.

The next example involves a linear system of three equations in three variables.

Example Solve the system

$$\begin{aligned} x - 3y + z &= -2 \\ 3x + y - z &= 8 \\ 2x - 2y + 3z &= -1. \end{aligned}$$

Solution The augmented matrix of the system is

$$\begin{bmatrix} 1 & -3 & 1 & -2 \\ 3 & 1 & -1 & 8 \\ 2 & -2 & 3 & -1 \end{bmatrix}.$$

Performing elementary transformations, we have:

$$\begin{array}{l} -3(\text{row } 1) + \text{row } 2 \\ -2(\text{row } 1) + \text{row } 3 \end{array} \quad \begin{bmatrix} 1 & -3 & 1 & | & -2 \\ 0 & 10 & -4 & | & 14 \\ 0 & 4 & 1 & | & 3 \end{bmatrix} \quad \begin{array}{l} x - 3y + z = -2 \\ 0x + 10y - 4z = 14 \\ 0x + 4y + z = 3 \end{array}$$

$$\tfrac{1}{2}(\text{row } 2) \quad \begin{bmatrix} 1 & -3 & 1 & | & -2 \\ 0 & 5 & -2 & | & 7 \\ 0 & 4 & 1 & | & 3 \end{bmatrix} \quad \begin{array}{l} x - 3y + z = -2 \\ 0x + 5y - 2z = 7 \\ 0x + 4y + z = 3 \end{array}$$

$$-\tfrac{4}{5}(\text{row } 2) + \text{row } 3 \quad \begin{bmatrix} 1 & -3 & 1 & | & -2 \\ 0 & 5 & -2 & | & 7 \\ 0 & 0 & {}^{13}\!/_5 & | & -{}^{13}\!/_5 \end{bmatrix} \quad \begin{array}{l} x - 3y + z = -2 \\ 0x + 5y - 2z = 7 \\ 0x + 0y + ({}^{13}\!/_5)z = -{}^{13}\!/_5 \end{array}$$

$$5(\text{row } 3) \quad \begin{bmatrix} 1 & -3 & 1 & | & -2 \\ 0 & 5 & -2 & | & 7 \\ 0 & 0 & 13 & | & -13 \end{bmatrix} \quad \begin{array}{l} x - 3y + z = -2 \\ 0x + 5y - 2z = 7 \\ 0x + 0y + 13z = -13 \end{array}$$

Solution continued overleaf

Note that the last matrix corresponds to the system

$$x - 3y + \quad z = -2$$
$$5y - \quad 2z = 7$$
$$13z = -13.$$

Since, from the last equation, $z = -1$, we can substitute -1 for z in the second equation to obtain $y = 1$. Finally, substituting -1 for z and 1 for y in the first equation, we have $x = 2$; so the solution set is $\{(2, 1, -1)\}$.

EXERCISE B.1

A   ■ *Use row transformations on the augmented matrix to solve each system.*

Example $x - 2y = -5$
$2x + 3y = 11$

Solution   Form the augmented matrix.

$$\begin{bmatrix} 1 & -2 & | & -5 \\ 2 & 3 & | & 11 \end{bmatrix}$$

Perform elementary row operations.

$$-2(\text{row 1}) + \text{row 2} \quad \begin{bmatrix} 1 & -2 & | & -5 \\ 0 & 7 & | & 21 \end{bmatrix}$$

$$\tfrac{1}{7}(\text{row 2}) \quad \begin{bmatrix} 1 & -2 & | & -5 \\ 0 & 1 & | & 3 \end{bmatrix}$$

The last matrix corresponds to the system

$$x - 2y = -5$$
$$y = \quad 3.$$

Substitute 3 for y in (1).

$$x - 2(3) = 5; \quad x = 1$$

The solution set is $\{(1, 3)\}$.

1. $x + 3y = 11$ **2.** $x - 5y = 11$ **3.** $x - 4y = -6$
$\quad\;\; 2x - y = 1$ $\quad\;\; 2x + 3y = -4$ $\quad\;\; 3x + y = -5$

4. $x + 6y = -14$ **5.** $2x + y = 5$ **6.** $3x - 2y = 16$
$\quad\;\; 5x - 3y = -4$ $\quad\;\; 3x - 5y = 14$ $\quad\;\; 4x + 2y = 12$

7. $x + 3y - z = 5$
$3x - y + 2z = 5$
$x + y + 2z = 7$

8. $x - 2y + 3z = -11$
$2x + 3y - z = 6$
$3x - y - z = 3$

9. $2x - y + z = 8$
$x - 2y - 3z = 4$
$3x + 3y - z = -4$

10. $x - 2y - 2z = 4$
$2x + y - 3z = 7$
$x - y - z = 3$

11. $2x - y - z = -4$
$x + y + z = -5$
$x + 3y - 4z = 12$

12. $x - 2y - 5z = 2$
$2x + 3y + z = 11$
$3x - y - z = 11$

B.2

LINEAR SYSTEMS IN TWO VARIABLES: SOLUTION BY DETERMINANTS

Associated with each square matrix A that has real-number entries is a real number called the **determinant** of A, which is denoted by δA or $\delta(A)$. The determinant is customarily displayed in the same form as the matrix, but with vertical bars instead of brackets.

2 × 2 determinants

In this section, we consider 2×2 determinants with values defined as follows:

▶ If $A = \begin{bmatrix} a_1 & b_1 \\ a_2 & b_2 \end{bmatrix}$, *then*

$$\delta(A) = \begin{vmatrix} a_1 & b_1 \\ a_2 & b_2 \end{vmatrix} = a_1 b_2 - a_2 b_1.$$

This value is obtained by multiplying the elements on the diagonals and adding the negative of the second product to the first product. The process can be shown schematically as

$$\begin{vmatrix} a_1 & b_1 \\ a_2 & b_2 \end{vmatrix} = a_1 b_2 - a_2 b_1.$$

For example,

$$\begin{vmatrix} 1 & 2 \\ -1 & 3 \end{vmatrix} = 3 - (-2) = 5 \quad \text{and} \quad \begin{vmatrix} 0 & -1 \\ -1 & 7 \end{vmatrix} = 0 - 1 = -1.$$

A determinant is simply another way to represent a single number.

Solutions of linear systems

Determinants can be used to solve linear systems. In this section, we shall confine our attention to linear systems of two equations in two variables of the form

$$a_1 x + b_1 y = c_1 \tag{1}$$

$$a_2 x + b_2 y = c_2. \tag{2}$$

If this system is solved by means of a linear combination, we have, upon multiplication of Equation (1) by $-a_2$ and Equation (2) by a_1, the equations

$$-a_1 a_2 x - a_2 b_1 y = -a_2 c_1 \tag{1a}$$

$$a_1 a_2 x + a_1 b_2 y = a_1 c_2. \tag{2a}$$

The sum of the members of (1a) and (2a) is

$$a_1 b_2 y - a_2 b_1 y = a_1 c_2 - a_2 c_1.$$

Now, factoring y from each term in the left-hand member, we have

$$(a_1 b_2 - a_2 b_1)y = a_1 c_2 - a_2 c_1,$$

from which

$$y = \frac{a_1 c_2 - a_2 c_1}{a_1 b_2 - a_2 b_1} \qquad (a_1 b_2 - a_2 b_1 \neq 0). \tag{3}$$

Now, the numerator of (3) is just the value of the determinant

$$\begin{vmatrix} a_1 & c_1 \\ a_2 & c_2 \end{vmatrix},$$

which we designate as D_y, and the denominator is the value of the determinant

$$\begin{vmatrix} a_1 & b_1 \\ a_2 & b_2 \end{vmatrix},$$

which we designate as D; so Equation (3) can be written:

$$y = \frac{D_y}{D} = \frac{\begin{vmatrix} a_1 & c_1 \\ a_2 & c_2 \end{vmatrix}}{\begin{vmatrix} a_1 & b_1 \\ a_2 & b_2 \end{vmatrix}}. \tag{4}$$

The elements of the determinant in the denominator of (4) are the coefficients of the variables in (1) and (2). The elements of the determinant in the numerator of (4) are identical to those in the denominator, except that *the elements in the column containing the coefficients of y have been replaced by c_1 and c_2*, the constant terms of (1) and (2).

By exactly the same procedure, we can show that:

$$x = \frac{D_x}{D} = \frac{\begin{vmatrix} c_1 & b_1 \\ c_2 & b_2 \end{vmatrix}}{\begin{vmatrix} a_1 & b_1 \\ a_2 & b_2 \end{vmatrix}}. \tag{5}$$

Equations (4) and (5) together yield the components of the solution of the system. The use of determinants in this way is known as **Cramer's rule** for the solution of a system of linear equations.

Example   Solve the following system using Cramer's rule:

$$2x + y = 4$$
$$x - 3y = -5.$$

Solution   We have

$$D = \begin{vmatrix} 2 & 1 \\ 1 & -3 \end{vmatrix} = -6 - 1 = -7,$$

$$D_x = \begin{vmatrix} 4 & 1 \\ -5 & -3 \end{vmatrix} = -12 + 5 = -7,$$

$$D_y = \begin{vmatrix} 2 & 4 \\ 1 & -5 \end{vmatrix} = -10 - 4 = -14.$$

Therefore,

$$x = \frac{D_x}{D} = \frac{-7}{-7} = 1 \quad \text{and} \quad y = \frac{D_y}{D} = \frac{-14}{-7} = 2.$$

The solution set is therefore $\{(1, 2)\}$.

If $D = 0$ when using Cramer's rule, the equations in the system are either dependent or inconsistent, depending upon whether or not D_y and D_x are both 0. This follows from the discussion on page 338, where these conditions are considered in terms of the coefficients a_1, b_1, a_2, b_2, and the constant terms c_1 and c_2.

EXERCISE B.2

A ■ *Evaluate.*

Examples   **a.** $\begin{vmatrix} 2 & -3 \\ 1 & 4 \end{vmatrix}$ **b.** $\begin{vmatrix} -1 & 2 \\ 5 & 3 \end{vmatrix}$

Solutions   **a.** $\begin{vmatrix} 2 & -3 \\ 1 & 4 \end{vmatrix} = (2)(4) - (1)(-3)$ **b.** $\begin{vmatrix} -1 & 2 \\ 5 & 3 \end{vmatrix} = (-1)(3) - (5)(2)$

$$= 11 \qquad\qquad\qquad\qquad = -13$$

1. $\begin{vmatrix} 1 & 0 \\ 2 & 1 \end{vmatrix}$ **2.** $\begin{vmatrix} 3 & -2 \\ 4 & 1 \end{vmatrix}$ **3.** $\begin{vmatrix} -5 & -1 \\ 3 & 3 \end{vmatrix}$ **4.** $\begin{vmatrix} 1 & -2 \\ -1 & 2 \end{vmatrix}$

5. $\begin{vmatrix} -1 & 6 \\ 0 & -2 \end{vmatrix}$ **6.** $\begin{vmatrix} 20 & 3 \\ -20 & -2 \end{vmatrix}$ **7.** $\begin{vmatrix} -2 & -1 \\ -3 & -4 \end{vmatrix}$ **8.** $\begin{vmatrix} -1 & -5 \\ -2 & -6 \end{vmatrix}$

■ *Find the solution set of each system by Cramer's rule.*

Example

$2x - 3y = 6$
$2x + y = 14$

Solution

$$D = \begin{vmatrix} 2 & -3 \\ 2 & 1 \end{vmatrix} = (2)(1) - (2)(-3) = 8$$

The elements in D_x are obtained from the elements in D by replacing the elements in the column containing the coefficients of x with the corresponding constants 6 and 14.

$$D_x = \begin{vmatrix} 6 & -3 \\ 14 & 1 \end{vmatrix} = (6)(1) - (14)(-3) = 48$$

The elements in D_y are obtained from the elements in D by replacing the elements in the column containing the coefficients of y with the corresponding constants 6 and 14.

$$D_y = \begin{vmatrix} 2 & 6 \\ 2 & 14 \end{vmatrix} = (2)(14) - (2)(6) = 16$$

Values for x and y can now be determined by Cramer's rule.

$$x = \frac{D_x}{D} = \frac{48}{8} = 6 \qquad y = \frac{D_y}{D} = \frac{16}{8} = 2$$

The solution set is $\{(6, 2)\}$.

9. $2x - 3y = -1$
   $x + 4y = 5$

10. $3x - 4y = -2$
   $x - 2y = 0$

11. $3x - 4y = -2$
   $6x + 12y = 36$

12. $2x - 4y = 7$
   $x - 2y = 1$

13. $\frac{1}{3}x - \frac{1}{2}y = 0$
   $\frac{1}{2}x + \frac{1}{4}y = 4$

14. $\frac{2}{3}x + y = 1$
   $x - \frac{4}{3}y = 0$

15. $x - 2y = 5$
   $\frac{2}{3}x - \frac{4}{3}y = 6$

16. $\frac{1}{2}x + y = 3$
   $-\frac{1}{4}x - y = -3$

17. $x - 3y = 1$
   $y = 1$

18. $2x - 3y = 12$
   $x = 4$

19. $ax + by = 1$
   $bx + ay = 1$

20. $x + y = a$
   $x - y = b$

B   ■ *Show that each statement is true for every real value of each variable.*

21. $\begin{vmatrix} a & a \\ b & b \end{vmatrix} = 0$

22. $\begin{vmatrix} a_1 & b_1 \\ a_2 & b_2 \end{vmatrix} = -\begin{vmatrix} a_2 & b_2 \\ a_1 & b_1 \end{vmatrix}$

23. $\begin{vmatrix} a_1 & b_1 \\ a_2 & b_2 \end{vmatrix} = -\begin{vmatrix} b_1 & a_1 \\ b_2 & a_2 \end{vmatrix}$

24. $\begin{vmatrix} ka_1 & b_1 \\ ka_2 & b_2 \end{vmatrix} = k\begin{vmatrix} a_1 & b_1 \\ a_2 & b_2 \end{vmatrix}$

25. $\begin{vmatrix} ka & a \\ kb & b \end{vmatrix} = 0$

26. $\begin{vmatrix} a_1 + ka_2 & b_1 + kb_2 \\ a_2 & b_2 \end{vmatrix} = \begin{vmatrix} a_1 & b_1 \\ a_2 & b_2 \end{vmatrix}$

27. Show that if both $D_y = 0$ and $D_x = 0$, it follows that $D = 0$ when c_1 and c_2 are not both 0, and the equations in the system

$$a_1 x + b_1 y = c_1$$
$$a_2 x + b_2 y = c_2$$

are dependent. [*Hint:* Show that the first two determinant equations imply that $a_1 c_2 = a_2 c_1$ and $b_1 c_2 = b_2 c_1$ and that the rest follows from the formation of a proportion with these equations.

28. Show that for the system given in Problem 27, if $D = 0$ and $D_x = 0$, then $D_y = 0$.

B.3

THIRD-ORDER DETERMINANTS

Associated with each 3×3 matrix A that has real-number entries is the determinant $\delta(A)$ with the value defined as follows:

▶ If $A = \begin{bmatrix} a_1 & b_1 & c_1 \\ a_2 & b_2 & c_2 \\ a_3 & b_3 & c_3 \end{bmatrix}$, *then*

$$\delta(A) = \begin{vmatrix} a_1 & b_1 & c_1 \\ a_2 & b_2 & c_2 \\ a_3 & b_3 & c_3 \end{vmatrix}$$

$$= a_1 b_2 c_3 - a_1 b_3 c_2 + a_3 b_1 c_2 - a_2 b_1 c_3 + a_2 b_3 c_1 - a_3 b_2 c_1. \tag{1}$$

Again, we note that a 3×3 determinant is simply a number, namely, that number represented by the expression in the right-hand member of (1). Fortunately, we can rewrite (1) in a simpler form in terms of 2×2 determinants formed from the elements of the 3×3 determinant.

Minor of a determinant   The **minor** of an element in a determinant is defined as the determinant that remains after deleting the row and column in which the element appears. In the determinant (1), for example,

the minor of the element a_1 is $\begin{vmatrix} b_2 & c_2 \\ b_3 & c_3 \end{vmatrix}$,

the minor of the element b_1 is $\begin{vmatrix} a_2 & c_2 \\ a_3 & c_3 \end{vmatrix}$,

the minor of the element c_1 is $\begin{vmatrix} a_2 & b_2 \\ a_3 & b_3 \end{vmatrix}$.

Expansion by minors

If, by suitably factoring pairs of terms in the right-hand member, (1) is rewritten in the form

$$\begin{vmatrix} a_1 & b_1 & c_1 \\ a_2 & b_2 & c_2 \\ a_3 & b_3 & c_3 \end{vmatrix} = a_1(b_2c_3 - b_3c_2) - b_1(a_2c_3 - a_3c_2) + c_1(a_2b_3 - a_3b_2), \quad (2)$$

we observe that the sums enclosed in parentheses in the right-hand member of (2) are the respective minors (second-order determinants) of the elements a_1, b_1, and c_1. Therefore, (2) can be written

$$\begin{vmatrix} a_1 & b_1 & c_1 \\ a_2 & b_2 & c_2 \\ a_3 & b_3 & c_3 \end{vmatrix} = a_1\begin{vmatrix} b_2 & c_2 \\ b_3 & c_3 \end{vmatrix} - b_1\begin{vmatrix} a_2 & c_2 \\ a_3 & c_3 \end{vmatrix} + c_1\begin{vmatrix} a_2 & b_2 \\ a_3 & b_3 \end{vmatrix}. \quad (3)$$

The right-hand member of (3) is called the **expansion** of the determinant by minors *about the first row*.

Suppose, instead of factoring the right-hand member of (1) into the right-hand member of (2), we factor it as

$$\begin{vmatrix} a_1 & b_1 & c_1 \\ a_2 & b_2 & c_2 \\ a_3 & b_3 & c_3 \end{vmatrix} = a_1(b_2c_3 - b_3c_2) - a_2(b_1c_3 - b_3c_1) + a_3(b_1c_2 - b_2c_1). \quad (4)$$

Then we have the expansion of the determinant by minors *about the first column*,

$$\begin{vmatrix} a_1 & b_1 & c_1 \\ a_2 & b_2 & c_2 \\ a_3 & b_3 & c_3 \end{vmatrix} = a_1\begin{vmatrix} b_2 & c_2 \\ b_3 & c_3 \end{vmatrix} - a_2\begin{vmatrix} b_1 & c_1 \\ b_3 & c_3 \end{vmatrix} + a_3\begin{vmatrix} b_1 & c_1 \\ b_2 & c_2 \end{vmatrix}. \quad (5)$$

With the proper use of signs, it is possible to expand a determinant by minors about *any* row or *any* column and obtain an expression equivalent to a factored form of the right-hand member of (1). A helpful device for determining the signs of the terms in an expansion of a third-order determinant by minors is the array of alternating signs

$$\begin{array}{ccc} + & - & + \\ - & + & - \\ + & - & + \end{array}$$

which we call the **sign array** for the determinant. To obtain an expansion of (1) about a given row or column, the appropriate sign from the sign array is prefixed to each term in the expansion.

As an example, let us first expand the determinant

$$\begin{vmatrix} 1 & 2 & -3 \\ 0 & 2 & -1 \\ 1 & 1 & 0 \end{vmatrix}$$

about the second row. We have

$$\begin{vmatrix} 1 & 2 & -3 \\ 0 & 2 & -1 \\ 1 & 1 & 0 \end{vmatrix} = -0\begin{vmatrix} 2 & -3 \\ 1 & 0 \end{vmatrix} + 2\begin{vmatrix} 1 & -3 \\ 1 & 0 \end{vmatrix} - (-1)\begin{vmatrix} 1 & 2 \\ 1 & 1 \end{vmatrix}$$

$$= 0 + 2(0 + 3) + 1(1 - 2)$$

$$= 6 - 1 = 5.$$

If we expand about the third row, we have

$$\begin{vmatrix} 1 & 2 & -3 \\ 0 & 2 & -1 \\ 1 & 1 & 0 \end{vmatrix} = 1\begin{vmatrix} 2 & -3 \\ 2 & -1 \end{vmatrix} - 1\begin{vmatrix} 1 & -3 \\ 0 & -1 \end{vmatrix} + 0\begin{vmatrix} 1 & 2 \\ 0 & 2 \end{vmatrix}$$

$$= 1(-2 + 6) - 1(-1 - 0) + 0$$

$$= 4 + 1 = 5.$$

You should expand this determinant about the first row and about each column to verify that the result is the same in each expansion.

The expansion of a higher-order determinant by minors can be accomplished in the same way. By continuing the pattern of alternating signs used for third-order determinants, the sign array extends to higher-order determinants. The determinants in each term in the expansion will be of order one less than the order of the original determinant.

EXERCISE B.3

A ■ *Evaluate.*

Example

$$\begin{vmatrix} 1 & 2 & 0 \\ 3 & -1 & 4 \\ -2 & 1 & 3 \end{vmatrix}$$

Solution

Expand about any row or column; the first row is used here.

$$\begin{vmatrix} 1 & 2 & 0 \\ 3 & -1 & 4 \\ -2 & 1 & 3 \end{vmatrix} = 1\begin{vmatrix} -1 & 4 \\ 1 & 3 \end{vmatrix} - 2\begin{vmatrix} 3 & 4 \\ -2 & 3 \end{vmatrix} + 0\begin{vmatrix} 3 & -1 \\ -2 & 1 \end{vmatrix}$$

$$= 1[(-1)(3) - (1)(4)] - 2[(3)(3) - (-2)(4)] + 0$$

$$= (-3 - 4) - 2(9 + 8)$$

$$= -7 - 34 = -41$$

1. $\begin{vmatrix} 2 & 0 & 1 \\ 1 & 1 & 2 \\ -1 & 0 & 1 \end{vmatrix}$ **2.** $\begin{vmatrix} 1 & 3 & 1 \\ -1 & 2 & 1 \\ 0 & 2 & 0 \end{vmatrix}$ **3.** $\begin{vmatrix} 2 & -1 & 0 \\ -3 & 1 & 2 \\ 1 & -3 & 1 \end{vmatrix}$

4. $\begin{vmatrix} 2 & 4 & -1 \\ -1 & 3 & 2 \\ 4 & 0 & 2 \end{vmatrix}$ **5.** $\begin{vmatrix} 1 & 2 & 3 \\ 3 & -1 & 2 \\ 2 & 0 & 2 \end{vmatrix}$ **6.** $\begin{vmatrix} 1 & 0 & 0 \\ 0 & 1 & 2 \\ 0 & 3 & 4 \end{vmatrix}$

7. $\begin{vmatrix} -1 & 0 & 2 \\ -2 & 1 & 0 \\ 0 & 1 & -3 \end{vmatrix}$ **8.** $\begin{vmatrix} 2 & 1 & 4 \\ 3 & 2 & 6 \\ 5 & -3 & 10 \end{vmatrix}$ **9.** $\begin{vmatrix} 2 & 5 & -1 \\ 1 & 0 & 2 \\ 0 & 0 & 1 \end{vmatrix}$

10. $\begin{vmatrix} 2 & 3 & 1 \\ 0 & 1 & 0 \\ -4 & 2 & 1 \end{vmatrix}$ **11.** $\begin{vmatrix} a & b & 1 \\ a & b & 1 \\ 1 & 1 & 1 \end{vmatrix}$ **12.** $\begin{vmatrix} a & a & a \\ 1 & 2 & 3 \\ 4 & 5 & 6 \end{vmatrix}$

13. $\begin{vmatrix} x & 0 & 0 \\ 0 & x & 0 \\ 0 & 0 & x \end{vmatrix}$ **14.** $\begin{vmatrix} 0 & 0 & x \\ 0 & x & 0 \\ x & 0 & 0 \end{vmatrix}$ **15.** $\begin{vmatrix} x & y & 0 \\ x & y & 0 \\ 0 & 0 & 1 \end{vmatrix}$

16. $\begin{vmatrix} 0 & a & b \\ a & 0 & a \\ b & a & 0 \end{vmatrix}$ **17.** $\begin{vmatrix} a & b & 0 \\ b & 0 & b \\ 0 & b & a \end{vmatrix}$ **18.** $\begin{vmatrix} 0 & b & 0 \\ b & a & b \\ 0 & b & 0 \end{vmatrix}$

■ *Solve for x.*

19. $\begin{vmatrix} x & 0 & 0 \\ 2 & 1 & 3 \\ 0 & 1 & 4 \end{vmatrix} = 3$ **20.** $\begin{vmatrix} x^2 & 0 & 1 \\ 2 & -1 & 3 \\ 3 & 2 & 0 \end{vmatrix} = 1$

21. $\begin{vmatrix} x^2 & x & 1 \\ 0 & 2 & 1 \\ 3 & 1 & 4 \end{vmatrix} = 28$ **22.** $\begin{vmatrix} x & 1 & 1 \\ 0 & x & 1 \\ 0 & x & 0 \end{vmatrix} = -4$

B   **23.** Show that for all values of x, y, and z,

$$\begin{vmatrix} x & x & a \\ y & y & b \\ z & z & c \end{vmatrix} = 0.$$

[*Hint:* Expand about the elements of the third column.] Make a conjecture about determinants that contain two identical columns.

24. Show that

$$\begin{vmatrix} 0 & 0 & 0 \\ a & b & c \\ d & e & f \end{vmatrix} = 0$$

for all values of a, b, c, d, e, and f. Make a conjecture about determinants that contain a row of 0 elements.

25. Show that

$$\begin{vmatrix} 1 & 2 & 3 \\ 4 & 5 & 6 \\ 0 & 0 & 1 \end{vmatrix} = - \begin{vmatrix} 4 & 5 & 6 \\ 1 & 2 & 3 \\ 0 & 0 & 1 \end{vmatrix}.$$

Make a conjecture about the result of interchanging any two rows of a determinant.

26. Show that

$$\begin{vmatrix} 2 & 0 & 1 \\ 4 & 1 & -2 \\ 6 & 1 & 1 \end{vmatrix} = 2 \begin{vmatrix} 1 & 0 & 1 \\ 2 & 1 & -2 \\ 3 & 1 & 1 \end{vmatrix}.$$

Make a conjecture about the result of factoring a common factor from each element of a column in a determinant.

B.4

LINEAR SYSTEMS IN THREE VARIABLES: SOLUTION BY DETERMINANTS

Consider the linear system in three variables

$$a_1 x + b_1 y + c_1 z = d_1 \tag{1}$$

$$a_2 x + b_2 y + c_2 z = d_2 \tag{2}$$

$$a_3 x + b_3 y + c_3 z = d_3. \tag{3}$$

Cramer's rule   By solving this system using the methods of Section B.2, it can be shown that Cramer's rule is applicable to such systems and, in fact, to all similar systems as well as to linear systems in two variables. That is:

▶
$$x = \frac{D_x}{D}, \qquad y = \frac{D_y}{D}, \qquad z = \frac{D_z}{D},$$

where

$$D = \begin{vmatrix} a_1 & b_1 & c_1 \\ a_2 & b_2 & c_2 \\ a_3 & b_3 & c_3 \end{vmatrix}, \qquad D_x = \begin{vmatrix} d_1 & b_1 & c_1 \\ d_2 & b_2 & c_2 \\ d_3 & b_3 & c_3 \end{vmatrix},$$

$$D_y = \begin{vmatrix} a_1 & d_1 & c_1 \\ a_2 & d_2 & c_2 \\ a_3 & d_3 & c_3 \end{vmatrix}, \qquad D_z = \begin{vmatrix} a_1 & b_1 & d_1 \\ a_2 & b_2 & d_2 \\ a_3 & b_3 & d_3 \end{vmatrix}.$$

Note that the elements of the determinant D in each denominator are the coefficients of the variables in (1), (2), and (3). Note also that the numerators are formed from D by replacing the elements in the x, y, or z column, respectively, by d_1, d_2, and d_3.

We illustrate the application of Cramer's rule by considering the system of equations used in the example on page 343:

$$x + 2y - 3z = -4$$
$$2x - y + z = 3$$
$$3x + 2y + z = 10.$$

The determinant D, with elements that are the coefficients of the variables, is given by

$$D = \begin{vmatrix} 1 & 2 & -3 \\ 2 & -1 & 1 \\ 3 & 2 & 1 \end{vmatrix}.$$

We can expand the determinant about the first column, obtaining

$$D = \begin{vmatrix} 1 & 2 & -3 \\ 2 & -1 & 1 \\ 3 & 2 & 1 \end{vmatrix} = 1 \begin{vmatrix} -1 & 1 \\ 2 & 1 \end{vmatrix} - 2 \begin{vmatrix} 2 & -3 \\ 2 & 1 \end{vmatrix} + 3 \begin{vmatrix} 2 & -3 \\ -1 & 1 \end{vmatrix}.$$

$$= -3 - 16 - 3 = -22.$$

Replacing the first column in D with -4, 3, and 10, we obtain

$$D_x = \begin{vmatrix} -4 & 2 & -3 \\ 3 & -1 & 1 \\ 10 & 2 & 1 \end{vmatrix}.$$

Expanding D_x about the third column, we have

$$D_x = \begin{vmatrix} -4 & 2 & -3 \\ 3 & -1 & 1 \\ 10 & 2 & 1 \end{vmatrix} = -3 \begin{vmatrix} 3 & -1 \\ 10 & 2 \end{vmatrix} - 1 \begin{vmatrix} -4 & 2 \\ 10 & 2 \end{vmatrix} + 1 \begin{vmatrix} -4 & 2 \\ 3 & -1 \end{vmatrix}$$

$$= -48 + 28 - 2 = -22.$$

In a similar fashion, we can compute D_y and D_z:

$$D_y = \begin{vmatrix} 1 & -4 & -3 \\ 2 & 3 & 1 \\ 3 & 10 & 1 \end{vmatrix} = -44, \qquad D_z = \begin{vmatrix} 1 & 2 & -4 \\ 2 & -1 & 3 \\ 3 & 2 & 10 \end{vmatrix} = -66.$$

We then have

$$x = \frac{D_x}{D} = \frac{-22}{-22} = 1, \qquad y = \frac{D_y}{D} = \frac{-44}{-22} = 2, \quad \text{and} \quad z = \frac{D_z}{D} = \frac{-66}{-22} = 3.$$

The solution set of the system is therefore $\{(1, 2, 3)\}$.

As noted on page 415 for a linear system in two variables, if $D = 0$ for a linear system in three variables, the system does not have a unique solution.

EXERCISE B.4

A ■ *Solve by Cramer's rule. If a unique solution does not exist (D = 0), so state.*

Example

$4x + 10y - z = 2$
$2x + 8y + z = 4$
$x - 3y - 2z = 3$

Solution

Determine values for D, D_x, D_y, and D_z. The elements of D are the coefficients of the variables in the order they occur. For D_x, D_y, and D_z, the respective column of elements in D is replaced by the constants 2, 4, and 3.

$$D = \begin{vmatrix} 4 & 10 & -1 \\ 2 & 8 & 1 \\ 1 & -3 & -2 \end{vmatrix} = 12 \qquad D_x = \begin{vmatrix} 2 & 10 & -1 \\ 4 & 8 & 1 \\ 3 & -3 & -2 \end{vmatrix} = 120$$

$$D_y = \begin{vmatrix} 4 & 2 & -1 \\ 2 & 4 & 1 \\ 1 & 3 & -2 \end{vmatrix} = -36 \qquad D_z = \begin{vmatrix} 4 & 10 & 2 \\ 2 & 8 & 4 \\ 1 & -3 & 3 \end{vmatrix} = 96$$

Use Cramer's rule to determine x, y, and z.

$$x = \frac{D_x}{D} = \frac{120}{12} = 10 \qquad y = \frac{D_y}{D} = \frac{-36}{12} = -3 \qquad z = \frac{D_z}{D} = \frac{96}{12} = 8$$

The solution set is $\{(10, -3, 8)\}$.

1. $x + y = 2$
   $2x - z = 1$
   $2y - 3z = -1$

2. $2x - 6y + 3z = -12$
   $3x - 2y + 5z = -4$
   $4x + 5y - 2z = 10$

3. $x - 2y + z = -1$
   $3x + y - 2z = 4$
   $y - z = 1$

4. $2x + 5z = 9$
   $4x + 3y = -1$
   $3y - 4z = -13$

5. $2x + 2y + z = 1$
   $x - y + 6z = 21$
   $3x + 2y - z = -4$

6. $4x + 8y + z = -6$
   $2x - 3y + 2z = 0$
   $x + 7y - 3z = -8$

7. $x + y + z = 0$
   $2x - y - 4z = 15$
   $x - 2y - z = 7$

8. $x + y - 2z = 3$
   $3x - y + z = 5$
   $3x + 3y - 6z = 9$

9. $x - 2y + 2z = 3$
   $2x - 4y + 4z = 1$
   $3x - 3y - 3z = 4$

10. $3x - 2y + 5z = 6$
   $4x - 4y + 3z = 0$
   $5x - 4y + z = -5$

11. $\frac{1}{4}x - z = -\frac{1}{4}$
   $x + y = \frac{2}{3}$
   $3x + 4z = 5$

12. $2x - \frac{2}{3}y + z = 2$
   $\frac{1}{2}x - \frac{1}{3}y - \frac{1}{4}z = 0$
   $4x + 5y - 3z = -1$

13. $x + 4z = 3$
   $y + 3z = 9$
   $2x + 5y - 5z = -5$

14. $2x + y = 18$
   $y + z = -1$
   $3x - 2y - 5z = 38$

APPENDIX B SUMMARY

[B.1]   A **matrix** is a rectangular array of elements or **entries**. Two matrices are said to be **row-equivalent** if one matrix can be transformed to the other by one or more of the following transformations:

1. Multiplying the entries of any row by a nonzero real number.

2. Interchanging two rows.

3. Multiplying the entries of any row by a real number and adding the results to the corresponding elements of another row.

We can use these transformations to solve linear systems of equations.

[B.2]   The order of a **determinant** is the number of rows or columns in the determinant. A second-order determinant is defined by

$$\begin{vmatrix} a_1 & b_1 \\ a_2 & b_2 \end{vmatrix} = a_1 b_2 - a_2 b_1.$$

Cramer's rule can be used to solve systems of two linear equations in two variables. If

$$a_1 x + b_1 y = c_1$$
$$a_2 x + b_2 y = c_2,$$

then, for $D \neq 0$,

$$x = \frac{D_x}{D} = \frac{\begin{vmatrix} c_1 & b_1 \\ c_2 & b_2 \end{vmatrix}}{\begin{vmatrix} a_1 & b_1 \\ a_2 & b_2 \end{vmatrix}} \quad \text{and} \quad y = \frac{D_y}{D} = \frac{\begin{vmatrix} a_1 & c_1 \\ a_2 & c_2 \end{vmatrix}}{\begin{vmatrix} a_1 & b_1 \\ a_2 & b_2 \end{vmatrix}}.$$

[B.3]   A third-order determinant is defined by

$$\begin{vmatrix} a_1 & b_1 & c_1 \\ a_2 & b_2 & c_2 \\ a_3 & b_3 & c_3 \end{vmatrix} = a_1 b_2 c_3 - a_1 b_3 c_2 + a_3 b_1 c_2 - a_2 b_1 c_3 + a_2 b_3 c_1 - a_3 b_2 c_1.$$

The **minor** of an element in a determinant D is the determinant that results when the row and column containing the element are deleted. A determinant of an order greater than or equal to three can be evaluated by expansion by minors about any row or column.

[B.4]   Cramer's rule can also be used to solve systems of three linear equations in three variables. If

$$a_1 x + b_1 y + c_1 z = d_1$$
$$a_2 x + b_2 y + c_2 z = d_2$$
$$a_3 x + b_3 y + c_3 z = d_3,$$

then, for $D \neq 0$,

$$x = \frac{D_x}{D}, \qquad y = \frac{D_y}{D}, \quad \text{and} \quad z = \frac{D_z}{D}.$$

The symbols introduced in this appendix are listed on the inside of the front cover.

REVIEW EXERCISES

[B.1] ■ *Use row transformations to solve each system.*

1. $x - 2y = 5$
$2x + y = 5$

2. $x + 2y - z = -3$
$2x - 3y + 2z = 2$
$x - y + 4z = 7$

[B.2] **3.** Evaluate $\begin{vmatrix} 3 & -2 \\ 1 & -5 \end{vmatrix}$.

4. Solve by Cramer's rule:

$$2x + 3y = -2$$
$$x - 8y = -39.$$

[B.3] ■ *Evaluate each determinant.*

5. $\begin{vmatrix} 2 & 1 & 3 \\ 0 & 4 & -1 \\ 2 & 0 & 3 \end{vmatrix}$

6. $\begin{vmatrix} 3 & -1 & 2 \\ -2 & 1 & 0 \\ 2 & 4 & 1 \end{vmatrix}$

[B.4] ■ *Solve each system by Cramer's rule.*

7. $x + y = 3$
$y + z = 5$
$x - y + 2z = 5$

8. $2x + 3y - z = -2$
$x - y + z = 6$
$3x - y + z = 10$

C. LINEAR INTERPOLATION

Only three digits for the number x and four for the number $\log_{10} x$ appear in Table II of common logarithms. By means of a process called **linear interpolation,** however, the table can be used to find approximations to logarithms for numbers with four digits.

Let us examine geometrically the concepts involved. A portion of the graph of

$$y = \log_{10} x$$

is shown in Figure C.1. The curvature is exaggerated to illustrate the principle involved. We propose to use the straight line joining the points P_1 and P_2 as an approximation to the curve passing through the points. If an enlarged graph of $y = \log_{10} x$ were available, the value of $\log_{10} 2.257$ could be found by using the ordinate (RT) to the curve for $x = 2.257$. Since there is no way to accomplish this with the table of values we have, we shall use the ordinate (RS) to the straight line as an approximation to the ordinate of the curve.

Consider Figure C.2, where line segments P_2P_3 and P_4P_5 are perpendicular to line segment P_1P_3. From geometry, we know that $\triangle P_1P_4P_5$ is similar to $\triangle P_1P_2P_3$. Hence, the lengths of corresponding sides are proportional:

$$\frac{x}{X} = \frac{y}{Y}. \tag{1}$$

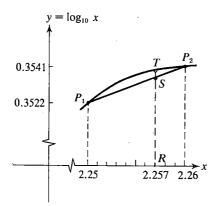

Figure C.1

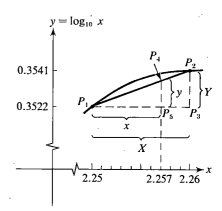

Figure C.2

If we know any three of the values in Equation (1), the fourth can be determined. For the purpose of interpolation, we assume that all the logarithmic function values noted in the table now have four-digit numerals; that is, we consider 2.250 instead of

2.25 and 2.260 instead of 2.26. We note in Figure C.2 that x (0.007) is the difference between 2.250 and 2.257, X (0.010) is the difference between 2.250 and 2.260, and Y (0.0019) is the difference between the logarithms 0.3522 and 0.3541. These data can be conveniently arranged as follows:

$$0.010\left\{0.007\left\{\begin{matrix}\log_{10} 2.250 = 0.3522\\ \log_{10} 2.257 = \quad ?\end{matrix}\right\}y\right\}0.0019$$
$$\log_{10} 2.260 = 0.3541$$

To find the value of y, we use (1):

$$\frac{0.007}{0.010} = \frac{y}{0.0019},$$

from which

$$y = \frac{7}{10}(0.0019) = 0.00133 \approx 0.0013$$

We can now add 0.0013 to 0.3522 to obtain a good approximation of the required logarithm. That is,

$$\log_{10} 2.257 = 0.3522 + 0.0013 = 0.3535.$$

The first example in Exercise C.1 shows another convenient arrangement for the calculations involved in the example presented here. The antilogarithm of a number can be found by a similar procedure. The second example in Exercise C.1 illustrates the process. However, with practice, it is possible to interpolate mentally in both procedures.

We can also use linear interpolation to find an approximation for e^x for a value of x that is between two entries in Table III and an approximation for $\ln x$ for a value of x that is between two entries in Table IV. Several examples are also shown in the exercise set.

·EXERCISE C

A   ■ *Find each logarithm.*

Example $\log_{10} 32.54$

Solution $\log_{10} 32.54 = \log_{10} (3.254 \times 10^1)$

Hence, the characteristic of the logarithm is 1. To obtain an approximation for the mantissa using linear interpolation, we arrange the data in tabular form as follows:

Solution continued overleaf

x	$\log_{10} x$
$10\left\{4\begin{cases}32.50\\32.54\\32.60\end{cases}\right.$	$\left.\begin{array}{c}1.5119\\?\\1.5132\end{array}\right\}y\right\}0.0013$

Note that for convenience we have written 4 and 10 for the differences for values of x instead of 0.04 and 0.10, respectively, since only their ratios are involved. We now set up a proportion and solve for y.

$$\frac{4}{10}=\frac{y}{0.0013}$$

$$y=\frac{4}{10}(0.0013)=0.00052\approx 0.0005$$

Adding this value of y to 1.5119, we have

$$\log_{10} 32.54 = 1.5119 + 0.0005 = 1.5124.$$

1. $\log_{10} 4.213$ **2.** $\log_{10} 8.184$ **3.** $\log_{10} 6.219$ **4.** $\log_{10} 10.31$

5. $\log_{10} 1522$ **6.** $\log_{10} 203.4$ **7.** $\log_{10} 37,110$ **8.** $\log_{10} 72.36$

9. $\log_{10} 0.5123$ **10.** $\log_{10} 0.09142$ **11.** $\log_{10} 0.008351$ **12.** $\log_{10} 0.03741$

■ *Find each antilogarithm.*

Example   antilog$_{10}$ 2.8472

Solution   We first note that antilog$_{10}$ 2.8472 = antilog$_{10}$(0.8472 + 2); hence, the characteristic is 2 and the mantissa is 0.8472. Since 0.8472 is not an entry in Table II, we note that the entries closest to this number are 0.8470 and 0.8476. The necessary data from the table can be arranged for interpolation as follows:

x	antilog$_{10}$ x
$6\left\{2\begin{cases}0.8470\\0.8472\\0.8476\end{cases}\right.$	$\left.\begin{array}{c}7.030\\?\\7.040\end{array}\right\}y\right\}0.010$

Then

$$\frac{2}{6}=\frac{y}{0.010},$$

where we have written 2 and 6 for 0.0002 and 0.0006, respectively. Thus,

$$y = \frac{2}{6}(0.010) \approx 0.003.$$

Adding this value of y to 7.030, we have

$$\text{antilog}_{10}\ 0.8472 = 7.030 + 0.003 = 7.033.$$

However, since the characteristic of the original number 2.8472 is 2, we have

$$\text{antilog}_{10}\ 2.8472 = 7.033 \times 10^2 = 703.3.$$

13. $\text{antilog}_{10}\ 0.5085$ **14.** $\text{antilog}_{10}\ 0.8087$ **15.** $\text{antilog}_{10}\ 1.9512$

16. $\text{antilog}_{10}\ 2.2620$ **17.** $\text{antilog}_{10}\ 1.0220$ **18.** $\text{antilog}_{10}\ 3.0759$

19. $\text{antilog}_{10}\ (0.7055 - 2)$ **20.** $\text{antilog}_{10}\ (0.6112 - 2)$ **21.** $\text{antilog}_{10}\ (0.8742 - 1)$

22. $\text{antilog}_{10}\ (0.9979 - 2)$ **23.** $\text{antilog}_{10}\ (0.8748 - 1)$ **24.** $\text{antilog}_{10}\ (0.7397 - 3)$

■ *Find each power.*

Example

$e^{3.24}$

Solution

Since 3.24 is between 3.2 and 3.3, we obtain $e^{3.2} = 24.533$ and $e^{3.3} = 27.113$ from Table III and set up a table as follows:

x	e^x
$10\begin{cases}4\begin{cases}3.20 \\ 3.24\end{cases} \\ \ \ 3.30\end{cases}$	$\begin{cases}24.533 \\ ? \\ 27.113\end{cases}$

We now set up a proportion and solve for y.

$$\frac{4}{10} = \frac{y}{2.58}$$

$$y = \frac{4}{10}(2.58) = 1.032$$

Adding this value of y to 24.533, we have

$$e^{3.24} = 24.533 + 1.032 = 25.565.$$

25. $e^{3.72}$ **26.** $e^{4.54}$ **27.** $e^{-2.78}$

28. $e^{-4.76}$ **29.** $e^{5.73}$ **30.** $e^{6.34}$

■ *Find each logarithm.*

Example   ln 2.24

Solution   Since 2.24 is between 2.2 and 2.3, we obtain ln 2.2 = 0.7885 and ln 2.3 = 0.8329
                   from Table IV and set up a table as follows:

x	ln x
$10\begin{cases} 4\begin{cases} 2.20 \\ 2.24 \\ \end{cases} \\ \quad 2.30 \end{cases}$	$\left.\begin{array}{l} 0.7885 \\ \quad ? \\ 0.8329 \end{array}\right\}{\scriptstyle y} \Big\} 0.0444$

We now set up a proportion and solve for y.

$$\frac{y}{0.0444} = \frac{4}{10}$$

$$y = \frac{4}{10}(0.0444) \approx 0.0178$$

Adding this value of y to 0.7885, we have

$$\ln 2.24 = 0.7885 + 0.0178 = 0.8063.$$

31. ln 3.16   **32.** ln 4.28   **33.** ln 6.72

34. ln 8.54   **35.** ln 0.73   **36.** ln 0.84

D. TABLES

TABLE I

SQUARES, SQUARE ROOTS, AND PRIME FACTORS

Number	Square	Square root	Prime factors	Number	Square	Square root	Prime factors
1	1	1.000		51	2,601	7.141	$3 \cdot 17$
2	4	1.414	2	52	2,704	7.211	$2^2 \cdot 13$
3	9	1.732	3	53	2,809	7.280	53
4	16	2.000	2^2	54	2,916	7.348	$2 \cdot 3^3$
5	25	2.236	5	55	3,025	7.416	$5 \cdot 11$
6	36	2.449	$2 \cdot 3$	56	3,136	7.483	$2^3 \cdot 7$
7	49	2.646	7	57	3,249	7.550	$3 \cdot 19$
8	64	2.828	2^3	58	3,364	7.616	$2 \cdot 29$
9	81	3.000	3^2	59	3,481	7.681	59
10	100	3.162	$2 \cdot 5$	60	3,600	7.746	$2^2 \cdot 3 \cdot 5$
11	121	3.317	11	61	3,721	7.810	61
12	144	3.464	$2^2 \cdot 3$	62	3,844	7.874	$2 \cdot 31$
13	169	3.606	13	63	3,969	7.937	$3^2 \cdot 7$
14	196	3.742	$2 \cdot 7$	64	4,096	8.000	2^6
15	225	3.873	$3 \cdot 5$	65	4,225	8.062	$5 \cdot 13$
16	256	4.000	2^4	66	4,356	8.124	$2 \cdot 3 \cdot 11$
17	289	4.123	17	67	4,489	8.185	67
18	324	4.243	$2 \cdot 3^2$	68	4,624	8.246	$2^2 \cdot 17$
19	361	4.359	19	69	4,761	8.307	$3 \cdot 23$
20	400	4.472	$2^2 \cdot 5$	70	4,900	8.367	$2 \cdot 5 \cdot 7$
21	441	4.583	$3 \cdot 7$	71	5,041	8.426	71
22	484	4.690	$2 \cdot 11$	72	5,184	8.485	$2^3 \cdot 3^2$
23	529	4.796	23	73	5,329	8.544	73
24	576	4.899	$2^3 \cdot 3$	74	5,476	8.602	$2 \cdot 37$
25	625	5.000	5^2	75	5,625	8.660	$3 \cdot 5^2$
26	676	5.099	$2 \cdot 13$	76	5,776	8.718	$2^2 \cdot 19$
27	729	5.196	3^3	77	5,929	8.775	$7 \cdot 11$
28	784	5.292	$2^2 \cdot 7$	78	6,084	8.832	$2 \cdot 3 \cdot 13$
29	841	5.385	29	79	6,241	8.888	79
30	900	5.477	$2 \cdot 3 \cdot 5$	80	6,400	8.944	$2^4 \cdot 5$
31	961	5.568	31	81	6,561	9.000	3^4
32	1,024	5.657	2^5	82	6,724	9.055	$2 \cdot 41$
33	1,089	5.745	$3 \cdot 11$	83	6,889	9.110	83
34	1,156	5.831	$2 \cdot 17$	84	7,056	9.165	$2^2 \cdot 3 \cdot 7$
35	1,225	5.916	$5 \cdot 7$	85	7,225	9.220	$5 \cdot 17$
36	1,296	6.000	$2^2 \cdot 3^2$	86	7,396	9.274	$2 \cdot 43$
37	1,369	6.083	37	87	7,569	9.327	$3 \cdot 29$
38	1,444	6.164	$2 \cdot 19$	88	7,744	9.381	$2^3 \cdot 11$
39	1,521	6.245	$3 \cdot 13$	89	7,921	9.434	89
40	1,600	6.325	$2^3 \cdot 5$	90	8,100	9.487	$2 \cdot 3^2 \cdot 5$
41	1,681	6.403	41	91	8,281	9.539	$7 \cdot 13$
42	1,764	6.481	$2 \cdot 3 \cdot 7$	92	8,464	9.592	$2^2 \cdot 23$
43	1,849	6.557	43	93	8,649	9.644	$3 \cdot 31$
44	1,936	6.633	$2^2 \cdot 11$	94	8,836	9.695	$2 \cdot 47$
45	2,025	6.708	$3^2 \cdot 5$	95	9,025	9.747	$5 \cdot 19$
46	2,116	6.782	$2 \cdot 23$	96	9,216	9.798	$2^5 \cdot 3$
47	2,209	6.856	47	97	9,409	9.849	97
48	2,304	6.928	$2^4 \cdot 3$	98	9,604	9.899	$2 \cdot 7^2$
49	2,401	7.000	7^2	99	9,801	9.950	$3^2 \cdot 11$
50	2,500	7.071	$2 \cdot 5^2$	100	10,000	10.000	$2^2 \cdot 5^2$

TABLE II

VALUES OF LOG$_{10}$ x
AND ANTILOG$_{10}$ x OR (10^x)

x	0	1	2	3	4	5	6	7	8	9
1.0	.0000	.0043	.0086	.0128	.0170	.0212	.0253	.0294	.0334	.0374
1.1	.0414	.0453	.0492	.0531	.0569	.0607	.0645	.0682	.0719	.0755
1.2	.0792	.0828	.0864	.0899	.0934	.0969	.1004	.1038	.1072	.1106
1.3	.1139	.1173	.1206	.1239	.1271	.1303	.1335	.1367	.1399	.1430
1.4	.1461	.1492	.1523	.1553	.1584	.1614	.1644	.1673	.1703	.1732
1.5	.1761	.1790	.1818	.1847	.1875	.1903	.1931	.1959	.1987	.2014
1.6	.2041	.2068	.2095	.2122	.2148	.2175	.2201	.2227	.2253	.2279
1.7	.2304	.2330	.2355	.2380	.2405	.2430	.2455	.2480	.2504	.2529
1.8	.2553	.2577	.2601	.2625	.2648	.2672	.2695	.2718	.2742	.2765
1.9	.2788	.2810	.2833	.2856	.2878	.2900	.2923	.2945	.2967	.2989
2.0	.3010	.3032	.3054	.3075	.3096	.3118	.3139	.3160	.3181	.3201
2.1	.3222	.3243	.3263	.3284	.3304	.3324	.3345	.3365	.3385	.3404
2.2	.3424	.3444	.3464	.3483	.3502	.3522	.3541	.3560	.3579	.3598
2.3	.3617	.3636	.3655	.3674	.3692	.3711	.3729	.3747	.3766	.3784
2.4	.3802	.3820	.3838	.3856	.3874	.3892	.3909	.3927	.3945	.3962
2.5	.3979	.3997	.4014	.4031	.4048	.4065	.4082	.4099	.4116	.4133
2.6	.4150	.4166	.4183	.4200	.4216	.4232	.4249	.4265	.4281	.4298
2.7	.4314	.4330	.4346	.4362	.4378	.4393	.4409	.4425	.4440	.4456
2.8	.4472	.4487	.4502	.4518	.4533	.4548	.4564	.4579	.4594	.4609
2.9	.4624	.4639	.4654	.4669	.4683	.4698	.4713	.4728	.4742	.4757
3.0	.4771	.4786	.4800	.4814	.4829	.4843	.4857	.4871	.4886	.4900
3.1	.4914	.4928	.4942	.4955	.4969	.4983	.4997	.5011	.5024	.5038
3.2	.5051	.5065	.5079	.5092	.5105	.5119	.5132	.5145	.5159	.5172
3.3	.5185	.5198	.5211	.5224	.5237	.5250	.5263	.5276	.5289	.5302
3.4	.5315	.5328	.5340	.5353	.5366	.5378	.5391	.5403	.5416	.5428
3.5	.5441	.5453	.5465	.5478	.5490	.5502	.5514	.5527	.5539	.5551
3.6	.5563	.5575	.5587	.5599	.5611	.5623	.5635	.5647	.5658	.5670
3.7	.5682	.5694	.5705	.5717	.5729	.5740	.5752	.5763	.5775	.5786
3.8	.5798	.5809	.5821	.5832	.5843	.5855	.5866	.5877	.5888	.5899
3.9	.5911	.5922	.5933	.5944	.5955	.5966	.5977	.5988	.5999	.6010
4.0	.6021	.6031	.6042	.6053	.6064	.6075	.6085	.6096	.6107	.6117
4.1	.6128	.6138	.6149	.6160	.6170	.6180	.6191	.6201	.6212	.6222
4.2	.6232	.6243	.6253	.6263	.6274	.6284	.6294	.6304	.6314	.6325
4.3	.6335	.6345	.6355	.6365	.6375	.6385	.6395	.6405	.6415	.6425
4.4	.6435	.6444	.6454	.6464	.6474	.6484	.6493	.6503	.6513	.6522
4.5	.6532	.6542	.6551	.6561	.6571	.6580	.6590	.6599	.6609	.6618
4.6	.6628	.6637	.6646	.6656	.6665	.6675	.6684	.6693	.6702	.6712
4.7	.6721	.6730	.6739	.6749	.6758	.6767	.6776	.6785	.6794	.6803
4.8	.6812	.6821	.6830	.6839	.6848	.6857	.6866	.6875	.6884	.6893
4.9	.6902	.6911	.6920	.6928	.6937	.6946	.6955	.6964	.6972	.6981
5.0	.6990	.6998	.7007	.7016	.7024	.7033	.7042	.7050	.7059	.7067
5.1	.7076	.7084	.7093	.7101	.7110	.7118	.7126	.7135	.7143	.7152
5.2	.7160	.7168	.7177	.7185	.7193	.7202	.7210	.7218	.7226	.7235
5.3	.7243	.7251	.7259	.7267	.7275	.7284	.7292	.7300	.7308	.7316
5.4	.7324	.7332	.7340	.7348	.7356	.7364	.7372	.7380	.7388	.7396
x	0	1	2	3	4	5	6	7	8	9

Table II   (*continued*)

x	0	1	2	3	4	5	6	7	8	9
5.5	.7404	.7412	.7419	.7427	.7435	.7443	.7451	.7459	.7466	.7474
5.6	.7482	.7490	.7497	.7505	.7513	.7520	.7528	.7536	.7543	.7551
5.7	.7559	.7566	.7574	.7582	.7589	.7597	.7604	.7612	.7619	.7627
5.8	.7634	.7642	.7649	.7657	.7664	.7672	.7679	.7686	.7694	.7701
5.9	.7709	.7716	.7723	.7731	.7738	.7745	.7752	.7760	.7767	.7774
6.0	.7782	.7789	.7796	.7803	.7810	.7818	.7825	.7832	.7839	.7846
6.1	.7853	.7860	.7868	.7875	.7882	.7889	.7896	.7903	.7910	.7917
6.2	.7924	.7931	.7938	.7945	.7952	.7959	.7966	.7973	.7980	.7987
6.3	.7993	.8000	.8007	.8014	.8021	.8028	.8035	.8041	.8048	.8055
6.4	.8062	.8069	.8075	.8082	.8089	.8096	.8102	.8109	.8116	.8122
6.5	.8129	.8136	.8142	.8149	.8156	.8162	.8169	.8176	.8182	.8189
6.6	.8195	.8202	.8209	.8215	.8222	.8228	.8235	.8241	.8248	.8254
6.7	.8261	.8267	.8274	.8280	.8287	.8293	.8299	.8306	.8312	.8319
6.8	.8325	.8331	.8338	.8344	.8351	.8357	.8363	.8370	.8376	.8382
6.9	.8388	.8395	.8401	.8407	.8414	.8420	.8426	.8432	.8439	.8445
7.0	.8451	.8457	.8463	.8470	.8476	.8482	.8488	.8494	.8500	.8506
7.1	.8513	.8519	.8525	.8531	.8537	.8543	.8549	.8555	.8561	.8567
7.2	.8573	.8579	.8585	.8591	.8597	.8603	.8609	.8615	.8621	.8627
7.3	.8633	.8639	.8645	.8651	.8657	.8663	.8669	.8675	.8681	.8686
7.4	.8692	.8698	.8704	.8710	.8716	.8722	.8727	.8733	.8739	.8745
7.5	.8751	.8756	.8762	.8768	.8774	.8779	.8785	.8791	.8797	.8802
7.6	.8808	.8814	.8820	.8825	.8831	.8837	.8842	.8848	.8854	.8859
7.7	.8865	.8871	.8876	.8882	.8887	.8893	.8899	.8904	.8910	.8915
7.8	.8921	.8927	.8932	.8938	.8943	.8949	.8954	.8960	.8965	.8971
7.9	.8976	.8982	.8987	.8993	.8998	.9004	.9009	.9015	.9020	.9025
8.0	.9031	.9036	.9042	.9047	.9053	.9058	.9063	.9069	.9074	.9079
8.1	.9085	.9090	.9096	.9101	.9106	.9112	.9117	.9122	.9128	.9133
8.2	.9138	.9143	.9149	.9154	.9159	.9165	.9170	.9175	.9180	.9186
8.3	.9191	.9196	.9201	.9206	.9212	.9217	.9222	.9227	.9232	.9238
8.4	.9243	.9248	.9253	.9258	.9263	.9269	.9274	.9279	.9284	.9289
8.5	.9294	.9299	.9304	.9309	.9315	.9320	.9325	.9330	.9335	.9340
8.6	.9345	.9350	.9355	.9360	.9365	.9370	.9375	.9380	.9385	.9390
8.7	.9395	.9400	.9405	.9410	.9415	.9420	.9425	.9430	.9435	.9440
8.8	.9445	.9450	.9455	.9460	.9465	.9469	.9474	.9479	.9484	.9489
8.9	.9494	.9499	.9504	.9509	.9513	.9518	.9523	.9528	.9533	.9538
9.0	.9542	.9547	.9552	.9557	.9562	.9566	.9571	.9576	.9581	.9586
9.1	.9590	.9595	.9600	.9605	.9609	.9614	.9619	.9624	.9628	.9633
9.2	.9638	.9643	.9647	.9652	.9657	.9661	.9666	.9671	.9675	.9680
9.3	.9685	.9689	.9694	.9699	.9703	.9708	.9713	.9717	.9722	.9727
9.4	.9731	.9736	.9741	.9745	.9750	.9754	.9759	.9763	.9768	.9773
9.5	.9777	.9782	.9786	.9791	.9795	.9800	.9805	.9809	.9814	.9818
9.6	.9823	.9827	.9832	.9836	.9841	.9845	.9850	.9854	.9859	.9863
9.7	.9868	.9872	.9877	.9881	.9886	.9890	.9894	.9899	.9903	.9908
9.8	.9912	.9917	.9921	.9926	.9930	.9934	.9939	.9943	.9948	.9952
9.9	.9956	.9961	.9965	.9969	.9974	.9978	.9983	.9987	.9991	.9996
x	0	1	2	3	4	5	6	7	8	9

TABLE III

VALUES OF e^x

x	e^x	e^{-x}		x	e^x	e^{-x}
0.00	1.0000	1.0000		0.50	1.6487	0.6065
0.01	1.0101	0.9901		0.51	1.6653	0.6005
0.02	1.0202	0.9802		0.52	1.6820	0.5945
0.03	1.0305	0.9705		0.53	1.6990	0.5886
0.04	1.0408	0.9608		0.54	1.7160	0.5827
0.05	1.0513	0.9512		0.55	1.7333	0.5769
0.06	1.0618	0.9418		0.56	1.7507	0.5712
0.07	1.0725	0.9324		0.57	1.7683	0.5655
0.08	1.0833	0.9231		0.58	1.7860	0.5599
0.09	1.0942	0.9139		0.59	1.8040	0.5543
0.10	1.1052	0.9048		0.60	1.8221	0.5488
0.11	1.1163	0.8958		0.61	1.8404	0.5434
0.12	1.1275	0.8869		0.62	1.8590	0.5380
0.13	1.1388	0.8781		0.63	1.8776	0.5326
0.14	1.1503	0.8694		0.64	1.8965	0.5273
0.15	1.1618	0.8607		0.65	1.9155	0.5220
0.16	1.1735	0.8521		0.66	1.9348	0.5169
0.17	1.1853	0.8437		0.67	1.9542	0.5117
0.18	1.1972	0.8353		0.68	1.9739	0.5066
0.19	1.2092	0.8270		0.69	1.9937	0.5016
0.20	1.2214	0.8187		0.70	2.0138	0.4966
0.21	1.2337	0.8106		0.71	2.0340	0.4916
0.22	1.2461	0.8025		0.72	2.0544	0.4868
0.23	1.2586	0.7945		0.73	2.0751	0.4819
0.24	1.2712	0.7866		0.74	2.0959	0.4771
0.25	1.2840	0.7788		0.75	2.1170	0.4724
0.26	1.2969	0.7711		0.76	2.1383	0.4677
0.27	1.3100	0.7634		0.77	2.1598	0.4630
0.28	1.3231	0.7558		0.78	2.1815	0.4584
0.29	1.3364	0.7483		0.79	2.2034	0.4538
0.30	1.3499	0.7408		0.80	2.2255	0.4493
0.31	1.3634	0.7334		0.81	2.2479	0.4449
0.32	1.3771	0.7261		0.82	2.2705	0.4404
0.33	1.3910	0.7190		0.83	2.2933	0.4360
0.34	1.4050	0.7118		0.84	2.3164	0.4317
0.35	1.4191	0.7047		0.85	2.3396	0.4274
0.36	1.4333	0.6977		0.86	2.3632	0.4232
0.37	1.4477	0.6907		0.87	2.3869	0.4190
0.38	1.4623	0.6839		0.88	2.4109	0.4148
0.39	1.4770	0.6771		0.89	2.4351	0.4107
0.40	1.4918	0.6703		0.90	2.4596	0.4066
0.41	1.5068	0.6636		0.91	2.4843	0.4025
0.42	1.5220	0.6570		0.92	2.5093	0.3985
0.43	1.5373	0.6505		0.93	2.5345	0.3946
0.44	1.5527	0.6440		0.94	2.5600	0.3906
0.45	1.5683	0.6376		0.95	2.5857	0.3867
0.46	1.5841	0.6313		0.96	2.6117	0.3829
0.47	1.6000	0.6250		0.97	2.6379	0.3791
0.48	1.6160	0.6188		0.98	2.6645	0.3753
0.49	1.6323	0.6126		0.99	2.6912	0.3716

Table III   (*continued*)

x	e^x	e^{-x}		x	e^x	e^{-x}
1.0	2.7183	0.3679		5.5	244.69	0.0041
1.1	3.0042	0.3329		5.6	270.43	0.0037
1.2	3.3201	0.3012		5.7	298.87	0.0034
1.3	3.6693	0.2725		5.8	330.30	0.0030
1.4	4.0552	0.2466		5.9	365.04	0.0027
1.5	4.4817	0.2231		6.0	403.43	0.0025
1.6	4.9530	0.2019		6.1	445.86	0.0022
1.7	5.4739	0.1827		6.2	492.75	0.0020
1.8	6.0496	0.1653		6.3	544.57	0.0018
1.9	6.6859	0.1496		6.4	601.85	0.0017
2.0	7.3891	0.1353		6.5	665.14	0.0015
2.1	8.1662	0.1225		6.6	735.10	0.0014
2.2	9.0250	0.1108		6.7	812.41	0.0012
2.3	9.9742	0.1003		6.8	897.85	0.0011
2.4	11.023	0.0907		6.9	992.27	0.0010
2.5	12.182	0.0821		7.0	1096.6	0.0009
2.6	13.464	0.0743		7.1	1212.0	0.0008
2.7	14.880	0.0672		7.2	1339.5	0.0007
2.8	16.445	0.0608		7.3	1480.3	0.0007
2.9	18.174	0.0550		7.4	1636.0	0.0006
3.0	20.086	0.0498		7.5	1808.0	0.0006
3.1	22.198	0.0450		7.6	1998.2	0.0005
3.2	24.533	0.0408		7.7	2208.4	0.0005
3.3	27.113	0.0369		7.8	2440.6	0.0004
3.4	29.964	0.0334		7.9	2697.3	0.0004
3.5	33.115	0.0302		8.0	2981.0	0.0003
3.6	36.598	0.0273		8.1	3294.5	0.0003
3.7	40.447	0.0247		8.2	3641.0	0.0003
3.8	44.701	0.0224		8.3	4023.9	0.0002
3.9	49.402	0.0202		8.4	4447.1	0.0002
4.0	54.598	0.0183		8.5	4914.8	0.0002
4.1	60.340	0.0166		8.6	5431.7	0.0002
4.2	66.686	0.0150		8.7	6002.9	0.0002
4.3	73.700	0.0136		8.8	6634.2	0.0002
4.4	81.451	0.0123		8.9	7332.0	0.0001
4.5	90.017	0.0111		9.0	8103.1	0.0001
4.6	99.484	0.0101		9.1	8955.3	0.0001
4.7	109.95	0.0091		9.2	9897.1	0.0001
4.8	121.51	0.0082		9.3	10938	0.0001
4.9	134.29	0.0074		9.4	12088	0.0001
5.0	148.41	0.0067		9.5	13360	0.0001
5.1	164.02	0.0061		9.6	14765	0.0001
5.2	181.27	0.0055		9.7	16318	0.0001
5.3	200.34	0.0050		9.8	18034	0.0001
5.4	221.41	0.0045		9.9	19930	0.0001

TABLE IV

VALUES OF ln x

x	ln x	x	ln x	x	ln x
		4.5	1.5041	9.0	2.1972
		4.6	1.5261	9.1	2.2083
0.1	−2.3026	4.7	1.5476	9.2	2.2192
0.2	−1.6094	4.8	1.5686	9.3	2.2300
0.3	−1.2040	4.9	1.5892	9.4	2.2407
0.4	−0.9163				
		5.0	1.6094	9.5	2.2513
0.5	−0.6931	5.1	1.6292	9.6	2.2618
0.6	−0.5108	5.2	1.6487	9.7	2.2721
0.7	−0.3567	5.3	1.6677	9.8	2.2824
0.8	−0.2231	5.4	1.6864	9.9	2.2925
0.9	−0.1054				
		5.5	1.7047	10	2.3026
1.0	0.0000	5.6	1.7228	11	2.3979
1.1	0.0953	5.7	1.7405	12	2.4849
1.2	0.1823	5.8	1.7579	13	2.5649
1.3	0.2624	5.9	1.7750	14	2.6391
1.4	0.3365				
		6.0	1.7918	15	2.7081
1.5	0.4055	6.1	1.8083	16	2.7726
1.6	0.4700	6.2	1.8245	17	2.8332
1.7	0.5306	6.3	1.8405	18	2.8904
1.8	0.5878	6.4	1.8563	19	2.9444
1.9	0.6419				
		6.5	1.8718	20	2.9957
2.0	0.6931	6.6	1.8871	25	3.2189
2.1	0.7419	6.7	1.9021	30	3.4012
2.2	0.7885	6.8	1.9169	35	3.5553
2.3	0.8329	6.9	1.9315	40	3.6889
2.4	0.8755				
		7.0	1.9459	45	3.8067
2.5	0.9163	7.1	1.9601	50	3.9120
2.6	0.9555	7.2	1.9741	55	4.0073
2.7	0.9933	7.3	1.9879	60	4.0943
2.8	1.0296	7.4	2.0015	65	4.1744
2.9	1.0647				
		7.5	2.0149	70	4.2485
3.0	1.0986	7.6	2.0281	75	4.3175
3.1	1.1314	7.7	2.0412	80	4.3820
3.2	1.1632	7.8	2.0541	85	4.4427
3.3	1.1939	7.9	2.0669	90	4.4998
3.4	1.2238				
		8.0	2.0794	100	4.6052
3.5	1.2528	8.1	2.0919	110	4.7005
3.6	1.2809	8.2	2.1041	120	4.7875
3.7	1.3083	8.3	2.1163	130	4.8676
3.8	1.3350	8.4	2.1282	140	4.9416
3.9	1.3610				
		8.5	2.1401	150	5.0106
4.0	1.3863	8.6	2.1518	160	5.0752
4.1	1.4110	8.7	2.1633	170	5.1358
4.2	1.4351	8.8	2.1748	180	5.1930
4.3	1.4586	8.9	2.1861	190	5.2470
4.4	1.4816				

FORMULAS FROM GEOMETRY

Plane Figures

1. Triangle ABC with sides of lengths a and c, base b, and altitude h.

Perimeter: $P = a + b + c$

Area: $A = \dfrac{1}{2}bh$

$\angle A + \angle B + \angle C = 180°$

a. Isoceles triangle.
Two sides of equal length.
Two angles with equal measure.

b. Equilateral triangle.
Three sides of equal length.
Three angles with equal measure.

c. Right triangle with hypotenuse c.

$c^2 = a^2 + b^2$

2. Square with side of length s.
Perimeter: $P = 4s$
Area: $A = s^2$

3. Rectangle with length l and width w.
Perimeter: $P = l + l + w + w$
$P = 2l + 2w$
Area: $A = lw$

4. Circle with radius r.
Diameter: $d = 2r$
Circumference: $C = 2\pi r$ or πd
Area: $A = \pi r^2$

Solid Figures

1. Right circular cylinder with height h and radius r of the base.
Volume: $V = \pi r^2 h$
Lateral area: $S = 2\pi r h$

2. Rectangular prism with length l, width w, and height h.
Volume: $V = lwh$

ANSWERS TO ODD-NUMBERED EXERCISES

Exercise 1.1 [page 4] **1.** $\{3, 4, 5\}$ **3.** $\{0, 1, 2\}$ **5.** $\{5, 6, 7, ...\}$ **7.** $\{5, 7, 9\}$
9. {natural numbers between 3 and 7} **11.** {even natural numbers between 7 and 13}
13. {natural numbers greater than 7} **15.** $\{0, 8\}$ **17.** $\{-\sqrt{15}\}$ **19.** $\{-5, -\sqrt{15}, -3.44, -\frac{2}{3}\}$
21. finite **23.** finite **25.** infinite **27.** $\in$ **29.** $\in$ **31.** $\notin$ **33.** no **35.** no

Exercise 1.2 [page 9] **1.** $3r$ **3.** 6 **5.** $n; t$ **7.** r **9.** $6+x$ **11.** $8>5$ **13.** $-6<-4$
15. $x+1<0$ **17.** $x-4\le0$ **19.** $-2<y<3$ **21.** $1\le x<7$ **23.** $-2<8$ **25.** $-7>-13$
27. $-6<-3$ **29.** $1\frac{1}{2}=\frac{3}{2}$ **31.** $3<5<7$ **33.** $-7<0<2$

35.

37.

39.

41.

43.

45.

47.

49.

51.

53.

Exercise 1.3 [page 15] **1.** 12 **3.** $3y$ **5.** $t\cdot4$ **7.** 1 **9.** 1 **11.** $3x; 3y$ **13.** 5 **15.** -3
17. x **19.** positive **21.** 3 **23.** 4 **25.** -2 **27.** -5 **29.** x, if $x\ge0$; $-x$, if $x<0$
31. $x-2$, if $x\ge2$; $-(x-2)$, if $x<2$ **33.** no; yes **35.** no; no **37.** yes **39.** no

Exercise 1.4 [page 18] **1.** $5+(-9)$ **3.** $-3+(-4)$ **5.** $-2+2$ **7.** $0+(-5)$ **9.** 11 **11.** 6
13. 3 **15.** -7 **17.** 5 **19.** -3 **21.** -9 **23.** 6 **25.** 2 **27.** 11 **29.** -6 **31.** 3
33. 4 **35.** -10 **37.** 10 **39.** 3 **41.** 7 **43.** 8 **45.** 11
47. Substituting 5 and 3 for a and b, we get $5-3\overset{?}{=}3-5$; $2\ne-2$. **49.** integers
Other values can be used for a and b.

Exercise 1.5 [page 23] **1.** -12 **3.** 12 **5.** 24 **7.** -20 **9.** 24 **11.** 0 **13.** -24
15. 24 **17.** -24 **19.** 24 **21.** $2 \cdot 2 \cdot 2$ **23.** $7 \cdot 7$ **25.** prime **27.** $-1 \cdot 2 \cdot 2 \cdot 3$
29. $2 \cdot 2 \cdot 2 \cdot 7$ **31.** $-1 \cdot 2 \cdot 2 \cdot 2 \cdot 2 \cdot 3$ **33.** -4 **35.** -13 **37.** 3 **39.** 0 **41.** undefined
43. 4 **45.** $15 = (-3)(-5)$ **47.** $-38 = 19(-2)$ **49.** $-52 = (-4)(13)$ **51.** $0 = (-7)(0)$
53. $7\left(\dfrac{1}{8}\right)$ **55.** $3\left(\dfrac{1}{8}\right)$ **57.** $82\left(\dfrac{1}{11}\right)$ **59.** $7\left(\dfrac{1}{100}\right)$ **61.** $\dfrac{3}{2}$ **63.** $\dfrac{2}{7}$ **65.** $\dfrac{5}{8}$ **67.** $\dfrac{9}{2}$
69. Substituting 1 for y gives $2(3 \cdot 1) \overset{?}{=} 2 \cdot 3(2 \cdot 1); \ 6 \neq 12.$ **71.** $x = 0, \ y \neq 0$ **73.** $x, \ y > 0$ or $x, \ y < 0$
75. Substituting 8 and 4 for a and b gives $8 \div 4 \overset{?}{=} 4 \div 8; \ 2 \neq \frac{1}{2}.$
77. 1. $b \neq 0$ (hypothesis)

2. $\dfrac{a}{b} = q$ implies $bq = a$ (definition of quotient)

3. $\dfrac{1}{b}(bq) = \dfrac{1}{b} \cdot a$ (multiplication property)

4. $\left(\dfrac{1}{b} \cdot b\right)q = a \cdot \dfrac{1}{b}$ (associative and commutative properties of multiplication)

5. $1 \cdot q = a \cdot \dfrac{1}{b}$ (reciprocal property)

6. $q = a \cdot \dfrac{1}{b}$ (identity element for multiplication)

7. $\dfrac{a}{b} = a \cdot \dfrac{1}{b}$ (substitution of $\dfrac{a}{b}$ for q)

Exercise 1.6 [page 26] **1.** 20 **3.** 0 **5.** -17 **7.** -13 **9.** -2 **11.** 6 **13.** -44 **15.** 5
17. -2 **19.** -2 **21.** 60 **23.** 100 **25.** 4 **27.** 1080 **29.** 56 centimeters **31.** \$1016
33. Substituting 2 for x gives $3 + 4 \cdot 2 \overset{?}{=} (3 + 4) \cdot 2; \ 11 \neq 14.$ **35.** 10 **37.** 2 **39.** undefined **41.** 1
43. 0

Review Exercises [page 29] **1.** $\{-3, 0, 1\}$ **2.** $\{0, 1\}$ **3.** $a < c$ **4.** $a \leq 7$ **5.** $a < b < c$
6. $y \leq 6$ **7.** **8.**

9. $2 + x$ **10.** $18 + t$ **11.** $\dfrac{1}{3}$ **12.** 1 **13. a.** 14 **b.** 8 **14. a.** -6 **b.** 5

15. a. -36 **b.** 0 **c.** 21 **d.** 24

16. a. $2 \cdot 2 \cdot 2 \cdot 3$ **b.** $2 \cdot 2 \cdot 7$ **c.** $-1 \cdot 2 \cdot 3 \cdot 7$ **d.** $-1 \cdot 2 \cdot 2 \cdot 2 \cdot 3 \cdot 3$
17. a. 12 **b.** 0 **c.** -8 **d.** -8 **18. a.** $-7 \cdot \dfrac{1}{5}$ **b.** $24 \cdot \dfrac{1}{7}$ **c.** $-4 \cdot \dfrac{1}{5}$ **d.** $8 \cdot \dfrac{1}{3}$
19. a. 3 **b.** 6 **20. a.** 5 **b.** 2 **21.** no **22.** yes **23.** yes **24.** $x \geq 0$
25. a. yes **b.** no **26. a.** $x > y$ **b.** $x < y$ **27.** 6 **28.** -26 **29.** 2 **30.** -3

Exercise 2.1 [page 35] **1.** binomial; degree 3 **3.** monomial; degree 4 **5.** trinomial; degree 2
7. trinomial; degree 3 **9.** -25 **11.** 9 **13.** -2 **15.** 50 **17.** -2 **19.** -5 **21.** 7

23. 13   **25.** 5   **27.** 0   **29.** 1   **31.** 64   **33.** 8   **35.** 54   **37.** -1; -21 **39.** -4; 16
41. 19; 7   **43.** 37; 11   **45.** 14   **47.** -4 **49.** 4   **51.** 8
53. Closure for multiplication; closure for both addition and multiplication.

Exercise 2.2 [page 40]   **1.** $7x^2$ **3.** $-3y$ **5.** z^2 **7.** $7x^2y - 2x$ **9.** $6r^2 + 4r$ **11.** $s^2 - 4s$
13. $t^2 + 2t - 1$ **15.** $-2u^2 - u - 3$ **17.** $3a^2 + 4a + 1$ **19.** $-5xy - xy^2$ **21.** $-x^2 - 4x + 7$
23. $-t^3 - 4t^2 + t + 1$ **25.** $-2x^2 + x - 3$ **27.** $b^2 + 3b + 1$ **29.** $-2y - 1$ **31.** $2 - x$ **33.** $x - 1$
35. $-x^2 - 3x - 1$ **37.** Substituting 1 for x gives $-(1+1) \overset{?}{=} -1 + 1$; $-2 \neq 0$ **39.** $2x - y$
41. $-4x - 5$ **43.** $2x - 1$ **45.** -1

Exercise 2.3 [page 44]   **1.** $-14t^3$ **3.** $-40a^3b^3c$ **5.** $44x^3y^4z^2$ **7.** $6x^5y^5$ **9.** $-2r^6s^4t^2$
11. $3x^2y^6z^4$ **13.** $6r^3t^4$ **15.** $6x^4$ **17.** $6z^5$ **19.** x^6 **21.** x^4y^4 **23.** y^6z^3 **25.** $8x^3z^6$
27. $4x^2y^4z^2$ **29.** $-8x^3y^9z^3$ **31.** $x^4y^2 + x^3y^3$ **33.** $4x^2y^4z^2 - x^2y^2z^4$ **35.** a^{3n-3} **37.** x^n
39. a^{3n+1} **41.** $x^{6n}y^3$ **43.** $x^{3n-6}y^3$ **45.** $x^{6n+3}y^{3n-3}$

Exercise 2.4 [page 47]   **1.** $4xy - 8y^2$ **3.** $-12x^3 + 6x^2 - 6x$ **5.** $-x^2 + 4x + 1$ **7.** $x^2 + 6x + 9$
9. $4y^2 - 20y + 25$ **11.** $x^2 - 9$ **13.** $n^2 + 10n + 16$ **15.** $r^2 + 3r - 10$ **17.** $y^2 - 7y + 6$
19. $2z^2 - 5z - 3$ **21.** $8r^2 + 2r - 3$ **23.** $4x^2 - a^2$ **25.** $y^3 - y + 6$ **27.** $x^3 + 2x^2 - 21x + 18$
29. $x^3 - 7x + 6$ **31.** $z^3 - 7z - 6$ **33.** $6x^3 + x^2 - 8x + 6$ **35.** $6a^4 - 5a^3 - 5a^2 + 5a - 1$ **37.** 6
39. $-2a^2 - 2a$ **41.** $4a + 4$ **43.** $4a + 4$ **45.** $-4x^2 - 11x$
47. $(x+a)(x+b) = x(x+a) + b(x+a)$ **49.** $(x+a)(x-a) = x(x+a) - a(x+a)$
$ = x^2 + ax + bx + ab$ $ = x^2 + ax - ax - a^2$
$ = x^2 + (a+b)x + ab$ $ = x^2 - a^2$
51. $(x+a)(x^2 - ax + a^2) = x(x^2 - ax + a^2) + a(x^2 - ax + a^2)$
$ = (x^3 - ax^2 + a^2x) + (ax^2 - a^2x + a^3)$
$ = x^3 - ax^2 + a^2x + ax^2 - a^2x + a^3$
$ = x^3 + a^3$
53. Substituting 1 for x and 2 for y gives $(1+2)^2 \overset{?}{=} 1^2 + 2^2$; $9 \neq 5$.   **55.** $2x^{2n} - x^n$ **57.** $a^{2n+1} - a^{n+1}$
59. $a^{3n+1} + a^{2n+2}$ **61.** $1 - a^{2n}$ **63.** $a^{6n} + a^{3n} - 2$ **65.** $2a^{2n} + 3a^nb^n - 2b^{2n}$

Exercise 2.5 [page 52]   **1.** $xy^2(x^2y^5)$ **3.** $3xz^2(-2x^3)$ **5.** $2xy^2(2x^3y^2z^2)$ **7.** $2(x+3)$
9. $4x(x+2)$ **11.** $3x(x - y + 1)$ **13.** $6(4a^2 + 2a - 1)$ **15.** $2x(x^3 - 2x + 4)$ **17.** $3z^2(4z^2 + 5z - 3)$
19. $a(y^2 + by + b)$ **21.** $3mn(m - 2n + 4)$ **23.** $3ac(5ac - 4 + 2c^2)$ **25.** $(a+b)(a+3)$
27. $(2x - y)(x + 3)$ **29.** $(2y - x)(a + b)$ **31.** $-(r - 7)$ **33.** $-(b - 2a)$ **35.** $-2(x - 1)$
37. $-a(b + c)$ **39.** $-(-2x + 1)$ **41.** $-(-x + y - z)$ **43.** $x^n(x^n - 1)$ **45.** $x^n(x^{2n} - x^n - 1)$
47. $x^n(x^2 + 1)$ **49.** $-x^n(x^n + 1)$ **51.** $-x^a(x + 1)$

Exercise 2.6 [page 57]   **1.** $(x+3)(x+2)$ **3.** $(y-4)(y-3)$ **5.** $(x-3)(x+2)$ **7.** $(y-5)(y+2)$
9. $(x-5)(x+5)$ **11.** $(xy-1)(xy+1)$ **13.** $(y^2-3)(y^2+3)$ **15.** $(x-2y)(x+2y)$
17. $(2x-5y)(2x+5y)$ **19.** $(2x-1)(x+2)$ **21.** $(4x-1)(x+2)$ **23.** $(3x-1)(x-1)$
25. $(3x-8)(3x+1)$ **27.** $(5x-3)(2x+1)$ **29.** $(2x+3)(2x+3)$ or $(2x+3)^2$ **31.** $(3x-a)(x-2a)$

33. $(4xy - 1)(4xy + 1)$ **35.** $(3xy + 1)(3xy + 1)$ or $(3xy + 1)^2$ **37.** $3(x + 2)(x + 2)$ or $3(x + 2)^2$
39. $2a(a - 5)(a + 1)$ **41.** $4(a - b)(a - b)$ or $4(a - b)^2$ **43.** $4y(x - 3)(x + 3)$ **45.** $x(4 + x)(3 - x)$
47. $x^2y^2(x - 1)(x + 1)$ **49.** $(y^2 + 1)(y^2 + 2)$ **51.** $(3x^2 + 1)(x^2 + 2)$ **53.** $(x^2 + 4)(x - 1)(x + 1)$
55. $(x - 2)(x + 2)(x - 1)(x + 1)$ **57.** $(2a^2 + 1)(a - 1)(a + 1)$ **59.** $(x^2 + 2a^2)(x - a)(x + a)$

Exercise 2.7 [page 60] **1.** $(a + b)(x + 1)$ **3.** $(ax + 1)(x + a)$ **5.** $(x + a)(x + y)$ **7.** $(3a - c)(b - d)$
 9. $(3x + y)(1 - 2x)$ **11.** $(a^2 + 2b^2)(a - 2b)$ **13.** $(x + 2y)(x - 1)$ **15.** $(2a^2 - 1)(b + 3)$
17. $(x^3 - 3)(y^2 + 1)$ **19.** $(x - 1)(x^2 + x + 1)$ **21.** $(2x + y)(4x^2 - 2xy + y^2)$
23. $(a - 2b)(a^2 + 2ab + 4b^2)$ **25.** $(xy - 1)(x^2y^2 + xy + 1)$ **27.** $(3a + 4b)(9a^2 - 12ab + 16b^2)$
29. $(2x - y)(x^2 - xy + y^2)$ **31.** $[(x + 1) - 1][(x + 1)^2 + (x + 1) + 1] = x(x^2 + 3x + 3)$
33. $[(x + 1) - (x - 1)][(x + 1)^2 + (x + 1)(x - 1) + (x - 1)^2] = 2(3x^2 + 1)$
35. $ac - ad + bd - bc$ $ac - ad + bd - bc$

$\quad\quad = (ac - ad) - (bc - bd)$ $= (bd - ad) - (bc - ac)$
$\quad\quad = a(c - d) - b(c - d)$ $= (b - a)d - (b - a)c$
$\quad\quad = (a - b)(c - d);$ $= (b - a)(d - c)$

Review Exercises [page 62] **1. a.** monomial; degree 3   **b.** trinomial; degree 2
 2. a. binomial; degree 5   **b.** trinomial; degree 4   **3.** $\frac{1}{5}$ **4.** 7   **5. a.** 8   **b.** 13
 6. a. -2 **b.** -14 **7. a.** $x + y - z$ **b.** $2x^2 - 2z^2 - x + 3y$ **8. a.** $2x - 5$ **b.** $-x^2 - 2x$
 9. a. $-6x^3y^4$ **b.** $-6x^2y^3z^3$ **10. a.** $-8x^6y^9z^3$ **b.** $x^3y^6 - 4x^6y^2$
11. a. $2x^3 - 4x^2 + 2x$ **b.** $4x^2 + 10x - 6$ **12. a.** $y^3 - 3y^2 + 3y - 2$ **b.** $z^3 + 2z^2 - z - 2$
13. a. $-3x - 15$ **b.** $7x^2 - 7x$ **14. a.** $4(3x^2 - 2x + 1)$ **b.** $x(x^2 - 3x - 1)$
15. a. $4y^2(y - 2)$ **b.** $2xy^2(2x^2 - x + 3)$ **16. a.** $-(y - 3x)$ **b.** $-(-2x + y - z)$
17. a. $-x(x - 2)$ **b.** $-3xy(2x - 1 + y)$ **18. a.** $(a + 2)(x - y)$ **b.** $(x - y)(2a + b)$
19. a. $(x - 7)(x + 5)$ **b.** $(y + 8)(y - 4)$ **20. a.** $(xy - 6)(xy + 6)$ **b.** $(a - 7b)(a + 7b)$
21. a. $(3y - 1)(y + 4)$ **b.** $x(x + 5)(x - 2)$ **22. a.** $9(x - 2)(x + 2)$ **b.** $3(2x - y)(2x + y)$
23. a. $(2x - y)(x + 2y)$ **b.** $(3x + y)(2x - y)$ **24. a.** $(3a + 2b)(5a + 6b)$ **b.** $6(2a - b)(a - b)$
25. a. $(3xy - 1)(3xy + 1)$ **b.** $x^2(2y - 1)(2y + 1)$ **26. a.** $(2x - a)(x + 3a)$ **b.** $(3x - 2a)(3x - 2a)$
27. a. $(x + y)(2x + 1)$ **b.** $(y - 3)(x - 1)$ **28. a.** $(x + y)(a - 2b)$ **b.** $(x - 2y)(2a + b)$
29. a. $(2x - y)(4x^2 + 2xy + y^2)$ **b.** $(x + 4y)(x^2 - 4xy + 16y^2)$
30. a. $(3y + z)(9y^2 - 3yz + z^2)$ **b.** $(x - 2a)(x^2 + 2ax + 4a^2)$ **31.** 17   **32.** 0   **33.** 4   **34.** 14
35. $x^{3n} - 3y^{3n}$ **36.** $2x^{2n} - 5x^n - 3$ **37.** $x^{2n}(x^{2n} + 1)$ **38.** $x^n(x - 1)(x + 1)$ **39.** $2x(x^2 + 3y^2)$
40. $2y(3x^2 + y^2)$

Exercise 3.1 [page 67] **1.** $\dfrac{-1}{4}$ **3.** $\dfrac{3}{5}$ **5.** $\dfrac{2}{5}$ **7.** $\dfrac{-3}{7}$ **9.** $\dfrac{-2x}{y}$ **11.** $\dfrac{3x}{4y}$

13. $\dfrac{-(x + 1)}{x}$ or $\dfrac{-x - 1}{x}$ **15.** $\dfrac{-(x - y)}{y + 2}$, $\dfrac{-x + y}{y + 2}$, or $\dfrac{y - x}{y + 2}$ **17.** $\dfrac{4}{y - 3}$ **19.** $\dfrac{-1}{y - x}$

21. $\dfrac{2 - x}{x - 3}$ **23.** Substituting 1 for x gives $-\dfrac{1 + 3}{2} \stackrel{?}{=} \dfrac{-1 + 3}{2};\; -2 \neq 1.$ **25.** $\dfrac{-x - 1}{y - x}$ **27.** $\dfrac{x - 2}{y - x}$

29. $\dfrac{a - 1}{3a + b}$

Exercise 3.2 [page 71]   **1.** $2x^2y$ **3.** $2tr^2$ **5.** $-2c^2$ **7.** ab^6c^3 **9.** -1 **11.** $2rt$ **13.** 5

15. $a-b$ **17.** $6(t+1)$ **19.** $2(y-2)$ **21.** $\dfrac{-1}{y+3}$ **23.** $\dfrac{-1}{x-y}$ **25.** $\dfrac{2x+3}{3}$ **27.** $\dfrac{3x-1}{3}$

29. $y-1$ **31.** a^2-3a+2 **33.** $y+7$ **35.** $\dfrac{x-4}{x+1}$ **37.** $\dfrac{2y-3}{y-1}$ **39.** $\dfrac{x+2y}{x+y}$ **41.** $4y^2+6y+9$

43. Substitute 0 for x and 1 for y: $\dfrac{2(0)+1}{1} \neq 2(0)$.

Exercise 3.3 [page 75]   **1.** $4a^2+2a+\dfrac{1}{2}$ **3.** $y^2-2+\dfrac{3}{7y^2}$ **5.** $6rs-5+\dfrac{2}{rs}$ **7.** $4ax-2x+\dfrac{1}{2}$

9. $-5m^3+3-\dfrac{7}{5m^3}$ **11.** $8m^2-5+\dfrac{7}{5m}$ **13.** $2y+5+\dfrac{2}{2y+1}$ **15.** $2t-1-\dfrac{6}{2t-1}$

17. $x^2+4x+9+\dfrac{19}{x-2}$ **19.** $a^3-3a^2+6a-16+\dfrac{47}{a+3}$ **21.** $4z^3-2z^2+3z+1+\dfrac{2}{2z+1}$

23. $x^3+2x^2+4x+8+\dfrac{15}{x-2}$ **25.** $x-1+\dfrac{-7x+12}{x^2-2x+7}$ **27.** $4a^2-9a+31+\dfrac{-104a+32}{a^2+3a-1}$

29. $t-1+\dfrac{-t^2-3t+3}{t^3-2t^2+t+2}$ **31.** $k=-2$

Exercise 3.4 [page 80]   **1.** $\dfrac{6}{9}$ **3.** $\dfrac{-30}{14}$ **5.** $\dfrac{20}{5}$ **7.** $\dfrac{6}{18x}$ **9.** $\dfrac{-a^2b}{b^3}$ **11.** $\dfrac{xy^2}{xy}$ **13.** $\dfrac{3a+3b}{a^2-b^2}$

15. $\dfrac{3xy-9x}{y^2-y-6}$ **17.** $\dfrac{-2x-4}{x^2+3x+2}$ **19.** 60 **21.** 120 **23.** 252 **25.** $6ab^2$ **27.** $24x^2y^2$

29. $a(a-b)^2$ **31.** $(a-b)(a+b)$ **33.** $(a+4)(a+1)^2$ **35.** $(x+4)(x-1)^2$ **37.** $x(x-1)^3$

39. $4(a+1)(a-1)^2$ **41.** $x^3(x-1)^2$ **43.** $\dfrac{4y}{6xy}$ and $\dfrac{1}{6xy}$ **45.** $\dfrac{9}{12y^2}$ and $\dfrac{8y}{12y^2}$

47. $\dfrac{2}{a^2-1}$ and $\dfrac{3a+3}{a^2-1}$ **49.** $\dfrac{a-1}{(a-4)(a-1)^2}$ and $\dfrac{2a-8}{(a-4)(a-1)^2}$

51. $\dfrac{3y^2+6y}{(y+1)(y+2)^2}$ and $\dfrac{y^2+y}{(y+1)(y+2)^2}$

Exercise 3.5 [page 84]   **1.** $\dfrac{x-3}{2}$ **3.** $\dfrac{a+b-c}{6}$ **5.** $\dfrac{2x-1}{2y}$ **7.** $\dfrac{1-2x}{x+2y}$ **9.** $\dfrac{6-2a}{a^2-2a+1}$ **11.** $\dfrac{2-2a}{ax}$

13. $\dfrac{-a-4}{6}$ **15.** $\dfrac{2x^2+xy+2y^2}{2xy}$ **17.** $\dfrac{2x+33}{(x-6)(x+3)}$ **19.** $\dfrac{-2ax-3a}{(3x+2)(x-1)}$ **21.** $\dfrac{-4}{15(x-2)}$

23. $\dfrac{-8x^2+2xy+6y^2}{(3x+y)(2x-y)}$ **25.** $\dfrac{6}{(x-2)(x+1)(x+1)}$ **27.** $\dfrac{-6y-4}{(y+4)(y-4)(y-1)}$ **29.** $\dfrac{y^2+6y-3}{(y+3)(y-3)(y+7)}$

31. $\dfrac{-1}{(z-4)(z-3)(z-2)}$ **33.** $\dfrac{x^3-2x^2+2x-2}{(x-1)(x-1)}$ **35.** $\dfrac{c}{s}+\dfrac{s}{c}=\dfrac{c\cdot c}{c\cdot s}+\dfrac{s\cdot s}{c\cdot s}$ **37.** $c+\dfrac{s^2}{c}=\dfrac{c\cdot c}{c\cdot 1}+\dfrac{s^2}{c}$

$$=\dfrac{c^2+s^2}{cs}=\dfrac{1}{sc} \qquad =\dfrac{c^2+s^2}{c}=\dfrac{1}{c}$$

39. $\dfrac{(c-1)(c+1)}{s^2}+1=\dfrac{c^2-1}{s^2}+\dfrac{s^2}{s^2}$

$\qquad =\dfrac{c^2+s^2-1}{s^2}$

$\qquad =\dfrac{1-1}{s^2}=0$

41. $\dfrac{s-c}{s+c}+\dfrac{2sc}{s^2-c^2}=\dfrac{(s-c)(s-c)}{(s-c)(s+c)}+\dfrac{(2sc)}{(s-c)(s+c)}$

$\qquad =\dfrac{s^2-2sc+c^2+2sc}{(s-c)(s+c)}$

$\qquad =\dfrac{s^2+c^2}{s^2-c^2}=\dfrac{1}{s^2-c^2}$

Exercise 3.6 [page 88] **1.** $\dfrac{2}{3}$ **3.** $\dfrac{7}{10}$ **5.** $\dfrac{10}{3}$ **7.** $\dfrac{1}{8x}$ **9.** $\dfrac{25p}{n}$ **11.** $\dfrac{-n}{2}$ **13.** $\dfrac{-b^2}{a}$

15. $\dfrac{3c}{35ab}$ **17.** $\dfrac{5}{ab}$ **19.** 5 **21.** $\dfrac{a(2a-1)}{a+4}$ **23.** $\dfrac{x+3}{x-5}$ **25.** $\dfrac{x-7}{x-5}$ **27.** $\dfrac{(x-2)(3x+1)}{(3x-1)(x-1)}$

29. $\dfrac{3-a}{a+1}$ **31.** $\dfrac{4}{3}$ **33.** $\dfrac{1}{ax^2y}$ **35.** $\dfrac{20ay}{3}$ **37.** $\dfrac{2}{9}$ **39.** $\dfrac{a+1}{a-2}$ **41.** $\dfrac{x-5}{x+5}$ **43.** $\dfrac{3x-1}{x-2}$

45. $3(x^2-xy+y^2)$ **47.** $(y-3)(x+2)$ **49.** $\dfrac{a-1}{a-3}$ **51.** $\dfrac{x+1}{x}$ **53.** $\dfrac{1}{x+1}$

Exercise 3.7 [page 91] **1.** $\dfrac{3}{2}$ **3.** $\dfrac{2}{21}$ **5.** $\dfrac{a}{bc}$ **7.** $\dfrac{4y}{3}$ **9.** $\dfrac{1}{5}$ **11.** $\dfrac{1}{10}$ **13.** $\dfrac{7}{2(5a+1)}$ **15.** x

17. $\dfrac{x}{x-1}$ **19.** $\dfrac{y}{y+2}$ **21.** $\dfrac{10}{7}$ **23.** $\dfrac{a(4a-3)}{4a+1}$ **25.** $\dfrac{-1}{y-3}$ **27.** $\dfrac{a-2b}{a+2b}$ **29.** $\dfrac{a+6}{a-1}$ **31.** 1

33. $\dfrac{\dfrac{(c+s)^2}{c-s}-\dfrac{1}{c-s}}{\dfrac{2cs}{c-s}}=\dfrac{\dfrac{c^2+2cs+s^2}{c-s}-\dfrac{2cs}{c-s}}{}$

$\qquad =\dfrac{\dfrac{c^2+s^2+2cs-2cs}{c-s}}{}$

$\qquad =\dfrac{1}{c-s}$

35. $\dfrac{\dfrac{1}{s-c}}{\dfrac{s}{s^2-c^2}}+\dfrac{s-c}{s}=\dfrac{\dfrac{1}{s-c}\cdot(s-c)(s+c)}{\dfrac{s}{(s-c)(s+c)}\cdot(s-c)(s+c)}+\dfrac{s-c}{s}$

$\qquad =\dfrac{s+c}{s}+\dfrac{s-c}{s}$

$\qquad =\dfrac{s+c+s-c}{s}=\dfrac{2s}{s}=2$

Review Exercises [page 94]

1. a. $\dfrac{1}{-(a-b)},\quad \dfrac{-1}{b-a},\quad -\dfrac{-1}{a-b},\quad -\dfrac{-1}{-(b-a)},\quad -\dfrac{1}{-(a-b)},\quad -\dfrac{1}{b-a}$

b. $\dfrac{1}{a-b}$ is a positive number if $a>b$ and a negative number if $a<b$. **2.** $\dfrac{1}{a-1},\quad \dfrac{-1}{1-a}$

3. a. $\dfrac{2x}{5y}$ **b.** $x-2$ **4. a.** $-2x-1$ **b.** $\dfrac{4x^2-2xy+y^2}{2x-y}$ **5. a.** $6x-3+\dfrac{3}{2x}$ **b.** $2y+1-\dfrac{1}{3y}$

6. a. $y-3+\dfrac{10}{2y+3}$ **b.** $bx^3+3x^2+3x+5+\dfrac{7}{x-1}$ **7. a.** $\dfrac{-18}{24}$ **b.** $\dfrac{x^2y}{2xy^2}$

8. a. $\dfrac{2x-6y}{x^2-9y^2}$ **b.** $\dfrac{y-y^2}{y^2-4y+3}$ **9. a.** $\dfrac{4x}{3}$ **b.** $\dfrac{-x-y}{4x}$

10. a. $\dfrac{8y-15x+7}{20xy}$ **b.** $\dfrac{3y+4}{6y-18}$ **11. a.** $\dfrac{2x+21}{(x-2)(x+3)}$ **b.** $\dfrac{x-2y+3}{x^2-4y^2}$

12. a. $\dfrac{5x+1}{(x^2-1)(x-1)}$ **b.** $\dfrac{-6y-4}{(y-4)(y+4)(y-1)}$ **13. a.** $\dfrac{x^2}{3}$ **b.** $\dfrac{2}{y}$ **14. a.** 1   **b.** $\dfrac{2y^2-2y}{y+1}$

15. a. $\dfrac{5xy}{3}$ **b.** 1   **16. a.** $\dfrac{(y+3)(y+2)}{(y-2)(y+1)}$ **b.** $\dfrac{1}{1-x}$ **17. a.** $\dfrac{5}{3}$ **b.** $\dfrac{2}{11}$

18. a. $\dfrac{6x+9}{3x-2}$ **b.** $\dfrac{y^2-1}{y^2+1}$ **19. a.** $\dfrac{2x}{x^2-1}$ **b.** 12   **20. a.** $\dfrac{3y+5}{y+1}$ **b.** $\dfrac{(x^2+1)(x-1)}{x(x-2)}$

21. $x-1+\dfrac{-2x+3}{x^2-x+2}$ **22.** $x+\dfrac{-x^2-x+4}{x^3-2x+1}$ **23.** x^2+3x+1 **24.** $x^2-5+\dfrac{18}{x^2+4}$ **25.** $\dfrac{13}{10}$

26. $\dfrac{3x-7y}{2x+2y}$ **27.** 1   **28.** $\dfrac{a^2-4a+1}{a^2+a-5}$ **29.** 2   **30.** $2b^2$

Exercise 4.1 [page 101]   **1.** {5}   **3.** {10}   **5.** {2}   **7.** {⁴⁄₃}   **9.** {⁸⁄₃}   **11.** {4}   **13.** {−4}
15. {²⁄₉}   **17.** {⁷⁄₂}   **19.** {−⁹⁄₂}   **21.** {−1}   **23.** {8}   **25.** {−20}   **27.** {−²⁷⁄₂}   **29.** {3}
31. {−30}   **33.** {−7}   **35.** ∅   **37.** {13}   **39.** ∅   **41.** $k=-7$ **43.** $k=11$

Exercise 4.2 [page 104]   **1.** 8   **3.** ⁵⁸⁄₃   **5.** 21.48   **7.** 153 feet per second   **9.** 12 years
11. 21,000 revolutions per minute   **13.** 66 inches   **15.** 132 feet   **17.** 30 years   **19.** 120 ohms
21. 8¾ years; 15 years   **23.** 62.6° Fahrenheit   **25. a.** $P=pn$ **b.** \$22
27. a. $A=\frac{1}{2}hb$ **b.** 10 centimeters   **29. a.** $S=2\pi rh$ **b.** 4 meters

Exercise 4.3 [page 109]   **1.** $x=\dfrac{a+b}{2}$ **3.** $y=\dfrac{b-c}{a}$ **5.** $x=\dfrac{c-b}{2a}$ **7.** $y=\dfrac{-bc}{a}$ **9.** $x=bc-a$

11. $y=\dfrac{c-ab}{2}$ **13.** $x=\dfrac{a}{a+1}$ **15.** $x=a+b$ **17.** $x=\dfrac{5b+8}{b+2}$ **19.** $x=\dfrac{a}{3a-1}$ **21.** $y=a$

23. $x=\dfrac{ab}{a+b}$ **25.** $k=v-gt$ **27.** $m=\dfrac{f}{a}$ **29.** $v=\dfrac{K}{p}$ **31.** $a=\dfrac{2s}{t^2}$ **33.** $t=\dfrac{v-k}{g}$ **35.** $h=\dfrac{V}{lw}$

37. $B=180-A-C$ **39.** $c=\dfrac{2A-hb}{h}$ **41.** $d=\dfrac{S-5\pi D}{3\pi}$ **43.** $n=\dfrac{l-a+d}{d}$ **45.** $l=\dfrac{S-2wh}{2w+2h}$

47. $t=\dfrac{15c-15V}{c}$ **49.** $n=\dfrac{125B-50w}{w}$

Exercise 4.4 [page 112]

1. a. Number of students that applied: x
    b. $\dfrac{3}{4}x=600$
    c. 800 students

3. a. Votes for loser: x
      Votes for winner: $x+122$
    b. $x+(x+122)=584$
    c. loser: 231; winner: 353

5. a. Total sales: x
    b. $0.02x=300$
    c. \$15,000

7. a. Number of bricks layed in 40 hours: x
    b. $\dfrac{x}{450}=\dfrac{40}{3}$
    c. 6000 bricks

9. a. Number of pounds of tin to make 90 pounds of alloy: x
  b. $\dfrac{x}{2.5} = \dfrac{90}{12}$
  c. 18¾ pounds

11. a. Number of dimes: d
  Number of quarters: $d + 12$
  b. $10d + 25(d + 12) = 1245$
  c. 27 dimes; 39 quarters

13. a. Number of first-class passengers: f
  Number of tourist passengers: $42 - f$
  b. $80f + 64(42 - f) = 2880$
  c. 12 first class; 30 tourist

15. a. Amount invested at 10%: x
  Amount invested at 12%: $8000 - x$
  b. $0.10x + 0.12(8000 - x) = 844$
  c. $5800 at 10% and $2200 at 12%

17. a. Money invested at 13%: x
  b. $0.08(3000) + 0.13x = 0.10(3000 + x)$
  c. $2000

19. a. Smallest angle: x
  Next larger angle: $2x$
  Largest angle: $3x + 12$
  b. $x + 2x + (3x + 12) = 180$
  c. 28°; 56°; 96°

21. a. Length of base: x
  Length of one of the equal sides: $3 + 2x$
  b. $x + (3 + 2x) + (3 + 2x) = 86$
  c. 16 centimeters; 35 centimeters; 35 centimeters

23. a. Distance from 32-gram weight: x
  Distance from 24-gram weight: $x + 2$
  b. $24(x + 2) = 32x$
  c. 24 grams at 8 centimeters; 32 grams at 6 centimeters

25. a. Distance from 2200-pound weight: x
  b. $200(6 - x) = 2200x$
  c. 6 inches from 2200-pound weight

27. a. Liters of 30% solution: x
  b. $0.30x + 0.12(40) = 0.20(x + 40)$
  c. 32 liters

29. a. Liters of pure alcohol: x
  b. $x + 0.45(12) = 0.60(x + 12)$
  c. 4.5 liters

31. a. Number of gallons to be drained: x
  b. $0.16(20) - 0.16x = 0.12(20)$
  c. 5 gallons

33. a. Rate of auto: r
  Rate of plane: $r + 120$
  b. $\dfrac{1260}{r + 120} = \dfrac{420}{r}$
  c. auto: 60 miles per hour; plane: 180 miles per hour

35. a. Time for second ship to reach first ship: t
  b. $20t + 5 = 30t$
  c. ½ hour

37. a. Time to fill tank with both pipes running together: t
  b. $\dfrac{1}{30}t + \dfrac{1}{45}t = 1$
  c. 18 hours

39. a. Time of the older machine: t
  b. $\left(\dfrac{1}{10}\right)6 + \left(\dfrac{1}{t}\right)6 = 1$
  c. 15 hours

Exercise 4.5 [page 127]

1. $\{x \mid x < 2\}$ or $(-\infty, 2)$

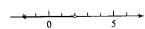

3. $\{x \mid x \leq 12\}$ or $(-\infty, 12]$

5. $\{x \mid x > 3\}$ or $(3, +\infty)$

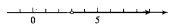

7. $\{x \mid x > 3\}$ or $(3, +\infty)$

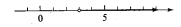

9. $\{x|x < -6\}$ or $(-\infty, -6)$

11. $\{x|x \geq -1\frac{1}{3}\}$ or $[-1\frac{1}{3}, +\infty)$

13. $\{x|x < -6\frac{6}{7}\}$ or $(-\infty, -6\frac{6}{7})$

15. $\{x|x \leq -4\}$ or $(-\infty, -4]$

17. $\{x|x > -11\}$ or $(-11, +\infty)$

19. $\{x|x \geq 0\}$ or $[0, +\infty)$

21. $\{x|x \leq -1\frac{8}{5}\}$ or $(-\infty, -1\frac{8}{5}]$

23. $\{x|x \leq -\frac{9}{2}\}$ or $(-\infty, -\frac{9}{2}]$

25. $\{x|6 < x < 10\}$ or $(6, 10)$

27. $\{x|-2 < x \leq 3\}$ or $(-2, 3]$

29. $\{x|-6 < x < -2\}$ or $(-6, -2)$

31. $\{x|-2 < x < 2\}$ or $(-2, 2)$

33. $\{x|-4 \leq x \leq 2\}$ or $[-4, 2]$

35. $\{x|-7 < x \leq -4\}$ or $(-7, -4]$

37. a. $80 \leq \dfrac{78 + 64 + 88 + 76 + x}{5} < 90$ **b.** 94% or more

39. a. $30 < \dfrac{5}{9}(F - 32) < 40$ **b.** between 86°F and 104°F   **41. a.** $90 + 18x < 24x$ **b.** after 15 days

43. a. $0.12x + 0.09(10,000 - x) \geq 1008$ **b.** at least \$3600

45. $\{x|-8 \leq x \leq 2\}$; $\{x|3 < x \leq 7\}$

47. $\{x|-7 \leq x \leq -3\}$; $\{x|0 < x \leq 4\}$

49. $\{x|-5 < x \leq -3\}$; $\{x|-2 < x \leq 0\}$; $\{x|1 < x < 3\}$

Exercise 4.6 [page 133]   **1.** $\{5, -5\}$ **3.** $\{13, -5\}$ **5.** $\{6, -7\}$ **7.** $\{1, \frac{5}{3}\}$ **9.** $\{-\frac{3}{4}\}$
11. $\{1, \frac{1}{2}\}$ **13.** $\{-\frac{1}{2}, -\frac{5}{6}\}$ **15.** $\{-\frac{1}{8}, \frac{7}{24}\}$ **17.** $\{\frac{3}{4}\}$
19. $\{x|-2 < x < 2\}$ or $(-2, 2)$ **21.** $\{x|-7 \leq x \leq 1\}$ or $[-7, 1]$

23. $\{x|1<x<4\}$ or $(1, 4)$

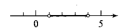

25. $\{x|-4\le x\le 12\}$ or $[-4, 12]$

27. $\{x|x<-3\}\cup\{x|x>3\}$ or $(-\infty, -3)\cup(3, +\infty)$

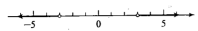

29. $\{x|x<-3\}\cup\{x|x>7\}$ or $(-\infty, -3)\cup(7, \infty)$

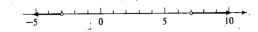

31. $\{x|x\le -2\}\cup\{x|x\ge 5\}$ or $(-\infty, -2]\cup[5, +\infty)$

Review Exercises [page 135] **1. a.** $\{-4\}$ **b.** $\{-\frac{7}{2}\}$ **2. a.** $\{-\frac{26}{21}\}$ **b.** $\{\frac{5}{2}\}$ **3.** 9 **4.** 17

5. 15 **6.** 51°C **7.** $x=-5y$ **8.** $y=-\dfrac{1}{5}x$ **9.** $d=\dfrac{l-a}{n-1}$ **10.** $a=\dfrac{2s-4t}{t^2}$ **11.** \$225

12. 140 hours **13.** \$2400 at 12% and \$4800 at 10% **14.** 10.5°; 31.5°; 138°

15. $\{x|x\le 27\}$ or $(-\infty, 27]$ **16.** $\{x|x\le \frac{3}{2}\}$ or $(-\infty, \frac{3}{2}]$ **17.** $\{x|2<x\le 5\}$ or $(2, 5]$

18. $\{x|1<x<3\}$ or $(1, 3)$

19. $\{x|-5<x<2\}$ or $(-5, 2)$

20. $\{x|0<x<4\}$ or $(0, 4)$ **21. a.** $\{3, -4\}$ **b.** $\{-\frac{5}{4}, \frac{11}{4}\}$

22. a. $\{x|x\le -8\}\cup\{x|x\ge 2\}$ or $(-\infty, -8]\cup[2, +\infty)$ **b.** $\{x|-3<x<7\}$ or $(-3, 7)$

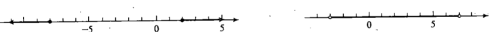

23. $k=-\frac{2}{3}$ **24.** $k=-9$ **25.** $\frac{2}{3}$ gallons

26. slower car: 50 miles per hour; faster car: 150 miles per hour **27.** 2½ hours

28. $2\frac{2}{9}$ days **29.** $49\frac{1}{11}$ minutes after 9 o'clock **30.** 10 inches

Exercise 5.1 [page 140] **1.** x^5 **3.** a^8 **5.** a^6 **7.** x^6 **9.** x^3y^6 **11.** $a^4b^4c^8$ **13.** $-16x^4$

15. $4a^7b^6$ **17.** x^2 **19.** xy^2 **21.** $\dfrac{x^3}{y^6}$ **23.** $\dfrac{8x^3}{y^6}$ **25.** $\dfrac{-8x^3}{27y^6}$ **27.** $\dfrac{16}{x}$ **29.** $\dfrac{y^4}{x}$

31. x^4y **33.** $\dfrac{8x}{9y^2}$ **35.** $36y^2$ **37.** $\dfrac{x^2s^{14}}{y^2t^2}$ **39.** $\dfrac{-1}{x^3a^3b}$ **41.** $\dfrac{a^2b^6}{x^8}$

43. Use 1 for x and 2 for y: $(1^2+2^2)^3 \overset{?}{=} 1^6+2^6$; $125\ne 1+64$. **45.** x^{2n} **47.** x^{3n} **49.** x^{2n}

Exercise 5.2 [page 143] **1.** $\frac{1}{2}$ **3.** 3 **5.** $-\frac{1}{8}$ **7.** $\frac{1}{5}$ **9.** $\frac{5}{3}$ **11.** $\frac{9}{5}$ **13.** $\frac{82}{9}$ **15.** $\frac{3}{16}$

17. $\frac{x^2}{y^3}$ **19.** $\frac{1}{x^6 y^3}$ **21.** $\frac{x^2}{y^6}$ **23.** $\frac{y^4}{x}$ **25.** x^4 **27.** $\frac{1}{x^6}$ **29.** $\frac{y}{x}$ **31.** $4x^5 y^2$ **33.** $\frac{z^2}{y^2}$ **35.** $\frac{xy}{z}$

37. $\frac{z}{y^2}$ **39.** $\frac{x^2 + y^2}{x^2 y^2}$ **41.** $\frac{x^2 y^2 + 1}{xy}$ **43.** $\frac{1}{(x-y)^2}$ **45.** $\frac{y^2 - x^2}{xy}$ **47.** $1 - xy$ **49.** $x + y$

51. Substituting 1 for x and 2 for y gives $(1+2)^{-2} \overset{?}{=} \frac{1}{1^2 + 2^2}$; $\frac{1}{9} \neq \frac{1}{5}$. **53.** a^{3-n} **55.** a^{2-2n}

57. $b^{-1} c^{-1}$ **59.** x^{-1} **61.** $\left(\frac{a}{b}\right)^{-n} = \frac{a^{-n}}{b^{-n}} = \frac{1}{a^n} \cdot b^n = \frac{b^n}{a^n} = \left(\frac{b}{a}\right)^n$

Exercise 5.3 [page 147] **1.** 2.85×10^2 **3.** 2.1×10 **5.** 8.372×10^6 **7.** 2.4×10^{-2}
9. 4.21×10^{-1} **11.** 4×10^{-6} **13.** 240 **15.** 687,000 **17.** 0.005 **19.** 0.0202 **21.** 12,270
23. 0.00235 **25.** 0.0005 **27.** 12.5 **29.** 0.00006 **31.** 10^{-5} or 0.00001 **33.** 10^0 or 1
35. 6×10^{-3} or 0.006 **37.** 1.8×10^6 or 1,800,000 **39.** four **41.** two **43.** three
45. three **47.** 72×10 or 720 **49.** 4×10^{-1} or 0.4 **51.** 8
53. a. 3×10^8 **b.** 1.18×10^{10} inches per second

Exercise 5.4 [page 152] **1.** 3 **3.** 2 **5.** -2 **7.** 9 **9.** 27 **11.** 16 **13.** $\frac{1}{4}$ **15.** $\frac{1}{8}$

17. $x^{2/3}$ **19.** $x^{1/3}$ **21.** $a^{3/2}$ **23.** $\frac{1}{x^{1/2}}$ **25.** $a^{1/3} b^{1/2}$ **27.** $\frac{a^4}{b^2}$ **29.** $\frac{r^2}{t^3}$ **31.** $\frac{t^2}{z}$ **33.** $\frac{yz}{x}$

35. $x^{3/2} + x$ **37.** $x - x^{2/3}$ **39.** $x^{-1} + 1$ **41.** $t + 1$ **43.** $b + b^{1/4}$ **45.** $x^{2/5}$ **47.** $x^{1/3}$

49. $x^{-2/3}$ **51.** Use 1 for a and 1 for b: $(1+1)^{1/2} \overset{?}{=} 1^{1/2} + 1^{1/2}$; $2^{1/2} \neq 1 + 1$. **53.** $x^{3n/2}$ **55.** $x^{3n/2}$

57. $x^{5n/2} y^{(3m+2)/2}$ **59.** $x^{n/3} y^n$ **61.** $x(x^{1/2} + 1)$ **63.** $x^{1/3}(x^{2/3} - x^{1/3})$

65. $16^{1/2} > 16^{1/4}$; $(\frac{1}{16})^{1/4} > (\frac{1}{16})^{1/2}$

Exercise 5.5 [page 156] **1.** $\sqrt{3}$ **3.** $\sqrt{x^3}$ **5.** $3\sqrt[4]{y}$ **7.** $x\sqrt[3]{y}$ **9.** $\sqrt[3]{xy}$ **11.** $-3\sqrt[5]{x^3}$

13. $\sqrt{x + 2y}$ **15.** $\sqrt[3]{(2x-y)^2}$ **17.** $\frac{1}{\sqrt[3]{4}}$ **19.** $\frac{1}{\sqrt[3]{x^2}}$ **21.** $5^{1/2}$ **23.** $x^{2/3}$ **25.** $a^{1/2} b^{1/2}$

27. $xy^{1/2}$ **29.** $2^{1/3} a^{1/3} b^{2/3}$ **31.** $(a-b)^{1/2}$ **33.** $a^{1/2} - 2b^{1/2}$ **35.** $a^{1/3} b^{1/2}$ **37.** $\frac{1}{x^{1/2}}$

39. $\frac{2}{(x+y)^{1/2}}$ **41.** 4 **43.** -5 **45.** 3 **47.** -4 **49.** x **51.** x^2 **53.** $2y^2$ **55.** $-x^2 y^3$

57. $\frac{2}{3} xy^4$ **59.** $\frac{-2}{5} x$ **61.** $2xy^2$ **63.** $2a^2 b^3$

65.

$$\begin{array}{c} -\sqrt{7} \quad -\sqrt{1} \quad\quad \sqrt{5} \;\; \sqrt{9} \\ \hline \bullet \quad\; \bullet \qquad\; \bullet \;\; \bullet \\ 0 \qquad\qquad\qquad 5 \end{array}$$

       67.

$$\begin{array}{c} -\sqrt{20} \;\; -\sqrt{6} \quad\; \sqrt{1} \qquad\quad 6 \\ \hline \bullet \quad\; \bullet \qquad\; \bullet \qquad\quad \bullet \\ -5 \qquad\; 0 \qquad\quad 5 \end{array}$$

69. $2|x|$ **71.** $|x+1|$ **73.** $\frac{2}{|x+y|}$ $(x + y \neq 0)$

Exercise 5.6 [page 161] **1.** $3\sqrt{2}$ **3.** $2\sqrt{5}$ **5.** $5\sqrt{3}$ **7.** x^2 **9.** $x\sqrt{x}$ **11.** $3x\sqrt{x}$ **13.** $2x^3\sqrt{2}$
15. $x\sqrt[4]{x}$ **17.** $xyz^2\sqrt[5]{x^2 y^4 z}$ **19.** $ab^2 c^2\sqrt[6]{ac^3}$ **21.** $3abc\sqrt[7]{ab^2 c^3}$ **23.** 6 **25.** $x^3 y$ **27.** 2

29. $10\sqrt{3}$ **31.** $100\sqrt{6}$ **33.** $\dfrac{\sqrt{5}}{5}$ **35.** $\dfrac{-\sqrt{2}}{2}$ **37.** $\dfrac{\sqrt{2x}}{2}$ **39.** $\dfrac{-\sqrt{xy}}{x}$ **41.** $\sqrt{x}$

43. $\dfrac{\sqrt{2xy}}{2x}$ **45.** $\dfrac{\sqrt[3]{x}}{x}$ **47.** $\dfrac{\sqrt[3]{18y^2}}{3y}$ **49.** $\dfrac{\sqrt[3]{2xy}}{2y}$ **51.** $\dfrac{\sqrt[5]{48x^2}}{2x}$ **53.** $a^2 b$ **55.** $7y\sqrt{2x}$

57. $\dfrac{2b}{a^2}\sqrt[3]{b}$ **59.** $\sqrt[5]{b}$ **61.** $\dfrac{1}{\sqrt{3}}$ **63.** $\dfrac{x}{\sqrt{xy}}$

65. Substitute two numbers such as 4 and 9 for a and b in the given equation and solve:

$$(\sqrt{a}+\sqrt{b})^2 \overset{?}{=} a+b$$
$$(\sqrt{4}+\sqrt{9})^2 \overset{?}{=} 4+9$$
$$(2+3)^2 \overset{?}{=} 13$$
$$25 \neq 13.$$

67. Substitute a negative number such as -2 for a in the given equation and solve:

$$\sqrt{a^2} \overset{?}{=} a$$
$$\sqrt{(-2)^2} \overset{?}{=} -2$$
$$\sqrt{4} \overset{?}{=} -2$$
$$2 \neq -2.$$

69. $\sqrt{3}$ **71.** $\sqrt{3}$ **73.** $\sqrt[3]{9}$ **75.** $\sqrt{x}$ **77.** $\sqrt[6]{72}$ **79.** $\sqrt[4]{20}$

Exercise 5.7 [page 166]   **1.** $5\sqrt{7}$ **3.** $\sqrt{3}$ **5.** $9\sqrt{2x}$ **7.** $-6y\sqrt{x}$ **9.** $15\sqrt{2a}$-   **11.** $5\sqrt[3]{2}$
13. $6-2\sqrt{5}$ **15.** $2\sqrt{3}+2\sqrt{5}$ **17.** $1-\sqrt{5}$ **19.** $x-9$ **21.** $-4+\sqrt{6}$ **23.** $7-2\sqrt{10}$
25. $2(1+\sqrt{3})$ **27.** $6(\sqrt{3}+1)$ **29.** $4(1+\sqrt{y})$ **31.** $\sqrt{2}(1-\sqrt{3})$ **33.** $1+\sqrt{3}$ **35.** $1+\sqrt{2}$

37. $1-\sqrt{x}$ **39.** $x-y$ **41.** $2\sqrt{3}-2$ **43.** $\dfrac{2(\sqrt{7}+2)}{3}$ **45.** $\dfrac{x(\sqrt{x}+3)}{x-9}$ **47.** $\dfrac{\sqrt{6}}{2}$

49. $\dfrac{3\sqrt{2}+2\sqrt{3}}{6}$ **51.** $\dfrac{\sqrt{3}}{3}$ **53.** $3\sqrt{x}$ **55.** $\dfrac{\sqrt{x+1}}{x+1}$ **57.** $\dfrac{-\sqrt{x^2+1}}{x(x^2+1)}$ **59.** $\dfrac{-1}{2(1+\sqrt{2})}$

61. $\dfrac{x-1}{3(\sqrt{x}+1)}$ **63.** $\dfrac{x-y}{x(\sqrt{x}+\sqrt{y})}$

Review Exercises [page 170]   **1. a.** $x^3 y^2$ **b.** $27x^6 y^3$ **2. a.** $\dfrac{1}{x^3 y}$ **b.** $\dfrac{y^6}{x^8}$ **3. a.** $\dfrac{11}{18}$ **b.** $\dfrac{x^2+y^2}{xy}$

4. a. $\dfrac{2x}{4x^2+4x+1}$ **b.** $\dfrac{(y^2-x^2)(x-y)^2}{x^2 y^2}$ **5. a.** 2.3×10^{-11} **b.** 3.07×10^{11} **6. a.** three   **b.** four

7. a. 10^2 or 100 **b.** 8×10^1 or 80 **8. a.** 9 **b.** $\tfrac{1}{2}$ **9. a.** x^2 **b.** $x^2 y^3$
10. a. $x-x^{5/3}$ **b.** $y-y^0$ or $y-1$ **11. a.** $x^{1/5}(x^{3/5})$ **b.** $y^{-1/2}(y^{-1/4})$

12. a. $\sqrt[3]{(1-x^2)^2}$ **b.** $\sqrt[3]{(1-x^2)^{-2}}$ or $\dfrac{1}{\sqrt[3]{(1-x^2)^2}}$ **13. a.** $(x^2 y)^{1/3}$ or $x^{2/3} y^{1/3}$ **b.** $(a+b)^{-2/3}$

14. a. $2y$ **b.** $-2xy^2$ **15. a.** $6\sqrt{5}$ **b.** $2xy\sqrt[4]{2y}$ **16. a.** $\dfrac{\sqrt{xy}}{y}$ **b.** $x\sqrt{3xy}$

17. a. $\dfrac{\sqrt[3]{4}}{1}$ **b.** $\dfrac{\sqrt[4]{54x^3}}{3x}$ **18. a.** $xy\sqrt[3]{x^2 y}$ **b.** xy **19. a.** $18\sqrt{3}$ **b.** $19\sqrt{2x}$

20. a. $3y\sqrt{2x}$ **b.** $4x\sqrt{3x}$ **21. a.** $3\sqrt{2}-2\sqrt{3}$ **b.** $5\sqrt{2}-5$ **22. a.** $12-7\sqrt{3}$ **b.** $x-4$

23. a. $8+4\sqrt{3}$ **b.** $2\sqrt{5}-4$ **24. a.** $\dfrac{y(\sqrt{y}+3)}{y-9}$ **b.** $\sqrt{x}-\sqrt{y}$ **25.** $x^{4n}y^{2n-2}$ **26.** $x^{n-1}y^{n-1}$

27. 0.2 **28.** $x(x^{-1/4}-x^{-3/4})$ **29.** $2|x-1|$ **30.** $\sqrt[3]{4}$

Exercise 6.1 [page 177]   **1.** $\{-2, 5\}$ **3.** $\{-\tfrac{5}{2}, 2\}$ **5.** $\{0, -\tfrac{1}{2}\}$ **7.** $\{6, -\tfrac{3}{2}\}$ **9.** $\{2, -\tfrac{1}{2}\}$
11. $\{\tfrac{5}{2}, -\tfrac{2}{3}\}$ **13.** $\{0, 3\}$ **15.** $\{0, 3\}$ **17.** $\{3, -3\}$ **19.** $\{3, -3\}$ **21.** $\{\tfrac{2}{3}, -\tfrac{2}{3}\}$ **23.** $\{\tfrac{3}{2}, -\tfrac{3}{2}\}$

25. $\{4, 1\}$ **27.** $\{7, -2\}$ **29.** $\{1\}$ **31.** $\{½, 1\}$ **33.** $\{3, -2\}$ **35.** $\{1, -6\}$ **37.** $\{1, -5\}$
39. $\{½, -3\}$ **41.** $\{6, -3\}$ **43.** $\{3/2, -2\}$ **45.** $\{1, -3\}$ **47.** $\{1, -1\frac{2}{3}\}$ **49.** $x^2 + x - 2 = 0$
51. $x^2 + 5x = 0$ **53.** $4x^2 - 13x - 12 = 0$ **55.** $10x^2 - 11x + 3 = 0$

Exercise 6.2 [page 183]   **1.** $\{10, -10\}$ **3.** $\{5/3, -5/3\}$ **5.** $\{\sqrt{7}, -\sqrt{7}\}$ **7.** $\{\sqrt{6}, -\sqrt{6}\}$
  9. $\{\sqrt{6}, -\sqrt{6}\}$ **11.** $\{9/2, -9/2\}$ **13.** $\{5, -1\}$ **15.** $\{5/2, -3/2\}$ **17.** $\{-2+\sqrt{3}, -2-\sqrt{3}\}$
19. $\{2+2\sqrt{3}, 2-2\sqrt{3}\}$ **21.** $\{5, 9\}$ **23.** $\{1, 4\}$ **25.** $\{-2/3, -8/3\}$ **27.** $\left\{\dfrac{1+\sqrt{15}}{7}, \dfrac{1-\sqrt{15}}{7}\right\}$
29. $\left\{\dfrac{7+2\sqrt{2}}{8}, \dfrac{7-2\sqrt{2}}{8}\right\}$ **31.** $\{2, -6\}$ **33.** $\{1\}$ **35.** $\{-5, -4\}$ **37.** $\{1+\sqrt{2}, 1-\sqrt{2}\}$
39. $\left\{\dfrac{-3+\sqrt{21}}{2}, \dfrac{-3-\sqrt{21}}{2}\right\}$ **41.** $\left\{\dfrac{-2+\sqrt{10}}{2}, \dfrac{-2-\sqrt{10}}{2}\right\}$ **43.** $\{5/2, -1\}$
45. $\left\{\dfrac{1+\sqrt{73}}{12}, \dfrac{1-\sqrt{73}}{12}\right\}$ **47.** $\{1, -3/5\}$ **49.** $\{\sqrt{a}, -\sqrt{a}\}$ **51.** $\left\{\sqrt{\dfrac{bc}{a}}, -\sqrt{\dfrac{bc}{a}}\right\}$
53. $\{a+4, a-4\}$ **55.** $\left\{\dfrac{3-b}{a}, \dfrac{-3-b}{a}\right\}$ **57.** $\{\sqrt{c^2-a^2}, -\sqrt{c^2-a^2}\}$ **59.** $\left\{-1+\sqrt{\dfrac{A}{P}}, -1-\sqrt{\dfrac{A}{P}}\right\}$
61. $y = \pm\dfrac{1}{3}\sqrt{9-x^2}$ **63.** $y = \pm\dfrac{2}{3}\sqrt{x^2-9}$ **65.** (See page 191.)

Exercise 6.3 [page 188]   **1.** $2i$ **3.** $4i\sqrt{2}$ **5.** $6i\sqrt{2}$ **7.** $6i\sqrt{6}$ **9.** $40i$ **11.** $-4i\sqrt{3}$
13. $4 + 2i$ **15.** $2 + 15i\sqrt{2}$ **17.** $2 + 2i$ **19.** $5 + 5i$ **21.** $-2 + i$ **23.** $-1 - 2i$ **25.** $8 + i$
27. $13 + 13i$ **29.** $21 - 18i$ **31.** $3 - 4i$ **33.** $5 + 0i$ or 5 **35.** $\dfrac{-i}{3}$ **37.** $\dfrac{-1}{5} - \dfrac{3}{5}i$
39. $1 + i$ **41.** $\dfrac{1}{2} - \dfrac{1}{2}i$ **43.** $\dfrac{12}{13} - \dfrac{5}{13}i$ **45.** $\dfrac{9}{34} + \dfrac{19}{34}i$ **47.** $4 + 2i$ **49.** $15 + 3i$ **51.** $-\dfrac{3}{2}i$
53. $\dfrac{3}{5} - \dfrac{4}{5}i$ **55.** $\{3i, -3i\}$ **57.** $\{3i\sqrt{2}, -3i\sqrt{2}\}$ **59.** $\left\{\dfrac{1+i\sqrt{7}}{2}, \dfrac{1-i\sqrt{7}}{2}\right\}$
61. $\{-1+i\sqrt{2}, -1-i\sqrt{2}\}$ **63. a.** -1 **b.** 1 **c.** $-i$ **d.** -1 **65.** $5 + 4i$ **67.** $x \geq 5$; $x < 5$

Exercise 6.4 [page 193]   **1.** $\{4, 1\}$ **3.** $\{1, -4\}$ **5.** $\left\{\dfrac{3+\sqrt{5}}{2}, \dfrac{3-\sqrt{5}}{2}\right\}$ **7.** $\left\{\dfrac{5+\sqrt{13}}{6}, \dfrac{5-\sqrt{13}}{6}\right\}$
  9. $\{3/2, -2/3\}$ **11.** $\{0, 5\}$ **13.** $\{i\sqrt{2}, -i\sqrt{2}\}$ **15.** $\left\{\dfrac{1+i\sqrt{7}}{4}, \dfrac{1-i\sqrt{7}}{4}\right\}$ **17.** $\left\{\dfrac{3+\sqrt{13}}{2}, \dfrac{3-\sqrt{13}}{2}\right\}$
19. $\left\{\dfrac{2+\sqrt{7}}{3}, \dfrac{2-\sqrt{7}}{3}\right\}$ **21.** $\left\{\dfrac{1+i\sqrt{23}}{6}, \dfrac{1-i\sqrt{23}}{6}\right\}$ **23.** $\left\{\dfrac{1+i\sqrt{3}}{4}, \dfrac{1-i\sqrt{3}}{4}\right\}$
25. 1; real and unequal   **27.** -16; imaginary and unequal   **29.** 0; one real   **31.** $x = 2k$; $x = -k$
33. $x = \dfrac{1 \pm \sqrt{1 - 4ac}}{2a}$ **35.** $x = -1 \pm \sqrt{1+y}$ **37.** $x = \dfrac{-y \pm \sqrt{24 - 11y^2}}{6}$ **39.** $y = \dfrac{-x \pm \sqrt{8 - 11x^2}}{2}$
41. $k = 4$ **43.** $\{k \mid k \leq -2\}$
45. If r_1 and r_2 are solutions of a quadratic equation, let

$$r_1 = \frac{-b + \sqrt{b^2 - 4ac}}{2a} \quad \text{and} \quad r_2 = \frac{-b - \sqrt{b^2 - 4ac}}{2a}.$$

Then,

$$r_1 + r_2 = \left(\frac{-b}{2a} + \frac{\sqrt{b^2 - 4ac}}{2a}\right) + \left(\frac{-b}{2a} - \frac{\sqrt{b^2 - 4ac}}{2a}\right) = \frac{-2b}{2a} = -\frac{b}{a}$$

and

$$r_1 \cdot r_2 = \left(\frac{-b}{2a} + \frac{\sqrt{b^2 - 4ac}}{2a}\right)\left(\frac{-b}{2a} - \frac{\sqrt{b^2 - 4ac}}{2a}\right) = \frac{b^2 - b^2 + 4ac}{4a^2} = \frac{c}{a}.$$

Exercise 6.5 [page 196] **1.** 5, 8 **3.** −9, −8 **5.** numerator is 3; denominator is 4 **7.** 4 or ¼
9. 7 meters by 9 meters **11.** 3 meters by 11 meters **13.** 3½ seconds **15.** 2.5 seconds
17. −13; −12; −11 **19.** 15 **21.** height is, 16 inches; width is 8 inches **23.** 3 meters
25. 3 miles per hour **27.** 20 miles per hour to the city; 30 miles per hour returning **29.** $1.20

Exercise 6.6 [page 201] **1.** {64} **3.** {−2} **5.** {−⅓} **7.** {2, −½} **9.** {12} **11.** {5} **13.** {4}
15. {−27} **17.** {17} **19.** {5} **21.** {0} **23.** {4} **25.** {1, 3} **27.** $A = \pi r^2$ **29.** $S = \dfrac{1}{R^3}$

31. $t = \pm\sqrt{r^2 + s^2}$ **33.** $E = \pm\sqrt{\left(\dfrac{A - B}{C}\right)^2 - D}$ **35.** $A \geq 0$ **37.** $2\sqrt{10}$ centimeters

Exercise 6.7 [page 204] **1.** {1, −1, 2, −2} **3.** $\left\{\dfrac{\sqrt{2}}{2}, \dfrac{-\sqrt{2}}{2}, 3i, -3i\right\}$ **5.** {25} **7.** {3, −3}
9. {64, −8} **11.** {64, −1} **13.** {9, 36} **15.** {¼, 16} **17.** {¼, −⅓} **19.** {626} **21.** {9, 16}
23. {4, 81}

Exercise 6.8 [page 209] **1.** $\{x | x < -1\} \cup \{x | x > 2\}$ or $(-\infty, -1); (2, +\infty)$ **3.** $\{x | 0 \leq x \leq 2\}$ or $[0, 2]$

5. $\{x | x < -1\} \cup \{x | x > 4\}$ or $(-\infty, -1); (4, +\infty)$ **7.** $\{x | -\sqrt{5} < x < \sqrt{5}\}$ or $(-\sqrt{5}, \sqrt{5})$

9. $\{x | x \in R\}$ or $(-\infty, +\infty)$ **11.** $\{x | -3 < x < 2\}$ or $(-3, 2)$

13. $\{x | -\frac{8}{3} < x < -2\}$ or $(-\frac{8}{3}, -2)$ **15.** $\{x | x \leq -\frac{1}{2}\} \cup \{x | x > 0\}$ or $(-\infty, -\frac{1}{2}]; (0, +\infty)$.

17. $\{x | x < 1\} \cup \{x \geq 2\}$ or $(-\infty, 1); [2, +\infty)$ **19.** $\{x | x < -1\} \cup \{x | 0 < x < 3\}$ or $(-\infty, -1); (0, 3)$

21. $\{x|x<0\} \cup \{x|2<x\le 4\}$ or $(-\infty, 0); (2, 4]$

23. $\{x|-3<x<-2\} \cup \{x|2<x<3\}$ or $(-3, -2); (2, 3)$ **25.** Between 4 seconds and 16 seconds.

Review Exercises [page 212]   1. a. $\{0, 2\}$ **b.** $\{2, 3\}$ **2. a.** $\{5, -3\}$ **b.** $\{2, -3\}$
3. a. $\{2\}$ **b.** $\{1, 3\}$ **4. a.** $x^2 - 3x - 10 = 0$ **b.** $12x^2 - x - 1 = 0$
5. a. $\{5, -5\}$ **b.** $\{\sqrt{7/3}, -\sqrt{7/3}\}$ **6. a.** $\{2, -8\}$ **b.** $\{4+\sqrt{15}, 4-\sqrt{15}\}$
7. a. $\{2+\sqrt{10}, 2-\sqrt{10}\}$ **b.** $\left\{\dfrac{-3+\sqrt{33}}{4}, \dfrac{-3-\sqrt{33}}{4}\right\}$ **8. a.** $4+6i$ **b.** $5-6i\sqrt{3}$
9. a. $6-i$ **b.** $5+3i$ **10. a.** $7+17i$ **b.** $4+3i$ **11. a.** $-\dfrac{4}{3}i$ **b.** $1-i$
12. a. $\{1, 2\}$ **b.** $\left\{\dfrac{3+i\sqrt{19}}{2}, \dfrac{3-i\sqrt{19}}{2}\right\}$ **13. a.** $\left\{\dfrac{3+\sqrt{5}}{2}, \dfrac{3-\sqrt{5}}{2}\right\}$ **b.** $\{1, -\frac{3}{2}\}$
14. 9 inches; 4 inches   **15.** $\frac{2}{3}$ or $\frac{3}{2}$ **16.** $\sqrt{6} \approx 2.45$ seconds   **17.** 3 seconds
18. a. $\{1, 4\}$ **b.** $\{8\}$ **19. a.** $t = \pm\sqrt{\dfrac{1-P^2 s}{2}}$ **b.** $p = \pm 2\sqrt{R^2 - R}$ **20. a.** $\{2, -2, i, -i\}$ **b.** $\{16\}$
21. a. $\{x|x<0\} \cup \{x|x>9\}$ or $(-\infty, 0); (9, +\infty)$ **b.** $\{x|-3<x<-2\}$ or $(-3, -2)$

22. a. $\{x|x\le -1\} \cup \{x|x>3\}$ or $(-\infty, -1]; (3, +\infty)$

b. $\{x|x\le -4\} \cup \{x|x>-2\}$ or $(-\infty, -4]; (-2, +\infty)$

23. $x = \dfrac{b \pm 2}{a}$ **24.** $\dfrac{1}{i^5}; -i$ **25.** $x = \dfrac{k \pm \sqrt{k^2 - 8}}{4}$ **26.** $y = \dfrac{-x \pm i\sqrt{7x^2}}{2}$ **27.** $k \ge 0$ **28.** $\{82\}$
29. $\{x|0<x<2\} \cup \{x|x\ge 4\}$ or $(0, 2); [4, +\infty)$

30. driving to work: 45 miles per hour; driving home: 30 miles per hour

Exercise 7.1 [page 216]   1. a. $(0, 7)$ **b.** $(2, 9)$ **c.** $(-2, 5)$
3. a. $(0, -\frac{3}{2})$ **b.** $(2, 0)$ **c.** $(-5, -2\frac{1}{4})$ **5.** $(-3, -7), (0, -4), (3, -1)$
7. $(1, 1), \quad (2, \frac{3}{4}), \quad (3, \frac{3}{5})$ **9.** $(1, 0), \quad (2, \sqrt{3}), \quad (3, 2\sqrt{2})$ **11.** $y = 6 - 2x; \{2, -2\}$
13. $y = \dfrac{x+2}{x}; \{0, 2\}$ **15.** $y = \dfrac{4}{x-1}; \{\frac{4}{3}, \frac{4}{7}\}$ **17.** $y = \dfrac{2}{x^2 - x - 4}; \{-1, -\frac{1}{2}\}$

19. $y = \dfrac{\pm\sqrt{x+8}}{2}$; $\left\{ \pm\dfrac{\sqrt{7}}{2}, \ \pm\dfrac{\sqrt{11}}{2} \right\}$ **21.** $y = \pm\dfrac{\sqrt{3x^2-4}}{2}$; $\left\{ \pm\sqrt{2}, \ \pm\dfrac{\sqrt{23}}{2} \right\}$ **23.** $(-4, 1)$

25. $(-1, 3)$ **27.** $(3, 8)$ **29.** $(3, \frac{1}{8})$ **31.** $k = \frac{7}{2}$

Exercise 7.2 [page 222]

1.

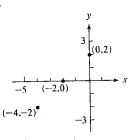

3.

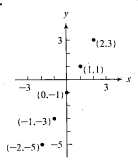

5.

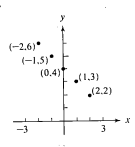

7.

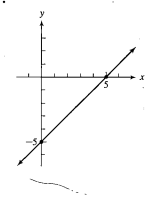

9.

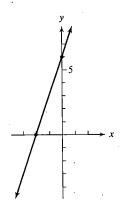

11.

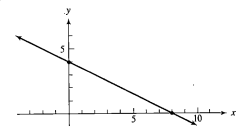

13.

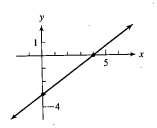

15.

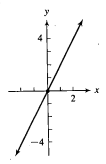

17.

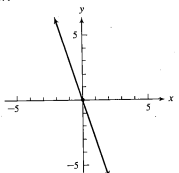

19.

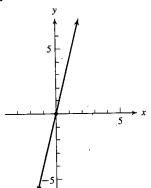

21.

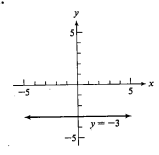

23.

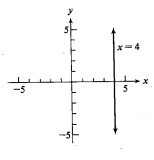

25.

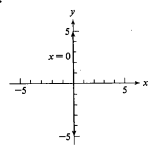

27.

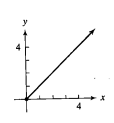

29.

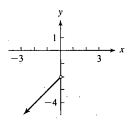

31.

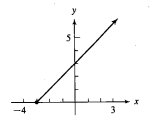

33.

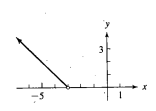

35.

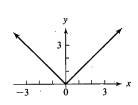

37.

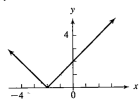

39.

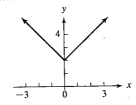

41.

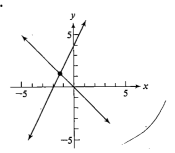

Exercise 7.3 [page 228] **1. a.** domain: $\{-2, -1, 0, 1\}$; range: $\{3, 4, 5, 6\}$ **b.** relation is a function

3. a. domain: $\{5, 6, 7, 8\}$; range $\{1\}$ **b.** relation is a function

5. a. domain: $\{2, 3\}$; range: $\{3, 4\}$ **b.** relation is not a function

7. a. domain: $\{0, 2, 3\}$; range: $\{0, 1, 2, 3\}$ **b.** relation is not a function

9. a. $y = 6 - 2x$ **b.** $x \in R$ **c.** function

11. a. $y = \dfrac{x - 8}{2}$ **b.** $x \in R$ **c.** function

13. a. $y = \dfrac{4}{x - 2}$ **b.** $x \in r, x \neq 2$ **c.** function

15. a. $y = \dfrac{6}{x - 4}$ **b.** $x \in R, x \neq 4$ **c.** function

17. a. $y = \pm\sqrt{8 - 2x^2}$ **b.** $-2 \leq x \leq 2$ **c.** not a function

19. a. $y = \pm 2\sqrt{x^2 - 4}$ **b.** $x \leq -2$ or $x \geq 2$ **c.** not a function

21. 9 **23.** 3 **25.** -6 **27. a.** $3x + 3h - 4$ **b.** $3h$ **c.** 3

29. a. $x^2 + 2hx + h^2 - 3x - 3h + 5$ **b.** $2hx + h^2 - 3h$ **c.** $2x + h - 3$

31. a. $x^3 + 3x^2h + 3xh^2 + h^3 + 2x + 2h - 1$ **b.** $3x^2h + 3xh^2 + h^3 + 2h$ **c.** $3x^2 + 3xh + h^2 + 2$

33.

$f(x)$

5

$f(3)$

-3

4

$f(-2)$

-4

x

35. $f(x + h) - f(x)$

37. $f(a + b) = 3(a + b) + 2$ $\quad f(a) + f(b) = (3a + 2) + (3b + 2)$
$\qquad\qquad = 3a + 3b + 2;$ $\qquad\qquad\qquad = 3a + 3b + 4;$
hence, $f(a + b) \neq f(a) + f(b)$.

39. $V = \left(\dfrac{1}{27}\right)\pi h^3$

41. $A = (4 - \pi)r^2$

Exercise 7.4 [page 235]

1. distance: 5; slope: $\dfrac{4}{3}$

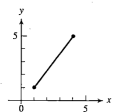

3. distance: 13; slope: $\dfrac{12}{5}$

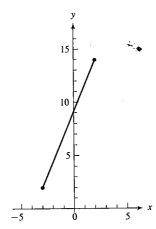

5. distance: $\sqrt{5}$; slope: $\dfrac{-1}{2}$

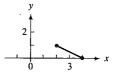

7. distance: $\sqrt{61}$; slope: $\dfrac{-5}{6}$

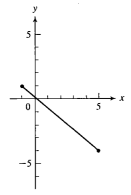

9. distance: 5; slope: 0

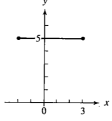

11. distance: 10; slope: not defined

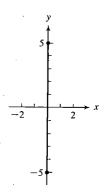

13. $15 + 9\sqrt{5}$

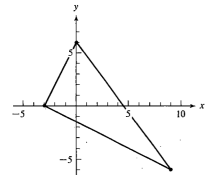

15. 48

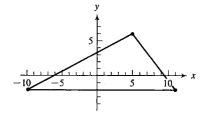

17. The slope of the segment with endpoints (5, 4) and (3, 0) is 2. The slope of the segment with endpoints (−1, 8) and (−4, 2) is 2. Therefore, the line segments are parallel.

19. Identify the given points as $A(0, -7)$, $B(8, -5)$, $C(5, 7)$, and $D(8, -5)$. Then, the slopes of AB and CD are

$$\text{slope } AB = \frac{-7-(-5)}{0-(8)} = \frac{1}{4} \quad \text{and} \quad \text{slope } CD = \frac{7-(-5)}{5-8} = -4;$$

and

$$\left(\frac{1}{4}\right)(-4) = -1.$$

21.

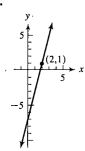

23.

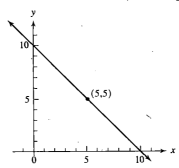

25.

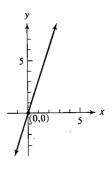

27.

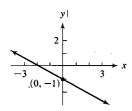

29.

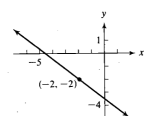

31.

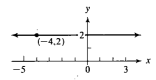

33. 1   **35.** 4   **37.** −1   **39.** −2

41. Identify the vertices as $A(0, 6)$, $B(9, -6)$, and $C(-3, 0)$. Then, by the distance formula, $(AB)^2 = 225$ and $(BC)^2 + (AC)^2 = 180 + 45 = 225$; the triangle is a right triangle.

43. Let the given points be $A(2, 4)$, $B(3, 8)$, $C(5, 1)$, and $D(4, -3)$. Since AB and CD have equal slopes $(m = 4)$ and BC and AD have equal slopes $(m = -7/2)$, there are two pairs of parallel sides. Hence, $ABCD$ is a parallelogram.

45. $k = -28$

Exercise 7.5 [page 240]   **1.** $2x - y + 5 = 0$ **3.** $x + y - 4 = 0$ **5.** $x - 2y + 6 = 0$ **7.** $3x + 2y - 1 = 0$

9. $y + 5 = 0$ **11.** $x + 3 = 0$ **13.** $y = -x + 3$; slope: -1; y-intercept: 3

15. $y = \frac{-3}{2}x + \frac{1}{2}$; slope: $\frac{-3}{2}$; y-intercept: $\frac{1}{2}$ **17.** $y = \frac{1}{3}x - \frac{2}{3}$; slope: $\frac{1}{3}$; y-intercept: $\frac{-2}{3}$

19. $y = \dfrac{8}{3}x$; slope: $\dfrac{8}{3}$; y-intercept: 0   **21.** $y = 0x - 2$; slope: 0; y-intercept: -2

23. $x - 7y = -18$ **25.** $x - y = 2$

27. $x - 2y = 0$ **29.** $2x + y = 0$

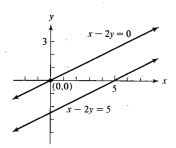

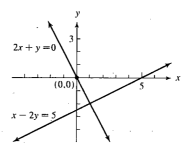

31. $F = \dfrac{9}{5}C + 32$ or $F = \dfrac{9C + 160}{5}$

Exercise 7.6 [page 245]

1.   **3.**

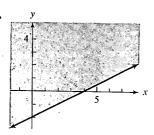

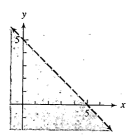

5.   **7.**

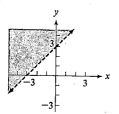   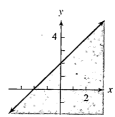

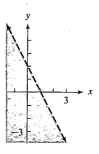   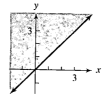

9.   **11.**

13.

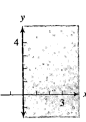

15.

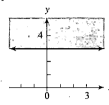

17.

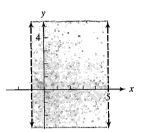

19.

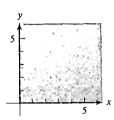

21.

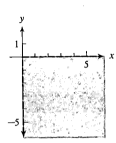

23.

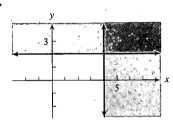

25.

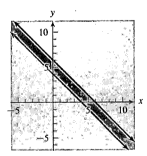

27. +

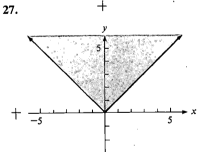

29.

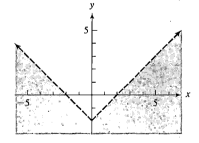

Review Exercises [page 248] **1. a.** $(0, -2)$ **b.** $(6, 0)$ **c.** $(3, -1)$ **2.** $(2, 1)$, $(4, 5)$, and $(6, 9)$

3. $y = \dfrac{3}{x - 2x^2}$ $(x \neq 0,\ x \neq \frac{1}{2})$

4. a.

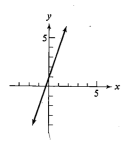

b.

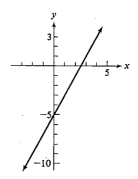

5. a.

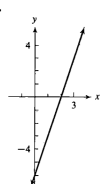

b.

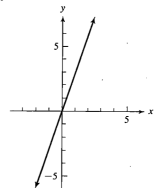

6. a. domain: $\{3, 4\}$; range: $\{5, 6\}$; not a function   **b.** domain: $\{2, 3, 4, 5\}$; range: $\{3, 4, 5\}$; function

7. a. domain: $x \in R$, $x \neq 1$; function   **b.** domain: $\{-2 \leq x \leq 2\}$; not a function   **8. a.** 2   **b.** h

9. a. 10   **b.** $4x + 2h - 3$ **10. a.** $\sqrt{178}$ **b.** $\dfrac{13}{3}$ **11. a.** $3\sqrt{5}$ **b.** -2

12. a. Slope of $P_1P_2 = 1$ and slope of $P_3P_4 = 1$; therefore, P_1P_2 and P_3P_4 are parallel.

    b. Slope of $P_1P_2 = 1$ and slope of $P_1P_3 = -1$; since $1(-1) = -1$, the lines are perpendicular.

13. $2x - y - 1 = 0$ **14.** $x + 2y = 0$ **15. a.** $y = 3x - 4$ **b.** slope: 3; y-intercept: -4

16. a. $y = -\dfrac{2}{3}x + 2$ **b.** slope: $-\dfrac{2}{3}$; y-intercept: 2

17. a.

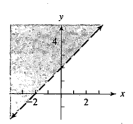

b.

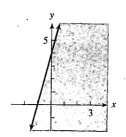

18. a.

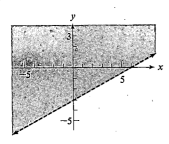

b.

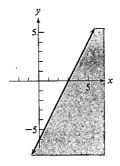

19. a.

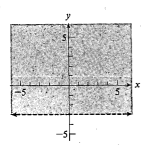

b.

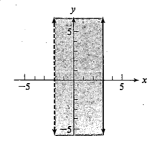

20. a.

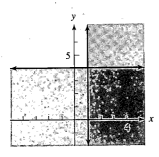

b.

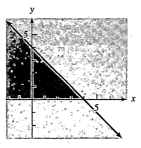

21.

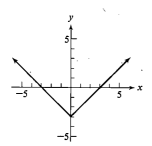

22.

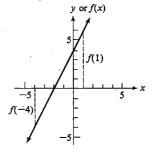

23. $k = \dfrac{-9}{7}$ **24.** $6x - y - 19 = 0$ **25.** $2x - y - 7 = 0$

26.   **27.** $A = \dfrac{\sqrt{3}}{4} s^2$ **28.** $A = \dfrac{7x - 2x^2}{4}$

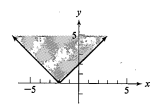

Exercise 8.1 [page 254]

1. a. $(-4, 17), (-3, 10),$
$(-2, 5), (-1, 2), (0, 1),$
$(1, 2), (2, 5), (3, 10),$
$(4, 17)$

b.

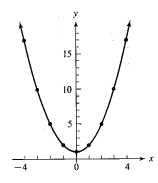

3. a. $(-4, 13), (-3, 6),$
$(-2, 1), (-1, -2),$
$(0, -3), (1, -2), (2, 1),$
$(3, 6), (4, 13)$

b.

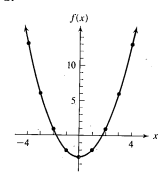

5. a. $(-4, -12), (-3, -5),$
$(-2, 0), (-1, 3), (0, 4),$
$(1, 3), (2, 0), (3, -5),$
$(4, -12)$

b.

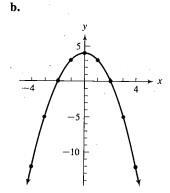

7. a. $(-4, 44), (-3, 24),$
$(-2, 10), (-1, 2), (0, 0),$
$(1, 4), (2, 14), (3, 30),$
$(4, 52)$

b.

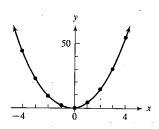

9. a. $(-4, 9), (-3, 4), (-2, 1),$
$(-1, 0), (0, 1), (1, 4),$
$(2, 9), (3, 16), (4, 25)$

b.

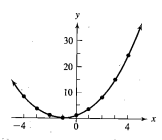

11. a. $(-4, -39), (-3, -24),$
$(-2, -13), (-1, -6),$
$(0, -3), (1, -4), (2, -9),$
$(3, -18), (4, -31)$

b.

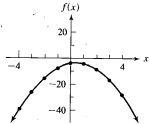

13. a. $(-4, -19), (-3, -11),$
$(-2, -5), (-1, -1),$
$(0, 1), (1, 1), (2, -1),$
$(3, -5), (4, -11)$

b.

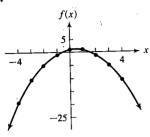

15. a. $(16, -4), (9, -3), (4, -2),$
$(1, -1), (0, 0), (1, 1),$
$(4, 2), (9, 3), (16, 4)$

b.

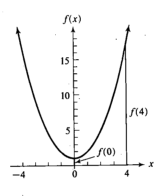

17. a. $(-60, -4), (-32, -3),$
$(-12, -2), (0, -1),$
$(4, 0), (0, 1), (-12, 2),$
$(-32, 3), (-60, 4)$

b.

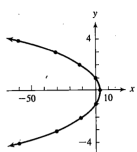

19. a. $(6, -4), (2, -3), (0, -2),$
$(0, -1), (2, 0), (6, 1),$
$(12, 2), (20, 3), (30, 4)$

b.

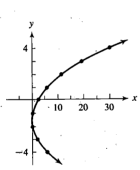

21.

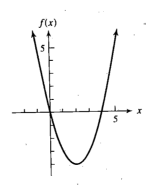

Exercise 8.2 [page 260]

1. x-intercepts: 1 and 4;
y-intercept: 4;
minimum point: $(5/2, -9/4)$

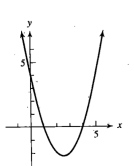

3. x-intercepts: 0 and 4;
$f(x)$-intercept: 0;
minimum point: $(2, -4)$

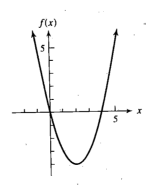

5. x-intercepts: -5 and 1;
$g(x)$-intercept: -5;
minimum point: $(-2, -9)$

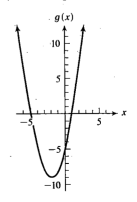

7. x-intercepts: -3 and $\frac{3}{2}$;
 y-intercept: -9;
 minimum point: $(-\frac{3}{4}, -8\frac{1}{8})$

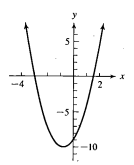

9. x-intercepts: 1 and 6;
 $f(x)$-intercept: -6;
 maximum point: $(\frac{7}{2}, \frac{25}{4})$

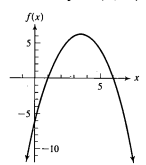

11. x-intercepts: -3 and $-\frac{1}{2}$;
 y-intercept: -3;
 maximum point: $(-\frac{7}{4}, \frac{25}{8})$

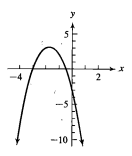

13.

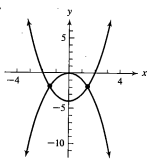

15.

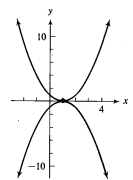

17.

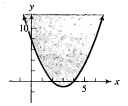

19.

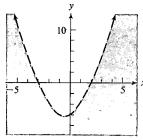

21.

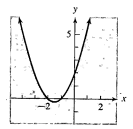

23.

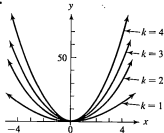

25.

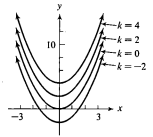

27. 6 and 6

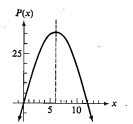

29. 2 seconds

31.

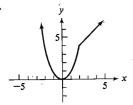

33.

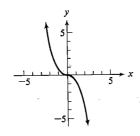

35.

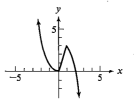

37. $f(x) = x^2$

39.

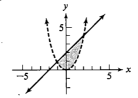

Exercise 8.3 [page 267]

1. a. $y = \pm \sqrt{4 - x^2}$
   b. domain: $\{x | -2 \le x \le 2\}$
   c.

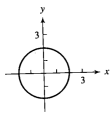

3. a. $y = \pm 3\sqrt{4 - x^2}$
   b. domain: $\{x | -2 \le x \le 2\}$
   c.

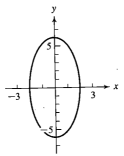

5. a. $y = \pm \frac{1}{2}\sqrt{16 - x^2}$
   b. domain: $\{x | -4 \le x \le 4\}$
   c.

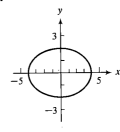

7. a. $y = \pm \sqrt{x^2 - 1}$
   b. domain: $\{x | x \le -1$ or
   $x \ge 1\}$
   c.

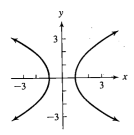

9. a. $y = \pm \sqrt{x^2 + 9}$
   b. domain: $\{x | x$ a real
   number$\}$
   c.

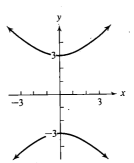

11. a. $y = \pm \sqrt{\dfrac{24 - 2x^2}{3}}$
   b. domain:
   $\{x | -2\sqrt{3} \le x \le 2\sqrt{3}\}$
   c.

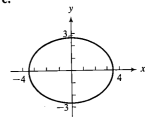

13. a. $y = \pm 3x$

 b. $\{x | x \text{ a real number}\}$

 c.

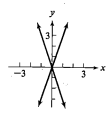

15. a. $y = \pm \dfrac{2}{3} x$

 b. $\{x | x \text{ a real number}\}$

 c.

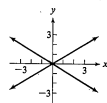

17. The graph of any equation of
the form $ax^2 + by^2 = c$
where $a, b > 0$ and $c = 0$ is
a point at the origin.

19.

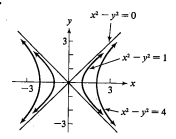

Exercise 8.4 [page 271]

1. a. circle

 b.

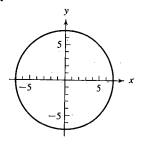

3. a. ellipse

 b.

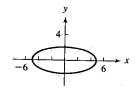

5. a. two intersecting lines

 b.

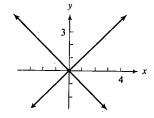

7. a. hyperbola

 b.

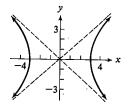

 c. $y = x$ or $y = -x$

9. a. ellipse

 b.

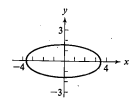

11. a. circle

 b.

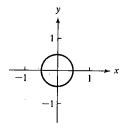

13. a. ellipse

b.

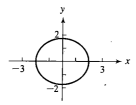

15. a. hyperbola

b.

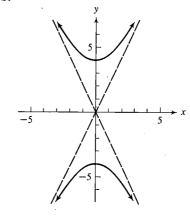

c. $y = 2x$ or $y = -2x$

17. no graph

19.

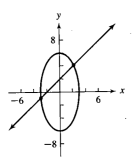

21.

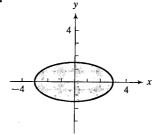

23.

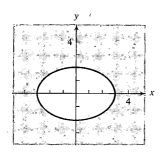

25. $by^2 = ax^2 - c$
$$y^2 = \frac{a}{b}x^2 - \frac{c}{b} = \frac{a}{b}x^2\left(1 - \frac{c}{ax^2}\right);$$
$$y = \pm\sqrt{\frac{a}{b}}x\sqrt{1 - \frac{c}{ax^2}}.$$

Exercise 8.5 [page 275] **1.** $d = kt$ **3.** $I = \dfrac{k}{R}$ **5.** $V = klw$ **7.** 3 **9.** 200 **11.** 2 **13.** $^{32}\!/_9$

15. 264 **17.** 16 **19.** 400 feet **21.** 160 pounds per square foot **23.** $^{3375}\!/_2$ pounds **25.** $4000

27. $\dfrac{9}{3^2} = \dfrac{y}{4^2}$; 16 **29.** $\dfrac{16}{2^2} = \dfrac{d}{10^2}$; 400 feet **31.** $\dfrac{40}{10} = \dfrac{P}{40}$; 160 pounds per square foot

33. $\dfrac{8 \cdot 750}{2 \cdot 4^2} = \dfrac{8 \cdot L}{2 \cdot 6^2}$; $\dfrac{3375}{2}$ pounds **35.** $\dfrac{2400}{6000 \cdot 4} = \dfrac{800}{P \cdot 2}$; $4000

37. Let D_1 and D_2 be the diameters and C_1 and C_2 be the corresponding circumferences of the two circles; then, since $\pi = \dfrac{C}{D}$, it follows that $\dfrac{C_1}{D_1} = \pi = \dfrac{C_2}{D_2}$ and hence $\dfrac{C_1}{C_2} = \dfrac{D_1}{D_2}$.

39. The intensity will decrease by ¼.   **41.** As k increases, the graph rises more steeply

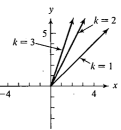

Exercise 8.6 [page 282]   1. $\{(-2, -2), (2, 2)\}$; a function

3. $\{(3, 1), (3, 2), (4, 3)\}$; not a function   **5.** $\{(1, 1), (2, 2), (3, 3)\}$; a function

7. a. $2y + 4x = 7$

b.

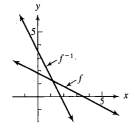

c. a function

9. a. $y - 3x = 6$

b.

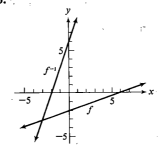

c. a function

11. a. $x = y^2 - 4y$

b.

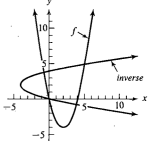

c. not a function

13. a. $x = \sqrt{4 + y^2}$

b.

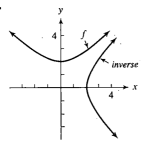

c. not a function

15. a. $x = |y|$

b.

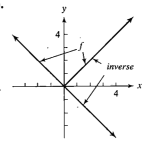

c. not a function

17. $f^{-1}: y = x;\quad f[f^{-1}(x)] = f[x] = x;$
$f^{-1}[f(x)] = f^{-1}[x] = x$

19. $f^{-1}: y = \dfrac{4 - x}{2};\quad f[f^{-1}(x)] = f\left[\dfrac{4 - x}{2}\right] = -2\left[\dfrac{4 - x}{2}\right] + 4 = -4 + x + 4 = x;$

$f^{-1}[f(x)] = f^{-1}[-2x + 4] = 2 - \dfrac{1}{2}[-2x + 4] = 2 + x - 2 = x$

21. f^{-1}: $y = \dfrac{4x + 12}{3}$; $f[f^{-1}(x)] = f\left[\dfrac{4x + 12}{3}\right] = \dfrac{3}{4}\left[\dfrac{4x + 12}{3}\right] - 3 = x + 3 - 3 = x$;

$f^{-1}[f(x)] = f^{-1}\left[\dfrac{3x - 12}{4}\right] = \dfrac{4}{3}\left[\dfrac{3x - 12}{4}\right] + 4 = x - 4 + 4 = x$

Review Exercises [page 284]   **1. a.** x-intercepts: 1 and 5; y-intercept: 5   **b.** $(3, -4)$

2.

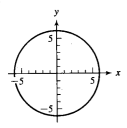

3.

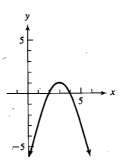

4.

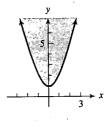

5. $\{x | x \geq 1 \quad \text{or} \quad x \leq -1\}$ **6.** $\{x | -z \leq x \leq 2\}$; ellipse   **7. a.** hyperbola   **b.** circle

8. a. parabola   **b.** ellipse

9.

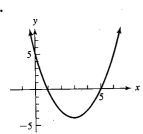

10.

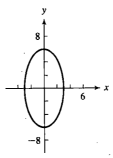

11.

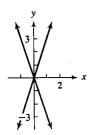

12.

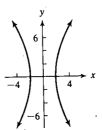

13. 9   **14.** $\dfrac{45}{2}$ **15.** 112 centimeters   **16.** $\dfrac{12,800}{81}$ pounds

17. a. $\{(7, 3), (8, 4), (4, 8)\}$; a function   **b.** $\{(6, 2), (8, 3), (8, 5)\}$; not a function

18. $x = y^2 + 9y$; not a function

19.

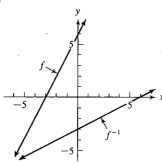

20.

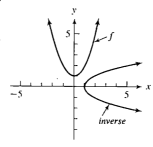

21.

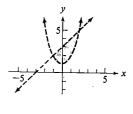

22.

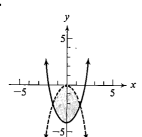

23.

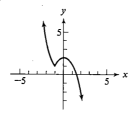

24.

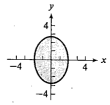

25.

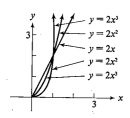

26. $y - 2x = 4$ can be written as $f(x) = 4 + 2x$.
The inverse defined by $x - 2y = 4$

can be written as $f^{-1}(x) = \dfrac{x - 4}{2}$. Hence,

$$f[f^{-1}(x)] = 4 + 2\left(\dfrac{x - 4}{2}\right) = x;$$

$$f^{-1}[f(x)] = \dfrac{(4 + 2x) - 4}{2} = x.$$

Exercise 9.1 [page 289] 1. (0, 1), (1, 3), (2, 9)   **3.** (−4, ¹⁄₁₆), (0, 1), (4, 16)
 5. (−4, 16), (0, 1), (4, ¹⁄₁₆)   **7.** (−2, ¹⁄₁₀₀), (−1, ¹⁄₁₀), (0, 1)

9.

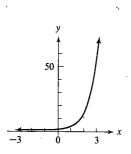

11.

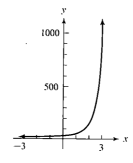

13.

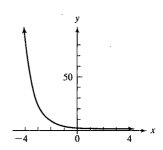

15.

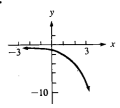

17.

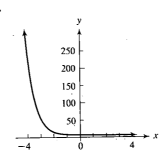

19.

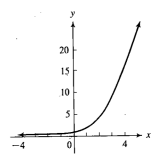

21.

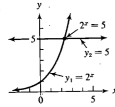

$x \approx 2.3$

23. An increasing function for all $a > 1$
and a decreasing function for all $0 < a < 1$.

Exercise 9.2 [page 292] **1.** $\log_4 16 = 2$ **3.** $\log_3 27 = 3$ **5.** $\log_{1/2} \frac{1}{4} = 2$ **7.** $\log_8 \frac{1}{2} = -\frac{1}{3}$

9. $\log_{10} 100 = 2$ **11.** $\log_{10} 0.1 = -1$ **13.** $2^6 = 64$ **15.** $3^2 = 9$ **17.** $\left(\frac{1}{3}\right)^{-2} = 9$

19. $10^3 = 1000$ **21.** $10^{-2} = 0.01$ **23.** 2 **25.** 3 **27.** ½ **29.** -1 **31.** 1 **33.** 2

35. -1 **37.** 2 **39.** 2 **41.** 64 **43.** -3 **45.** 100 **47.** 4 **49.** 1 **51.** 0 **53.** 1

55. 0 **57.** $x > 9$

Exercise 9.3 [page 296] **1.** $\log_b 2 + \log_b x$ **3.** $\log_b 3 + \log_b x + \log_b y$ **5.** $\log_b x - \log_b y$

7. $\log_b x + \log_b y - \log_b z$ **9.** $3 \log_b x$ **11.** $\frac{1}{2}\log_b x$ **13.** $\frac{2}{3}\log_b x$ **15.** $2 \log_b x + 3 \log_b y$

17. $\frac{1}{2}\log_b x + \log_b y - 2 \log_b z$ **19.** $\frac{1}{3}\log_{10} x + \frac{2}{3}\log_{10} y - \frac{1}{3}\log_{10} z$ **21.** $\frac{3}{2}\log_{10} x - \frac{1}{2}\log_{10} y$

23. $\log_{10} 2 + \log_{10} \pi + \frac{1}{2}\log_{10} l - \frac{1}{2}\log_{10} g$ **25.** $\frac{1}{2}\log_{10}(s - a) + \frac{1}{2}\log_{10}(s - b)$ **27.** $\log_b xy$ **29.** $\log_b \frac{x^2}{y^3}$

31. $\log_b \frac{x^3 y}{z^2}$ **33.** $\log_{10} \frac{\sqrt{xy}}{z}$ **35.** $\log_b x^{-2}$ or $\log_b \frac{1}{x^2}$ **37.** {500} **39.** {4} **41.** {3}

43. Substitute arbitrary numbers, for example, 1 and 10, for x and y in the given equation.

$$\log_{10}(1 + 10) \overset{?}{=} \log_{10} 1 + \log_{10} 10$$
$$\log_{10} 11 \overset{?}{=} 0 + 1$$
$$\log_{10} 11 \neq 1$$

Since $\log_{10} 11 \neq 1$, $\log_{10}(x + y) \neq \log_{10} x + \log_{10} y$ for all $x, y > 0$.

45. 0.7781 **47.** -0.3980 **49.** 0.9542 **51.** 0.8751 **53.** -8.0970 **55.** 1.8751

57. Show that the left and right members of the equality reduce to the same quantity.
For the left side:

$$\log_b 4 + \log_b 8 = \log_b 2^2 + \log_b 2^3$$
$$= 2 \log_b 2 + 3 \log_b 2 = 5 \log_b 2.$$

For the right side:

$$\log_b 64 - \log_b 2 = \log_b 2^6 - \log_b 2$$
$$= 6 \log_b 2 - \log_b 2 = 5 \log_b 2.$$

59. For the left side:

$$2 \log_b 6 - \log_b 9 = \log_b 6^2 - \log_b 9$$
$$= \log_b 36 - \log_b 9 = \log_b \frac{36}{9} = \log_b 4.$$

For the right side:

$$2 \log_b 2 = \log_b 2^2 = \log_b 4.$$

61. For the left side:

$$\frac{1}{2}\log_b 12 - \frac{1}{2}\log_b 3 = \frac{1}{2}(\log_b 12 - \log_b 3)$$

$$= \frac{1}{2}\left(\log_b \frac{12}{3}\right) = \frac{1}{2}(\log_b 2^2) = \log_b 2.$$

For the right side:

$$\frac{1}{3}\log_b 8 = \log_b 8^{1/3} = \log_b 2.$$

Exercise 9.4 [page 305]   1. 2   **3.** -4 **5.** 0   **7.** 0.8280   **9.** 1.9227   **11.** 2.5011
13. $0.9101 - 1$ **15.** $0.9031 - 2$ **17.** 2.3945   **19.** 4.10   **21.** 36.7   **23.** 0.0642   **25.** 0.00718
27. 0.297   **29.** 0.0503   **31.** 0.00205   **33.** 7.52   **35.** 784   **37.** 0.0357   **39.** 1.5373
41. 0.5655   **43.** 4.4817   **45.** 0.0907   **47.** 1.3610   **49.** 2.7726   **51.** $0.0837 - 1$ or -0.9163
53. 1.1735   **55.** 6.0496   **57.** 90.017

Exercise 9.5 [page 311]   1. 2   **3.** -4 **5.** 1   **7.** 4   **9.** 1.7348   **11.** 3.3700   **13.** -1.1367
15. -0.15851 **17.** 41.687   **19.** 0.13490   **21.** 129.72   **23.** 0.58157   **25.** 25.704
27. 3.3113   **29.** 0.050119   **31.** 1.3610   **33.** 2.7726   **35.** -0.91629 **37.** 1.4918
39. 10.381   **41.** 0.30119   **43.** 0.67032   **45.** 4.1371   **47.** 1.8776   **49.** 0.074274

Exercise 9.6 [page 316]   1. 0.693   **3.** 1.37   **5.** 3.83   **7.** 0.642   **9.** 3.81   **11.** -1.20
13. 0.977   **15.** -3.72 **17.** 0.768   **19.** -0.684 **21.** 0.828   **23.** 7.52   **25.** -2.97
27. 1.65   **29.** -3.07 **31.** 2.63   **33.** 2.89   **35.** 2.06   **37.** 2.81   **39.** 0.89   **41.** ± 1.40

43. -2.10 **45.** $t = \dfrac{1}{k}\ln y$ **47.** $t = 2(\ln T - \ln R)$ **49.** $k = e^{T/T_0} - 10$

51. $\ln N = \dfrac{\log_{10} N}{\log_{10} e} \approx \dfrac{\log_{10} N}{0.43429} \approx 2.303 \log_{10} N$

53. $\log_{10} y^x = x \log_{10} y$; hence, in exponential form $y^x = 10^{x \log_{10} y}$.

Exercise 9.7 [page 320] **1.** 9.60 inches of mercury   **3.** 1.91 miles   **5.** 3.34 miles

7. a. 19,960,000   **b.** 25,349,000; 32,193,000; 40,884,000   **9.** 1.56%   **11.** 18.7 years

13. 12.1 grams   **15.** 99.3 volts   **17.** 2.77 seconds   **19.** 34.7 hours   **21.** 17.5%   **23.** 13.5%

25. $15,529.25; $16,035.68   **27.** 16 years

Exercise 9.8 [page 328] **1.** 4.01   **3.** 2.30   **5.** 0.000461   **7.** 64.4   **9.** 2.01   **11.** 3.44×10^{-10}

13. 0.0458   **15.** 0.278   **17.** 9.87   **19.** 4.75   **21.** 1.39   **23.** 3.48   **25.** 57.8   **27.** 2.21

29. 4.6 kilowatt-hours   **31.** 18.7 horsepower   **33.** 1.11 seconds

Review Exercises [page 331]

1. a.

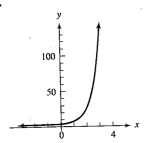

b.

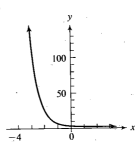

2. a. $\log_9 27 = \dfrac{3}{2}$ **b.** $\log_{4/9} \dfrac{2}{3} = \dfrac{1}{2}$ **3. a.** $5^4 = 625$ **b.** $10^{-4} = 0.0001$ **4. a.** 2   **b.** 8

5. a. $\log_b 3 + 2 \log_b x + \log_b y$ **b.** $\log_b y + \dfrac{1}{2} \log_b x - 2 \log_b z$ **6. a.** $\log_b \dfrac{x^5}{y^2}$ **b.** $\log_b \sqrt[3]{\dfrac{xz^2}{y^4}}$

7. a. {1}   **b.** {5}   **8. a.** 1.0791   **b.** 0.6276   **9. a.** -0.147 **b.** 3.26

10. a. 431   **b.** 0.0379   **11. a.** 17.2   **b.** 0.285   **12. a.** 2.29   **b.** 0.273

13. a. 1.95   **b.** -0.916 **14. a.** 2.08   **b.** 14.9   **15.** 0.114   **16.** 4.13   **17.** 0.453

18. 7.05   **19.** 4.61   **20.** 8.05   **21.** 4.53   **22.** 1.78   **23.** 2.46   **24.** 7.32

25. 0 mg; 3.9 mg   **26.** 461.6 lumens   **27.** 13.5 years   **28.** 1.7 seconds   **29.** 26.1   **30.** 1.39

31. 1.22   **32.** 1.63   **33.** 0.280   **34.** 0.000249   **35.** 0

36. Left side: $2 \log_b 8 - \log_b 4 = \log_b \dfrac{8^2}{4} = \log_b 16$.

Right side: $4 \log_b 2 = \log_b 2^4 = \log_b 16$.

37. $t = -\dfrac{1}{k}\ln\!\left(\dfrac{N}{N_0}\right)$ **38.** $t = ke^{(N - C)/N_0}$ **39.** 6.2   **40.** 2.5×10^{-6}

Exercise 10.1 [page 339]

1. $\{(3, 2)\}$

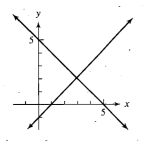

3. $\{(2, 1)\}$

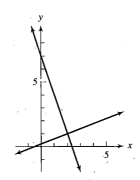

5. $\{(-5, 4)\}$

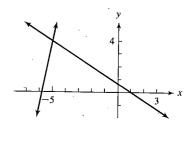

7. $\{(1, 2)\}$

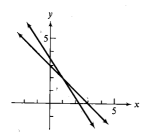

9. $\{(1, 0)\}$

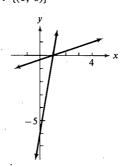

11. $\{(5, -2)\}$

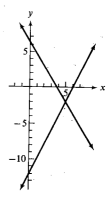

13. $\{(0, \frac{3}{2})\}$

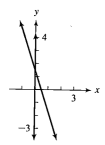

15. $\{(\frac{2}{3}, -1)\}$

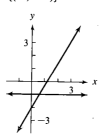

17. $\{(1, 2)\}$ **19.** $\{(-\frac{19}{5}, -\frac{18}{5})\}$ **21.** infinitely many solutions **23.** $\varnothing$; no solutions **25.** 9; 15
27. 39; 40 **29.** 46 children; 36 adults **31.** \$1200 at 10%; \$800 at 8% **33.** white, 1965; red, 1969
35. a. $C = 20 + 0.40x$ **b.** $R = 1.20x$ **c.** 25 records **37.** $\{(\frac{1}{5}, \frac{1}{2})\}$ **39.** $\{(-5, 3)\}$
41. $\{(\frac{1}{6}, \frac{1}{2})\}$ **43.** $a = 1$; $b = -1$ **45.** $x = \dfrac{c_1b_2 - c_2b_1}{a_1b_2 - a_2b_1}$, $\quad y = \dfrac{a_1c_2 - a_2c_1}{a_1b_2 - a_2b_1}$ $(a_1b_2 - a_2b_1 \neq 0)$

Exercise 10.2 [page 344]   **1.** $\{(1, 2, -1)\}$ **3.** $\{(2, -2, 0)\}$ **5.** $\{(2, 2, 1)\}$ **7.** no unique solution
  9. $\{(3, -1, 1)\}$ **11.** $\{(4, -2, 2)\}$ **13.** 3; 6; 6   **15.** 60 nickels; 20 dimes; 5 quarters
  17. $x^2 + y^2 + 2x + 2y - 23 = 0$ **19.** $a = 1$; $b = 2$; $c = 3$

Exercise 10.3 [page 348]

1. $\{(-1, -4), (5, 20)\}$

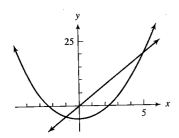

3. $\{(2, 3), (3, 2)\}$

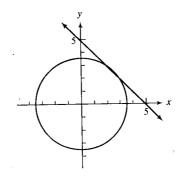

5. $\{(-3, 4), (4, -3)\}$

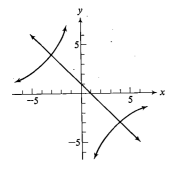

7. $\{(2, 2), (-2, -2)\}$

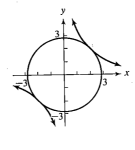

9. $\{(i\sqrt{7}, 4), (-i\sqrt{7}, 4)\}$

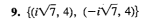
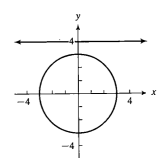

11. $\left\{\left(\dfrac{-1+\sqrt{13}}{2}, \dfrac{1+\sqrt{13}}{2}\right), \left(\dfrac{-1-\sqrt{13}}{2}, \dfrac{1-\sqrt{13}}{2}\right)\right\}$

13. $\{(3, 1), (2, 0)\}$

15. $\{(-1, -3)\}$

17. 2; 3

19. length, 12 inches; width, 1 inch

21. $\{(2, 2), (-2, -2)\}$

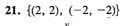
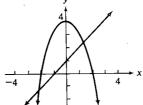
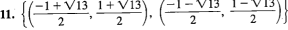

Exercise 10.4 [page 352] **1.** $\{(1, 3), (-1, 3), (1, -3), (-1, -3)\}$
3. $\{(1, 2), (-1, 2), (1, -2), (-1, -2)\}$ **5.** $\{(3, \sqrt{2}), (-3, \sqrt{2}), (3, -\sqrt{2}), (-3, -\sqrt{2})\}$
7. $\{(2, 1), (-2, 1), (2, -1), (-2, -1)\}$ **9.** $\{(\sqrt{3}, 4), (-\sqrt{3}, 4), (\sqrt{3}, -4), (-\sqrt{3}, -4)\}$
11. $\{(2, -2), (-2, 2), (2i\sqrt{2}, i\sqrt{2}), (-2i\sqrt{2}, -i\sqrt{2})\}$ **13.** $\{(1, -1), (-1, 1), (i, i), (-i, -i)\}$
15.   **17.**

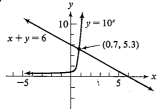

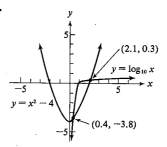

19. a. 1   **b.** 2   **c.** 4

Review Exercises [page 354] **1.** $\{(\tfrac{1}{2}, \tfrac{7}{2})\}$ **2.** $\{(1, 2)\}$ **3.** $\{(12, 0)\}$ **4.** $\{(\tfrac{22}{35}, \tfrac{11}{14})\}$
5. a. unique solution   **b.** dependent
6. a. inconsistent   **b.** unique solution   **7.** $\{(2, 0, -1)\}$ **8.** $\{(2, 1, -1)\}$
9. $\{(2, -5, 3)\}$ **10.** $\{(\tfrac{4}{17}, -\tfrac{2}{17}, -\tfrac{84}{17})\}$ **11.** $\{(-2, 1, 3)\}$ **12.** $\{(2, -1, 0)\}$
13. $\left\{\left(3, \dfrac{\sqrt{3}}{3}\right), \left(3, -\dfrac{\sqrt{3}}{3}\right)\right\}$ **14.** $\{(4i, -2), (-4i, -2)\}$ **15.** $\{(4, -13), (1, 2)\}$
16. $\{(\tfrac{6}{7}, -\tfrac{37}{7}), (2, -3)\}$ **17.** $\{(1, \sqrt{5}), (1, -\sqrt{5}), (-1, \sqrt{5}), (-1, -\sqrt{5})\}$
18. $\{(2, 3), (2, -3), (-2, 3), (-2, -3)\}$ **19.** $\left\{(1, -2), (-1, 2), \left(2\sqrt{3}, \dfrac{-\sqrt{3}}{3}\right), \left(-2\sqrt{3}, \dfrac{1}{\sqrt{3}}\right)\right\}$
20. $\{(2, 1), (-2, -1), (i, -2i), (-i, 2i)\}$ **21.** width: 7 centimeters; length: 10 centimeters
22. width: 2 feet; length: 7 feet   **23.** $y = x^2 - x - 6$ **24.** $x^2 + y^2 - 4x - 2y - 5 = 0$
25. $b = \dfrac{-a^2 - 16}{4}$ **26.**

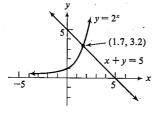

Exercise 11.1 [page 360] **1.** $-4, -3, -2, -1$ **3.** $-\tfrac{1}{2}, 1, \tfrac{7}{2}, 7$ **5.** $2, \tfrac{3}{2}, \tfrac{4}{3}, \tfrac{5}{4}$ **7.** $0, 1, 3, 6$
9. $-1, 1, -1, 1$ **11.** $1, 0, -\tfrac{1}{3}, \tfrac{1}{2}$ **13.** $1 + 4 + 9 + 16$ **15.** $3 + 4 + 5$ **17.** $2 + 6 + 12 + 20$

19. $-\tfrac{1}{2} + \tfrac{1}{4} - \tfrac{1}{8} + \tfrac{1}{16}$ **21.** $1 + 3 + 5 + \cdots$ **23.** $1 + \tfrac{1}{2} + \tfrac{1}{4} + \cdots$ **25.** $\displaystyle\sum_{i=1}^{4} i$

27. $\displaystyle\sum_{i=1}^{4} x^{2i-1}$ **29.** $\displaystyle\sum_{i=1}^{5} i^2$ **31.** $\displaystyle\sum_{i=1}^{\infty} \dfrac{i}{i+1}$ **33.** $\displaystyle\sum_{i=1}^{\infty} \dfrac{i}{2i-1}$ **35.** $\displaystyle\sum_{i=1}^{\infty} \dfrac{2^{i-1}}{i}$

Exercise 11.2 [page 365]   **1.** 11, 15, 19   **3.** $-9, -13, -17$ **5.** $x+2, x+3, x+4$
$s_n = 4n - 1$ $s_n = 3 - 4n$ $s_n = x + n - 1$

7. $x+5a, x+7a, x+9a$ **9.** $2x+7, 2x+10, 2x+13$ **11.** $3x, 4x, 5x$ **13.** 31   **15.** $15\frac{1}{2}$
$s_n = x + 2an - a$ $s_n = 2x + 3n - 2$ $s_n = nx$

17. -92 **19.** 2; 3; 41   **21.** twenty-eighth term   **23.** 1; 8   **25.** 16   **27.** 9; 14; 19

29. 63   **31.** 806   **33.** -6 **35.** 1938   **37.** 196 bricks   **39.** 2, 7, 12   **41.** $\frac{14}{15}$

43. The first n odd natural numbers form an arithmetic sequence with $a = 1$ and $d = 2$. Hence, the sum of the first n odd natural numbers is

$$S_n = \frac{n}{2}[2a + (n-1)d] = \frac{n}{2}[2 + (n-1)2]$$

$$= \frac{n}{2}[2 + 2n - 2] = \frac{n}{2}[2n] = n^2.$$

Exercise 11.3 [page 371]   **1.** 128, 512, 2048   **3.** $\frac{16}{3}, \frac{32}{3}, \frac{64}{3}$ **5.** $-\frac{1}{2}, \frac{1}{4}, -\frac{1}{8}$
$s_n = 2(4)^{n-1}$ $s_n = \frac{2}{3}(2)^{n-1}$ $s_n = 4(-\frac{1}{2})^{n-1}$

7. $-\frac{x^2}{a^2}, \frac{x^3}{a^3}, -\frac{x^4}{a^4}$ **9.** 1536   **11.** $-243a^{20}$ **13.** 3   **15.** 3, 9   **17.** 18 or -18
$s_n = \frac{a}{x}\left(-\frac{x}{a}\right)^{n-1}$

19. 16, 8, 4 or $-16, 8, -4$ **21.** 1092   **23.** $\frac{31}{32}$ **25.** $\frac{364}{729}$

27.

29. 1 hour: 20 $(10 \cdot 2)$;
2 hours: 40 $(10 \cdot 2^2)$;
3 hours: 80 $(10 \cdot 2^3)$;
4 hours: 160 $(10 \cdot 2^4)$;
n hours: $(10 \cdot 2^n)$.

Exercise 11.4 [page 377]   **1.** 24   **3.** does not exist   **5.** $\frac{9}{20}$ **7.** $\frac{8}{49}$ **9.** 2   **11.** $\frac{1}{3}$

13. $\frac{31}{99}$ **15.** $\frac{2408}{999}$ **17.** $\frac{29}{225}$ **19.** \$220.80; \$222.55   **21.** 20 centimeters   **23.** 30 feet

25. After 2 years: $P(1 + r)(1 + r) = P(1 + r)^2$;
after 3 years: $P(1 + r)^2(1 + r) = P(1 + r)^3$;
after n years: $P(1 + r)^n$.

Exercise 11.5 [page 381]   **1.** $8 \cdot 7 \cdot 6 \cdot 5 \cdot 4 \cdot 3 \cdot 2 \cdot 1$ **3.** $2 \cdot 4 \cdot 3 \cdot 2 \cdot 1$ **5.** $6 \cdot 5 \cdot 4 \cdot 3 \cdot 2 \cdot 1$ **7.** 120

9. 72   **11.** 15   **13.** 28   **15.** 3!   **17.** $\frac{6!}{2!}$ **19.** $\frac{8!}{5!}$ **21.** $n(n-1)(n-2) \cdot \cdots \cdot 3 \cdot 2 \cdot 1$

23. $3n(3n-1)(3n-2) \cdot \cdots \cdot 3 \cdot 2 \cdot 1$ **25.** $(n-2)(n-3)(n-4) \cdot \cdots \cdot 3 \cdot 2 \cdot 1$

27. $x^5 + 15x^4 + 90x^3 + 270x^2 + 405x + 243$ **29.** $x^4 - 12x^3 + 54x^2 - 108x + 81$

31. $8x^3 - 6x^2y + \dfrac{3}{2}xy^2 - \dfrac{1}{8}y^3$ **33.** $\dfrac{1}{64}x^6 + \dfrac{3}{8}x^5 + \dfrac{15}{4}x^4 + 20x^3 + 60x^2 + 96x + 64$

35. $x^{20} + 20x^{19}y + \dfrac{20 \cdot 19}{2!}x^{18}y^2 + \dfrac{20 \cdot 19 \cdot 18}{3!}x^{17}y^3$

37. $a^{12} + 12a^{11}(-2b) + \dfrac{12 \cdot 11}{2!}a^{10}(-2b)^2 + \dfrac{12 \cdot 11 \cdot 10}{3!}a^9(-2b)^3$

39. $x^{10} + 10x^9(-\sqrt{2}) + \dfrac{10 \cdot 9}{2!}x^8(-\sqrt{2})^2 + \dfrac{10 \cdot 9 \cdot 8}{3!}x^7(-\sqrt{2})^3$ **41.** $-3003a^{10}b^5$

43. $3360x^6y^4$ **45.** $21x^{10}y^4$ **47.** $63a^5b^4$

49. a. $1^{-1} + (-1)(1^{-2})x + 1(1^{-3})x^2 + (-1)(1^{-4})x^3$ or $1 - x + x^2 - x^3$
   b. $1 - x + x^2 - x^3$ **c.** results are equal

Exercise 11.6 [page 387] **1.** 24   **3.** 362,880   **5.** 2058   **7.** 720   **9.** 648   **11.** 448
13. 17,576   **15.** 676,000   **17.** 60   **19.** 120   **21.** 45,360   **23.** 6720   **25.** 30

Exercise 11.7 [page 391] **1.** 1365   **3.** 7   **5.** 2,598,960   **7.** 15   **9.** 84

11. $\dbinom{13}{6} \cdot \dbinom{13}{3} \cdot \dbinom{13}{4} = 350{,}904{,}840$ **13.** 7425

15. $\dbinom{n}{n-r} = \dfrac{n!}{(n-r)![n-(n-r)]!} = \dfrac{n!}{(n-r)!r!} = \dfrac{n!}{r!(n-r)!} = \dbinom{n}{r}$ **17.** 7

19. From Section 11.5, we see that the coefficients of the binomial expansion of $(a+b)^4$ are 1, 4, 6, 4, and 1. Since

$$\binom{4}{0} = \frac{4!}{0!4!} = 1, \qquad \binom{4}{1} = \frac{4!}{1!3!} = 4, \qquad \binom{4}{2} = \frac{4!}{2!2!} = 6,$$

$$\binom{4}{3} = \frac{4!}{3!1!} = 4, \quad \text{and} \quad \binom{4}{4} = \frac{4!}{4!0!} = 1,$$

the combinations are equal to the respective coefficients of the binomial expansion $(a+b)^4$.

Review Exercises [page 394] **1.** $1, -\tfrac{1}{2}, \tfrac{1}{3}, -\tfrac{1}{4}$ **2.** $2 + 6 + 12 + 20$ **3.** $4n + 1$
 4. a. 94   **b.** 138   **5.** 68   **6.** $5\left(\dfrac{9}{5}\right)^{n-1}$ **7. a.** $-\tfrac{8}{1}$ **b.** $-\tfrac{26}{27}$ **8.** $12\tfrac{1}{243}$ **9.** $\tfrac{1}{2}$
10. $\tfrac{4}{9}$ **11.** 56   **12.** $x^{10} - 20x^9y + 180x^8y^2 - 960x^7y^3$ **13.** $-15{,}360x^3y^7$ **14. a.** 125   **b.** 60
15. 360   **16.** 1260   **17.** 15   **18.** 126   **19.** 16   **20.** 24   **21.** 1539   **22.** 2, 5, 8
23. $2^{(n/4)+1}$ **24.** 56 feet   **25.** $\dfrac{2}{(n+2)(n+1)}$ **26.** $(2n-1)(2n-2)(n+1)$

Exercise A.1 [page 399] **1.** $x - 2$ $(x \neq 6)$ **3.** $x + 2$ $(x \neq -2)$ **5.** $x^3 - x^2 + \dfrac{-1}{x-2}$ $(x \neq 2)$

7. $2x^2 - 2x + 3 + \dfrac{-8}{x+1}$ $(x \neq -1)$ **9.** $2x^3 + 10x^2 + 50x + 249 + \dfrac{1251}{x-5}$ $(x \neq 5)$

11. $x^2 + 2x - 3 + \dfrac{4}{x+2}$ $(x \neq -2)$ 13. $x^5 + x^4 + 2x^3 + 2x^2 + 2x + 1 + \dfrac{1}{x-1}$ $(x \neq 1)$

15. $x^4 + x^3 + x^2 + x + 1$ $(x \neq 1)$ 17. $x^5 + x^4 + x^3 + x^2 + x + 1$ $(x \neq 1)$

Exercise A.2 [page 401] **1.** 74; -46 **3.** 5; 44

5.

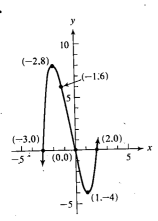

7.

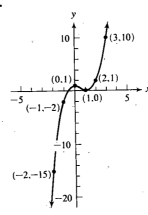

9.

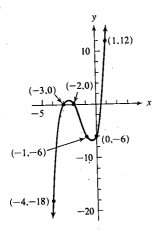

11.

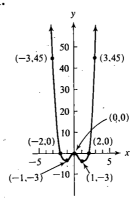

13. no   **15.** yes   **17.** -1 and -2 **19.** 0, 2, and 4

Exercise A.3 [page 406] **1.** $x = -3$ **3.** $x = 2, x = -3$ **5.** $x = -2, x = 3$

7. vertical asymptotes: $x = 3, x = -3$
horizontal asymptote: $y = 0$

9. vertical asymptote: $x = \frac{1}{2}$
horizontal asymptote: $y = \frac{1}{2}$

11. vertical asymptotes: $x = -1$, $x = 4$
  horizontal asymptote: $y = 2$

13. horizontal asymptote: $y = 0$
  vertical asymptote: $x = -3$
  y-intercept: $\frac{1}{3}$

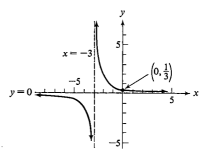

15. horizontal asymptote: $y = 0$
  vertical asymptotes: $x = 4$, $x = -1$
  y-intercept: $-\frac{1}{2}$

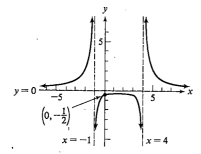

17. horizontal asymptote: $y = 0$
  vertical asymptotes: $x = 4$, $x = 1$
  y-intercept: $\frac{1}{2}$

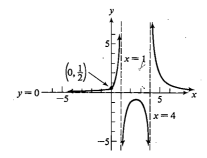

19. horizontal asymptote: $y = 1$
  vertical asymptote: $x = -3$
  intercepts: $x = 0$, $y = 0$

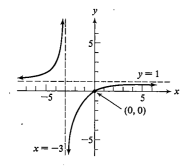

21. horizontal asymptote: $y = 1$
  vertical asymptote: $x = -2$
  y-intercept: $\frac{1}{2}$
  x-intercept: -1

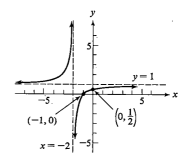

23. horizontal asymptote: $y = 0$
vertical asymptotes: $x = 2$, $x = -2$
intercepts: $(0, 0)$

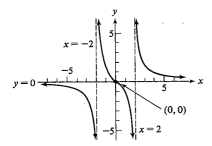

25. horizontal asymptote: $y = 0$
vertical asymptotes: $x = -1$, $x = -4$
y-intercept: $-\frac{1}{2}$
x-intercept: 2

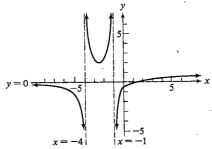

27.

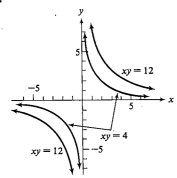

Review Exercises [page 408]   **1.** $y^2 + 2y - 4$ **2.** $y^6 + y^5 + y^4 + y^3 + y^2 + y + 1$ **3.** 19; -25 **4.** 4; 8

5.

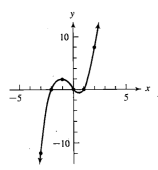

6.

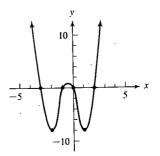

7. no   **8.** $(x + 1); (x - 3)$

9.

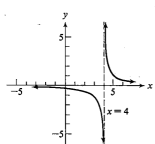

10.

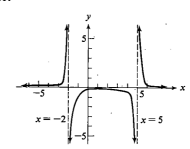

11.

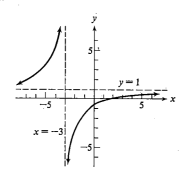

12.

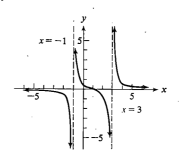

Exercise B.1 [page 412]   **1.** $\{(2, 3)\}$ **3.** $\{(-2, 1)\}$ **5.** $\{(3, -1)\}$ **7.** $\{(1, 2, 2)\}$
9. $\{(\frac{76}{39}, -\frac{112}{39}, \frac{48}{39})\}$ **11.** $\{(-3, 1, -3)\}$

Exercise B.2 [page 415]   **1.** 1   **3.** -12 **5.** 2   **7.** 5   **9.** $\{(1, 1)\}$ **11.** $\{(2, 2)\}$ **13.** $\{(6, 4)\}$

15. inconsistent, $\varnothing$ **17.** $\{(4, 1)\}$ **19.** $\left\{\left(\dfrac{1}{a+b}, \dfrac{1}{a+b}\right)\right\}$ **21.** $\begin{vmatrix} a & a \\ b & b \end{vmatrix} = ab - ab = 0$

23. $\begin{vmatrix} a_1 & b_1 \\ a_2 & b_2 \end{vmatrix} = a_1b_2 - a_2b_1;$ $-\begin{vmatrix} b_1 & a_1 \\ b_2 & a_2 \end{vmatrix} = -(b_1a_2 - b_2a_1) = -b_1a_2 + b_2a_1 = a_1b_2 - a_2b_1$

25. $\begin{vmatrix} ka & a \\ kb & b \end{vmatrix} = kab - kba = 0$

27. Since $D_x = \begin{vmatrix} c_1 & b_1 \\ c_2 & b_2 \end{vmatrix} = c_1b_2 - c_2b_1 = 0,$ then $b_1c_2 = b_2c_1.$

Since $D_y = \begin{vmatrix} a_1 & c_1 \\ a_2 & c_2 \end{vmatrix} = a_1c_2 - a_2c_1 = 0,$ then $a_1c_2 = a_2c_1.$

Forming the proportion $\dfrac{b_1c_2}{a_1c_2} = \dfrac{b_2c_1}{a_2c_1},$ if c_1 and c_2 are not both 0, then $\dfrac{b_1}{a_1} = \dfrac{b_2}{a_2},$

from which $a_1b_2 = a_2b_1$ and $a_1b_2 - a_2b_1 = 0.$ Since $D = \begin{vmatrix} a_1 & b_1 \\ a_2 & b_2 \end{vmatrix} = a_1b_2 - a_2b_1,$ it follows that $D = 0.$

Exercise B.3 [page 419] **1.** 3 **3.** 9 **5.** 0 **7.** -1 **9.** -5 **11.** 0 **13.** x^3 **15.** 0

17. $-2ab^2$ **19.** {3} **21.** $\{2, -17/7\}$

23. Expanding about the elements of the third column of the given determinant produces

$$a\begin{vmatrix} y & y \\ z & z \end{vmatrix} - b\begin{vmatrix} x & x \\ z & z \end{vmatrix} + c\begin{vmatrix} x & x \\ y & y \end{vmatrix} = 0 + 0 + 0 = 0.$$

25. For the left side: $\begin{vmatrix} 1 & 2 & 3 \\ 4 & 5 & 6 \\ 0 & 0 & 1 \end{vmatrix} = 1\begin{vmatrix} 1 & 2 \\ 4 & 5 \end{vmatrix} = 5 - 8 = -3.$

For the right side: $-\begin{vmatrix} 4 & 5 & 6 \\ 1 & 2 & 3 \\ 0 & 0 & 1 \end{vmatrix} = -1\begin{vmatrix} 4 & 5 \\ 1 & 2 \end{vmatrix} = -1(8 - 5) = -1(3) = -3.$

Exercise B.4 [page 423] **1.** {(1, 1, 1)} **3.** {(1, 1, 0)} **5.** {(1, -2, 3)} **7.** {(3, -1, -2)}

9. no unique solution **11.** $\{(1, -\frac{1}{3}, \frac{1}{2})\}$ **13.** {(-5, 3, 2)}

Review Exercises [page 425] **1.** {(3, -1)} **2.** {(-1, 0, 2)} **3.** -13 **4.** {(-7, 4)} **5.** -2

6. -19 **7.** {(1, 2, 3)} **8.** {(2, -1, 3)}

Exercise C [page 427] **1.** 0.6246 **3.** 0.7937 **5.** 3.1824 **7.** 4.5695 **9.** $0.7095 - 1$

11. $0.9218 - 3$ **13.** 3.225 **15.** 89.38 **17.** 10.52 **19.** 0.05076 **21.** 0.7485 **23.** 0.7495

25. 41.298 **27.** 0.0621 **29.** 308.30 **31.** 1.1505 **33.** 1.9051 **35.** -0.3166

INDEX